THE
FOUNDATION
ENGINEERING
HANDBOOK

THE
FOUNDATION
ENGINEERING
HANDBOOK

Edited by
Manjriker Gunaratne

Taylor & Francis
Taylor & Francis Group
Boca Raton   London   New York

A CRC title, part of the Taylor & Francis imprint, a member of the
Taylor & Francis Group, the academic division of T&F Informa plc.

MER
624.15 GUN
7dy

Published in 2006 by
CRC Press
Taylor & Francis Group
6000 Broken Sound Parkway NW, Suite 300
Boca Raton, FL 33487-2742

International Standard Book Number-10: 0-8493-1159-4 (Hardcover)
International Standard Book Number-13: 978-0-8493-1159-8 (Hardcover)
Library of Congress Card Number 2005050886

Library of Congress Cataloging-in-Publication Data

The foundation engineering handbook / edited by Manjriker Gunaratne.
    p. cm.
  Includes bibliographical references and index.
  ISBN 0-8493-1159-4
  1. Foundations--Handbooks, manuals, etc. 2. Soil mechanics--Handbooks, manuals, etc. I. Gunaratne, Manjriker.

TA775.F677 2006
624.1'5--dc22   2005050886

Taylor & Francis Group
is the Academic Division of T&F Informa plc.

Visit the Taylor & Francis Web site at
http://www.taylorandfrancis.com

and the CRC Press Web site at
http://www.crcpress.com

Preface

A genuine need existed for an updated foundation engineering handbook that incorporates, in addition to classical principles of foundation designs, significant contributions made to the art of foundation design by practitioners and researchers during the last two decades. Of special significance in this regard is the knowledge of (1) innovative *in situ* testing and site improvement techniques that have evolved recently; (2) cost-effective design methods that make use of geogrids for mechanically stabilized earth retaining structures; (3) concepts involved in ground deformation modeling using finite elements; and (4) latest modifications in the ACI codes applicable to structural design of foundations. This handbook largely fulfills the above needs, since the editor and the contributors have focused on discussing the state of the art of theoretical and applied foundation engineering and concrete design in a concise and simple fashion.

Reliability-based design concepts that have been incorporated in most up-to-date structural and pavement design guidelines are making inroads into foundation engineering as well. Hence, the editor decided to include reliability-based design and LRFD (load resistance factor design) concepts along with relevant illustrative examples in this handbook. This step not only makes this handbook somewhat unique among other currently available foundation engineering literature, but also it provides an opportunity for practitioners and students alike to familiarize themselves with the basics of limit state design applied to foundation engineering.

Furthermore, the editor's extensive experience as an engineering educator has constantly indicated that, in spite of the availability of a number of excellent textbooks in foundation engineering, a quick reference that mostly focuses on significant and commonly-used foundation engineering principles and illustrative examples has been in demand. This handbook also addresses such a need, since it can be adopted conveniently as a textbook, both at the undergraduate and graduate levels.

It is indeed my pleasure to have worked with a distinguished set of contributors who took time off of their extremely busy professional careers and produced their best in keeping with their usual professional performance. My appreciation is conveyed to Ingrid Hall of the Civil and Environmental Engineering Department, University of South Florida's civil engineering graduate students Alex Mraz, Ivan Sokolic, Mathiyaparanam and Kalyani Jeyisankar, Dumina Randeniya, and undergraduate student Mercedes Quintas for their help in preparing the manuscript. The support of my children, Ruwan and Aruni, and my wife, Prabha, during the arduous task of making this project a reality is also gratefully acknowledged. I wish to extend my special thanks to Cindy Renee Carelli, former engineering acquisitions editor; Matt Lamoreaux, current engineering acquisitions editor; Elizabeth Spangenberger; and other staff of Taylor & Francis for their meticulous work on publishing this handbook. Thanks are also due to the relevant publishers who permitted the use of material from other references.

I also express my profound gratitude to late Professor Alagiah Thurairajah, former dean of the Faculty of Engineering, Peradeniya University, Sri Lanka, and prominent member of the Cambridge University's *Cam Clay* group for introducing me to North America and postgraduate studies in geotechnics.

Finally, it is to my mother, Jeannette Gunaratne, and my late father, Raymond Gunaratne, that I dedicate this book.

Manjriker Gunaratne
University of South Florida
Tampa

Abstract

This handbook contains some of the most recent developments in theoretical and applied foundation engineering in addition to classical foundation design methods. The inclusion of recent developments mostly enriches the classical design concepts in Chapters 3–7, 10, and 11. It also enables the reader to update his or her knowledge of new modeling concepts applicable to foundation design. Most recently developed *in situ* testing methods discussed in detail in Chapter 2 certainly familiarize the reader with state-of-the-art techniques adopted in site testing. In addition, modern ground stabilization techniques introduced in Chapter 12 by an experienced senior engineer in Hayward-Baker Inc., a leading authority in site improvement work across North America, provides the reader with the knowledge of effective site improvement techniques that are essential for foundation design. Innovative and widely used methods of testing pile foundations are introduced with numerical illustrations in Chapters 2 and 7. LRFD designs in Chapters 3 and 6 and the design of retaining structures with geogrids included in Chapter 10 are unique features of this foundation engineering handbook. For the benefit of the reader, the basic and advanced soil mechanics concepts needed in foundation design are elaborated with several numerical examples in Chapter 1.

Editor

Manjriker Gunaratne is a professor of civil engineering at the University of South Florida. He completed his pre-engineering education at Ananda College, Colombo, Sri Lanka, receiving the S.A. Wijetileke prize for the highest ranking student. Thereafter, he obtained his bachelor of science in engineering (Honors) degree from the Faculty of Engineering, University of Peradeniya, Sri Lanka, in 1978. In 1977, he was awarded the Professor E.O.E. Pereira prize for the highest ranking student at the Part (II) examination in the overall engineering class. Subsequently, he pursued postgraduate education in North America, earning master of applied science and doctoral degrees in civil engineering from the University of British Columbia, Vancouver, Canada, and Purdue University, West Lafayette, Indiana, respectively. During his 18 years of service as an engineering educator, he has authored 25 papers in a number of peer-reviewed journals, such as the American Society of Civil Engineering (geotechnical, transportation, civil engineering materials, and infrastructure systems) journals, *International Journal of Numerical and Analytical Methods in Geomechanics, Civil Engineering Systems*, and others. In addition, he has made a number of presentations at various national and international forums in geotechnical and highway engineering.

He has held fellowships at the United States Air Force (Wright-Patterson Air Force Base) and the National Aeronautics and Space Administration (Robert Goddard Space Flight Center) and a consultant's position with the United Nations Development Program in Sri Lanka. He has also been a panelist for the National Science Foundation and a member of the task force for investigation of dam failures in Florida, U.S.A.

Contributors

Dr. Austin Gray Mullins is an associate professor of civil engineering at the University of South Florida, Tampa, Florida, who specializes in geotechnical and structural engineering. He obtained B.S., M.S., and Ph.D. degrees in civil engineering from the University of South Florida. Prior to joining USF's department of civil and environmental engineering, he worked as an engineer at Greiner Inc. Roadway Group, Tampa. His most recent research work has been in the areas of statnamic testing of building foundations and drilled shafts as well as structural testing of bridges. He is a professional engineer registered in the state of Florida.

Dr. Alaa Ashmawy is an associate professor of civil engineering at the University of South Florida, Tampa, Florida, with specialization in geotechnical and geoenvironmental engineering. He obtained the B.S. degree in civil engineering from Cairo University, Egypt, and M.S. and Ph.D. degrees from Purdue University. Prior to joining USF's department of civil and environmental engineering, he was a postdoctoral research associate at the Georgia Institute of Technology. His most recent research work has been in the areas of hydraulic and diffusion characteristics of surface amended clays, evaluation of the Purdue TDR Method for soil water content and density measurement, and discrete element modeling of angular particles. He is a professional engineer registered in the state of Florida.

Dr. Panchy Arumugasaamy graduated with first class honors bachelor degree in civil engineering from the University of Sri Lanka, Katubedde Campus, and is the recipient of the 1973 gold medal from the UNESCO Team for ranking first in the Faculty of Engineering and Architecture of that year. He earned his Ph.D. degree in structural engineering in 1978 from the University of Sheffield, England. In 1998, he earned his Executive M.B.A. graduate degree from Ohio University. He has over 25 years of extensive experience in engineering consulting (civil and structural engineering), project management, teaching, advanced research, and product development. He is well respected by his peers for his competencies in the analysis and design of complex structural systems for buildings, bridges, and other structures for different types of applications, and assessment of behavior of elements using both classical and computer aided methods. He is familiar with many codes of practices including American Codes (ACI, AISC, ASCE, SEAOC, and AASHTO), CEP-FIP codes, BSI (for bridges), and CSA. He has hands-on experience in computer modeling, computer aided design including 2D and 3D frame analysis, grillage analysis for bridges, 2D and 3D finite element analysis, and plate analysis to optimize the structural system (steel and concrete structures). He is also proficient in 3D computer modeling. He has also specialized in optical engineering and holds many patents for his inventions. He has published many papers on national and international journals as a coauthor and has received the following awards for the best designs and research papers. He is currently working with MS Consultants Inc. as the head of the structural division in Columbus, Ohio.

He has been a research scholar and senior adjunct faculty at University of West Indies, St. Augustine, Trinidad and Tobago (WI), Florida Atlantic University, Boca Raton, and research associate professor at the University of Nebraska, Lincoln-Omaha.

James D. Hussin received his B.S. in civil engineering from Columbia University and M.S. in geotechnical engineering from California Institute of Technology (CalTech). Dr. Hussin has been with Hayward-Baker Inc. for 20 years and in his current position of director is responsible for the company's national business development and marketing efforts and oversees engineering for the southeast U.S. and the Caribbean. Before joining Hayward-Baker, Dr. Hussin was a geotechnical consultant in Florida and South Carolina. Dr. hussin is a member and past chairman of the American Society of Civil Engineers (ASCE) Geoinstitute National Soil Improvement Committee and is a current board member of the National ASCE Technical Coordination Council that oversees the technical committees. Dr. Hussin has over 20 publications, including associate editor of the ASCE Special Publication No. 69, "Ground Improvement, Ground Reinforcement, Ground Treatment, Developments 1987–1997."

Contents

1

Review of Soil Mechanics Concepts and Analytical Techniques Used in Foundation Engineering

Manjriker Gunaratne

CONTENTS

1.1   Introduction

Geotechnical engineering is a branch of civil engineering in which technology is applied in the design and construction of structures involving geological materials. Earth's surface material consists of soil and rock. Of the several branches of geotechnical engineering, soil and rock mechanics are the fundamental studies of the properties and mechanics of soil and rock, respectively. Foundation engineering is the application of the principles of soil mechanics, rock mechanics, and structural engineering to the design of structures associated with earthen materials. On the other hand, rock engineering is the corresponding application of the above-mentioned technologies in the design of structures associated with rock. It is generally observed that most foundation types supported by intact bedrock present no compressibility problems. Therefore, when designing common foundation types, the foundation engineer's primary concerns are the strength and compressibility of the subsurface soil and, whenever applicable, the strength of bedrock.

1.2   Soil Classification

1.2.1   Mechanical Analysis

According to the texture or the "feel," two different soil types can be identified. They are: (1) coarse-grained soil (gravel and sand) and (2) fine-grained soil (silt and clay). While the engineering properties (primarily strength and compressibility) of coarse-grained soils depend on the size of individual soil particles, the properties of fine-grained soils are mostly governed by the moisture content. Hence, it is important to identify the type of soil at a given construction site since effective construction procedures depend on the soil type. Geotechnical engineers use a universal format called the unified soil classification system (USCS) to identify and label different types of soils. The system is based on the results of common laboratory tests of mechanical analysis and Atterberg limits.

In classifying a given soil sample, mechanical analysis is conducted in two stages: (1) sieve analysis for the coarse fraction (gravel and sand) and (2) hydrometer analysis for the fine fraction (silt and clay). Of these, sieve analysis is conducted according to American Society for Testing and Materials (ASTM) D421 and D422 procedures, using a set of U.S. standard sieves (Figure 1.1) the most commonly used sieves are U.S. Standard numbers 20, 40, 60, 80, 100, 140, and 200, corresponding to sieve openings of 0.85, 0.425, 0.25, 0.18, 0.15, 0.106, and 0.075 mm, respectively. During the test, the percentage (by weight) of the soil sample retained on each sieve is recorded, from which the percentage of soil ($R\%$) passing through a given sieve size (D) is determined.

On the other hand, if a substantial portion of the soil sample consists of fine-grained soils ($D < 0.075$ mm), then sieve analysis has to be followed by hydrometer analysis

FIGURE 1.1
Equipment used for sieve analysis. (Courtesy of the University of South Florida.)

(Figure 1.2). The hydrometer analysis test is performed by first treating the "fine fraction" with a deflocculating agent such as sodium hexametaphosphate (Calgon) or sodium silicate (water glass) for about half a day and then allowing the suspension to settle in a hydrometer jar kept at a constant temperature. As the heavier particles settle, followed by the lighter ones, a calibrated ASTM 152H hydrometer is used to estimate the fraction (percentage, $R\%$) that is still settling above the hydrometer bottom at any given stage. Further, the particle size (D) that has settled past the hydrometer bottom at that stage in

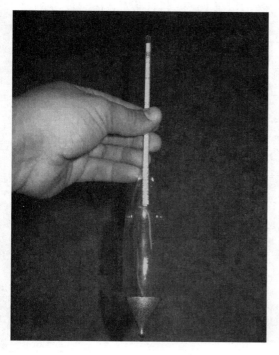

FIGURE 1.2
Equipment used for hydrometer analysis. (Courtesy of the University of South Florida.)

time can be estimated from Stokes' law. Then, it can be seen that $R\%$ is the weight percentage of soil finer than D.

Complete details of the above-mentioned tests such as the correction to be applied to the hydrometer reading and determination of the effective length of the hydrometer are provided in Bowles (1986) and Das (2002). For soil samples that have significant coarse and fine fractions, the sieve and hydrometer analysis results ($R\%$ and D) can be logically combined to generate grain (particle) size distribution curves such as those indicated in Figure 1.3. As an example, from Figure 1.3, it can be seen that 30% of soil type A is finer than 0.075 mm (U.S. Standard no. 200 sieve), with $R\% = 30$ and $D = 0.075$ mm being the last pair of results obtained from sieve analysis. In combining sieve analysis data with hydrometer analysis data, one has to convert $R\%$ (based on the fine fraction only) and D (size) obtained from hydrometer analysis to $R\%$ based on the weight of the entire sample in order to ensure continuity of the curve. As an example, let the results from one hydrometer reading of soil sample A be $R\% = 90$ and $D = 0.05$ mm. To plot the curve, one requires the percentage of the entire sample finer than 0.05 mm. Since what is finer than 0.05 mm is 90% of the fine fraction (30% of the entire sample) used for hydrometer analysis, the converted $R\%$ for the final plot can be obtained by multiplying 90% by the fine fraction of 30%. Hence, the converted data used to plot Figure 1.3 are $R\% = 27$ and $D = 0.05$ mm.

1.2.2 Atterberg Limits

As mentioned earlier, properties of fine-grained soils are governed by water. Hence, the effect of water has to be considered when classifying fine-grained soils. This is achieved

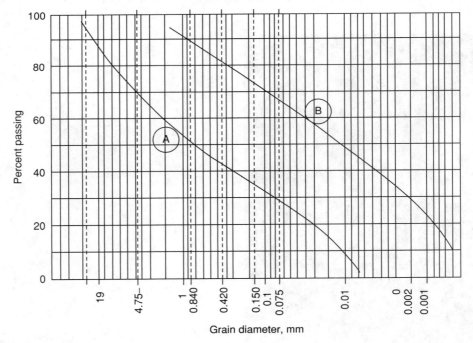

FIGURE 1.3
Grain (particle) size distribution curves. (From *Concrete Design Handbook*, CRC Press. With permission.)

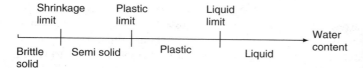

Shrinkage limit	Plastic limit	Liquid limit	
Brittle solid	Semi solid	Plastic	Liquid	Water content

FIGURE 1.4
Variation of the fine-grained soil properties with the water content.

by employing the Atterberg limits or consistency limits. The physical state of a fine-grained soil changes from brittle to liquid state with increasing water content, as shown in Figure 1.4.

Theoretically, the plastic limit (PL) of a soil is defined as the water content at which the soil changes from "semisolid" to "plastic" (Figure 1.4). For a given soil sample, this is an inherent property of the soil that can be determined by rolling a plastic soil sample into a worm shape to gradually reduce its water content by exposing more and more of an area until the soil becomes semisolid. This change can be detected by cracks appearing on the sample. According to ASTM 4318, the PL is the water content at which cracks develop on a rolled soil sample at a diameter of 3 mm. Thus, the procedure to determine the PL is one of trial and error. Although the apparatus (ground glass plate and moisture cans) used for the test is shown in Figure 1.5, the reader is referred to Bowles (1986) and Das (2002) for more details.

On the other hand, the liquid limit (LL), which is visualized as the water content at which the state of a soil changes from "plastic" to "liquid" with increasing water content, is determined in the laboratory using the Casagrande liquid limit device (Figure 1.5). This device is specially designed with a standard brass cup on which a standard-sized soil paste is applied during testing. In addition, the soil paste is grooved in the middle by a standard grooving tool thereby creating a "gap" with standard dimensions. When the brass cup is made to drop through a distance of 1 cm on a hard rubber base, the number of drops (blows) required for the parted soil paste to come back into contact through a

FIGURE 1.5
Equipment for the plastic limit/liquid limit tests. (Courtesy of the University of South Florida.)

distance of 0.5 in. is counted. Details of the test procedure can be found in Bowles (1986) and Das (2002). ASTM 4318 specifies the LL as the water content at which the standard-sized gap is closed in 25 drops of the cup. Therefore, one has to repeat the experiment for different trial water contents, each time recording the number of blows required to fulfill the closing condition of the soil gap. Finally, the water content corresponding to 25 blows (or the LL) can be interpolated from the data obtained from all of the trials. The plasticity index (PI) is defined as follows:

$$PI = LL - PL \qquad (1.1)$$

1.2.3 Unified Soil Classification System

In the commonly adopted USCS shown in Table 1.1, the aforementioned soil properties are effectively used to classify soils. Example 1.1 illustrates the classification of the two soil samples shown in Figure 1.3. Definitions of the following two curve parameters are necessary to accomplish the classification:

Coefficient of uniformity $(C_u) = D_{60}/D_{10}$

Coefficient of curvature $(C_c) = D_{30}^2/D_{60}D_{10}$

where D_i is the diameter corresponding to the ith percent passing.

Example 1.1
Classify soils A and B shown in Figure 1.3.

Solution
Soil A. The percentage of coarse-grained soil is equal to 70%. Therefore, A is a coarse-grained soil. The percentage of sand in the coarse fraction is equal to $(70 - 30)/70 \times 100 = 57\%$. Thus, according to the USCS (Table 1.1), soil A is sand. If one assumes a clean sand, then

$C_c = (0.075)^2/(2 \times 0.013) = 0.21$ does not meet criterion for SW (well-graded)

$C_u = (2)/(0.013) = 153.85$ meets criterion for SW

Hence, soil A is a poorly graded sand, or SP (poorly graded).

Soil B. The percentage of coarse-grained soil is equal to 32%. Hence, soil B is a fine-grained soil. Assuming that LL and PL are equal to 45 and 35, respectively (then PI is equal to 10 from Equation (1.1)), and using Casagrande's plasticity chart (Table 1.1), it can be concluded that soil B is a silty sand with clay (ML or lean clay).

1.3 Effective Stress Concept

Pores (or voids) within the soil skeleton contain fluids such as air, water, or other contaminants. Hence, any load applied on a soil is partly carried by such pore fluids in addition to being borne by the soil grains. Therefore, the total stress at any given location

TABLE 1.1

Unified Soil Classification System

Division		Description	Group Symbol	Identification	Laboratory Classification Criteria
More than 50% soil retained in US 200 sieve (0.075 mm)	More than 50% retained in US No. 4 (4.75 mm)	Clean gravels	GW	Well graded gravels	$C_u > 4, 1 < C_c < 3$
			GP	Poorly graded gravels	Not meeting GW criteria
		Gravel with fines	GM	Silty gravel	Falls below A line in the plasticity chart, or PI less than 4
			GC	Clayey gravel	Falls above A line in the plasticity chart, or PI greater than 7
	More than 50% passing US No. 4 (4.75 mm)	Clean sand	SW	Well graded sand	$C_u > 4, 1 < C_c < 3$
			SP	Poorly graded sand	Not meeting SW criteria
		Sand with fines	SM	Silty sand	Falls below A line in the plasticity chart, or PI less than 4
			SC	Clayey sand	Falls above A line in the plasticity chart, or PI greater than 7
More than 50% soil passing US 200 sieve (0.075 mm)	Fine grained soils (LL < 50)		ML	Inorganic silts with low plasticity	
			CL	Inorganic clays with low plasticity	
			OL	Organic clays/silts with low plasticity	
	Fine grained soils (LL > 50)		MH	Inorganic silts with high plasticity	
			CH	Inorganic clays with high plasticity	
			OH	Organic clays/silts with low plasticity	
Highly organic soils			Pt		Use the Casagrande Plasticity chart shown above

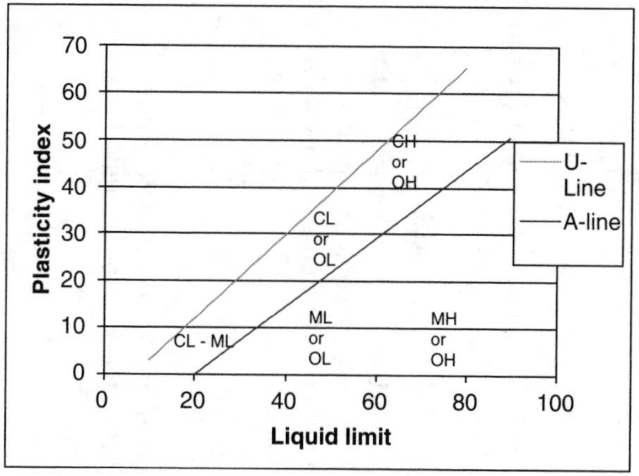

Casagrande Plasticity chart — Plasticity Index vs. Liquid limit, showing U-Line and A-line; regions labeled CL-ML, ML or OL, CL or OL, MH or OH, CH or OH.

within a soil mass can be expressed as the summation of the stress contributions from the soil skeleton and the pore fluids as

$$\sigma = \sigma' + u_p \tag{1.2}$$

where σ is the total stress (above atmospheric pressure), σ' is the stress in the soil skeleton (above atmospheric pressure), and u_p is the pore (fluid) pressure (above atmospheric pressure).

The stress in the soil skeleton or the intergranular stress is also known as the effective stress since it indicates that portion of the total stress carried by grain to grain contacts.

In the case of dry soils in which the pore fluid is primarily air, if one assumes that all pores anywhere within the soil are open to the atmosphere through interporous connectivity, from Equation (1.2) the effective stress would be the same as the total stress:

$$\sigma' = \sigma \tag{1.3}$$

On the other hand, in completely wet (saturated) soils, the pore fluid is mostly water and the effective stress is completely dependent on the pore water pressure (u_w). Then, from Equation (1.2):

$$\sigma' = \sigma - u_w \tag{1.4a}$$

Using the unit weights of soil (γ) and water (γ_w), Equation (1.4a) can be modified to a more useful form as shown in Equation (1.4b):

$$\sigma'_{v0} = \gamma z - \gamma_w d_w \tag{1.4b}$$

where z is the depth of the location from the ground surface (Figure 1.6) and d_w is the depth of the location from the groundwater table (Figure 1.6). A detailed discussion of the unit weights of soil is provided in Section 1.6.

Finally, in partly saturated soils, the effective stress is governed by both the pore water and pore air pressures (u_a). For unsaturated soils that contain both air and water with a high degree of saturation (85% or above), Bishop and Blight (1963) showed that

$$\sigma = \sigma' + u_a - \chi(u_a - u_w) \tag{1.5}$$

where ($u_a - u_w$) is the soil matrix suction that depends on the surface tension of water and χ is a parameter in the range of 0 to 1.0 that depends on the degree of saturation. One can verify the applicability of Equation (1.4a) for saturated soils based on Equation (1.5), since $\chi = 1$ for completely saturated soils.

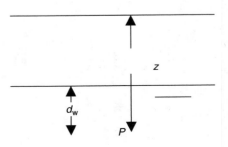

FIGURE 1.6
Illustration of *in situ* stresses.

1.4   Strength of Soils

The two most important properties of a soil that a foundation engineer must be concerned with are strength and compressibility. Since earthen structures are not designed to sustain tensile loads, the most common mode of soil failure is shearing. Hence, the shear strength of the foundation medium constitutes a direct input to the design of structural foundations.

1.4.1   Drained and Undrained Strengths

The shear strength of soils is assumed to originate from the strength properties of cohesion (c) and internal friction (ϕ). Using Coulomb's principle of friction, the shear strength of a soil, τ_f, can be expressed as

$$\tau_f = c + \sigma_n \tan \phi \qquad (1.6)$$

where σn is the effective normal stress on the failure plane. More extensive studies on stress–strain relations of soils (Section 1.8) indicate that more consistent and reliable strength parameters are obtained when Equation (1.6) is expressed with respect to the intergranular or the effective normal stress. Hence, c and ϕ are also known as the effective strength parameters and sometimes indicated as cN and NN. It is obvious that the strength parameters obtained from a shear strength test conducted under drained conditions would yield effective strength parameters due to the absence of pore water pressure. Hence, the effective strength parameters cN and NN are also termed the drained strength parameters. Similarly, failure loads computed based on effective or drained strength parameters are applicable in construction situations that either do not involve development of pore water pressures or where an adequate time elapses for dissipation of any pore pressures that could develop.

Effective strength parameters can also be obtained from any shear strength test conducted under undrained conditions if the pore water pressure developed during shearing is monitored accurately and Equation (1.6) is applied to estimate the shear strength in terms of the effective normal stress σ_n. On the other hand, during any shear strength test conducted under undrained conditions, if Equation (1.6) is applied to estimate the shear strength in terms of the total normal stress σ, one would obtain an entirely different set of strength parameters c and N, which are called the total stress-based strength parameters. Using the concepts provided in the Section 1.7 and relevant stress paths, it can be shown that the total stress-based strength parameters are generally lower in magnitude than the corresponding effective stress parameters.

From the discussion of soil strength it is realized that the measured shear strength of a soil sample depends on the extent of pore pressure generation and therefore the drainage condition that prevails during a shearing test. Hence, the type of soil and the loading rate expected during construction have an indirect bearing on the selection of the appropriate laboratory drainage condition that must be set up during testing.

A wide variety of laboratory and field methods is used to determine the shear strength parameters of soils, c and ϕ. The laboratory triaxial and discrete shear testing, the *in situ* standard penetration testing (SPT), static cone penetration testing (CPT), and vane shear testing (VST) are the most common tests used to obtain foundation design parameters. The determination of the strength parameters using SPT and CPT is addressed in detail in Chapter 2. Hence, only method of evaluating strength parameters based on the triaxial test will be discussed in this chapter.

1.4.2   Triaxial Tests

In this test, a sample of undisturbed soil retrieved from a site is tested under a range of pressures that encompasses the expected field stress conditions imposed by the building foundation. Figure 1.7(a) shows the schematic of the important elements of a triaxial setup; the actual testing apparatus is shown in Figure 1.7(b).

The pore pressure increase that can be expected during triaxial loading of a soil can be expressed using Skempton's pore pressure parameters, A and B, for that particular soil as

$$\Delta u = B\Delta\sigma_3 + A[\Delta\sigma_1 - \Delta\sigma_3] \tag{1.7}$$

where $\Delta\sigma_1$ and $\Delta\sigma_3$ are the increments of the major and the minor principal stresses, respectively.

When A and B for a given soil type are determined using a set of preliminary triaxial tests, one would be able to predict the magnitude of the pore pressure that would be generated in that soil under any triaxial stress state. It can be shown that, for saturated soils, $B = 1.0$.

An alternative way of expressing the pore pressure increase due to triaxial loading is as follows:

$$\Delta u = \Delta\sigma_{\text{oct}} + 3a\,\Delta\tau_{\text{oct}} \tag{1.8}$$

where a is the Henkel pore pressure parameter and σ_{oct} and τ_{oct} are octahedral normal and shear stresses defined, respectively, as

$$\sigma_{\text{oct}} = [\sigma_1 + \sigma_2 + \sigma_3]/3 \tag{1.9a}$$

$$\tau_{\text{oct}} = [(\sigma_1 - \sigma_2)^2 + (\sigma_2 - \sigma_3)^2 + (\sigma_3 - \sigma_1)^2]^{1/2}/3 \tag{1.9b}$$

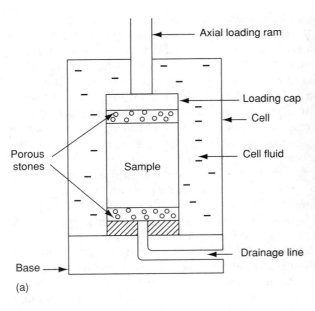

(a)

(b)

FIGURE 1.7
(a) Schematic diagram of triaxial test. (From *Concrete Design Handbook*, CRC Press. With Permission.) (b) Triaxial testing apparatus for soils. (Courtesy of the University of South Florida.)

where σ_2 is the intermediate principal stress. Under the triaxial state of stress, Equations (1.9a) and (1.9b) simplify to

$$\sigma_{\text{oct}} = [\sigma_1 + 2\sigma_3]/3 \tag{1.10a}$$

$$\tau_{\text{oct}} = \sqrt{2}\,(\sigma_1 - \sigma_3)/3 \tag{1.10b}$$

With respect to the drainage condition that is employed during testing, three types of triaxial tests can be conducted: (1) consolidated drained tests (CD), (2) consolidated undrained tests (CU), and (3) unconsolidated undrained tests (UU). In CU and CD tests, prior to applying the axial compression, the pressure of the cell fluid is used to consolidate the soil sample back to the *in situ* effective stress state that existed prior to sampling. On the other hand, in the UU tests, the cell pressure is applied with no accompanying drainage or consolidation, simply to provide a confining pressure.

1.4.2.1 Triaxial Testing of Rocks

When foundations are designed on rocks, as in the case of pile foundations driven to bedrock and pile and drilled shaft foundations cast on bedrock, an accurate estimate of the shear strength of the *in situ* rock is essential. A variety of methods is available in the literature (Goodman, 1989) to determine the shear strength of rock. Of them, the most accurate method of shear strength estimation is perhaps through triaxial testing. Triaxial testing is even more reliable for rock samples than in soils since sample disturbance is not a major issue in the case of rocks. Moreover, correlations that have been developed between the shear strength of rock and the unconfined compression strength (Section 1.4.3) and the rock quality designation (RQD) also provide convenient means of estimating the shear strength parameters of rocks. Further details of such correlations are provided in Section 6.10. Triaxial testing of rock samples is performed using a special apparatus that can sustain the relatively large confining pressures and deviator stresses that must be applied on rock samples to induce shear failure. A set of such apparatus is illustrated in Figure 1.8(a) and (b).

1.4.2.2 Selection of Triaxial Test Type Based on the Construction Situation

The CD strength is critical when considering long-term stability. Examples of such situations are:

1. Slowly constructed embankment on a soft clay deposit
2. Earth dam under steady-state seepage
3. Excavation of natural slopes in clay

On the other hand, CU strength is more relevant for the following construction conditions:

1. Raising of an embankment subsequent to consolidation under its original height
2. Rapid drawdown of a reservoir of an earthen dam previously under steady-state seepage
3. Rapid construction of an embankment on a natural slope

(a)

FIGURE 1.8
(a) Triaxial cell and membrane
used in testing of rock samples.
(b) Triaxial testing of rocks.   (b)

TABLE 1.2

Measured CU Triaxial Test Data

Test	Cell Pressure (kPa)	Deviator Stress at Failure (kPa)	Pore Pressure at Failure (kPa)
1	20	20.2	5.2
2	40	30.4	8.3

Finally, the UU strength is applicable under the following conditions:

1. Rapid construction of an embankment over a soft clay
2. Large dam constructed with no change in water content in the clay core
3. Footing placed rapidly on a clay deposit

1.4.2.3 Computation of Strength Parameters Based on Triaxial Tests

Computations involving CU and UU tests are given in Examples 1.2 and 1.3, and the reader is referred to Holtz and Kovacs (1981) for more details of the testing procedures.

Example 1.2

Assume that one conducts two CU triaxial tests on a sandy clay sample from a tentative site in order to determine the strength properties. The applied cell pressures, deviator stresses, and measured pore pressures at failure are given in Table 1.2. The strength parameters can be estimated using the Mohr circle method as follows:

Solution

Total strength parameters. The total stresses (σ_1 and σ_3) acting on both test samples at failure are indicated in Figure 1.9(a). Accordingly, the Mohr circles for the two stress states can be drawn as shown in Figure 1.10. Then the total strength parameters (also referred to as the undrained strength parameters) can be evaluated from the slope of the direct common tangent, which is the Coulomb envelope (Equation (1.6)), plotted on the Mohr circle diagram as $c = 4.0 \, \text{kPa}$ and $\phi = 13.2°$. It is obvious that the generated pore pressure has been ignored in the above solution. The most appropriate applications of c and ϕ obtained above are cases where foundations are rapidly constructed on a well-consolidated ground.

Effective strength parameters. The effective stresses on both (saturated) test samples at failure are computed by subtracting the pore pressure from the total stress (Equation (1.4a)), as indicated in Figure 1.9(b). The Mohr circles corresponding to the two stress

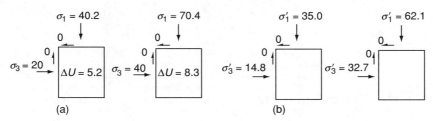

(a)   (b)

FIGURE 1.9

Stress states at failure for Example 1.2: (a) total stress (kPa); (b) effective stress (kPa). (From *Concrete Design Handbook*, CRC Press. With permission.)

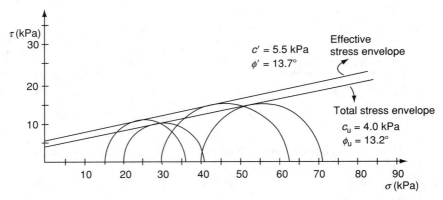

FIGURE 1.10
Mohr circle diagram for a CU test in Example 1.2. (From *Concrete Design Handbook*, CRC Press. With permission.)

states are shown in Figure 1.10. The effective strength parameters (also referred to as the drained strength parameters) can be found from the slope of the Coulomb envelope for effective stresses plotted on the Mohr circle diagram as

$$c' = 5.5 \text{ kPa} \quad \text{and} \quad \phi' = 13.7°$$

The most appropriate applications of the c' and ϕ' are cases where foundations are constructed rather slowly on a well-consolidated ground.

Example 1.3
Assume that one wishes to determine the strength properties of a medium stiff clayey foundation under short-term (undrained) conditions. The most effective method for achieving this is to conduct a UU (quick) test. For the results presented in Table 1.3, estimate the undrained strength parameters.

Solution
In these tests, since the pore pressure generation is not typically monitored the total stresses can be plotted, as shown in Figure 1.11. From Table 1.3, it can be seen that the deviator stress at failure does not change with the changing cell pressure during UU tests. This is because, in UU tests, since no drainage is permitted the soil samples are not consolidated to the corresponding cell pressures. Therefore, the soil structure is largely unaffected by the change in cell pressure. Hence, the following strength parameters can be obtained from Figure 1.11:

$$c_u = 50.6 \text{ kPa} \quad \text{and} \quad \phi_u = 0°$$

TABLE 1.3

Measured UU Triaxial Test Data

Test	Cell Pressure (kPa)	Deviator Stress at Failure (kPa)	Pore Pressure at Failure (kPa)
1	40	102.2	NA
2	60	101.4	NA

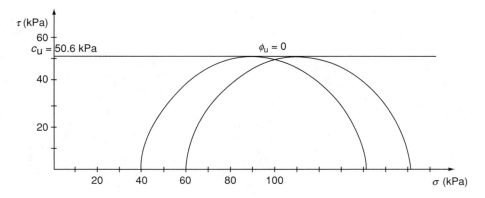

FIGURE 1.11
Mohr circle diagram for a UU test for Example 1.3. (From *Concrete Design Handbook*, CRC Press. With permission.)

It should be noted that the subscript "u" is used to distinguish the UU test parameters. Under UU conditions, if Equation (1.6) is applied, then the undrained shear strength $s_u = c_u$.

The most critical foundation design scenario presented by saturated, slow draining soils such as clays and silts involve undrained conditions prevailing immediately after the foundation is constructed. Therefore, the undrained shear strength (s_u) is typically used to design foundations on soils where the predominant soil type is clay or silt.

1.4.3 Unconfined Compression Test

Very often, it is convenient to use the unconfined compression strength to express the undrained shear strength of clayey soils especially when *in situ* tests are used for such determinations. An unconfined compression test can be used to determine the c_u values based on the measured unconfined compression strength (q_u). Since this test can be visualized as an undrained triaxial test with no confining pressure (hence unconsolidated), the Mohr circle for stress conditions at sample failure can be shown as in Figure 1.12. Then, it can be seen that

$$c_u = \frac{1}{2}q_u \qquad (1.11)$$

The same triaxial apparatus including the loading frame shown in Figure 1.8 can be used to test a clayey soil sample under unconfined compression conditions as well.

Example 1.4
Determine the unconfined compression strength and the undrained shear strength of the soil tested in unconfined compression conditions as shown in Table 1.4.

Solution
The compression test data in Table 1.4 are plotted in Figure 1.13. From Figure 1.13, the unconfined compression strength is determined to be 320 kPa. Therefore, from Equation (1.11), the undrained strength of the clay is estimated to be 160 kPa.

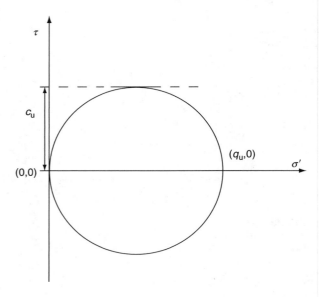

FIGURE 1.12
Mohr circle plot for failure stress condition in
unconfined compression test.

1.5 Compressibility and Settlement

Soils, like any other material, deform under loads. Hence, even if the condition of
structural integrity or bearing capacity of a foundation is satisfied, the ground supporting
the structure can undergo compression, leading to structural settlement. In most dry soils,
this settlement will cease almost immediately after the particles readjust in order to attain
an equilibrium with the structural load. For convenience, this immediate settlement is
evaluated using the theory of elasticity although it is very often nonelastic in nature.

TABLE 1.4

Data for Example 1.8 (Height of Sample — 7.5 cm; Cross-Sectional Area of Sample — 10.35 cm^2)

Vertical Displacement (mm)	Axial Force (N)	Strain (%)	Stress (kPa)
0.030	23.478	0.04	22.68
0.315	52.174	0.39	50.22
0.757	71.739	0.95	68.66
1.219	90.000	1.52	85.64
1.666	106.957	2.08	101.20
2.179	127.826	2.72	120.15
2.682	143.478	3.35	133.99
3.152	163.043	3.94	151.34
3.612	211.304	4.51	194.96
4.171	240.000	5.21	219.82
4.740	260.870	5.92	237.14
5.291	280.435	6.61	253.06
5.850	300.000	7.31	268.69
6.340	314.348	7.92	279.68
7.224	358.696	9.03	315.30
7.991	365.217	9.99	317.65
8.623	349.565	10.78	301.37
9.360	290.870	11.70	248.18

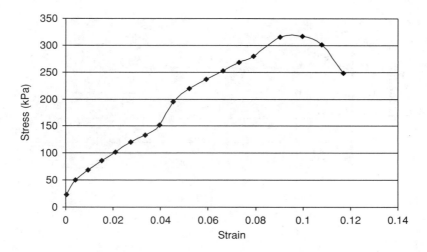

FIGURE 1.13
Plot of the unconfined compression test results in Example 1.4.

However, if the ground material consists of wet, fine-grained (low permeability) soil, the settlement will continue for a long period of time with slow drainage of water accompanied by the readjustment of the soil skeleton until the excess pore water pressure completely dissipates. This is usually evaluated by Terzaghi's consolidation theory. In some situations involving very fine clays and organic soils, settlement continues to occur even after the pore water pressure in the foundation vicinity attains equilibrium with that of the far field. Secondary compression concepts introduced later in this chapter are needed to estimate this prolonged secondary settlement.

1.5.1 Estimation of Immediate Settlement in Soils

The most commonly adopted analytical methods for immediate settlement evaluation in soils are based on the elastic theory. However, one must realize that reliable estimates of elastic moduli and Poisson ratio values for soils are not easily obtained. This is mainly because of the sampling difficulty and, particularly, the dependency of the elastic modulus on the stress state. On the other hand, reliable field methods for obtaining elastic moduli are also scarce. Very often, settlement of footings founded on granular soils or unsaturated clays is determined on the basis of plate load tests (Chapter 4). The following expression can be used to determine the immediate settlement (Bowles, 1896):

$$s_e = f \frac{B q_0}{E_s}(1 - \nu_s^2)\frac{\alpha}{2} \tag{1.12}$$

where α is a factor to be determined from Figure 1.14, B is the width of the foundation, L is the length of the foundation, q_0 is the contact pressure (P/BL), s_e is the immediate settlement, E_s is the elastic modulus of soil, ν_s is the Poisson ratio of soil, and f is equal to 0.5 or 1.0 (depending on whether s_e is evaluated at the corner or center of the foundation).

Another widely used method for computing granular soil settlements is the Schmertmann and Hartman (1978) method based on the elastic theory as well:

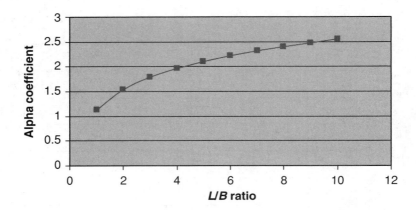

FIGURE 1.14
Chart for obtaining the α factor.

$$S_e = C_1 C_2 (\Delta\sigma - q) \sum_0^z \frac{I_z}{E_s} \Delta z \qquad (1.13)$$

where I_z is the strain influence factor in Figure 1.15 (Schmertmann and Hartman, 1978), C_1 is the foundation depth correction factor ($= 1 - 0.5[q/(\Delta\sigma - q)]$), C_2 is the correction factor for creep of soil ($= 1 + 0.2 \log[\text{time in years}/0.1]$), $\Delta\sigma$ is the stress at the foundation level ($= P/BL$), and q is the overburden stress at the foundation level ($= \gamma z$).

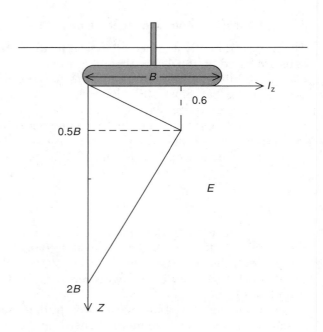

FIGURE 1.15
Strain influence factor.

TABLE 1.5

Poisson Ratios (μ) for Geomaterials

Type of Soil	μ
Clay, saturated	0.4–0.5
Clay, unsaturated	0.1–0.3
Sandy clay	0.2–0.3
Silt	0.3–0.35
Sand, gravelly sand	−0.1 to 1.00
Commonly used	0.3–0.4
Rock	0.1–0.4 (depends somewhat on type of rock)
Loess	0.1–0.3
Ice	0.36
Concrete	0.15
Steel	0.33

Source: From Bowles, J.E., 2002, *Foundation Analysis and Design*, McGraw-Hill, New York. With permission.

The elastic properties needed to manipulate the above expressions are provided in Tables 1.5 (Bowles, 1995) and Table 1.6, where the author, based on his experience, has extracted approximate values from Bowles (1995) for most common soil types.

1.5.1.1 Elastic Properties and In Situ Test Parameters

The most commonly used *in situ* tests that can be used to determine elastic properties of soil are the SPT and CPT tests (discussed in Chapter 2). Some useful relationships that can provide the elastic properties from *in situ* test results are given in Table 1.7. However, in

TABLE 1.6

Approximate Elastic Moduli of Geomaterials

Soil Type	Elastic Modulus (MPa)
Soft clay	2–25
Medium clay	15–50
Stiff clay	50–100
Loose sand	10–20
Medium dense sand	20–50
Dense sand	50–80
Loose gravel (sandy)	50–150
Dense gravel (sandy)	100–200
Silt	2–20

TABLE 1.7

Soil Elastic Moduli from *In Situ* Test Data

Soil	SPT	CPT
Sand (normally consolidated)	$E_s = 500(N + 15)$ $= 7{,}000\sqrt{N}$ $= 6{,}000N$	$E_s = (2\text{--}4)q_u$ $= 8{,}000\sqrt{q_c}$
	— —	— —
		$E_s = 1.2(3D_r^2 + 2)q_c$
	$\ddagger E_s = (15{,}000\text{--}22{,}000)\ln N$	$^*E_s = (1 + D_r^2)q_c$
Sand (saturated)	$E_s = 250(N + 15)$	$E_s = Fq_c$ $e = 1.0, F = 3.5$ $e = 0.6, F = 7.0$
Sands, all (norm. consol.)	$\P E_s = (2{,}600\text{--}2900)N$	
Sand (overconsolidated)	$\dagger E_s = 40{,}000 + 1{,}050N$ $E_{s(\text{OCR})} \approx E_{s,\text{nc}}\sqrt{\text{OCR}}$	$E_s = (6\text{--}30)q_c$
Gravelly sand	$E_s = 1{,}200(N + 6)$ $= 600(N + 6)\quad N < 15$ $= 600(N + 6) + 2{,}000\quad N > 15$	
Clayey sand	$E_s = 320(N + 15)$	$E_s = (3\text{--}6)q_c$
Silts, sandy silt, or clayey silt	$E_s = 300(N + 6)$ If $q_c < 2{,}500$ kPa use$\quad\quad {}^sE_s^r = 2.5qc$ $2{,}500 < q_c < 5{,}000$ use$\quad E_s^t = 4qc + 5000$ where $E_s^t =$ constrained modulus $= \dfrac{E_3(1 - \mu)}{(1 + \mu)(1 - 2\mu)} = \dfrac{1}{m_\mu}$	$E_s = (1\text{--}2)q_c$
Soft clay or clayey silt		$E_s = (3\text{--}8)q_c$

$^*E_s$ (elastic modulus) for SPT (Standard penetration test) and units q_c for CPT (Cone penetration test).
Notes: E_s in kPa for SPT and units of q_c for CPT; divide kPa by 50 to obtain ksf. The N values should be estimated as N_{55} and not N_{70}.
Source: From Bowles, J.E., 2002, *Foundation Analysis and Design*, McGraw-Hill, New York. With permission.

foundation engineering, it is also common to assume the following approximate relations with respect to granular soils:

$$E_s(tsf) = 8N \tag{1.14a}$$

$$E_s(\text{kPa}) = 768N \tag{1.14b}$$

where N is the SPT blow count, and

$$E_s = 2q_c \tag{1.15}$$

where q_c is the cone resistance in CPT measured in units of stress; E_s and q_c have the same units.

A comprehensive example illustrating the use of the above relations is provided in Section 3.3.

1.5.2 Estimation of Foundation Settlement in Saturated Clays

When the foundation load is applied on a saturated fine-grained soil, it is immediately acquired by the pore water, as illustrated in Figure 1.16(a). However, with the gradual dissipation of pore pressure accompanied by drainage of water, the applied stress (total stress, $\Delta\sigma$) is gradually transerred to the sortskeleton as an effective (Figure 1.16b). The longterm rearrangement of the soil skeleton and the consequent foundation settlement that take place during this process is known as the phenomenon of consolidation settlement.

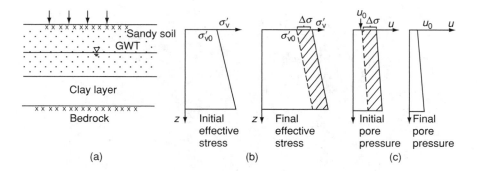

FIGURE 1.16
Illustration of consolidation settlement: (a) subsurface profile; (b) effective stress distribution; and (c) pore pressure distribution. (From *Concrete Design Handbook*, CRC Press. With permission.)

The soil properties required for estimation of the magnitude and rate of consolidation settlement can be obtained from the one-dimensional (1D) laboratory consolidation test. Figure 1.17 shows the consolidometer apparatus where a saturated sample (typically 2.5 in. or 62.5 mm diameter and 1.0 in. or 25.0 mm height) is subjected to a constant load while the deformation and (sometimes) the pore pressure are monitored until the primary consolidation process is complete, resulting in what is known as the "ultimate primary settlement." A detailed description of this test can be found in Das (2002). The sample is tested in this manner for a wide range of stresses that encompass the expected average pressure increase produced by the foundation at the clay layer.

Figure 1.18 shows the results of a consolidation test conducted on a clay sample. The coefficient of consolidation (C_v) for the soil can be obtained from the above results using Casagrande's logarithm-of-time method (Holtz and Kovacs, 1981). The coefficient of consolidation, C_v, is defined based on Equation (1.16):

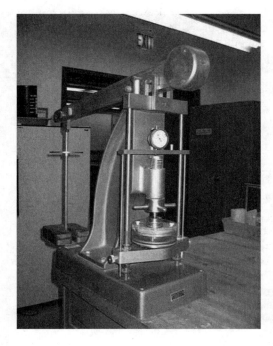

FIGURE 1.17
Laboratory consolidometer apparatus. (Courtesy of the University of South Florida.)

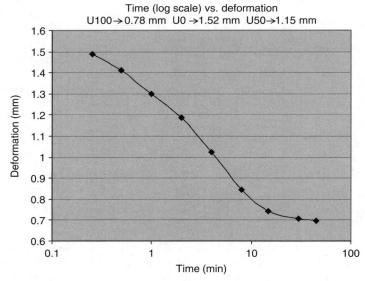

FIGURE 1.18
Settlement versus logarithm-of-time curve.

$$T = \frac{C_v t}{H_{dr}^2} \tag{1.16}$$

where H_{dr} is the longest drainage path in the consolidating soil layer and T is the nondimensional time factor. It should be noted that water is permitted to drain from both sides of the laboratory soil sample during consolidation. Hence, $H_{dr} = 0.5$ in. or 12.5 mm.

Furthermore, for a clay layer that is subjected to a constant or linear pressure increment throughout its depth, the relationship between the average degree of consolidation, U (settlement at any time t as a percentage of the ultimate primary settlement) and the nondimensional time factor, T, shown in Table 1.8, can be derived using Terzaghi's 1D consolidation theory.

Example 1.5
Compute the value of C_v using Figure 1.18.

Solution

From Figure 1.18, when $U = 50\%$, $t = 135$ sec
However, from Table 1.8, when $U = 50\%$, $T = 0.197$
Substitute in Equation (1.16), $C_v = 5.96 \times 10^{-2}\,\text{mm}^2/\text{sec}$

When the above consolidation test is repeated for several different pressure increments, each time doubling the pressure, the variation of the postconsolidation (equilibrium) void ratio (e) with pressure (p) can be plotting the following relations:

$$\frac{\Delta e}{1 + e_0} = \frac{\Delta H}{H} \tag{1.17a}$$

TABLE 1.8

Degree of Consolidation versus Time Factor

U_{avg}	T
0.1	0.008
0.2	0.031
0.3	0.071
0.4	0.126
0.5	0.197
0.6	0.287
0.7	0.403
0.8	0.567
0.9	0.848
0.95	1.163
1.0	∞

$$e = e_0 - \Delta e \qquad (1.17b)$$

where e_0 and H are the initial void ratio and the sample height, respectively, while ΔH and Δe are their respective changes. It must be noted that for an applied pressure of p when the primary consolidation is over with complete dissipation of pore pressure and the equilibrium void ratio is reached, the effective stress in the soil (p') is equal to p. Hence, when e values corresponding to the applied pressure p are plotted, it is realized that, in effect, the resulting plot is an e versus p' plot. A typical laboratory consolidation curve (e versus log p') for a clayey soil sample is shown in Figure 1.19. The following important parameters can be obtained from Figure 1.19:

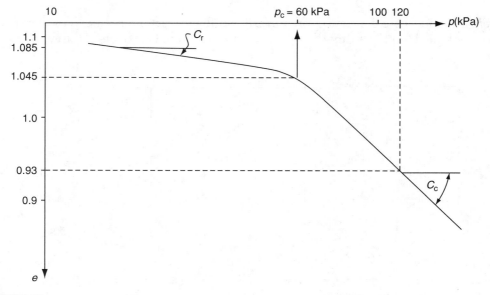

FIGURE 1.19
Laboratory consolidation curve (e versus log p'). (From *Concrete Design Handbook*, CRC Press. With permission.)

Recompression index, $C_r = (1.095 - 1.045)/(\log 60 - \log 10) = 0.064$

Compression index, $C_c = (1.045 - 0.93)/(\log 120 - \log 60) = 0.382$

Preconsolidation pressure, $p_c = 60\,\text{kPa}$

All of the above information can be used to estimate the ultimate consolidation settlement of a saturated clay layer (of thickness H) due to an average pressure increase of Δp. The ultimate consolidation settlement (s_{con}) can be determined by the following expressions, depending on the initial effective stress state (σ'_{vo}) and the load increment Δp, as illustrated in Figure 1.20.

Case 1 ($\sigma'_{vo} > p_c$)

$$s_{con} = \frac{C_c H}{1 + e_0} \log \frac{\sigma'_{vo} + \Delta p}{\sigma'_{vo}}$$ (1.18a)

Case 2 ($\sigma'_{vo} + \Delta p < p_c$)

$$s_{con} = \frac{C_r H}{1 + e_0} \log \frac{\sigma'_{vo} + \Delta p}{\sigma'_{vo}}$$ (1.18b)

Case 3 ($\sigma'_{vo} + \Delta p > p_c > p_0$)

$$s_{con} = \frac{C_r H}{1 + e_0} \log \frac{p_c}{\sigma'_{v_0}} + \frac{C_c H}{1 + e_0} \log \frac{\sigma'_{vo} + \Delta p}{p_c}$$ (1.18c)

Equations (1.18) can be derived easily based on Equation (1.17a) and Figure 1.19. The average pressure increase in the clay layer due to the foundation can be accurately determined by using Newmark's chart, as shown in Figure 1.21. When the footing is drawn on the chart to a scale of $OQ = d_c$, the depth of the mid-plane of the clay layer from the bottom of footing, Δp, can be evaluated by

$$\Delta p = qIM$$ (1.19)

where q, I, and M are the contact pressure, the influence factor (specific to the chart), and when the scaled footing is drawn on the chart, the number of elements of the chart covered

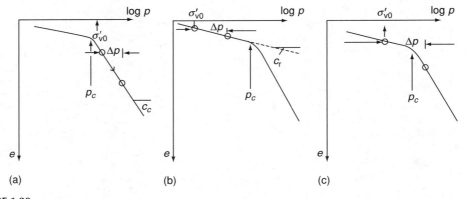

(a)   (b)   (c)

FIGURE 1.20
Illustration of the use of consolidation equation: (a) case 1, (b) case 2, and (c) case 3. (From *Concrete Design Handbook*, CRC Press. With permission.)

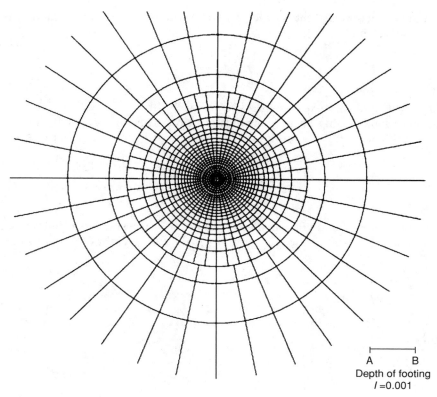

A   B
Depth of footing
$I = 0.001$

FIGURE 1.21
Newmark's influence chart.

by the drawn footing, respectively. The footing must be drawn so that the vertical projection of the location where the settlement is desired coincides with the center of the chart.

1.6 Soil Densities and Compaction

It is essential for designers of foundations and retaining structures to possess knowledge of the density of soils under different moisture states. In addition, sound knowledge of how to determine and improve soil densities is vital as well. For this purpose, commonly used soil densities and corresponding density, water content, and void ratio relations are introduced in the following sections.

1.6.1 Bulk Unit Weight

The bulk or moist unit weight (γ_b) is the total weight (W_T) of a unit volume of soil that includes water and air. In order to determine γ_b, one has to accurately estimate the volume (V_T) of a soil mass (Equation (1.20a)). Hence, the estimation of *in situ* γ_b becomes somewhat of a difficult task that is generally addressed by specially designed tests like the sand-cone test (Section 1.6.5.2):

$$\gamma_b = \left(\frac{W_T}{V_T}\right) \tag{1.20a}$$

Typically unit weights are expressed in kN/m³ or lbf/ft³.

On the other hand, using basic quantification properties of soil (such as the moisture content and the void ratio), the bulk unit weight of a soil can also be expressed conveniently as

$$\gamma_b = \gamma_w G_s \left(\frac{1+w}{1+e}\right) \tag{1.20b}$$

where w is the moisture content, e is the void ratio, G_s is the specific gravity of solids that typically ranges between 2.5 and 2.75, and γ_w is the unit weight of water (9.8 kN/m³ or 62.4 lbf/ft³).

1.6.2   Dry Unit Weight

The dry unit weight (γ_d) is the weight of solids (W_S) of a unit volume of soil that includes water and air:

$$\gamma_d = \left(\frac{W_S}{V_T}\right) \tag{1.21a}$$

Similarly, by setting the water content in Equation (1.20b) to zero, it can be seen that the dry unit weight of a soil can be expressed conveniently as

$$\gamma_d = \gamma_w G_s \left(\frac{1}{1+e}\right) \tag{1.21b}$$

Then, using Equations (1.20b) and (1.21b), one can derive the relationship that enables the dry unit weight of a soil to be determined conveniently from the bulk unit weight:

$$\gamma_d = \left(\frac{\gamma_b}{1+w}\right) \tag{1.22}$$

1.6.3   Saturated Unit Weight

The subsurface soil beneath the groundwater table or within the capillary zone is saturated with water. The bulk unit weight under saturated conditions is conveniently expressed by the saturated unit weight (γ_{sat}), which implies a degree of saturation (s) of 100%. The relationship between the basic quantification properties of soil furnishes a valuable computational tool in unit weight estimations:

$$se = wG_s \tag{1.23}$$

Then, using an s value of 100%, one can use Equations (1.20b) and (1.23) to express the saturated unit weight as

$$\gamma_{\text{sat}} = \gamma_{\text{w}} \left(\frac{G_{\text{s}} + w}{1 + e} \right) \tag{1.24}$$

1.6.4 Submerged (Buoyant) Unit Weight

In foundation stress computations (Equation (1.4b)) involving "under-water" soil conditions, the buoyant effect due to the water table can be included directly by using the submerged or buoyant unit weight γ' (Equation (1.25)), in place of the saturated unit weight. This is especially useful in effective stress computations because the need for separate consideration of pore pressure (Equations (1.4b)) can be precluded:

$$\gamma' = \gamma_{\text{sat}} - \gamma_{\text{w}} \tag{1.25}$$

1.6.5 Soil Compaction

Prior to construction of building foundations newly constructed embankments and natural subgrades must be compacted to density specifications within limitations of water content. One has to generally perform a laboratory compaction test on the foundation soil in advance, in order to set the appropriate compaction specifications. The two commonly performed tests are: (1) standard Proctor compaction test and (2) modified Proctor compaction test. This section provides a summary of the laboratory compaction computations. The reader is referred to Das (2002) for experimental details of these tests.

1.6.5.1 Laboratory Compaction

During laboratory compaction tests, a sample from the foundation soil is compacted at different water contents using the standard compaction equipment shown in Figure 1.22. The weight of the compacted soil filling the standard mold and its water content are

FIGURE 1.22
Laboratory soil compaction equipment. (Courtesy of the University of South Florida.)

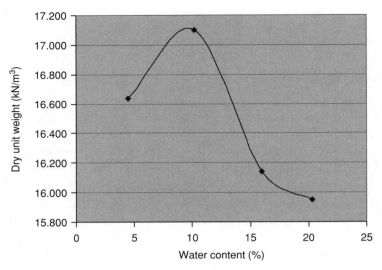

FIGURE 1.23
Laboratory compaction plot for Example 1.6.

recorded in each trial as shown in Table 1.9. Then, the laboratory compaction curve is plotted based on these data as indicated in Figure 1.23.

Example 1.6
Plot the compaction curve for the data provided in the first two columns of Table 1.9 for a standard Proctor compaction test and determine the maximum dry unit weight achievable under standard compaction conditions and the corresponding optimum water content. *Note:* The volume of the standard compaction mold (Figure 1.22) is 940 cc.

Solution
Table 1.9 also shows the computational procedure used to obtain the bulk unit weight and the dry unit weight for each trial based on Equations (1.20a) and (1.22). Figure 1.23 illustrates the plot of dry unit weight of the compacted soil versus the water content. It can be seen from Figure 1.23 that

(i) the maximum dry unit weight $= 17.1 \text{ kN/m}^3$

(ii) optimum water content $= 10\%$

TABLE 1.9

Data for Example 1.6

Mass of Compacted Soil and Mold (g)	Compacted Water Content (%)	Unit Weight (kN/m³), Equation (1.20a)	Dry Unit Weight (kN/m³), Equation (1.22)	Unit Weight (kN/m³) for 100% Saturation, Equation (1.24)
5873	4.48	17.383	16.637	23.214
6012	10.18	18.840	17.100	20.452
6001	15.99	18.722	16.142	18.241
6046	20.33	19.193	15.951	16.877

It is also quite common to plot the 100% degree of saturation line on the same plot (Table 1.9, column 5). Based on the above laboratory results, the specifications for field compaction of the particular soil can be set as follows.

The compacted dry unit weight in the field must be at least 98% of the laboratory maximum dry unit weight (i.e., 17.1 kN/m^3). The field water content must be within 2% of the optimum water content found in the laboratory (i.e., between 8 and 12%).

1.6.5.2 Evaluation of Field Compaction

Unit weights of compacted *in situ* soils and newly laid embankments can be evaluated using many methods. The most common ones are: (1) the sand-cone test, (2) the rubber balloon test, and (3) the nuclear gage method. Experimental details of the first two methods are found in Holtz and Kovacs (1981) while use of the nuclear gage is described in detail in Wray (1986). In this chapter, a numerical example illustrating the sand-cone test procedure is presented. The equipment used in the sand-cone test is shown in Figure 1.24.

Example 1.7
Estimate the field dry unit weight of a given embankment fill based on the sand-cone test readings provided below. For the benefit of the reader, the recorded data are italicized to differentiate them from the computations.

Solution
Step 1. Determination of density of sand in the laboratory
A uniformly graded sand (typically Ottawa sand), which is not very sensitive to compaction, is used for the calibration. Hence, the density of this sand, which is assumed to be invariant, can be first established based on measurements made with a mold of a known volume (e.g., standard compaction mold of volume 940 cc).

Diameter of mold $=$ 10.13 cm

Height of mold $=$ 11.65 cm

Mass of mold and sand $=$ 5602 g

Mass of empty mold $=$ 4252 g

FIGURE 1.24
Equipment for sand-cone test. (Courtesy of the University of South Florida.)

Volume of mold $= 938.94\,cm^3$

Mass of sand in mold $= 1350\,g$

Density of sand $= 1.4378\,g/cm^3 = 1437.8\,kg/m^3$

Step 2. Determination of mass of sand to fill cone

The next step is to determine the volume of the cone (Figure 1.24) by filling it with the calibrated sand (Ottawa sand):

Mass of jar and cone before filling the cone $= 3516\,g$

Mass of jar, cone, and sand after filling the cone $= 1934\,g$

Mass of sand filling the cone $= 1582\,g$

Step 3. In place measurements

Then, the sand-cone apparatus is placed on a previously dug hole in the field compacted layer and Ottawa sand is poured in gently until it completely fills the hole. On the other hand, the *in situ* soil removed from the hole is collected into a pan and weighed. The volume of the hole created by the removal of soil is estimated by knowing the amount of calibrated sand required to fill the hole and the cone:

Mass of jar and sand before use $= 6538\,g$

Mass of jar and sand after use $= 4325\,g$

Mass of collected soil $= 870\,g$

Mass of sand in hole + cone $= 2213\,g$

Mass of sand in hole $= 2213 - 1582 = 631$

Volume of sand in hole $= 631\,g/1.4378 = 438.9\,cm^3$

From Equation (1.20a),

Bulk density of soil $= 870/438.9 = 1.9824\,g/cm^3 = 1982.4\,kg/m^3$

Step 4. In place moisture content measurements

Finally, a simple water content test is performed for the soil fill as indicated in Table 1.10.

Average moisture content $= 1/3\,(6.16\% + 4.52\% + 6.13\%) = 5.6\%$

Based on Equation (1.22),

Dry density $= 1982.4/1.056 = 1877\,kg/m^3$

Dry unit weight $= 1877(9.8)/10^{-3} = 18.39\,kN/m^3$

TABLE 1.10

Water Content for Example 1.7

	Trial 1	Trial 2	Trial 3
Mass of container + wet soil (g)	146.54	142.52	147.32
Mass of container + dry soil (g)	144.63	140.89	144.83
Mass of container (g)	113.65	104.89	104.18

1.7 Finite Element Concepts Used in Modeling of Earthen Structures

1.7.1 Finite Element Approach

The finite element method (FEM) is widely used to model the load–deformation behavior of foundations, piles, retaining walls, and other earthen structures and derive important parameters relating to their design. With the availability of sophisticated and efficient computational facilities, finite element analysis can facilitate effective design criteria, even on a case-by-case basis, with the aid of parametric studies that involve design parameters relevant to each case. It is particularly attractive for situations that involve the design of irregular and relatively complex earthen structures. The basic philosophy involved in modeling an earthen structure with the FEM can be summarized by the following basic principles that form the framework of FEM formulation:

1. Satisfy the force equilibrium of each finite soil or structural element.
2. Satisfy the deformation compatibility at nodal points of each finite element considered.
3. Incorporate an appropriate stress–strain behavior model for each soil or structural material that composes the structure.

The mathematical techniques used to achieve the above-mentioned tasks will be summarized later in this chapter.

However, because of its very nature, finite element solutions also suffer from all drawbacks characteristic of numerical approximations. Based on the above discussion one realizes that the two most important steps that require the special attention of the analyst are:

1. Discretization of the soil–structure influence zone into finite soil or structural elements that could capture all of the load, deformation, geometric, and boundary effects that determine the overall behavior of the particular earthen structure under the given loading conditions.
2. Selection of the appropriate constitutive models that would describe, as accurately as possible, the stress–strain behavior of different soil and structural materials that make up the earthen structure being analyzed.

1.7.2 Finite Element Formulation

The first step involved in the formulation is the determination of the type of element to be used in modeling. Then, the strain field at any point on the selected soil (or structural) element must be expressed in terms of the nodal deflections. Analysts have employed a variety of different elements such as linear triangular elements, bilinear quadrilateral elements, trilinear hexahedral elements depending on their applicability to model different situations (Hughes, 1987). The use of isoparametric quadrilateral elements has been common in geotechnical modeling because they can be designed to take on convenient shapes, such as curved boundaries, often encountered in geotechnical problems. Standard nodal shape functions (N) for many elements are available in the literature (Zienkiewich, 1977; Hughes, 1987).

The significance of the nodal shape function of a quadrilateral element, N_j, where j ($j = $ 1 to 4) denotes the local node number, is that the coordinates, displacement, velocity, or the acceleration of any point within a given soil (or structural) element can be expressed in terms of corresponding nodal values using the shape functions as follows:

$$u_i(t) = \sum_{j=1}^{4} N_j(\xi, \eta) u_i^j(t) \tag{1.26}$$

where $N_j(\xi,\eta)$ is the shape function for the local node j in the local isoparametric coordinates (ξ,η), $u_i^j(t)$ is the time variation of any physical quantity i (coordinates, displacement, velocity, or acceleration) of the nodal point j, and $u_i(t)$ is the time variation of the corresponding physical quantity i at any other point within the element.

An example of shape functions for isoparametric bilinear quadrilateral elements is provided below:

$$N_1(\xi,\eta) = \frac{1}{4}(1 - \xi)(1 - \eta) \tag{1.27a}$$

$$N_2(\xi,\eta) = \frac{1}{4}(1 + \xi)(1 - \eta) \tag{1.27b}$$

$$N_3(\xi,\eta) = \frac{1}{4}(1 + \xi)(1 + \eta) \tag{1.27c}$$

$$N_4(\xi,\eta) = \frac{1}{4}(1 - \xi)(1 + \eta) \tag{1.27d}$$

It is realized that if one is only concerned with the static behavior of the earthen structure, then the time variation need not be considered and the modeling problem becomes far less complicated. Then, the quantity u will only represent the displacement of the points (or the nodes) of interest.

The strain field ($\underline{\varepsilon}$) at any location can be expressed in terms of the differential form of the displacement field ($\underline{u}$) of that point as

$$\bar{\varepsilon} = [B]\bar{u} \tag{1.28}$$

Following are some examples of strain–displacement matrices [B]. For 2D (plane stress or plane strain) situations (e.g., retaining wall and dams) that involve two displacements (u and v) and three strains (ε_x, ε_y, γ_{xy}),

$$[B] = \begin{bmatrix} \partial/\partial x & 0 \\ 0 & \partial/\partial y \\ \partial/\partial y & \partial/\partial x \end{bmatrix} \tag{1.29}$$

For 3D axisymmetric situations (e.g., axial loading of piles), which can be described by only two displacements (u and w) and three strains (ε_r, ε_z, ε_θ, γ_{rz}),

$$[B] = \begin{bmatrix} \partial/\partial r & 0 \\ 0 & \partial/\partial z \\ 1/r & 0 \\ \partial/\partial z & \partial/\partial r \end{bmatrix} \tag{1.30}$$

On the other hand, strain components can be related to the corresponding stresses using the constitutive relations matrix as follows:

$$\bar{\sigma} = [D]\bar{\varepsilon} \tag{1.31}$$

For 2D (plane stress or plane strain) situations,

$$\bar{\sigma} = \begin{bmatrix} \sigma_x & \sigma_y & \tau_{xy} \end{bmatrix}^T \tag{1.32}$$

For 3D axisymmetric situations,

$$\bar{\sigma} = \begin{bmatrix} \sigma_r & \sigma_z & \sigma_\theta & \gamma_{xy} \end{bmatrix}^T \tag{1.33}$$

In the case of structural elements that exhibit elastic behavior, the $[D]$ matrix can be expressed as follows for plane strain conditions:

$$[D] = \frac{E}{(1+\nu)(1-2\nu)} \begin{bmatrix} 1-\nu & \nu & 0 \\ \nu & 1-\nu & 0 \\ 0 & 0 & (1-2\nu)/2 \end{bmatrix} \tag{1.34}$$

where E and ν are the elastic modulus and the Poisson ratio of the element material, respectively. Similarly, the incremental form of Equation (1.31) can be used to model the nonlinear elastic behavior that is predominantly exhibited by soils at low stress levels as follows:

$$[\Delta\sigma] = [D][\Delta\varepsilon] \tag{1.35}$$

$$[D] = \frac{E_t}{(1+\nu_t)(1-2\nu_t)} \begin{bmatrix} 1-\nu_t & \nu_t & 0 \\ \nu_t & 1-\nu_t & 0 \\ 0 & 0 & (1-2\nu_t)/2 \end{bmatrix} \tag{1.36}$$

where E_t and ν_t are the respective instantaneous (tangential) elastic modulus and the Poisson ratio applicable to the current stress level.

However, it is the yielding of soils that governs the behavior of soils in most loading cases. Therefore, some soil constitutive relations that are commonly used to model the yielding (plastic) behavior of soil are discussed separately in Section 1.8.

By combining Equations (1.31) and (1.28), the stress vector can be expressed in terms of the displacement as

$$\bar{\sigma} = [D][B]\bar{u} \tag{1.37}$$

1.7.3 Equilibrium and Compatibility Conditions

When the FEM is used to solve geomechanics problems under static or dynamic conditions, one must satisfy the force equilibrium equations or the equations of motion, respectively, for both soil skeleton and pore water phases. On the other hand, under transient conditions, the equations of motion must be satisfied invariably for both soil skeleton and pore water phases of any general foundation system. In addition, the volumetric compatibility between the pore water and the soil skeleton must be assured as well. Therefore, the equations of equilibrium (or motion) can be combined with those of volumetric compatibility between the pore water and the soil skeleton. As an example, in the case of axisymmetric problems, the resulting equation can be expressed for any given point in the radial (r) direction as (Zienkiewich et al., 1977)

$$\frac{\partial \varepsilon_v}{\partial t} + \frac{\partial}{\partial r}\left[-\frac{k}{\gamma_w}\frac{\partial p}{\partial z} - \frac{k}{g}\ddot{u}\right] - \frac{\partial p}{\partial t}\frac{1-n}{K_s} + \frac{\partial \sigma_m}{\partial t}\frac{1}{K_s} + \frac{\partial p}{\partial t}\frac{n}{K_w} = 0 \tag{1.38}$$

where $\ddot{u}$ is the displacement of a point in the r direction, ε_v is the average volumetric strain in the element, k is the coefficient of permeability assumed to be the same in all directions (isotropic condition), p is the pore water pressure, σ_m is the mean normal stress ($= \sigma_r + \sigma_z + \sigma_\theta$), n is the porosity of the soil skeleton ($=$ void ratio/[1 + void ratio]) (Section 1.6), K_s is the compressibility of the solid grains, and K_w is the compressibility of water.

Similar equations can be written in the other directions (i.e., z and θ) as well. Then, by using the standard Galerkin method (Zienkiewich, 1977), Equation (1.38) for a single point can be generalized for an entire isoparametric finite element in the following integral form:

$$\int_V [N]^T\left[\frac{\partial \varepsilon_v}{\partial t} + \frac{\partial}{\partial r}\left[-\frac{k}{\gamma_w}\frac{\partial p}{\partial z} - \frac{k}{g}\ddot{u}\right] - \frac{\partial p}{\partial t}\frac{1-n}{K_s} + \frac{\partial \sigma_m}{\partial t}\frac{1}{K_s} + \frac{\partial p}{\partial t}\frac{n}{K_w}\right] = 0 \tag{1.39}$$

where

$$[N] = \begin{bmatrix} N_1 & 0 & N_2 & 0 & N_3 & 0 & N_4 & 0 \\ 0 & N_1 & 0 & N_2 & 0 & N_3 & 0 & N_4 \end{bmatrix} \tag{1.40}$$

where N_i are defined by Equation (1.27) and V is the volume of the finite element.

Finally, by applying the standard finite element procedure and summing up Equation (1.39) over m number of elements covering the entire domain that is modeled, the following stiffness relationship can be derived (Zienkiewich, 1977):

$$\sum_{n=1}^{m}\int_{V_n}(\rho[N]^T[N]\ddot{u})\,dV + \int_{V_n}([B]^T[D][B]u)\,dV + \int_{V_n}([N]^T[N]p)\,dV - \int_{b_n}([N]^Tt)\,ds = 0 \tag{1.41}$$

where V_n is the volume of the nth finite element, b_n is the boundary of nth finite element, and t is the traction (force per unit length) along the boundary of the nth element.

It is also seen how Equation (1.31) has been used in substituting for $[\sigma]$ in Equation (1.39). Equation (1.41) is usually integrated within each element using the Gauss quadrature method. The integration points used in the 2×2 Gauss quadrature technique are indicated in Figure 1.25. After the integration operation and mathematical manipulation, Equation (1.41) can be expressed in terms of the nodal force vector $[F]$ and displacement vector $[\delta]$ and arranged in a format similar to Equation (1.42). Then the corresponding stiffness matrix, $[K]$, of the foundation structure can be obtained by comparison with Equation (1.42):

$$[F] = [K][\delta] \tag{1.42}$$

When the stiffness matrix, $[K]$, of the foundation structure is available, one can use a linear algebraic solution scheme to solve for the unknown deflections and forces (such as reactions) with the known boundary conditions (deflections and forces). The solution technique is illustrated in Section 8.3.2.1.

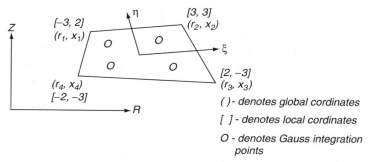

FIGURE 1.25
Illustration of an isoparametric finite element and integration points.

1.8 Common Methods of Modeling the Yielding Behaviour of Soils

Analytical solutions to foundation engineering problems, irrespective of whether they are numerical or deterministic in their approach, must incorporate criteria for modeling the stress–strain behavior of foundation material as accurately as possible. Most geomaterials exhibit nonlinear elastic properties at low stress levels, and therefore elastic methods described in Section 1.5, such as the Schmertmann method, would provide reasonable estimates of settlements at relatively low stress levels only. However, when relatively large loads are applied on a foundation, the foundation soils generally start to yield under these loads producing irrecoverable deformation. For instance, it was shown in Section 1.5 that low permeability soils like clay and silt exhibit time-dependent irrecoverable consolidation, which cannot be modeled using the elastic or the incremental elastic theory. Furthermore, excessive settlements undergone by uncompacted coarse-grained soils like loose and medium dense sand and gravel cannot be analytically predicted satisfactorily using the elastic or the incremental elastic theory alone.

In addition, if the analyst is equipped with a comprehensive stress–strain theory that could model the complete behavior of a loaded earthen structure from initial small deformation stages through large deformation yielding to ultimate failure, the analyst would be able to extract vital design parameters that would be useful in the design of that earthen structure not only to satisfy strength limits but also the desired serviceability limits. Hence, foundation engineers have to employ sophisticated constitutive (stress–strain) models that can model earthen structures, analytically or numerically. Two such popular models, (1) modified Cam-clay model for clays and (2) cap model that is typically used for granular soils, will be discussed in the subsequent sections. However, the effective application of any constitutive model to predict the behavior of an earthen structure accurately depends on the fulfillment of the following experimental tasks:

1. Appropriate laboratory testing to determine the specific material parameters needed to execute the theoretical model.

2. Field pilot testing based on scale models or prototypes themselves to verify the analytical or numerical predictions for actual field applications, and perhaps to further calibrate the analytical model.

1.8.1 Modified Cam-Clay Model

The modified Cam-clay model is based on research performed by Roscoe and Burland (1968). It has been applied successfully in many field applications involving deformation of soft clays. In this theory, the isotropic consolidation behavior of clays is approximated by the following relationships.

1.8.1.1 Isotropic Consolidation of Clays

The following terminology applies to Figure 1.26. The mean normal effective stress, p, is defined by

$$p = \frac{1}{3}(\sigma'_1 + 2\sigma'_3) \tag{1.43}$$

where σ'_1 and σ'_3 are the major and minor principal effective stresses, respectively. It is seen that under laboratory triaxial conditions (Section 1.4), the above stresses correspond, respectively, to axial and cell pressures. The specific volume is defined as $1 +$ void ratio or $1 + e$.

Then, the standard equation of the normal consolidation line is

$$v = N - \lambda \ln(p) \tag{1.44}$$

Similarly, the standard equation of any recompression line (RCL) is

$$v = v_\kappa - \kappa \ln(p) \tag{1.45}$$

where N, κ, and λ are model parameters that can be obtained from laboratory isotropic consolidation tests performed using the triaxial cell. During isotropic consolidations tests, the clay sample is compressed in 3D conditions as compared to conventional 1D tests. Figure 1.26 also shows that there are an infinite number of RCLs that form a family of RCLs for different v_κ values. Figure 1.26 also shows how v_κ can be related to the over-consolidation ratio (OCR) that is characteristic of each RCL. OCR is defined as the

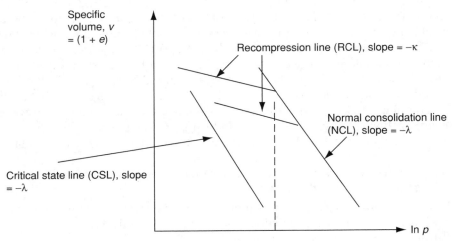

FIGURE 1.26
Isotropic consolidation parameters for the modified Cam-clay model.

ratio of the maximum past effective pressure (p_c in Figure 1.26) to the current effective pressure as

$$v_\kappa = N - (\lambda - \kappa) \ln [(OCR)p_0] \tag{1.46}$$

or

$$v_\kappa = N - (\lambda - \kappa) \ln [p_c] \tag{1.47}$$

It is important to realize that the deformation behavior of overly consolidated (OC) clay samples within a given RCL is elastic in nature. However, plastic deformations occur if the stress path corresponding to a specific loading situation displaces the stress–volume state of an OC sample onto a different RCL.

1.8.1.2 Critical State of Deformation of Clay

Shear strength testing of clays under triaxial (Section 1.4) or direct shear conditions shows that the ultimate failure occurs at a critical state where excessive shear deformation occurs with no further change in the stress conditions, i.e., shear or mean normal effective stress. This final state reached is a unique state for a given clay type independent of the initial consolidation state of the clay sample, i.e., normally consolidated (NC) or OC, and the drainage condition that exists during shearing, i.e., drained (CD) or undrained (CU). If the shear stress is defined by the following expression:

$$q = \sigma'_1 - \sigma'_3 \tag{1.48}$$

then the critical state line can be depicted on a q–p plot as in Figure 1.27. The equation of the CSL can be expressed as

$$q = Mp \tag{1.49}$$

where M is another modified Cam-clay model parameter that can be obtained from any triaxial test performed on a sample from that clay.

In order to better visualize the gradual deformation of a field clay sample starting under K_0 conditions (Figure 1.27) until it ultimately reaches the critical state line and fails, its combined stress and volumetric strain path can be plotted on a 3D q–p–v plot shown in Figure 1.28. Figure 1.28 also shows that the stress and volumetric deformation states of clays are bound by a convex surface known as the state boundary surface (SBS) and a

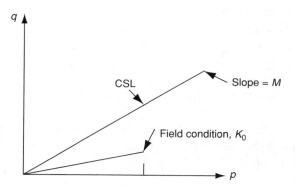

FIGURE 1.27
Illustration of the critical state line and the field stress state of a clay layer.

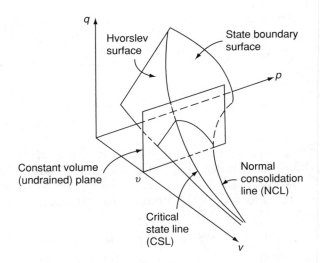

FIGURE 1.28
Three dimensional representation of the deformation of clay up to failure.

ruled Hvorslev surface, on both of which yielding can take place. The CSL defines the boundary of the above two surfaces. Therefore, CSL is, in fact, a line in 3D space whose corresponding q–p projection is shown in Figure 1.27.

The equation of the CSL projection on v–p is given by (Figure 1.27)

$$v = N - (\lambda - \kappa)\ \ln 2 - \lambda\ \ln(p) \tag{1.50}$$

Therefore, in 3D space, the CSL is a line represented by Equations (1.49) and (1.50). On the other hand, the equation of the Hovrslev surface (Figure 1.28) can be expressed as follows:

$$q = (M - h)e^{[N-(\lambda-\kappa)\ln 2 - v/\lambda]} + hp \tag{1.51}$$

where h, which defines the slope of the Hvorslev surface on the q–p plane, is another material constant that can be determined from triaxial tests. Furthermore, the equation of the SBS (Figure 1.28) can be expressed as

$$v = N - (\lambda - \kappa)\ \ln\left(p + \frac{q^2}{M^2 p}\right) - \kappa\ \ln(p) \tag{1.52}$$

The following observations are made based on Figure 1.28:

1. Stress states of NC clays plot on the state boundary surface.
2. Under undrained shearing, NC clays start from K_0 conditions on SBS and exhibit strain hardening behavior (path 1 in Figure 1.29) until failure on the CSL.
3. Under undrained shearing, slightly OC clays start from K_0 conditions immediately inside the SBS and exhibit strain hardening behavior (path 2 in Figure 1.29) up to the SBS and subsequently strain hardens further until it reaches failure on the CSL.
4. Under undrained shearing, highly OC clays start from K_0 conditions well inside the SBS and exhibit strain hardening behavior (path 3 in Figure 1.29) up to the Hvorslev surface and subsequently strain hardens further until failure is reached on the CSL.

5. Under shearing, OC samples exhibit elastic behavior until they approach either the SBS or the Hovrslev surface depending on the over-consolidation ratio. Therefore, during undrained shearing with the specific volume v remaining constant, the only way in which they can retain their elastic properties without moving onto a different RCL is to retain the same mean effective stress, p. Hence undrained stress paths of OC clays remain vertical until they approach either the SBS or the Hovrslev surface (Figure 1.29).

This discussion illustrates that the SBS and Hovrslev surfaces can be considered as yield surfaces in the study of plastic behavior of clays.

1.8.1.3 Stress–Strain Relations for Yielding Clays

The normality condition or the associated flow rule in plastic theory states that the plastic flow vector (for 2D yield surfaces) is normal to the yield surface at the stress point corresponding to any stress state. By assuming the normality condition, Roscoe and Burland (1968) derived the following plastic stress–strain relationship for clay yielding on the SBS:

$$\begin{bmatrix} \Delta\varepsilon_p \\ \Delta\varepsilon_q \end{bmatrix} = \frac{\lambda - \kappa}{vp(M^2 + \eta^2)} \begin{bmatrix} M^2 - \eta^2 & 2\eta \\ 2\eta & 4\eta^2/(M^2 - \eta^2) \end{bmatrix} \begin{bmatrix} \Delta p \\ \Delta q \end{bmatrix} \tag{1.53}$$

where $\eta = q/p$ represents the current stress state on the SBS. Equation (1.53) amply illustrates the shear-volume coupling phenomenon or the occurrence of volumetric strains due to shearing stresses. This is particularly noticeable in the case of granular material, such as medium dense and dense sands, for which more applicable stress–strain models are discussed in the Sections 1.8.2 and 1.8.3.

The usefulness of relationships such as Equation (1.53) is that they can be used conveniently in the [D] matrix of the finite element formulation (Section 1.7) providing a convenient mechanism to incorporate plastic deformation in finite element modeling of foundation problems. Table 1.11 shows the model parameters that must be evaluated for calibration of the modified Cam-clay model and the appropriate laboratory tests that can be used for the evaluation.

1.8.2 Cap Model

The cap model, more appropriately considered as a collection of many models, is based on concepts introduced by Drucker et al. (1957) and further developed by DiMaggio and Sandler (1971). This model has been used to represent both high-pressure and

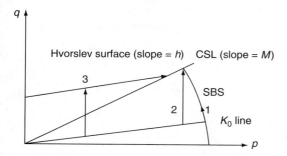

FIGURE 1.29
Undrained stress paths of field clay samples.

TABLE 1.11

Evaluation of Modified Cam-Clay Model Parameters

Model Parameter	Laboratory Test	Type of Sample
M	Triaxial (CU or CD)	NC or OC
λ	Isotropic consolidation	NC
N	Isotropic consolidation	NC
K	Isotropic consolidation	OC
h	Triaxial (CU)	OC

low-pressure mechanical behaviors of many geologic materials, including sand, clay, and rock. As shown in Figure 1.30 and mathematically represented in Equations (1.56)–(1.63), the cap model consists of a strain hardening elliptical cap and an elastic perfectly plastic Drucker–Prager failure surface plotted on a p (mean normal effective stress) versus q (deviator stress) plot. Although the p versus q plot in Figure 1.30 is adequate to represent triaxial stress states, the generalized cap model is plotted with I_1 versus $\sqrt{J_2}$ where I_1 and J_2 are the first invariant of the stress tensor (Equation (1.54)) and the second invariant of the deviatoric stress tensor (Equation (1.55)), respectively:

$$I_1 = \sigma_x + \sigma_y + \sigma_z \tag{1.54}$$

$$J_2 = (\sigma_x - \sigma_y)^2 + (\sigma_y - \sigma_z)^2 + (\sigma_z - \sigma_x)^2 + 6(\tau_{xy}^2 + \tau_{yz}^2 + \tau_{zx}^2) \tag{1.55}$$

where σ and τ represent the normal and shear stresses on the x, y, and z planes.

Then, for triaxial (axisymmetric) stress states, from Equations (1.54) and (1.55), it is seen that $I_1 = 3p$ and $\sqrt{J_2} = q$.

In the cap model, the soil is assumed to be linear elastic inside the yield surface while on the yield surface it is assumed to deform plastically based on an associative flow rule as in the case of the Cam-clay model discussed in Section 1.8.1. The cap model has widespread use as a constitutive model in many finite element computational methodologies that are used in earthen structure design. This is primarily because the presence of the strain hardening cap provides a facility to model the shear-induced dilatancy behavior of over-consolidated clays and dense sands and similarly the shear-induced compression behavior of normally consolidated clays and loose sands.

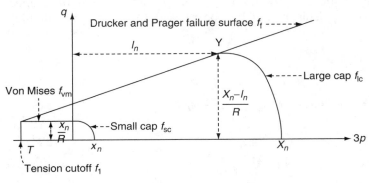

FIGURE 1.30
Illustration of the cap model.

The mathematical representation of the cap model is described below:
Drucker and Prager failure surface

$$-3\alpha p + q - k = 0 \tag{1.56}$$

Von Mises failure surface

$$q - x_n/R = 0 \tag{1.57}$$

Elliptical hardening caps
Large cap

$$(3p - l_n)^2 + R^2 q^2 - (X_n - l_n)^2 = 0 \tag{1.58}$$

Small cap

$$(3p)^2 + R^2 q^2 - x_n^2 = 0 \tag{1.59}$$

Tension cutoff

$$3p - T = 0 \tag{1.60}$$

where

$$X_n = \frac{1}{D} \ln\left(1 + \frac{\varepsilon_v^P}{W}\right) \tag{1.61}$$

$$\alpha = \frac{2\sin\phi}{\sqrt{3}(3 - \sin\phi)} \tag{1.62}$$

$$k = \frac{6c(\cos\phi)}{\sqrt{3}(3 - \sin\phi)} \tag{1.63}$$

D, W, and R are soil parameters, and

$$\varepsilon_v^P \text{ is the accumulated plastic volumetric strain} = \sum d\varepsilon_v^P \tag{1.64}$$

It is noted that the plastic volumetric strain is accumulated only if the stress (p) and strain increments are both compressive. The parameter l_n is limited to negative values to avoid development of tension. For positive l_n values, a modified small cap is assumed in the tensile stress range together with the Von Mises failure surface as shown in Figure 1.30. If the maximum tensile stress that the soil can take, T, is known, then tension cutoff can also be introduced as seen in Figure 1.30.

The above relationships can be used to express the plastic normal and shear strains $\Delta\varepsilon_p$ and $\Delta\varepsilon_q$ in terms of the stress increments Δp and Δq and form the corresponding [D] matrix in a relevant finite element formulation (Section 1.7) as in the case of the Cam-clay model (Equation (1.53)). The main advantage of using such stress–strain models in finite element modeling of foundations is that they provide a convenient mechanism to include the plastic deformation of soils in design considerations enabling the formulation of more realistic and economic design methodologies. Table 1.12 shows the model parameters that must be evaluated for calibration of the cap model and the appropriate laboratory tests that can be used for the evaluation.

TABLE 1.12

Evaluation of Cap Model Parameters

Model Parameter	Laboratory Test	Type of Soil
D	Isotropic consolidation	NC clay, loose sand
α	Triaxial (CU or CD)	NC clay, loose sand
K	Triaxial (CU or CD)	NC clay, loose sand
R	Triaxial (CU)	OC/NC clay, loose/dense sand
W	Isotropic consolidation	NC clay, loose sand
T	Triaxial extension	OC clay, dense sand

1.8.3 Nonlinear Elastic Stress–Strain Relations

Another stress–strain relationship popularly used to model foundation soils in finite element formulations is the nonlinear elastic shear stress–shear strain model developed by Hardin and Drnevich (1972). This can be mathematically expressed by Equation (1.65) and plotted on a stress–strain plot as shown in Figure 1.31:

$$\tau = \frac{\gamma}{1/G_{max} + \gamma/\tau_{max}} \tag{1.65}$$

where G_{max} and τ_{max} are the initial shear modulus and the maximum shear stress, respectively, as shown in Figure 1.31. It can be seen from Figure 1.31 that the shear modulus decreases in magnitude with increasing shear and G_{max} can also be interpreted as the shear modulus under very low strain levels. Equation (1.65) is of special significance when the response of soils is analyzed under dynamic loading conditions.

1.8.3.1 Evaluation of Nonlinear Elastic Parameters

The nonlinear elastic stress–strain parameters G_{max} and τ_{max} can be determined from direct shear and simple shear tests under static loading conditions and cyclic simple shear tests under dynamic loading conditions. While τ_{max} can be evaluated using the Mohr–Coulomb criterion as presented in Equation (1.6),

$$\tau = c + \sigma \tan \phi \tag{1.6}$$

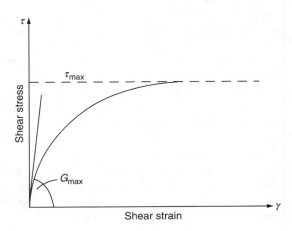

FIGURE 1.31
Nonlinear elastic relationship.

Analytical expressions are also available for evaluating G_{max}. For example, G_{max} for sandy soils can be expressed by the following expression:

$$G_{max} = A \frac{(B - e)^2}{1 + e} \bar{\sigma}_0^{1/2} \tag{1.66}$$

where $\bar{\sigma}_0$ is the mean (octahedral) normal stress (Equation (1.9)) and e is the void ratio. When G_{max} and $\bar{\sigma}_0$ are measured in kPa, the values of A and B are as follows:

For round-grained sands, $A = 6908$ and $B = 2.17$

For angular-grained sands, $A = 3230$ and $B = 2.97$

On the other hand, in the case of clays, the following modified form of Equation (1.66), presented by Hardin and Drnevich (1972), can be used to evaluate G_{max} from basic index and consolidation properties as well as the stress state:

$$G_{max} = A \frac{(B - e)^2}{1 + e} (OCR)^K \bar{\sigma}_0^{1/2} \tag{1.67}$$

where both G_{max} and σ_0 are expressed in kPa and OCR is the over-consolidation ratio. Furthermore,

$A = 3230, B = 2.97$

$K = 0.4 + 0.007(PI)$ for $0 < PI < 40$

$K = 0.68 + 0.001(PI - 40)$ for $40 < PI < 80$

when PI is the plasticity index (Section 1.2.2).

1.8.3.2 Evaluation of G_{max} from Standard Penetration Tests

Seed (1986) presented the following correlation between G_{max} and the SPT blow count:

$$G_{max} = 5,654.1(N_{60})^{0.34} \bar{\sigma}_0^{0.4} \tag{1.68}$$

where both G_{max} and σ_0 are expressed in kPa. N_{60} is the SPT blow count obtained with a test setup that delivers 60% of the theoretical free-fall energy of the hammer to the drill rod.

1.8.4 Concepts of Stress Dilatancy Theory for Granular Soils

In contrast to the popular approach to stress–strain relations for soils assuming a linear elastic (Section 1.5), nonlinear elastic (Section 1.8.3), and elasto-plastic (Sections 1.8.1 and 1.8.2) continua, particulate mechanics has also been used to explain the behavior of soils, especially in the case of sands. Stress dilatancy theory (Rowe, 1963) is a result of the research conducted toward understanding the deformation of soils as discrete particulate matter. The following assumptions form the framework of this theory:

1. During deformation particles tend to slide much more frequently than rolling.
2. The individual particles are rigid and cannot contribute an elastic component toward deformation.
3. Instant sliding is confined to some preferred angle.

The following expression can be used to describe the stress–strain relationship at a given stress level:

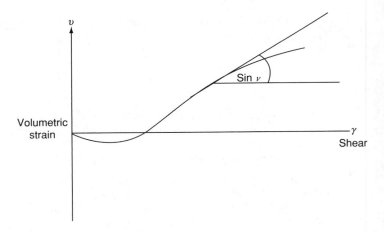

FIGURE 1.32
Shear dilatation of granular soils.

$$\frac{\sigma'_1}{\sigma'_3} = \left[\frac{1 - (dv/d\gamma)}{1 + (dv/d\gamma)}\right][\tan^2(45 + \phi_{cv}/2)] \tag{1.69}$$

where ϕ_{cv} is the angle of friction under large shear strains (residual friction) where shearing would occur under constant volume.

The principal stress ratio under triaxial conditions, σ'_1/σ'_3, can be related to the developed angle of friction ϕ as

$$\frac{\sigma'_1}{\sigma'_3} = \tan^2(45 + \phi/2) \tag{1.70}$$

and the rate of volume change at a constant confining pressure with respect to shear strains, $dv/d\gamma$, can be expressed in terms of the dilation angle, v, as

$$\frac{dv}{d\gamma} = -\tan^2(45 + \nu/2) \tag{1.71}$$

Inspection of Equation (1.69) shows that when shearing occurs in granular soils without any volumetric strains ($v = 0$), then $\phi = \phi_{cv}$. On the other hand, according to this theory, in the case of dense or medium dense sands, shear failure would occur at maximum dilation conditions (Figure 1.32). Under those conditions, the developed angle of friction will be equal to the angle of interparticle friction, or $\phi = \phi_f$.

This condition will also correspond to the occurrence of the maximum principal stress ratio, $[\sigma'_1/\sigma'_3]_{max}$. Equation (1.69) can also be rearranged to derive the $[D]$ matrix of the finite element formulation (Section 1.7) and provides a mechanism to incorporate plastic deformation of granular soils in finite element modeling of foundation problems.

References

Bishop, A.W. and Blight, G.E., 1963, Some aspects of effective stress in saturated and partly saturated soils, *Géotechnique*, 13(3): 177–197.
Bowles, J.E., 1986, *Engineering Properties of Soils and their Measurements*, McGraw-Hill, New York.

Bowles, J.E., 2002, *Foundation Analysis and Design*, McGraw-Hill, New York.

Das, B.M., 2002, *Soil Mechanics Laboratory Manual*, Oxford University Press, New York.

DiMaggio, F.L. and Sandler, I.S., 1971, Material models for granular soils, *Journal of the Engineering Mechanics Division, ASCE*, 97 (EM3): 935–950.

Drucker, D.C., Gibson, R.E., and Henkel, D.J., 1957, Soil mechanics and work-hardening theories of plasticity, *Transactions of the ASCE*, 122: 338–346.

Goodman, R.E., 1989, *Introduction to Rock Mechanics*, John Wiley, New York.

Hardin, B.O. and Drnevich, V.P., 1972, Shear modulus and damping in soils; design equations and curves, *Journal of the Soil Mechanics and Foundations Division, ASCE*, 98(SM7): 667–692.

Harr, M., 1962. *Groundwater and Seepage*, McGraw-Hill, New York. Chapter 13

Holtz, R.D. and Kovacs, W.D., 1981, *An Introduction to Geotechnical Engineering*, Prentice Hall, Englewood Cliffs, NJ.

Hughes, T.J.R., 1987, *The Finite Element Method*, Prentice Hall, Englewood Cliffs, NJ.

Roscoe, K.H. and Burland, J.B., 1968, *On the Generalized Stress–Strain Behavior of Wet Clay*, Engineering Plasticity, Cambridge University Press, Cambridge.

Rowe, P.W., 1963, The stress dilatancy relation for static equilibrium of an assembly of particles in contact, *Proceedings of the Royal Society of Soil Mechanics and Foundation Engineers*.

Schmertmann, J.H. and Hartman, J.P., 1978, Improved strain influence factor diagrams. *Journal of the Geotechnical Engineering Division, American Society of Civil Engineers*, 104(GT8): 1131–1135.

Schofield, A.N. and Wroth, C.P., 1968, *Critical State Soil Mechanics*, McGraw-Hill, New York.

Seed, H.B., Wong, R.T., Idriss, I.M., and Tokimatsu, K., 1986, Moduli and damping factors for dynamic analysis of cohesive soils, *Journal of Geotechnical Engineering, ASCE*, 112(GT 11): 1016–1032.

Wray, W., 1986, *Measuring Engineering Properties of Soils*, Prentice Hall, Englewood Cliffs, NJ.

Zienkiewich, O.C., Humpheson, C., and Lewis, R.W., 1977, A unified approach to soil mechanics problems (including plasticity and visco-plasticity) In *Finite Element in Geomechanics*, Gudehus, D., ed., John Wiley & Sons, New York.

2

In Situ *Soil Testing*

Gray Mullins

CONTENTS

2.1   Introduction to Subsurface Exploration

2.1.1   Preliminary Site Exploration

The designer of a super-structure foundation must invariably perform a detailed surface and subsurface (soil) exploration of the potential site prior to deciding on the nature and type of the foundation. The subsurface investigation program for a given site should account for the type, size, and importance of the proposed structure. These parameters help focus the design of the site exploration program by determining the quantity and depth of soil soundings (or borings) required. Planning for a site investigation should also include the location of underground utilities (i.e., phone, power, gas, etc.). As such, a local "call before you dig" service should be notified several days prior to the anticipated investigation. These services are usually subsidized by the various local utilities and have no associated cost.

Subsurface exploration and testing specifically serve the following purposes (FHWA, 1998):

1. Aid in the preliminary selection of substructure types that are viable for a particular site and the type of superstructure to be supported.
2. Provide a basis for selecting soil and rock properties needed for substructure design.
3. Identify special substructure conditions requiring special provisions to supplement standard construction specifications.

For most projects, the following types of subsurface information are needed for the selection, design, and construction of substructures:

1. Definition of soil–rock stratum boundaries
2. Variation of groundwater table
3. Location of adequate foundation-bearing layers
4. Magnitude of structure settlement and heave
5. Dewatering requirements during construction
6. Potential problems including building cracks, slope instability, etc.
7. Construction access

In developing site exploration programs, the geotechnical engineer should qualitatively assess the effects of variables such as the expected type and importance of the structure, magnitude and rate of loading, and foundation alternatives with respect to technical, economic, and constructability considerations (FHWA, 1998). An exhaustive subsurface exploration can be separated into two distinct phases: (1) preliminary investigation and (2) detailed investigation. In the preliminary investigation, one would attempt to obtain as much valuable information about the site as possible with least expense. In this respect, a wealth of useful information can be collected using the following sources:

1. Topographic maps: landforms, ground slopes and shapes, and stream locations
2. Aerial photographs: landforms, soil types, rock structure, and stream types
3. U.S. Department of Agriculture (USDA) Agronomy soil maps: landforms and soil descriptions close to the ground surface

4. Well drilling logs: identification of soil and rock as well as groundwater levels at the time
5. Existing boring logs
6. Local department of transportation (DOT) soil manuals
7. Local U.S. Geological Survey (USGS) soil maps
8. Local U.S. Army Corps of Engineers hydrological data
9. Local university research publications

In addition to screening of possible sites based on information from documentation of previous studies, a thorough site visit can provide vital information regarding the soil and groundwater conditions at a tentative site, leading to more efficient selection of foundation depth and type as well as other construction details. Hence, a site inspection can certainly aid in economizing the time and cost involved in foundation construction projects. During site visits (or reconnaissance surveys) one can observe such site details as topography, accessibility, groundwater conditions, and nearby structures (especially in the case of expected pile driving or dynamic ground modification). Firsthand inspection of the performance of existing buildings can also add to this information. A preliminary investigation can be an effective tool for screening all alternative sites for a given installation.

2.1.2 Site Exploration Plan

A detailed investigation has to be conducted at a given site only when that site has been chosen for the construction, since the cost of such an investigation is enormous. Guidelines for planning a methodical site investigation program are provided in Table 2.1 (FHWA, 1998).

This stage of the investigation invariably involves heavy equipment for boring. Therefore, at first, it is important to set up a definitive plan for the investigation, especially in terms of the bore-hole layout and the depth of boring at each location. In addition to planning boring locations, it is also prudent on the part of the engineer to search for any subsurface anomalies or possible weak layers that can undermine construction. As for the depth of boring, one can use the following criteria:

If the bedrock is in the vicinity, continue boring until a sound bedrock is reached, as verified from rock core samples. If bedrock is unreachable, one can seek depth guidelines for specific buildings such as those given by the following expressions (Das, 1995):

$D = 3S^{0.7}$ (for light steel and narrow concrete buildings)

$D = 6S^{0.7}$ (for heavy steel and wide concrete buildings)

If none of the above conditions is applicable, one can explore up to a depth at which the foundation stress attenuation reduces the applied stress by 90%. This generally occurs around a depth of $2B$, where B is the minimum foundation dimension.

2.1.2.1 Soil Boring

The quantity of borings is largely dependent on the overall acreage of the project, the number of foundations, or the intended use of the site. For foundations, the depth of borings depends on the zone of soil influenced by the foundation geometry and the given

TABLE 2.1

Guideline Minimum Boring and Sampling Criteria

Geotechnical Feature	Minimum Number of Borings	Minimum Depth of Borings
Structure foundation	One per substructure unit for width ≤30 m	Advance borings: (1) through unsuitable foundation soils (e.g., peat, highly organic soils, soft fine-grained soils) into competent material of suitable bearing capacity; (2) to a depth where stress increases due to estimated footing load is less than 10% of the existing effective soil overburden stress; or (3) a minimum of 3 m into bedrock if bedrock is encountered at shallower depth
	Two per substructure unit for width >30 m	
Retaining walls	Borings alternatively spaced every 30 to 60 m in front of and behind wall	Extend borings to depth of two times wall height or a minimum of 3 m into bedrock
Culverts	Two borings depending on length	See structure foundations
Bridge approach embankments over soft ground	For approach embankments placed over soft ground, one boring at each embankment to determine problems associated with stability and settlement of the embankment (*note:* borings for approach embankments are usually located at proposed abutment locations to serve a dual function)	See structure foundations
		Additional shallow explorations at approach embankment locations are an economical means to determine depth of unsuitable surface soils
Cuts and embankments	Borings typically spaced every 60 (erratic conditions) to 150 m (uniform conditions) with at least one boring taken in each separate landform	*Cut*: (1) in stable materials, extend borings a minimum of 3 to 5 m below cut grade
	For high cuts and fills, two borings along a straight line perpendicular to centerline or planned slope face to establish geologic cross section for analysis	(2) in weak soils, extend borings below cut grade to firm materials, or to the depth of cut below grade whichever occurs first
		Embankment: extend borings to firm material or to depth of twice the embankment height

Source: Modified after FHWA, 1993, *Soils and Foundations, Workshop Manual*, 2nd ed., FHWA HI-88-009, National Highway Institute, NHI Course No. 13212, Revised, July. With permission.

loading. For instance, a proposed roadway alignment typically requires a hand-auger investigation every 100 ft along the centerline to a depth of 5 ft to define uniformity of the subgrade material as well as spatial variability. Therein, the importance of the structure, in the form of causal effects should a failure occur, is somewhat minimal. Further, if undesirable soil conditions were identified, a follow-up investigation could be requested. In contrast, preliminary borings along the alignment of a proposed bridge foundation can be more frequent and are much deeper depending on the depth to a suitable bearing stratum. At a minimum, one boring should be performed at each pier location to a depth of 3 to 5 foundation diameters below the anticipated foundation. Likewise, buildings with large column loads often require a boring at each column location unless the soil shows extremely consistent behavior. For extremely important structures, the designer or client not only requires more scrutiny from the subsurface investigation, but also requires an amplification factor (or importance factor) be applied to the load to assure a low probability of failure (FHWA, 1998).

In virtually all cases, the additional cost of a thorough subsurface investigation can be reconciled with a cost-effective foundation design. Uncertainty in subsurface conditions in most instances promotes needless over-design. Depending on the type of design to be considered, the designer must recognize the effect of site variability as well as the type of testing that can be conducted to increase confidence and reduce the probability of failure.

Hand augers and continuous flight augers (Figure 2.1) can be used for boring up to a depth of about 3 m in loose to moderately dense soil. For extreme depths, a mechanized auger (Figure 2.2) can be used in loose to medium dense sands or soft clays. When the cut soil is brought to the surface, a technically qualified person should observe the texture, color, and the type of soil found at different depths and prepare a bore-hole log laying out soil types at different depths. This type of boring is called dry sample boring.

On the other hand, if relatively hard strata are encountered, the investigators have to resort to a technique known as wash boring. Wash boring is carried out using a mechanized auger and a water-circulation system that aids in cutting and drawing the cut material to the surface. A schematic diagram of the wash-boring apparatus is shown in Figure 2.2(a), and the Florida Department of Transportation drill rig, which utilizes the above technique, is shown in Figure 2.2(b).

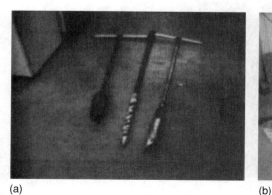
(a)

(b)

FIGURE 2.1
Drilling equipment: (a) hand augers; (b) mechanized auger. (Courtesy of the University of South Florida.)

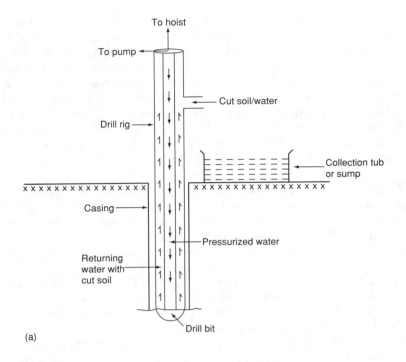

(a)

(b)

FIGURE 2.2
(a) Schematic diagram of the Florida Department of Transportation's CME-75 drill rig. (b) Wash boring.

2.2   Need for *In Situ* Testing

In addition to visual classification, one has to obtain soil type and strength and deformation properties for a foundation design. Hence, the soil at various depths has to be sampled as the bore holes advance. Easily obtained disturbed samples suffice for classification, index, and compaction properties in this regard. However, more sophisticated laboratory or *in situ* tests are needed for determining compressibility and shear strength parameters.

2.2.1   Sample Disturbance

In situ testing is the ultimate phase of the site investigation where foundation design parameters can be evaluated to a relatively higher degree of reliability than under laboratory conditions. This is because the reliability and the accuracy of the design parameters obtained in the laboratory depend on the disturbance undergone by the retrieved samples during the retrieval, transport, extrusion, and sample preparation processes. The predominant factors causing soil sample disturbance are as follows:

1. Use of samplers that have a relatively high metal percentage in the cross section. For this purpose, the area ratio of a sampler is defined as

$$A_r = \frac{D_o^2 - D_i^2}{D_o^2}(100)\% \tag{2.1a}$$

   where D_o and D_i are, respectively, the external and internal diameters of the sampler. The common samplers that are used for collecting disturbed samples are known as standard split-spoon samplers (described in Section 2.4.1) in relation to standard penetration tests. For these samplers, the value of A_r exceeds 100%. On the other hand, Shelby tubes (another sampler type) (Figure 2.3) have a relatively small metal cross section and hence an A_r value of less than 15%.

2. Friction between the internal sampler wall and the collected sample causing a compression or shortening of the sample. This can be addressed by introducing a minute inward protrusion of the cutting edge of the sampler.

3. Loosening of the sampling due to upheaval of roots, escape of entrapped air, etc.

Effects of causes 2 and 3 can be evaluated by the recovery ratio of the collected sample defined as

$$L_r = \frac{l_r}{d_p}(100)\% \tag{2.1b}$$

where d_p and l_r are the depth of penetration of the sampler and length of the collected sample, respectively. Furthermore, it is realized that L_r values close to 100% indicate minimum sample disturbance.

4. Evaporation of moisture from the sample causing particle reorientation and change in density. This effect can be eliminated by wrapping and sealing the soil sample during transport to the testing laboratory.

In terms of foundation engineering, *in situ* (or in place) testing refers to those methods that evaluate the performance of a geotechnical structure or the properties of the soils or rock

FIGURE 2.3
Shelby tube samplers. (Courtesy of the University of South Florida.)

used to support that structure. This testing can range from a soil boring at a surveyed location to monitoring the response of a fully loaded bridge pier, dam, or other foundation element. The reliability of a given structure to function as designed is directly dependent on the quality of the information obtained from such testing. Therein, it is imperative that the design engineer be familiar with the types of tests and the procedures for proper execution as well as the associated advantages and disadvantages.

Methods of *in situ* evaluation can be invasive or noninvasive, destructive or nondestructive, and may or may not recover a specimen for visual confirmation or laboratory testing. Invasive tests (e.g., soil borings or penetration tests) tend to be more time consuming, expensive, and precise, whereas noninvasive tests (e.g., ground penetrating radar or seismic refraction) provide a large amount of information in a short period of time that is typically less quantifiable. However, when used collectively, the two methods can complement each other by: (1) defining areas of concern from noninvasive techniques and (2) determining the foundation design parameter from invasive techniques. This is particularly advantageous on large sites where extreme variations in soil strata may exist. All of the relevant *in situ* tests except the plate load test are discussed in this chapter. A description of the plate load test is provided in Chapter 4 along with the methodology for designing combined footings.

2.3 Geophysical Testing Methods

2.3.1 Ground Penetrating Radar

Ground penetrating radar (GPR) is a geophysical exploration tool used to provide a graphical cross section of subsurface conditions. This cross-sectional view is created from the reflections of repeated short-duration electromagnetic (EM) waves that are generated by an antenna in contact with the ground surface as the antenna traverses across the ground surface. The reflections occur at the interfaces between materials with differing electrical properties. The electrical property from which variations cause these reflections is the dielectric permittivity, which is directly related to the electrical conductivity of the material. GPR is commonly used to identify underground utilities, underground storage tanks, buried debris, or geological features. The information from GPR can be used to make recommendations for more invasive techniques such as borings. Figure 2.4 shows a ground-launch GPR system being pushed along a predetermined transect line.

The higher the electrical contrast between the surrounding earth materials and the target of interest, the higher the amplitude of the reflected return signal. Unless the buried object or target of interest is highly conductive, only part of the signal energy is reflected back to the antenna located on the ground surface with the remaining portion of the signal continuing to propagate downward to be reflected by deeper features. If there is little or no electrical contrast between the target of interest and the surrounding earth materials, it would be very difficult if not impossible to identify the object using GPR.

The GPR unit consists of a set of integrated electronic components that transmits high-frequency (100 to 1500 MHz) EM waves into the ground and records the energy reflected

FIGURE 2.4
GPR field device. (Courtesy of Universal Engineering, Inc.)

back to the ground surface. The GPR system comprises an antenna, which serves as both a transmitter and a receiver, and a profiling recorder that processes the data and provides a graphical display of the data.

The depth of penetration of the GPR is very site specific and is controlled by two primary factors: subsurface soil conditions and antenna frequency. The GPR signal is attenuated (absorbed) as it passes through earth materials. As the energy of the GPR signal is diminished due to attenuation, the energy of the reflected waves is reduced, eventually to a level where the reflections can no longer be detected. In general, the more conductive the earth materials, the higher the GPR signal attenuation. In Florida, typical soil conditions that severely limit the GPR signal penetration are near-surface clays, organic materials, and the presence of sea water in the soil pore water space.

A GPR survey is conducted along survey lines (transects), which are measured paths along which the GPR antenna is moved. Known reference points (i.e., building corners, driveways, etc.) are placed on a master map, which includes traces of the GPR transects overlying the survey geometry. This survey map allows for correlation between the GPR data and the position of the GPR antenna on the ground.

For geological characterization surveys, the GPR survey is conducted along a set of perpendicularly oriented transects. The survey is conducted in two directions because subsurface features are often asymmetric for residential surveys. Spacing between the survey lines is initially set at 10 ft. More closely spaced grids may be used to further characterize a recorded anomaly. The features observed on the GPR data that are most commonly associated with potential sinkhole activity are:

1. A down-warping of GPR reflector sets, which are associated with suspected lithological contacts, toward a common center. Such features typically have a bowl- or funnel-shaped configuration and are often associated with deflection of overlying sediment horizons caused by the migration of sediments into voids in the underlying limestone (Figure 2.5). In addition, buried depressions caused by

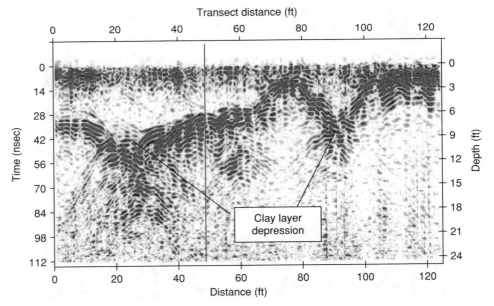

FIGURE 2.5
GPR image. (Courtesy of Universal Engineering, Inc.)

differential subsidence over buried organic deposits and debris may also cause these observed features.

2. A localized significant increase in the depth of penetration or amplitude of the GPR signal response. The increase in GPR signal penetration depth or amplitude is often associated with a localized increase in sand content at depth.

3. An apparent discontinuity in GPR reflector sets that are associated with suspected lithological contacts. The apparent discontinuities or disruption of the GPR reflector sets may be associated with the downward migration of sediments.

The greater the severity of the above-mentioned features or a combination of these features, the greater the likelihood that the identified feature is related to past or present sinkhole activity.

Depth estimates to the top of the lithological contacts or targets of interest are derived by dividing the time of travel of the GPR signal from the ground surface to the top of the feature by the velocity of the GPR signal. The velocity of the GPR signal is usually obtained for a given geographic area and earth material from published sources. In general, the accuracy of the GPR-derived depth estimates ranges from $\pm 25\%$ of the total depth.

Although the GPR is very useful in locating significant lithological soil changes, strata thickness, and inferred subsurface anomalies, the GPR cannot identify the nature of earth materials or their condition (i.e., loose versus dense sand, soft versus stiff clay). The GPR data are best used in conjunction with other geotechnical and physical tests to constrain the interpretation of the virtual cross-section profiles.

2.3.2 Resistivity Tests

Electrical resistivity imaging (ERI) (Figure 2.6) is a geophysical method that maps the differences in the electrical properties of geologic materials. These changes in electrical properties can result from variations in lithology, water content, pore-water chemistry, and the presence of buried debris. The method involves transmitting an electric current into the ground between two electrodes and measuring the voltage between two other electrodes. The direct measurement is an apparent resistivity of the area beneath the electrodes that includes deeper layers as the electrode spacing is increased. The spacing of

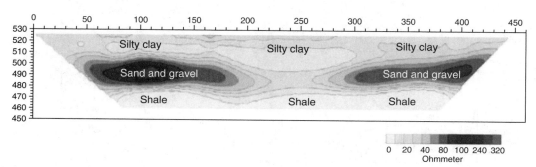

FIGURE 2.6
Rendering of soil cross section from ERI output. (Courtesy of Universal Engineering, Inc.)

electrodes can be increased about a central point, resulting in a vertical electric sounding that is modeled to create a 1D geoelectric cross section. Recent advances in technology allow for rapid collection of adjacent multiple soundings along a transect that are modeled to create a 2D geoelectric pseudo-cross-section. The cross section is useful for mapping both the vertical and horizontal variations in the subsurface (see Figure 2.6).

Although the results from this method are not absolute, the resistivity trends obtained are useful for mapping stratigraphy such as aquatards, bedrock, faults, and fractures. It can delineate anomalous formations or voids in karstic material, the presence of salt water intrusion in coastal regions, and detect leaks in dams as well as other applications. It is most successful in large cleared areas without severe changes in topography; it is not recommended for small congested or urban sites. Buried utilities or other highly conductive anomalies can adversely affect the results.

This method is fast, noninvasive, and relatively inexpensive when compared to drilling. When compared to EM methods, it is less susceptible to interference from overhead power lines. It is easily calibrated to existing boreholes to allow for correlations between measured resistivity and estimated soil properties. As with other geophysical test methods, it is best suited for environmental or water resources disciplines that require stratigraphy or soil property mapping of large land parcels.

2.3.2.1 Seismic Refraction

The seismic refraction technique measures the seismic velocity of subsurface materials and models the depth to interfaces with a velocity increase. Soil conditions and geologic structure are inferred from the results, since changes in material type, or soil conditions, are often associated with changes in seismic velocity. Seismic energy, which is introduced into the subsurface using a drop weight or explosive source, propagates through the earth as a wave front that is refracted by the material through which it passes. As illustrated in Figure 2.7, the wave front intersects a high-velocity interface, creating a "head wave" that travels in the high-velocity material nearly parallel to the interface. The energy in this head wave leaves the interface and passes back through the low-velocity material to the surface. Geophones placed at selected intervals along the ground surface detect the ground motion and send an electrical signal, via a cable, to a recording seismograph.

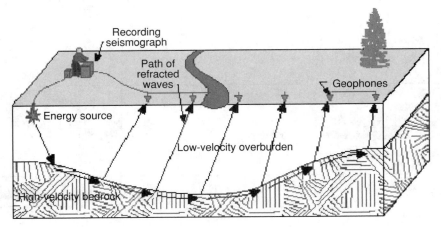

FIGURE 2.7
Conceptual sketch of seismic refraction layout and wave paths. (Courtesy of Universal Engineering, Inc.)

The objective is to determine the arrival times of these refracted waves in order to calculate the velocity of the material and model the depth to various interfaces.

This test is used to determine stratigraphy such as the depth to bedrock. It is best suited for stratigraphy that increases in density (or seismic velocity) with depth. In such cases, it can estimate the depth of borrow materials, aid in mapping faults and fractured rock zones, locate the water table, and estimate material elastic properties such as shear modulus and Poisson's ratio. The depth of exploration is limited by the energy source and the maximum length of geophone spacing. The test is less expensive when compared to other soil exploration methods and other comparable seismic reflection methods. The vertical resolution is usually better than electrical, magnetic, or gravity methods of site investigation.

2.4 Physical Sampling and Penetration Tests

2.4.1 Standard Penetration Test

The standard penetration test (SPT) is undoubtedly the most common method of soil exploration for foundation design. It is an invasive test that not only provides information from which soil strength can be estimated, but also provides a physical sample that can be visually inspected or used for laboratory classification. Although the test method has undergone several iterations with respect to upgrading equipment, it is sensitive to operator and equipment variability. Regardless, the general concept of penetration resistance and the hands on soil sample recovery make it the choice of many designers.

The SPT is described by the American Society for Testing and Materials (ASTM) as test number D-1586, entitled "Standard Method for Penetration Test and Split-Barrel Sampling of Soils." This standard defines the appropriate manner in which the test should be conducted which involves drilling techniques, penetration and sampling methods, proper equipment, and the reporting of results. In general, a 2 in. outer diameter split-spoon sampler is driven into the ground with a 140 lb (0.622 kN) drop hammer dropped 30 in. (0.77 m) repeatedly until a penetration of 18 in. is achieved. The number of blows of the hammer is recorded for each of three 6-in. (15.24 cm) intervals (totaling 18 in. or 45.72 cm). The number of blows required for advancing the sampler to the last 12 in. or 30.48 cm (second and third intervals) is defined as the SPT N-value. Upon extraction of the sampler, the soil retrieved is visually inspected, documented, and placed in jars for more elaborate testing (if so determined by the engineer). At best, continuous sampling produces a single SPT N-value every 1.5 ft. At minimum, a sample should be taken every 5 ft (1.54 m) of depth.

Between each penetration test, a boring should be advanced to permit the next sample without interference from side shear resistance along the length of the drill rod. Several boring techniques are acceptable: one-hole rotary drilling, continuous flight hollow stem augering, wash boring, or continuous flight solid stem augering. However, under no circumstance should the soil beneath the advanced borehole be disturbed by jetting or suction action caused by improper drilling techniques. For instance, extracting a continuous flight auger from submerged soils will reduce the *in situ* stresses and produce lower N-values.

2.4.1.1 *SPT Correlations with Shear Strength Properties*

Apart from the visual and physical classifications that can be obtained from an SPT, correlations have been established that provide estimates of *in situ* soil properties based on the soil type and blow count. The basic principle underlying the SPT test is the relation between the penetration resistance and shear strength of the soil, which can be visualized as a unique relationship. These correlations can be based on the corrected or uncorrected SPT blow count N' or N, respectively.

Corrected blow counts provide a method of accounting for the *in situ* state of stress surrounding a soil sample while it was being tested. For instance, sands with identical structure which appear stronger (higher blow counts) at greater depths than when at shallower depths. As such, soil properties such as unit weight may be better estimated if overburden effects are removed or normalized. However, soil properties such as shear strength or available end bearing are enhanced by greater *in situ* stresses and are generally correlated to uncorrected blow counts. The following expression is used to correct SPT N-values by normalizing it to a 1 tsf (95.5 kPa) overburden *in situ* state:

$$N' = C_N N \tag{2.2a}$$

where N is the measured (field) SPT value, N' is the SPT value corrected for the overburden stress, and

$$C_N = \left[\frac{95.76}{\sigma'_v}\right]^{1/2} \tag{2.2b}$$

where σ'_v is the effective overburden pressure of the test location (in kPa) expressed by Equation (1.4b):

$$\sigma'_v = \gamma_b z - \gamma_w d_w \tag{1.4b}$$

where z is the depth of the test location and d_w the depth of the test location from the ground water table.

Table 2.2–Table 2.6 provide estimated values for corrected and uncorrected blow counts. It must be noted from Equation (2.3) that the unconfined compression strength and the undrained cohesion (strength) are related by

TABLE 2.2

Determination of the Frictional Shear Strength of Sands and Clays from SPT Blow Count

		γ_{moist}		ϕ
Corrected SPT-N'	**Description**	**pcf**	**kN/m³**	**Degree**
Sands				
0	Very loose	70–100	11.0–15.7	25–30
4	Loose	90–115	14.1–18.1	27–32
10	Medium	110–130	17.3–20.4	30–35
30	Dense	120–140	18.8–22.0	35–40
50	Very dense	130–150	20.4–23.6	38–43
Clay				q_u, ksf (47.92 kPa)
0	Very soft	100–120	15.7–18.8	0
2	Soft			0.5
4	Medium	110–130	17.3–20.4	1.0
8	Stiff			2.0
16	Very stiff	120–140	18.8–22.0	4.0
32	Hard			8.0

Source: Modified after FHWA, 1993, *Soils and Foundations, Workshop Manual*, 2nd edn, FHWA HI-88-009, National Highway Institute, NHI Course No. 13212, Revised, July.

TABLE 2.3

Determination of the Frictional Shear Strength of Sands and Clays from SPT Blow Count

	γ_{sat}		γ_{sub}		ϕ
SPT-*N*	pcf	kN/m^3	pcf	kN/m^3	Degree
Sands					
0–2	100	15.7	37.6	5.9	26
3–4	100	15.7	37.6	5.9	28
4–10	105	16.5	42.6	6.7	29
10–20	110	17.3	47.6	7.5	30
20–30	115	18.1	52.6	8.3	32
30–40	120	18.9	57.6	9.1	33
>40	125	19.6	62.6	9.8	34
Clay					
0–2	105	16.5	42.6	6.7	0
2–4	110	17.3	47.6	7.5	0
4–8	115	18.1	52.6	8.3	0
8–15	120	18.9	57.6	9.1	0
15–30	125	19.6	62.6	9.8	0
>30	125	19.6	62.6	9.8	0

Notes: Clay shear strength $C = N/T_i$ in ksf (47.92 kPa, where T_i is the soil type factor); $T_i = 8$ for most clay, $T_i = 10$ for low plasticity, $T_i = 12$ for peat.
Source: From Kulhawy, F.H. and Mayne, P.W., 1990, *Manual on Estimating Soil Properties for Foundation Design*, EPRI EL-6800 Research Project 1493-6, Electric Power Research Institute, August. With permission.

$$c_u = \frac{1}{2}q_u \tag{2.3}$$

Alternatively, the frictional properties of granular soils can be obtained using the following simple expression (Bowles, 2002):

$$\phi = 4.5N_{70} + 20 \tag{2.4a}$$

The standard penetration value can also be used estimate the over-consolidation ratio of a soil based on Equation (2.4b) (Bowles, 2002):

$$OCR = 0.193 \left(\frac{N}{\sigma\delta}\right)^{0.689} \tag{2.4b}$$

TABLE 2.4

Determination of the Frictional Shear Strength of Limestone from SPT Blow Count

	Shear Strength	
SPT-*N*	psf	kN/m^2
10–20	4,000	190
20–50	8,000	380
50–100	15,000	720

Notes: $\gamma_{sat} = 135$ pcf (21.2 kN/m^3); $\gamma_{sub} = 72.6$ pcf (11.4 kN/m^3); $\phi = 0°$; $K_a = 1.0$; $K_p = 1.0$.
Source: From Kulhawy, F.H., and Mayne, P.W., 1990, *Manual on Estimating Soil Properties for Foundation Design*, EPRI EL-6800 Research Project 1493-6, Electric Power Research Institute, August. With permission

TABLE 2.5

Empirical Values for ϕ, D_r, and Unit Weight of Granular Soils Based on the SPT at about 6 m Depth and Normally Consolidated (Approximately, $\phi = 28° + 15°D_r [\pm 2°]$)

Description	Very Loose	Loose	Medium	Dense	Very Dense
Relative density D_r	0	0.15	0.35	0.65	0.85
SPT N'_{70}					
Fine	1–2	3–6	7–15	16–30	?
Medium	2–3	4–7	8–20	21–40	>40
Coarse	3–6	5–9	10–25	26–45	>45
ϕ					
Fine	26–28	28–30	30–34	33–38	
Medium	27–28	30–32	32–36	36–42	<50
Coarse	28–30	30–34	33–40	40–50	
γ_{wet}, kN/m^3	11–16[a]	14–18	17–20	17–22	20–23

[a]Excavated soil or material dumped from a truck has a unit weight of 11 to 14 kN/m^3 and must be quite dense to weigh much over 21 kN/m^3. No existing soil has a $D_r = 0.00$ nor a value of 1.00. Common ranges are from 0.3 to 0.7.
Source: From Bowles, J.E., 2002, *Foundation Analysis and Design*, McGraw-Hill, New York. With permission.

As discussed in Section 2.4.1.2, the subscript 70 indicates 70% efficiency in energy transfer from the hammer to the sampler. This value has been shown to be relevant for the North American practice of SPT.

2.4.1.2 Efficiency of Standard Penetration Testing

The actual energy effective in the driving of the SPT equipment varies due to many factors. Hence, in addition to the effective overburden stress at the tested location, the SPT parameter depends on the following additional factors:

1. Hammer efficiency
2. Length of drill rod
3. Sampler
4. Borehole diameter

TABLE 2.6

Consistency of Saturated Cohesive Soils[a]

Consistency	N'_{70}	q_u, kPa	Remarks
	Increasing NC Young clay		
Very soft	0–2	<25	Squishes between fingers when squeezed
Soft	3–5	25–50	Very easily deformed by squeezing
Medium	6–9	50–100	??
	OCR Aged/cemented		
Stiff	10–16	100–200	Hard to deform by hand squeezing
Very stiff	17–30	200–400	Very hard to deform by hand squeezing
Hard	>30	>400	Nearly impossible to deform by hand

[a]Blow counts and OCR division are for a guide — in clay "exceptions to the rule" are very common.
Source: From Bowles, J.E., 2002, *Foundation Analysis and Design*, McGraw-Hill, New York. With permission.

Accordingly, the following equation has been suggested for obtaining an appropriate standard SPT parameter to be used in foundation designs:

$$N'_E = N'\eta_i \qquad (2.5a)$$

where N'_E is the standard hammer efficiency ($= 70\%$), N' is the SPT value corrected for the effective overburden stress (Equation (2.2a)), and η_i are the factors that account for the variability due to factors 1–4 mentioned above.

The hammer used to drive the sampler can be either manual or automatic. Numerous configurations of both hammer types have been manufactured. The safety type is the most common manual hammer as it is equally suited to both drive and extract the split spoon. This type of hammer is lifted by the friction developed between a rope and a spinning cathead power take-off. The number of wraps around the cathead as well as the diameter of cathead are specified as well as the condition of the rope and cathead surface (Figure 2.8). Due to the incomplete release of the drop weight from the cathead, the total potential energy of the drop is not available to advance the sampler. A recent study showed that manual hammers transfer anywhere between 39% and 93% of the energy (average 66%), while automatic hammers transfer between 52% and 98% (average 79%). Although the reproducibility of an automatic hammer is better than manual hammer, the variation in energy efficiency cited is dependent on the upward velocity of the hammer as controlled by the revolutions per minute (rpm) of the drive chain motor (Figure 2.9). To this end,

(a) (b) (c)

FIGURE 2.8
SPT apparatus with manual hammer: (a) manual hammer; (b) hammer drop onto the cathead; (c) pull rope wrapped around spinning cathead.

(a)   (b)

FIGURE 2.9
Automated SPT apparatus: (a) truck-mount drill rig; (b) chain-driven automatic SPT hammer.

a given machine should be calibrated to produce an exact 30-in. drop height and the rpm required to produce that drop recorded and maintained.

As the standard hammer efficiency is 70%, it must be noted that for an SPT system with a hammer efficiency of 70% ($E = 70$), $\eta_1 = 1.0$. However, the hammer efficiencies of most commonly used SPT apparatus are 55% and 60%. Therefore, it is common for foundation engineers to encounter correlation equations for design parameters where the SPT blow count is expressed as N'_{55} or N'_{60}. However, the standard N'_{70} can easily be converted to the equivalent N'_{55} or N'_{60} using the corresponding η_1 factors in Equation (2.5a) as follows:

$$N'_{70} = N' \eta_{1,70} \prod_{i=2}^{4} \eta_i \tag{2.5a}$$

$$N'_{60} = N' \eta_{1,60} \prod_{i=2}^{4} \eta_i \tag{2.5b}$$

If it is assumed that the only difference between N'_{70} and N'_{60} is due to the differences in the corresponding η_1 factors, then one can simplify Equations (2.5a) and (2.5b) to

$$N'_{60} = N'_{70} \frac{\eta_{1,60}}{\eta_{1,70}} \tag{2.5c}$$

Since the η_1 values would be directly proportional to the corresponding efficiencies, the following relationship holds:

$$N'_{60} = N'_{70} \frac{60}{70} \tag{2.5d}$$

One realizes the reason for this conversion based on the logic that the penetration must be directly proportional to the energy available for the penetration action and hence the efficiency of the system since the energy input into any SPT apparatus is fixed under standard operating conditions. Therefore, N'_{55} or N'_{60} can be deduced from N'_{70} using the factors of 55/70 (or 0.787) and 60/70 (or 0.857), respectively, when the available correlations demand such a conversion. Furthermore, information that can be used to obtain η_2 to η_4 is given in Bowles (2002).

Although it is relatively easy to perform, SPT suffers because it is crude and not repeatable. Generally, a variation up to 100% is not uncommon in the SPT value when using different standard SPT equipment in adjacent borings in the same soil formation. On the other hand, a variation of only 10% is observed when using the same equipment in adjacent borings in the same soil formation.

2.4.2 *In Situ* Rock Testing

The design of rock-socketed drilled shafts is highly dependent on the integrity of the rock core samples obtained from field investigation. When sufficient samples are recovered, laboratory tests can be conducted to determine the splitting tensile strength, q_t (ASTM D 3967), and the unconfined compressive strength, q_u (ASTM D 2938). The shear strength of the shaft–limestone interface, f_{su}, is then expressed as a function of q_t and q_u (McVay et al., 1992). This value is typically adjusted by rock quality indicators such as the rock quality designation, RQD (ASTM D 6032), or the percent recovery, REC. For example, the State of Florida outlines a method using the percent recovery to offset the highly variable strength properties of the Florida limestone formation. Therein, a design value, $(f_{su})_{\text{DESIGN}}$, is expressed as $REC * f_{su}$ (Lai, 1999). These methods work well in consistent, competent rock but are subject to coring techniques, available equipment, and driller experience. Sampling problems are compounded in low-quality rock formations as evidenced by the occurrence of zero RQD and low REC values.

2.4.2.1 *Timed Drilling*

To counter poor-quality samples (or no sample at all), some designers with extensive local experience use timed drilling techniques to estimate rock quality and shaft design values in addition to, or in lieu of, the previous methods. With this technique, the driller must record the time to advance a wash boring through a bearing stratum while maintaining a constant "crowd" pressure, fluid flow, and rotational bit speed. Advance times would typically need to be greater than 2 to 3 min/ft to be useful. Lower advance times are common in weaker soils that are more effectively tested by standard penetration testing. Like SPT and CPT, the equipment should be maintained in reasonably consistent physical dimensions (i.e., the bit should stay in good working condition). Although this method is very simple, it is highly empirical and largely dependent on the uniformity of the drilling techniques. Additionally, the designer must have developed a large enough database (with load test calibration) to design with confidence. Such databases exist, but are proprietary and not common knowledge.

2.4.2.2 *Coring Methods*

When designing from rock core samples, it is important to consider the factors affecting sample retrieval and hence their quality. The recovered samples can range in diameter from 0.845 to 6 in. (2.15 to 15.25 cm) where larger samples are preferred in soft limestone. The State of Florida requires a minimum core diameter of 2.4 in. (6.1 cm) but recommends

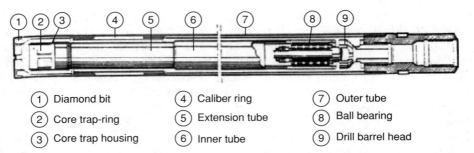

① Diamond bit	④ Caliber ring	⑦ Outer tube
② Core trap-ring	⑤ Extension tube	⑧ Ball bearing
③ Core trap housing	⑥ Inner tube	⑨ Drill barrel head

FIGURE 2.10
Schematic for double-tube core barrel. (After Wittke, W., 1990, *Rock Mechanics, Theory and Application with Case Histories*, Springer-Verlag, Berlin. With permission.)

4 in. (10.2 cm). The drill core samples can be obtained from three different types of core barrels: single tube, double tube, and triple tube. The simplest is the single tube in which the drill core and flushing fluid occupy the same space and consequently can lead to erosion of low strength or fragmented rock samples. As a result, this type of core barrel is not permitted for use with Florida limestone (FDOT, 1999).

Double-tube core barrels differ from single-tube barrels by essentially isolating the drill core from the flushing fluid (Figure 2.10). Simple versions of this type of core barrel use a rotating inner tube that requires a small fraction ($\approx 10\%$) of the drilling fluid to circulate around the drill core to prevent binding and direct contact of the sample with the tube. Most double-tube systems now use a fixed inner tube that requires no flush fluid around the drill core and thus causes fewer disturbances to the sample. During extraction of the entire barrel assembly, a core trap-ring at the leading edge of the inner barrel snares the drill core preventing its loss (see Figure 2.11). Recovering the sample from the inner tube without disturbing it is difficult in soft, fragmented, or interlayered rock deposits. Both fixed and rotating inner core barrels are permitted by FDOT but significant variations in recovery values should be expected.

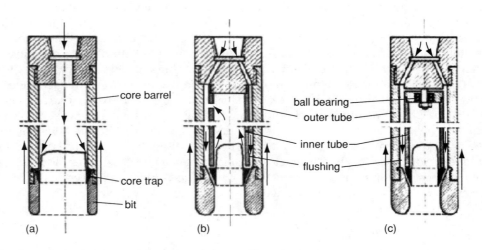

FIGURE 2.11
In situ rock coring apparatus: (a) single tube; (b) double tube with rotating inner tube; (c) double tube with fixed inner tube. (After Wittke, W., 1990, *Rock Mechanics, Theory and Application with Case Histories*, Springer-Verlag, Berlin. With permission.)

FIGURE 2.12
Triple-tube core barrel components.

The triple-tube core barrel, in concept, is essentially the same as the double tube (with the fixed inner tube). It differs in the way the specimen is recovered in that the inner tube is fitted with yet a third sleeve or split tubes in which the drill core is housed. The entire sleeve or split tube is extruded from the inner barrel using a plunger and pressure fitting that pushes directly on the split tubes. The extrusion process is similar to that of Shelby tube samples except the sample is not stressed. In this manner, the sample is not compressed or shaken loose. Figure 2.12–Figure 2.15 show the components of the triple-tube core barrel and sample extruder.

Further variables affecting core drilling results include: the type of drill bit, the flow rate of the flushing fluid, the end gap between the inner and outer barrels, the crowd pressure, and the advance rate through softer interlayered deposits. With so many variables controlling sample recovery, methods of investigating the remaining borehole for the *in situ* limestone characteristics could have significant merit.

2.4.2.3 In Situ *Rock Strength Tests*

Direct measurements of the *in situ* bond or shear strength of the drilled shaft-to-rock interface can be obtained through small-scale anchor pull-out tests or full-scale load tests. Anchor pull-out tests are purported to have produced reasonable correlations with full-scale results (Bloomquist et al., 1991). The test method involves simply grouting a high-strength post-tensioning rod into a borehole, and measuring the load required to pull the grout plug free. (*Note*: load is directly applied to the base of the plug to produce compression and the associated Poisson expansion in the specimen.) Attention must be given to the surface area formed by the volume of grout actually placed. This test is an attractive option in that it is relatively inexpensive, requires minimal equipment mobilization, and can be conducted at numerous locations throughout a site. However, it has received little attention as a whole and remains comparatively unused.

Design-phase, full-scale, *in situ* testing of the shaft–limestone interface is by far the surest method to determine the design parameters of a drilled shaft. This can be accomplished by several means: top down static loading, bi-directional static loading, statnamic loading, or drop-hammer dynamic loading (discussed later). However, due to the

FIGURE 2.13
Sample extruder.

associated costs, only a small fraction of rock sockets are tested in this fashion, and rarely at the design phase. Additionally, a single test may not adequately account for the spatial variability of rock formations without correlation to standard site investigation methods. As such, a host of *in situ* borehole devices have been developed to aid in estimating soil and rock strength parameters.

In situ borehole modulus devices are classified into two categories based on their loading apparatus: (1) rotationally symmetric borehole loading devices and (2) diametrically arranged lateral loading plates (Wittke, 1990). Figure 2.16 shows the loading scheme of the two conceptual mechanisms.

Type 1 probes apply load via a rubber diaphragm that is pressurized by either gas or liquid. In general, measurements of displacement are made directly when using gas pressure, and indirectly through change in volume when using fluid pressure. Table 2.7 lists Type 1 devices that have been developed by various manufacturers.

Type 2 probes use two semicylindrical thrust plates diametrically aligned to apply loads to the arc of the borehole. Measurements of displacement are obtained directly at a minimum of two locations along the longitudinal axis of the plates. Whereas Type 1 devices produce uniformly distributed radial stresses in the borehole, the stress distribution in Type 2 devices is dependent on the relative stiffness of the rock and the plate. Table 2.8 lists Type 2 devices that have been developed by various manufacturers.

Another mechanism (not originally intended for rocks) that has interesting features with respect to weak rocks is the Iowa borehole shear device (Section 2.10). The test

FIGURE 2.14
Cutting bits used by Florida Department of Transportation District I.

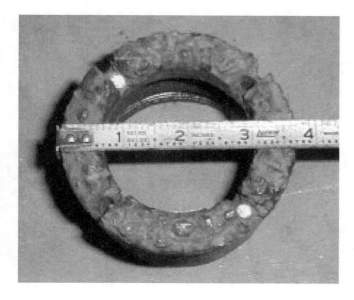

FIGURE 2.15
HQ3 triple-tube cutting tip, "Devil's Nightmare" 3.78 in. (9.6 cm) of outer diameter.

scheme for this device is a combination of both the anchor pull-out test and the borehole modulus test. The device is expanded into the walls of the borehole and is then pulled to determine the shear strength of the soil. Typically, several lateral pressures are investigated (Demartinecourt, 1981).

2.5   Cone Penetration Test

The cone penetration test (CPT) is an invasive soil test that defines soil strata type, soil properties, and strength parameters. It is highly repeatable, insensitive to operators, and best suited for uncemented soils, sands, or clay. Although this test retrieves no sample for

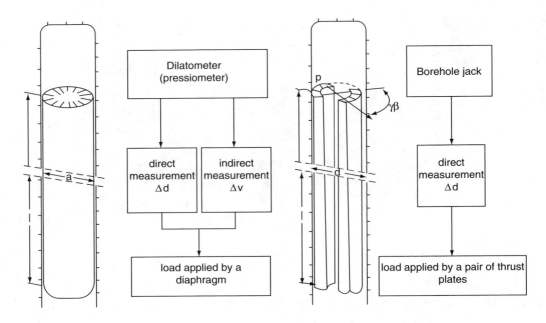

FIGURE 2.16
Loading scheme of the two types of modulus devices. (After Wittke, W., 1990, *Rock Mechanics, Theory and Application with Case Histories*, Springer-Verlag, Berlin. With permission.)

laboratory testing or visual inspection, it has the capability of producing enormous amounts of physical information based on correlations with side-by-side tests with other test methods such as SPT. Further, as the test provides direct measurements of ultimate end bearing and side shear, it is directly applicable for design of foundations of all kinds.

TABLE 2.7

List of Type 1 Borehole Devices

Name	Method of Measuring	Number of Measuring Devices	Max. Applied Pressure, P_{max} (MN/m²)	Borehole Diameter, d (mm)	Test Length, l (mm)	l/d
Menard pressuremeter	Indirect (Δv)	—	10	34–140	502–910	65
CSM cell	Indirect (Δv)	—	70	38	165	4.3
Janod Mermin probe	Direct (Δd)	3	15	168	770	4.6
Sounding dilatometer	Direct (Δd)	2	4/7.5	200/300	1000/1200	5/4
Comes probe	Direct (Δd)	3	15	160	1600	10
LNEC dilatometer	Direct (Δd)	4	15	76	540	7.1
Tube deformeter	Direct (Δd)	4	4	297	1300	4.4
Prigozhin pressuremeter	Direct (Δd)	2	20	46	680	14.8
Atlas dilatometer	Direct (Δd)	8	10	144	890	6.2
BGR dilatometer	Direct (Δd)	4	40	86	1000	11.6
Dilatometer 95	Direct (Δd)	3	12	100	1000	10
Dilatometer 112	Direct (Δd)	3	12	116	1000	8.6
Elastometer 100	Direct (Δd)	2	10	62	520	8.4
Elastometer 200	Direct (Δd)	3	20	62	520	8.4

Source: Wittke, W., 1990, *Rock Mechanics, Theory and Application with Case Histories*, Springer-Verlag, Berlin.

TABLE 2.8

List of Type 2 Borehole Devices

Name	Angle of Opening of Thrust Plates	Max. Applied Pressure, P_{max} (MN/m²)	Borehole Diameter, d (mm)	Test Length, l (mm)	l/d
Geoextensometer	$2\beta = 1430$	34	76	306	4.0
Goodman jack	$2\beta = 900$	64	74–80	204	2.6–2.8
CSIRO pressiometer	$2\beta = 1200$	35	76	280	3.7

Source: Wittke, W., 1990, *Rock Mechanics, Theory and Application with Case Histories*, Springer-Verlag, Berlin.

The CPT is described in ASTM test number D-3441, entitled "Standard Test Method for Deep, Quasi-Static, Cone and Friction-Cone Penetration Tests of Soil." This is to include cone penetration type tests that use mechanical or electronic load detection, tip or tip and friction stress delineation, and those tests where the penetration into the soil is slow and steady in a vertically aligned orientation. Those tests conducted with mechanical load detection are typically denoted as "Dutch cone tests" and those using electronic detection are simply called "cone penetration tests." The term "quasi-static" refers to a steady rate of penetration where the acceleration is zero, but the velocity of penetration is constant (1 to 2 cm/sec $\pm$ 25%).

The test apparatus consists of a 60° conical tip of known cross-sectional area that is thrust into the soil at a near constant rate. Behind the cone tip, a friction sleeve of known surface area is also included that is used to detect the side shear or adhesion between the steel sleeve and the surrounding soil. The force required to advance the tip through the soil is divided by the cross-sectional area to determine the tip stress, q_c. Similarly, the force required to advance the friction sleeve is divided by the sleeve surface area to produce the local friction value, f_s. The tip area and sleeve area vary from device to device but the most common areas are 10 and 150 cm², respectively.

The tip area (diameter) can influence the magnitude of the resulting q_c value similar to the effects of foundation diameter on capacity. This is due to the increased zone of influence beneath the tip as the cone diameter increases for various devices. Therefore, in relatively uniform soils, the tip diameter has little effect. In layered or more heterogeneous strata, a smaller tip diameter will better register the minute changes in soil type and strength. Larger diameter cones physically average the effects of thin layers. Figure 2.17 shows two different-sized cone tip and sleeve assemblies.

Another feature that CPT equipments usually incorporate is the capability of monitoring the pore-water pressure while advancing the cone–sleeve assembly. There can be a significant amount of excess pore pressure developed by forcing the volume of the cone into a somewhat fixed volume of a poorly draining material. In contrast, in dense fine-grained soils, the cavity expansion can cause a decrease in pore pressure. The assemblies shown in Figure 2.17 have a pressure transducer within the tip body that registers the pore pressure directly behind the tip (between the tip and the sleeve). This information can be used to convert the total stress registered by the tip to effective stress similar to a consolidated undrained triaxial compression test. When pore pressure measurements are taken the test is denoted as CPTU. Smaller cones tend to induce less cavity expansion and therefore fewer effects on total stress.

The thrust required to advance the cone assembly is dependent on the strength of the soil as well as the size of the tip. Given the disparity between the cone sizes in Figure 2.17, it is not surprising that the size of the equipment required to advance these cones is also disparate. Figure 2.18 shows the associated truck-mounted CPT rigs that use these

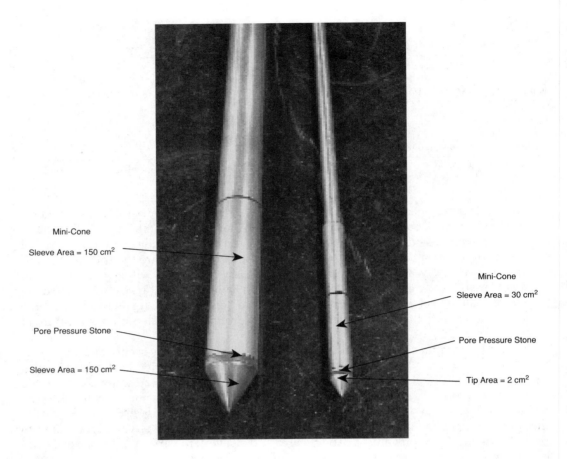

FIGURE 2.17
Two-cone tip and sleeve assemblies of different sizes.

devices. The larger diameter cone requires a 20-ton (178 kN) thrust mechanism and can reach depths of 100 m; the mini-cone requires about one fifth the thrust and can be mounted on the front of a standard truck or utility vehicle. Mini-CPTs are limited to a practical maximum depth 20 m.

The thrust mechanisms also vary between the two systems in Figure 2.18. The standard CPT system uses 1-m segmental rods to advance the cone tip–sleeve assembly. The thrusting ram is designed to grip, thrust downward 1 m, release and stroke upward 1 m, re-grip, and repeat. At a penetration rate of 2 cm/sec, the process progresses at an average advance rate of 1 cm/sec to account for the re-gripping. An average sounding to 30 m should take on the order of 1.5 to 2 h (in and out). An SPT to a similar depth could take multiple days. The mini-cone uses a more time efficient method of advancing the cone. It uses a contiguous length of mild stainless steel tubing (10 to 20 m long) that is continuously fed at a constant rate without having to re-stroke and re-grip. Therein, the tubing is straightened from a large diameter coil as it is continuously gripped by

(a)   (b)

FIGURE 2.18

Field cone penetration devices: (a) standard 20-ton cone truck (Courtesy of the Florida Department of Transportation.); (b) 4-ton mini-cone vehicle. (Courtesy of the University of South Florida.)

side-by-side opposing chains specially designed to mate to the coil diameter. Figure 2.19 shows both thrust mechanisms.

The data collection during cone penetration testing is typically performed at 5-cm intervals but can be as frequent as permitted by the data acquisition system. This gives a virtually continuous sounding of tip and sleeve stresses. As both of these devices use strain gage-based load cells, the instrumentation leads are routed through the center of the thrusting rod–tube. The data are processed to produce the soil type as well as other parameters in real time. The basis for the data regression is based on correlations developed by Robertson and Campanella (1983). Although many correlations exist, the most significant uses a calculated parameter that defines the ratio of measured side to the measured tip stress. This ratio is defined as the friction ratio R_f. To aid in classifying various soils, 12 soil types were defined that could be readily identified given the cone bearing stress, q_c, and the friction ration, R_f. Figure 2.20 shows the classification chart used to identify soils from CPT data. Further, correlations from CPT to SPT test data were developed to elevate the comfort of designers more familiar with SPT data. Therein, the q_c/N ratio was defined for each of the 12 soil types (also shown in Figure 2.20).

Furthermore, the undrained strength of clay, S_u, can be obtained using CPT data as follows:

$$S_u = \frac{q_c - p_0}{N_k} \tag{2.6}$$

where p_0 is the total overburden pressure and the N_k factor ranges between 15 and 20 (Bowles, 2002). It must be noted that S_u is expressed in the same units as q_c.

2.5.1 Cone Penetration Testing with Pore Pressure Measurements (Piezocone)

Currently it is commonplace to have cone tips fitted with pore pressure transducers that can produce a continuous record of the ground pore pressures at various depths. It is typical to install the piezometer that consists of a porous ring attached to a pore pressure transducer onto the sleeve of the CPT equipment immediately below the cone tip. If such a piezocone is used to obtain foundation strength parameters, the following modification for evaluating the corrected cone resistance, q_T, has been suggested by Robertson and Campanella (1983):

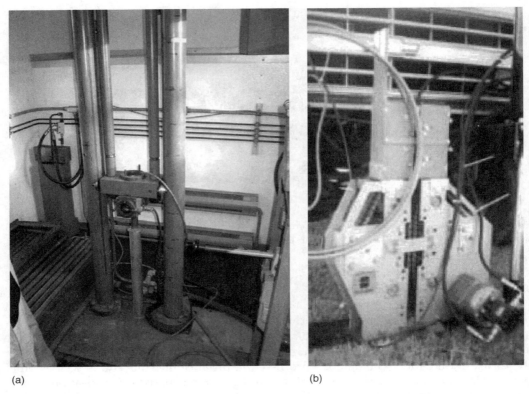

(a)   (b)

FIGURE 2.19
CPT driving mechanisms: (a) twin 10-ton rams used to thrust standard cone rod; (b) continuous feed chain drive used to thrust mini-cone.

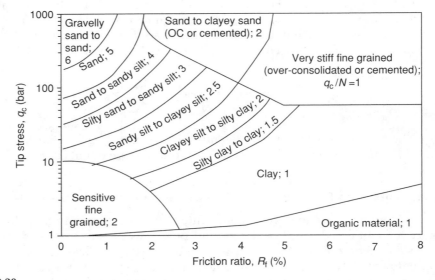

FIGURE 2.20
CPT data correlated to soil type and equivalent SPT-N. (After Robertson, P.K. and Campanella, R.E., 1983, *Canadian Geotechnical Journal*, Vol. 20, No. 4. With permission.)

$$q_T = q_c + u(1 - a) \tag{2.7}$$

where μ is the measured pore pressure and a is taken as the approximate area ratio $(d_1/D)^2$ of the cone; d_1 and D are internal and external diameters of the sleeve and the pore pressure sensor, respectively. The value of parameter a is typically in the range of 0.6 to 0.9 (Bowles, 2002).

2.6 The Field Vane Shear Test

The field vane shear test (ASTM D 2573) is the most common test for evaluating the undrained shear strength of soft to stiff clays because of its speed of performance and repeatability. During the test, a standard-size vane (Figure 2.21) is placed in the borehole and pushed to the depth where the evaluation of the undrained shear strength is required. Then it is twisted carefully and the torsional force required to cause shearing *in situ* is measured. The blade is rotated at a specified rate that should not exceed 0.1°/sec (practically 1° every 10 sec). If it is assumed that the undrained shear strength (S_u) of the tested clay is the same in both the horizontal and vertical planes, then S_u can be obtained from the following equation:

$$T = \frac{\pi S_u d^2}{2}\left(h + \frac{d}{2}\right) \tag{2.8}$$

where T is the maximum torque required for shearing the clay with the vane, d is the diameter of the vane, and h is the height of the vane.

2.7 The Pressuremeter Test

The basic concept behind the pressuremeter test (PMT) (ASTM D 4719) is that the uniform pressure required for monitored expansion of a cylindrical cavity in the ground would indicate not only the compressibility characteristics of the tested ground but also the

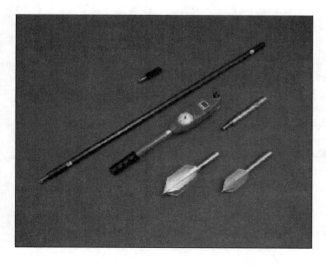

FIGURE 2.21
Field vane apparatus. (Courtesy of Geonor Corp.)

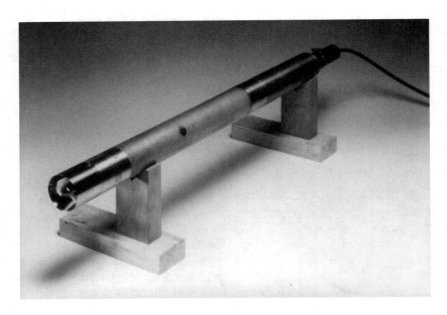

FIGURE 2.22
Self-boring pressuremeter. (From www.cambridge-insitu.com. With permission.)

ultimate pressure that it can sustain before complete shear failure due to lateral stressing. The main exercise involved in the test is the monitoring of the relationship between pressure and deformation of the tested soil. In practice, the test is conducted by first drilling a hole down to the desired elevation. The PMT probe (Figure 2.22) is then inserted inside the cavity and inflated to expand the cavity while recording the resulting volume change (ΔV) versus the applied pressure (p) in the probe. The test results are plotted on a ΔV versus p plot.

The ultimate objective of pressuremeter tests is to characterize the stress–strain relationship of the *in situ* soil up to the ultimate shear failure. As mentioned above, the testing mechanism permits the inference of both the deformation properties and the coefficient of lateral earth pressure that would model the perfectly plastic shear failure according to the Coulomb shear (frictional) failure theory. The deformation properties are deduced using the pressuremeter modulus that can be linked to the elastic modulus through Equation (2.9):

$$E = 2(1 + \mu)(V_0 + v_{\mathrm{m}})\frac{\mathrm{d}p}{\mathrm{d}v} \tag{2.9}$$

where E is the elastic modulus of the soil, μ is the Poisson ratio of the soil associated with 2D elastic deformation, V_0 is the initial volume of the measuring cell (typically $535\,\mathrm{cm}^3$), v_{m} is the expanded volume of the measuring cell at the mid-point of the linear portion of the V versus P curve (Figure 2.23), and $\mathrm{d}p/\mathrm{d}v$ is the pressuremeter modulus, which is equal to the slope of the linear portion of the V versus P curve (Figure 2.23).

On the other hand, the shear strength of the soil can be evaluated approximately through the lateral pressure, p_{h}, at which the pressuremeter membrane comes into complete contact with the surrounding soil. Under these conditions, the following expression for lateral earth pressure at rest can be applied:

$$K_0 = \frac{p_{\mathrm{h}}}{\sigma'_{\mathrm{v}}} \tag{2.10}$$

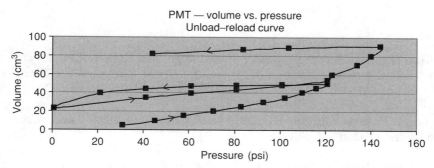

FIGURE 2.23
Typical results from a pressuremeter test. (Courtesy of the University of South Florida.)

Results obtained from a pressuremeter test performed at a depth of 30 ft (10 m) at the University of South Florida campus is illustrated in Figure 2.23. This plot shows a partial expansion of the pressuremeter up to a pressure of 120 psi without reaching limiting conditions, initial unloading, and reloading until a limiting pressure of 150 psi. It is also seen that the final unloading curve is parallel to the initial unloading and the reloading curves. From the initial unloading and reloading portions of the plot in Figure 2.23, it can be inferred that the linear portion of the curve as beginning around the point (20, 35) and terminating around the point (100, 45).

Alternatively, the pressuremeter results can also be plotted on a p (corrected inflation pressure) versus r (radius of the pressuremeter probe) plot. Hughes et al. (1977) used the elastoplastic theory of cavity expansion to model the pressuremeter inflation to obtain the following useful relations that can be applied to a p versus r plot, when a pressuremeter test is conducted in a sandy soil.

The pressuremeter inflation pressure at the elastic limit, p_e, can be expressed as

$$p_e = p_h(1 + \sin \phi) \tag{2.11}$$

In the range of plastic deformation, the gradient of the log (radial deflection) versus log (inflation pressure) curve is given by

$$m = [1 + \tan^2 (45 + \nu/2)]\frac{(1 + \sin \phi)}{2 \sin \phi} \tag{2.12}$$

where ν is the dilation angle that is the slope of the shear stress (τ) versus shear strain (γ) plot for the sand.

Example 2.1
Based on the plot in Figure 2.23, estimate the elastic deformation and shear strength parameters of the tested soil.

From Figure 2.23, the lateral pressure for complete contact with soil = 20 psi = *in situ* lateral soil pressure at rest.

Assuming that there was no groundwater up to a depth of 20 ft, the overburden pressure = 120 pcf × 30 ft = 3600 psf = 172.5 kPa. By applying Equation (2.10),

$$K_0 = 20(144)/3600 = 0.8$$

Assuming that the tested soil is a well-drained soil and the following common relationship holds for the coefficient of lateral earth pressure at rest:

$$K_0 \approx 1 - \sin \phi \qquad (2.13)$$

$\therefore \phi = 11°$

Also, from Figure 2.23, one can derive the value of v_m as $[\frac{1}{2}(45 + 35) - 20]$ cm^3 or 20 cm^3 (0.0007 ft^3), and the pressuremeter modulus, dp/dv, as $(100 - 20)$ psi/$(45 - 35)$ cm^3 = 80 psi/ 10 $(0.394)^3$ in.^{3} = 130.8 lb/in.^{5} = 32,546,744 lb/ft^5 = 80(6.9) kPa/10(10 − 6) m^3 = 55.2 (10^6) kN/m^5.

Assuming a Poisson ratio of 0.33 and V_0 of 535 cm^3 (0.0189 ft^3) and substituting for dp/dv in Equation (2.9),

$$E = 2(1.3)(0.0189 \text{ ft}^3 + 0.0007 \text{ ft}^3)(32{,}546{,}744) \text{ lb/ft}^5 = 1658 \text{ ksf}$$

In SI units, $E = 2(1.3)(535 + 20)(10^{-6} \text{ m}^3)(55.2(10^6)) \text{ kN/m}^5) = 79.65 \text{ MPa}$

Hence, the elastic soil modulus for lateral deformation is about 1658 ksf or 80 MPa.

Example 2.2
Table 2.9 provides the data obtained from a pressuremeter test in sand estimate the angle of internal friction and the dilation angle for the sand. From Figure 2.24, $p_h = 1.5 \text{ kg/cm}^2$ and $p_e = 2.5 \text{ kg/cm}^2$.

Applying Equation (2.11), $2.5 = 1.5(1 + \sin \phi)$
$\phi = 42°$

The gradient of the $\log r$ versus $\log p$ plot within the plastic range ($p > 2.5 \text{ kg/m}^2$) is found to be 3.1. Then, by applying Equation (2.12),

$$3.1 = [1 + \tan^2 (45 + \nu/2)]\frac{(1 + \sin 42)}{2 \sin 42}$$

the dilation angle (ν) for sand is found to be around 11.5°.

TABLE 2.9

Data for Example 2.2

Corrected Pressure (kg/cm^2)	Expanded Radius (cm)
0	3.417
0.352	3.635
0.394	4.192
0.922	4.705
1.408	4.744
1.9	4.768
2.394	4.789
2.837	4.814
3.326	4.85
3.811	4.901
4.294	4.961
4.737	5.051
5.183	5.162
5.641	5.295
6.091	5.452
6.531	5.635
6.955	5.859
7.143	6.176

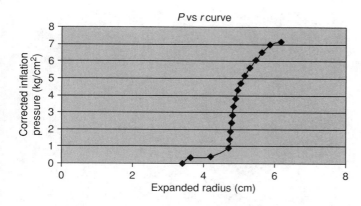

FIGURE 2.24
Illustration for Example 2.2.

Using Equation (1.69), one can also estimate the residual friction angle ϕ_{cv} using $\phi = 42°$ and $\nu = 11.5°$:

$$\tan^2 (45 + \phi/2) = [\tan^2 (45 + \nu/2)][\tan^2 (45 + \phi_{cv}/2)]$$

Hence, $\phi_{cv} = 33°$.

2.8 The Dilatometer Test

The dilatometer test (DMT) (ASTM D 6635) presents an alternative to the pressuremeter in terms of using measuring foundation design parameters using an inflatable membrane as opposed to the expansion of a cylindrical cavity. It measures the lateral defection of the tested soil by applying gas pressure through a vertical plate inserted at the desired level. DMT is more versatile than the pressuremeter in that in addition to the deformation modulus of the soil various other soil parameters, such as soil type, shear strength, *in situ* pore water pressure, OCR, k_o, and coefficent of consolidation, C_v, can also be deduced through many different correlations with the DMT measurements. In addition, a great deal of other information can be obtained from the DMT such as *in situ* stratigraphy, compressibility, and stress history.

The instrument is a paddle-shaped stainless steel plate (Figure 2.25) with a 60-mm thin high strength, expandable, circular steel membrane (0.2-mm thick) mounted at the center of one face. The membrane is expanded by pressurizing with nitrogen gas through a tube connected to the blade. The tip of the blade is sharpened to facilitate easy penetration in the soil (Figure 2.26). An electropneumatic tube (gas tube and electrical cable) runs through the hollow rod, which connects the blade to a pressure control and gauge readout unit. Before the actual test, the dilatometer must be calibrated to monitor the response of the electro-pneumatic readout unit. Gas pressure is applied to the diaphragm of the dilatometer to hear the initial buzzer sound, which will indicate that the membrane is in the seating position. This initial reading must be deducted from the actual readings taken during testing.

The DMT boasts of high reproducibility and is relatively quick to perform. Also, there are direct design parameters that can be obtained from the DMT. In terms of the limitations, like the CPT, no soil sample can be obtained from this test and there is slight disturbance of soil to be tested. Moreover, this test is not as straightforward as the CPT or

FIGURE 2.25
Calibration of the flat plate dilatometer. (Courtesy of the University of South Florida.)

SPT and, therefore, may be more difficult for a nontechnical crew to perform. Hence, the quality of the data obtained from this test depends on calibration, which can be a source of error. This test measures the soil pressure in only one direction compared to the pressuremeter test. Furthermore, DMT cannot be performed in soils that contain gravel or rock since the membrane is fairly delicate and can be susceptible to damage.

2.8.1 Measurement Procedure

1. Obtain p_1 pressure readings corresponding to the instant when the membrane is just flush with the plate.

FIGURE 2.26
Field application of the flat plate dilatometer (From www.marchetti-dmt.it. With permission.)

2. Take the p_2 pressure reading when the membrane expands by 1.1 mm into the surrounding soil.

3. Release the inflation pressure until the membrane returns to the original position that is flat with the plate.

4. Estimate pore pressure (u) and vertical effective stress (p'_0) prior to blade insertion.

5. Determine DMT parameters as follows:

I_D (material index)

$$I_D = \frac{p_2 - p_1}{p_2 - u} \tag{2.14}$$

K_D (horizontal stress index)

$$K_D = \frac{p_1 - u}{p'_0} \tag{2.15}$$

E_D (dilatometer modulus)

$$E_D = 34.7(p_2 - p_1) \tag{2.16}$$

2.8.2 Determination of Foundation Design Parameters

The material index, I_D, and the dilatometer modulus, E_D, can be used in Figure 2.27 to identify the type of soil. E_D and an assumed Poisson's ratio, μ_s can also be used to determine the elastic modulus using the following relationship:

$$E_D = \frac{E_s}{1 - \mu_s^2} \tag{2.17}$$

Furthermore, K_D, the horizontal stress index, can be used in estimating the *in situ* coefficient of lateral stress using the following approximate relationship:

$$K_0 = \left(\frac{K_D}{1.5}\right)^{0.47} - 0.6 \tag{2.18}$$

Design parameters specific for fine-grained soils ($I_D \leq 1.2$ from Figure 2.27) are given in Equation (2.19)–Equation (2.20)

$$OCR = [K_D]^{1.56} \tag{2.19}$$

$$C_u = 0.22[\sigma'_{vo}][0.5K_D]^{1.25} \tag{2.20}$$

Design parameters specific for granular soils ($I_D \geq 1.8$ from Figure 2.27) is given by Equation (2.21)

$$\phi = 28° + 14.6° \ \log(K_0) - 2.61° \ \log^2(K_0) \tag{2.21}$$

Table 2.10 provides a comparison of approximate costs involved with the SPT, CPT, and dilatometer testing and Table 2.11 illustrates the correlations between the elastic modulus of soil and the SPT and CPT test parameters.

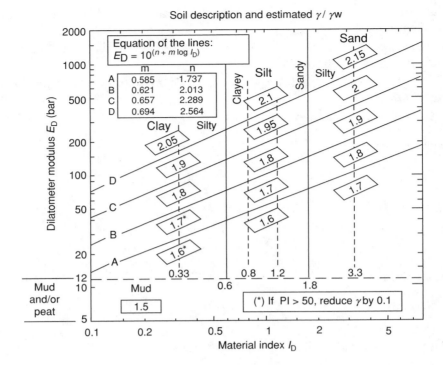

FIGURE 2.27
Use of the dilatometer in the identification of soil type. (From Marchetti, S., Monaco, P., Totani, G., and Calabrese, M., 2001, The Flat Dilatometer Test (DMT) in Soil Investigations, A Report by the ISSMGE Committee TC16, Proc. INSITU 2001, International Conference on In situ Measurement of Soil Properties, Bali, Indonesia, May. With permission.)

2.9 California Bearing Ratio Test

California bearing ratio (CBR) test (Figure 2.28) is a penetration test in which a standard piston having an area of $3\,in.^2$ is used to penetrate a potential roadbed or road-base soil sample. The soil sample is first compacted using a standard Proctor compaction test at the optimum moisture content and soaked for a specified period to simulate critical wet conditions. A surcharge load is applied on the compacted soil sample to simulate the *in situ* stress due to the pavement structure (Table 2.12). The standard penetration rate is 0.05 in./min and the pressure value at each 0.1-in. penetration is recorded up to 0.5 in. Results from a typical test are illustrated in Figure 2.29.

TABLE 2.10

Comparison of Approximate Costs Involved with SPT, CPT, and Dilatometer Testing

Test	Cost	Unit	Mobilization
SPT	$10–20	Per linear ft	$150–300
CPT	$9–15	Per linear ft	$300
DMT	$30–85	Per linear ft	$150–300

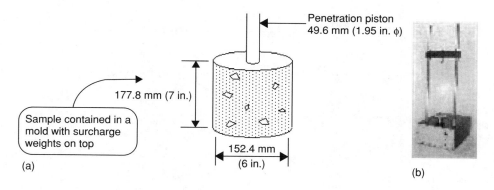

FIGURE 2.28

CBR testing equipment: (a) schematic diagram of CBR test apparatus (Courtesy of WSDOT.); (b) CBR compression tester. (From www.qcqa.com. With permission.)

In order to determine the empirical CBR parameter, the pressure value corresponding to 0.1-in. penetration of the tested soil sample is expressed as a percentage of the corresponding pressure value for standard high-quality crushed rock (Table 2.12). If the ratio obtained from the comparison of the corresponding pressure values for a penetration of 0.2 in. is greater, the latter is used as the CBR value. Therefore, as shown in Table 2.13, the CBR of the tested soil is computed to be 9.

TABLE 2.11

Soil Elastic Moduli from *In Situ* Test Data

Soil	SPT	CPT
Sand (normally consolidated)	$E_s = 500(N + 15)$ $= 7,000 \sqrt{N}$ $= 6,000N$	$E_s = (2\text{–}4)q_u$ $= 8,000 \sqrt{q_c}$ —
Sand (saturated)	— $\ddagger E_s = (15,000\text{–}22,000) \ln N$ $E_s = 250(N + 15)$	$E_s = 1.2(3D_r^2 + 2)q_c$ $*E_s = (1 + D_r^2)q_c$ $E_s = Fq_c$ $e = 1.0, F = 3.5$ $e = 0.6, F = 7.0$
Sands, all (norm. consol.) Sand (overconsolidated)	$\P E_s = (2,600\text{–}2,900)N$ $\dagger E_s = 40,000 + 1,050N$ $E_{s(OCR)} \approx E_{s,nc}\sqrt{OCR}$	$E_s = (6\text{–}30)q_c$
Gravelly sand	$E_s = 1,200(N + 6)$ $= 600(N + 6) \quad N < 15$ $= 600(N + 6) + 2,000 \quad N > 15$	
Clayey sand Silts, sandy silt, or clayey silt	$E_s = 320(N + 15)$ $E_s = 300(N + 6)$ If $q_c < 2,500$ kPa use $2,500 < q_c < 5,000$ use where $E_s^t = \text{constrained modulus} = \dfrac{E_3(1 - \mu)}{(1 + \mu)(1 - 2\mu)} = \dfrac{1}{m_\mu}$	$E_s = (3\text{–}6)q_c$ $E_s = (1\text{–}2)q_c$
Soft clay or clayey silt		$E_s = (3\text{–}8)q_c$

Notes: E_s in kPa for SPT and units of q_c for CPT; divide kPa by 50 to obtain ksf. The N values should be estimated as N_{55} and not N_{70}.

Source: From Bowles, J.E., 2002, *Foundation Analysis and Design*, McGraw-Hill, New York. With permission.

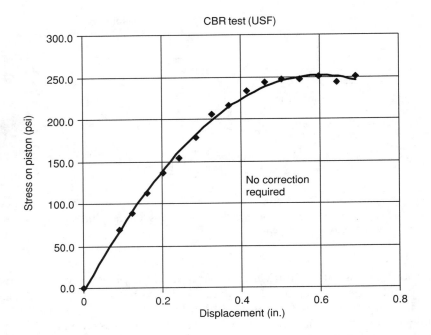

FIGURE 2.29
Typical results from a CBR test. (Courtesy of the University of South Florida.)

2.10 Borehole Shear Test

Borehole shear test (BST) can be adopted to rapidly measure the soil shear strength parameters of fine- to medium-grained soils, *in situ*. The BST is analogous to a laboratory direct shear test or a simple shear test with free drainage. Hence, the shear strength parameters obtained are consolidated drained ones. The most common BST technique used is the Iowa borehole shear test apparatus (Figure 2.30) (Miller et al., 1998).

2.10.1 Test Procedure

The expandable shear head of the BST is first lowered into the borehole to the desired depth. Next, the shear head is expanded against the walls of the hole under a known

TABLE 2.12

Penetration Results for Standard High-Quality Crushed Rock

Penetration (in.)	Pressure (psi)
0.1	1000
0.2	1500
0.3	1900
0.4	2300
0.5	2600

where q_{ult} is the ultimate bearing capacity of the foundation (kN/m², kPa, or ksf), P is the total load at the footing level (structural + refill soil load) (kN or kips), A is the footing area (m² or ft²), F is the an appropriate safety factor that accounts for the uncertainties involved in the determination of the structural loads (P) and the ultimate bearing capacity (q_{ult}).

3.1.2 Settlement Criterion

The designer must also ensure that the footing does not undergo either excessive total settlement or differential settlement within the footing. Excessive settlement of the foundation generally occurs due to irreversible compressive deformation taking place immediately or in the long term. Excessive time-dependent settlement occurs in saturated compressible clays with prior warning through cracking, tilting, and other signs of building distress. On the other hand, significant immediate settlement can occur in loose sands or compressible clays and silts. Therefore, the footing must be proportioned to limit its estimated settlements (δ_{est}) within tolerable settlements (δ_{tol}):

$$\delta_{est} \leq \delta_{tol} \tag{3.1b}$$

3.2 Evaluation of Bearing Capacity

Based on the discussion in Section 3.1.1, a foundation derives its bearing capacity from the shear strength of the subsoil within the influence area (Figure 3.1) and the embedment of the footing (D). Over the years, many eminent geotechnical engineers have suggested expressions for the ultimate bearing capacity of foundations that have also been verified on various occasions by load tests (e.g., plate load test). Some common expressions for the ultimate bearing capacity are provided next.

3.2.1 Bearing Capacity Evaluation in Homogeneous Soil

Terzaghi's bearing capacity expression

$$q_{ult} = cN_c s_c + qN_q + 0.5B\gamma N_\gamma s_\gamma \tag{3.2}$$

Meyerhoff's bearing capacity expression
For vertical loads

$$q_{ult} = cN_c s_c d_c + qN_q s_q d_q + 0.5B\gamma N_\gamma s_\gamma d_\gamma \tag{3.3}$$

For inclined loads

$$q_{ult} = cN_c d_c i_c + qN_q d_q i_q + 0.5B\gamma N_\gamma d_\gamma i_\gamma \tag{3.4}$$

Hansen's bearing capacity expression

$$q_{ult} = cN_c s_c d_c i_c g_c b_c + qN_q s_q d_q i_q g_q b_q + 0.5B\gamma N_\gamma s_\gamma d_\gamma i_\gamma g_\gamma b_\gamma \tag{3.5}$$

For undrained conditions

$$q_{ult} = 5.14 s_u (1 + s'_c + d'_c - i'_c - g'_c - b'_c) + q \tag{3.6}$$

Vesic's bearing capacity expression

$$q_{ult} = cN_c s_c d_c i_c g_c b_c + qN_q s_q d_q i_q g_q b_q + 0.5B\gamma N_\gamma s_\gamma d_\gamma i_\gamma g_\gamma b_\gamma \tag{3.7}$$

TABLE 3.1

Bearing Capacity Factors

ϕ	Terzaghi's (1943) Expression			Hansen, Meyerhoff, and Vesic's Expressions		Hansen (1970) N_γ	Meyerhoff (1951, 1963) N_γ	Vesic (1973, 1975) N_γ
	N_c	N_q	N_γ	N_c	N_q			
0	5.7	1.0	0.0	5.14	1.0	0.0	0.0	0.0
5	7.3	1.6	0.5	6.49	1.6	0.1	0.1	0.4
10	9.6	2.7	1.2	8.34	2.5	0.4	0.4	1.2
15	12.9	4.4	2.5	11.0	3.9	1.2	1.1	2.6
20	17.7	7.4	5.0	14.8	6.4	2.9	2.9	5.4
25	25.1	12.7	9.7	20.1	10.7	6.8	6.8	12.5
30	37.2	22.5	19.7	30.1	18.4	15.1	15.7	22.4
35	57.8	41.4	42.4	46.4	33.5	34.4	37.6	48.1
40	95.7	81.3	100	75.3	64.1	79.4	93.6	109.3
45	172	173	298	134	135	201	262.3	271.3

where c is the cohesive strength, ϕ is the friction angle, N_i are the bearing capacity factors (Table 3.1), q is the effective vertical stress at the footing base level, γ is the unit weight of the surcharge soil, s is the shape factor (Table 3.2; Figure 3.2), d is the depth factor (Table 3.2), i is the inclination factor (Table 3.2 and Table 3.3), g is the ground slope factor (Table 3.3), and b is the base tilt factor (Table 3.3).

Finally, appropriate safety factors recommended for various construction situations are given in Table 3.4.

Example 3.1

For the column shown in Figure 3.3, design a suitable footing to carry a column load of 400 kN, in a subsoil that can be considered as a homogenous silty clay with the following properties: unit weight $= \gamma = 17\,\text{kN/m}^3$; internal friction $= \phi = 15°$; cohesion $= c = 20\,\text{kPa}$.

Case 1. Assume that the ground water table is not in the vicinity.

Case 2. Assume that the ground water table is 0.5 m above the footing.

TABLE 3.2A

Shape and Depth Factors for Hansen's Expression (Hansen, 1970)

Shape Factors	Depth Factors
$s'_c = 0.2\dfrac{B}{L}$ for $\phi = 0°$	$d'_c = 0.4k$
	$d_c = 1.0 + 0.4k$
$s_c = 1.0 + \dfrac{N_q}{N_c}\dfrac{B}{L}$	$k = D/B$ for $D/B \leq 1$
	$k(\text{radians}) = \tan^{-1}(D/B)$ for $D/B > 1$
$s_q = 1.0 + \dfrac{B}{L}\sin\phi$	$d_q = 1 + 2\tan\phi(1 - \sin\phi)^2 k$
$s_\gamma = 1.0 - 0.4\dfrac{B}{L}$	$d_\gamma = 1.00$

Source: From Bowles, J.E., 2002, *Foundation Analysis and Design*, McGraw-Hill, New York. With permission.

TABLE 3.2B

Shape, Depth, and Inclination Factors for Meyerhoff's Expression (Meyerhoff, 1951, 1963) (Figure 3.2)

Shape factors	$s_c = 1 + 0.2K_P \dfrac{B}{L}$
	$s_q = s_\gamma = 1 + 0.1K_P \dfrac{B}{L}, \ \phi > 10°$
	$s_q = s_\gamma = 1, \ \phi = 0°$
Depth factors	$d_c = 1 + 0.2\sqrt{K_P}\dfrac{D}{B}$
	$d_q = d_\gamma = 1 + 0.1\sqrt{K_P}\dfrac{D}{B}, \ \phi > 10°$
	$d_q = d_\gamma = 1, \ \phi = 0°$
Inclination factors	$i_c = i_q = \left(1 - \dfrac{\theta°}{90°}\right)^2$
	$i_\gamma = \left(1 - \dfrac{\theta°}{\phi°}\right)^2, \ \phi > 0°$
	$i_\gamma = 0$ for $\theta > 0, \ \phi = 0°$

Note: Where θ is the load inclination to the vertical and $K_P = \tan^2(45 + \phi/2)$.
Source: From Bowles, J.E., 2002, *Foundation Analysis and Design*, McGraw-Hill, New York. With permission.

Solution

First one must decide on a suitable footing shape and depth. In the case of the footing shape, unless there are limitations in spacing such as the close proximity to the property line, there is generally no reason for one not to use a square or a circular footing. Hence, in this design, one can assume a circular footing.

As for the foundation depth, typically one would seek some significant embedment that does not reach the ground water table or a weak layer known to be underlying the foundation soil. In the current case, obviously none of these can be used as a criterion to select the footing depth. Therefore, one could assume a depth approximately equal to the minimum footing dimension (diameter) of the footing. However, once the design

TABLE 3.2C

Shape and Depth Factors for Vesic's Expression (Vesic, 1973, 1975)

Shape Factors	Depth Factors
$s_c = 1.0 + \dfrac{N_q}{N_c}\dfrac{B}{L}$	$d'_c = 0.4k$ for $\phi = 0°$
	$d_c = 1.0 + 0.4k$
	$k = D/B$ for $D/B \le 1$
	$k(\text{radians}) = \tan^{-1}(D/B)$ for $D/B > 1$
$s_q = 1.0 + \dfrac{B}{L}\tan\phi$	$d_q = 1 + 2\tan\phi(1 - \sin\phi)^2 k$
$s_\gamma = 1.0 - 0.4\dfrac{B}{L}$	$d_\gamma = 1.00$

Source: From Bowles, J.E., 2002, *Foundation Analysis and Design*, McGraw-Hill, New York. With permission.

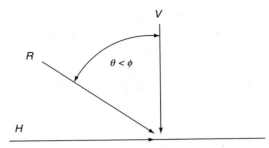

FIGURE 3.2
Guide for obtaining inclination factors.

parameters are obtained, one can reevaluate this criterion to verify that the depth is realistic from a construction point of view.

Tables indicate the following bearing capacity parameters:

Terzaghi's factors (Table 3.1)

$$N_c = 12.9, N_q = 4.4, N_\gamma = 2.5$$
$$s_c = 1.3, s_\gamma = 0.6$$

Hansen's factors (Table 3.1)

$$N_c = 10.97, N_q = 3.9, N_\gamma = 1.2$$
$$s_c = 1.359, s_q = 1.26, s_\gamma = 0.6$$
$$d_c = 1.4, d_q = 1.294, d_\gamma = 1.0$$

TABLE 3.3A

Inclination, Ground Slope, and Base Tilt Factors for Hansen's Expression (Hansen, 1970) (Figure 3.2). Primed Factors are for $\phi = 0$

Load Inclination Factors	Factors for Base on Slope (β)
$i'_c = 0.5 - 0.5\sqrt{1 - \frac{H_i}{A_f c_a}}$	$g'_c = \frac{\beta°}{147°}$
$i_c = i_q - \frac{1 - i_q}{N_q - 1}$	$g_c = 1.0 - \frac{\beta°}{147°}$
$i_q = \left[1 - \frac{0.5H_i}{V + A_f c_a \cot \phi}\right]^{\alpha_1}, 2 \le \alpha_1 \le 5$	$g_q = g_\gamma = (1 - 0.5 \tan \beta)^5$ ($\beta°$ measured clockwise from horizontal)
$i_\gamma = \left[1 - \frac{(0.7 - \theta°/450°)H_i}{V + A_f c_a \cot \phi}\right]^{\alpha_2}, 2 \le \alpha_2 \le 5$	
	Factors for tilted base (η)
	$b'_c = \frac{\eta°}{147°}$
	$b_c = 1 - \frac{\eta°}{147°}$
	$b_q = \exp(-0.0349\eta° \tan \phi)$
	$b_\gamma = \exp(-0.0471\eta° \tan \phi)$ (η is measured counter-clockwise from horizontal)

Source: From Bowles, J.E., 2002, *Foundation Analysis and Design*, McGraw-Hill, New York. With permission.

TABLE 3.3B

Inclination, Ground Slope, and Base Tilt Factors for Vesic's Expression
(Vesic 1973, 1975) (Figure 3.2). Primed Factors are for $\phi = 0$

Load Inclination Factors Slope	Factors for Base on Slope (β)

$$i_c = 1 - \frac{mH_i}{A_f c_a N_c}$$

$$g'_c = \frac{\beta}{5.14}$$

$$i_c = i_q - \frac{1 - i_q}{N_q - 1}$$

$$g_c = i_q - \frac{1 - i_q}{5.14 \tan \phi}$$

$$i_q = \left[1.0 - \frac{H_i}{V + A_f c_a \cot \phi}\right]^m$$

$$g_q = g_\gamma = (1.0 - \tan \beta)^2 \ (\beta° \text{ measured clockwise from horizontal})$$

$$i_\gamma = \left[1.0 - \frac{H_i}{V + A_f c_a \cot \phi}\right]^{m+1}$$

When H is parallel to B

$$m = m_B = \frac{2 + B/L}{1 + B/L}$$

Factors for tilted base (η)

When H is parallel to L

$$b'_c = g'_c$$

$$m = m_L = \frac{2 + L/B}{1 + L/B}$$

$$b_c = 1 - \frac{2\eta}{5.14 \tan \phi}$$

When H has components parallel to both B and L

$$b_q = b_\gamma = (1.0 - \eta \tan \phi)^2 \ (\eta \text{ is measured counter-clockwise from horizontal})$$

$$m^2 = m_B^2 + m_L^2$$

Notes: c, cohesion, that is, attraction between the same material; c_a, adhesion, that is, attraction between two different materials (e.g., concrete and soil).
Hence, $c_a < c$. Bowles (2002) suggests $c_a = 0.6$–$1.0c$. The actual value depends on the concrete finish. If concrete foundation base is smooth, then c_a would be higher than that of a rough base.
Source: From Bowles, J.E., 2002, *Foundation Analysis and Design*, McGraw-Hill, New York. With permission.

Case (1)

The vertical effective stress at the footing base level $= q = (17)(\text{depth}) = 17B$. Then, the following expressions can be written for the ultimate bearing capacity:
Terzaghi method (Equation (3.2))

$$q_{ult} = 20(12.9)(1.3) + (17B)(4.4) + 0.5(17)(B)(2.5)(0.6)$$
$$= 335.4 + 87.55B$$

Hansen method (Equation (3.5))

$$q_{ult} = 20(10.97)(1.359)(1.4) + (17B)(3.9)(1.26)(1.294) + 0.5(17)(B)(1.2)(0.6)(1.0)$$
$$= 417.4 + 114.22B$$

Contact stress at the foundation level $= 4 \times 400/(AB^2) + 17B =$ stresses imposed by the column and the re-compacted soil (Figure 3.3).

The following criterion can be applied to compare the contact stress and the ultimate bearing capacity with a safety factor of 2.5:

$$4 \times 400/(AB^2) + 17B = q_{ult}/(2.5)$$

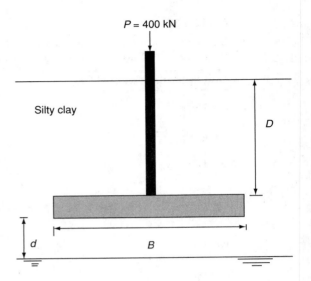

FIGURE 3.3
Illustration for Example 3.1.

From Terzaghi's expression

$$509.3/B^2 + 17B = (335.4 + 87.55B)/2.5$$
$$B = 1.75\,\text{m}$$

From Hansen's expression

$$509.3/B^2 + 17B = (417.4 + 114.22B)/2.5$$
$$B = 1.55\,\text{m}$$

Although the two solutions are different, one realizes that the disparity is insignificant from a construction point of view. Furthermore, in both cases, the footing depth obtained is within practical limits.

Case 2. Assume that the water table is 0.5 m above the footing.
Using Hansen's expression (Equation 3.5)

$$q_{\text{ult}} = 20(10.97)(1.359)(1.4) + [17B - (9.8)(0.5)](3.9)(1.26)(1.294) + 0.5(17 - 9.8)(B)(1.2)(0.6)(1.0)$$
$$= 386.27 + 110.69B$$

$$509.3/B^2 + 17B = (386.27 + 110.69B)/2.5$$
$$B = 1.62\,\text{m}$$

It is noted that a slightly larger area is needed to counteract the loss of foundation strength due to the groundwater table.

3.2.2 Net Ultimate Bearing Capacity

If the structural (column) load is to be used in the bearing capacity criterion (Equation (3.1)) to design the footing, then one has to strictly use the corresponding bearing capacity that excludes the effects of the soil overburden. This is known as the net ultimate bearing capacity of the ground and it is expressed as

$$q_{n,\text{ult}} = q_{\text{ult}} - q \tag{3.8}$$

where q is the total overburden stress.

On the other hand, the net load increase on the ground would be the structural load only, if it is assumed that concrete counteracts the soil removed to lay the footing. Then, Equation (3.1) can be modified as

$$P_{\text{structural}}/A \le \frac{q_{n,\text{ult}}}{F} \tag{3.9}$$

3.2.3 Foundations on Stiff Soil Overlying a Soft Clay Stratum

One can expect a punching type of bearing capacity failure if the surface layer is relatively thin and stiffer than the underlying softer layer. In this case, if one assumes that the stiff stratum (i.e., stiff clay, medium dense, or dense sand), where the footing is founded to satisfy the bearing capacity criterion with respect to the surface layers, then the next most critical criterion is that the stress induced by the footing (Figure 3.4) at the interface of the stiff soil–soft clay must meet the relatively low bearing capacity of the soft layer. The distributed stress can be computed by the following equations:

For rectangular spread footings

$$\Delta p = q \left[\frac{BL}{(B + d_c)(L + d_c)} \right] \tag{3.10}$$

For square or circular spread footings

$$\Delta p = q \left[\frac{B}{(B + d_c)} \right]^2 \tag{3.11}$$

For strip footings

$$\Delta p = q \left[\frac{B}{(B + d_c)} \right] \tag{3.12}$$

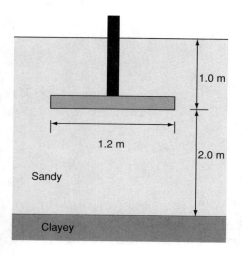

1.0 m

1.2 m

2.0 m

Sandy

Clayey

FIGURE 3.4
Illustration for Example 3.2.

Example 3.2

Assume that the square footing shown in Figure 3.2 has been well designed to carry a 500 kN load and to be founded in the sand layer overlying the soft clay layer. Check the bearing capacity criterion in the clay layer (undrained cohesion $= 20\,\text{kPa}$).

If Hansen's bearing capacity equation (Equation (3.5)) is used to estimate the *net* ultimate bearing capacity of the clay layer,

$$q_{n,\text{ult}} = cN_cs_cd_ci_cg_cb_c + q(N_q - 1)s_qd_qi_qg_qb_q + 0.5B\gamma N_\gamma s_\gamma d_\gamma i_\gamma g_\gamma b_\gamma \qquad (3.5)$$

Under undrained conditions, since $\phi_u = 0$,

$N_c = 5.14$, $N_q = 1.0$, $N_\gamma = 0$ (Table 3.1)
$q_{n,\text{ult}} = cN_cs_cd_ci_cg_cb_c$ (Equation (3.5))
$s_c = 1.195$ (square footing)

$$d_c = 1.0 + 0.4\left(\frac{3.0}{1.2}\right)$$

$$q_{n,\text{ult}} = (20)(5.14)(1.195)(2.0)$$
$$= 245.69 \text{ kPa}$$

Alternatively, from Equation (3.6)

$$q_{n,\text{ult}} = 5.14s_u(1 + s'_c + d'_c - i'_c - g'_c - b'_c) + q - q$$

From Table 3.2(a)
    $s'_c = 0.2$
    $d'_c = 0.4k = 0.4(3.0/1.2)$ (since $d/b = 3.0/1.2$ when one considers that the bearing capacity of the clay layer with respect to the distributed load from the footing)
Also, $i'_c = -0$, $g'_c = 0$, and $b'_c = 0$
Hence

$$q_{n,\text{ult}} = 5.14s_u(1 + s'_c + d'_c - i'_c - g'_c - b'_c) + q - q = 5.14(20)(1 + 0.2 + 1) = 226.16 \text{ kPa}$$

The net stress applied on the soft clay can be estimated as

$$\Delta p = q\left[\frac{B}{(B + d_c)}\right]^2$$
$$= [500/(1.2)^2][(1.2/3.2)^2] \qquad (3.11)$$
$$= 48.8 \text{ kPa}$$

$$\text{FOS} = 226/123/48.8 = 2.53 \text{ (satisfactory)}$$

3.2.4 Foundations on Soft Soil Overlying a Hard Stratum

When foundations are constructed on thin clayey surface layers overlying relatively hard strata (Figure 3.5), the mechanism of bearing capacity failure transforms into one in which

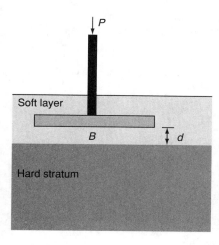

FIGURE 3.5
Soft surface layer overlying a harder layer.

the footing tends to squeeze the soft layer away while sinking in. In such cases, the net ultimate bearing capacity of the surface layer can be obtained from the following expressions (Tomlinson et al., 1995):

Circular/square footings

$$q_{n,\text{ult}} = \left(\frac{B}{2d} + \pi + 1\right) S_u \quad \text{for } \frac{B}{d} \geq 2 \tag{3.13}$$

Strip footings

$$q_{n,\text{ult}} = \left(\frac{B}{3d} + \pi + 1\right) S_u \quad \text{for } \frac{B}{d} \geq 6 \tag{3.14}$$

where B is the footing dimension, d is the thickness of the surface layer, and S_u is the undrained strength of the surface layer.

It must be noted that if the criteria $B/d \geq 2$ and $B/d \geq 6$ are not satisfied for circular and strip footings, respectively, the foundation can be treated as one placed in a homogeneous clay layer. For homogeneous cases, the bearing capacity estimation can be performed based on the methods discussed in the Section 3.2.1.

3.2.5 Bearing Capacity in Soils Mixed in Layers

When the subsurface constitutes an alternating (sandwiched) mixture of two distinct soil types as shown in Figure 3.6, one can use engineering judgment to estimate the bearing capacity. As an example, Figure 3.6 has the following layers as identified by the cone penetration test (CPT) results:

1. SM (silty sand), which is sand contaminated with a significant portion of silt. As expected the cone resistance q_c profile peaks out for sand.
2. CL or ML (clay and silt). As one would expect, the q_c profile drops for clay or silt (if the shaft friction, f_s, profile was provided, it would be relatively high for these layers).

In order to estimate the bearing capacity, the q_c values have to be averaged within the influence zone (Section 3.2.7.1). Since the soil types are not physically separated into two

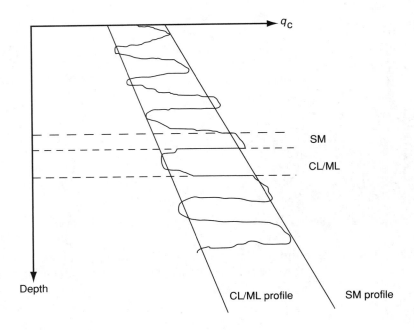

FIGURE 3.6
Bearing capacity of soils mixed in layers.

distinct layers, and because SM and CL (or ML) have very different engineering properties, it is conceptually incorrect to average the q_c values across the entire influence zone. Hence, the only way to address this is to assume one soil type at a time and obtain two bearing capacity estimates, an upper bound and a lower bound for the actual bearing capacity:

Step 1. Assume SM type only with a continuous linear q_c profile (with depth) defined by the peaks in Figure 3.6, thus ignoring the presence of clay and silt (CL or ML). Then, one deals with a silty sand only and the corresponding bearing capacity estimate would be $Q_{ult)1}$ (the upper bound).

Step 2. Assume CL or ML type only with a continuous linear q_c profile (with depth) defined by the troughs (indentations), thus ignoring the presence of sand (SM) and assuming undrained conditions. Then, one deals with clay or silt only and the corresponding bearing capacity estimate would be $Q_{ult)2}$ (the lower bound).

Then, the effective bearing capacity could be estimated from the following inequality:

$$Q_{ult)2} < Q_{ult} < Q_{ult)1} \tag{3.15}$$

3.2.6 Bearing Capacity of Eccentric Footings

The pressure distribution on the bottom of an eccentric footing can be determined from combined axial and bending stresses, as seen in Figure 3.7. One also realizes that, in order to prevent tensile forces at the bottom that tend to uplift the footing, the following conditions must be satisfied:

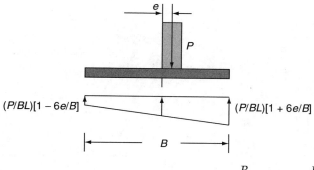

$(P/BL)[1 - 6e/B]$ $(P/BL)[1 + 6e/B]$

FIGURE 3.7
Bottom pressure distribution on rigid eccentric footings.

$$e_x \leq \frac{B}{6}, \quad e_y \leq \frac{L}{6} \tag{3.16a}$$

The above conditions are modified for rock as follows:

$$e_x \leq \frac{B}{4}, \quad e_y \leq \frac{L}{4} \tag{3.16b}$$

For the load and resistance factor design (LRFD) method (Section 3.4), the following modifications are made in the maximum eccentricity criteria (for no tension at the footing–soil interface) in view of load factoring:

$$e_x \leq \frac{B}{4} \tag{3.17a}$$

and

$$e_y \leq \frac{L}{4} \tag{3.17b}$$

The above conditions are modified for rock as follows:

$$e_x \leq \frac{3B}{8} \tag{3.17c}$$

$$e_y \leq \frac{3L}{8} \tag{3.17d}$$

Since the contact pressure is nonuniform at the bottom of the footing (Figure 3.7), Meyerhoff (1963) and Hansen (1970) suggested the following effective footing dimensions to be used in order to compute the bearing capacity of an eccentrically loaded rectangular footing. For eccentricities in both X and Y directions (Figure 3.8):

$$B' = B - 2e_x \tag{3.18a}$$

$$L' = L - 2e_y \tag{3.18b}$$

At times, a horizontal load that has two components, i.e., H_B parallel to B and H_L parallel to L, can act on the column producing two eccentricities e_x and e_y on the footing. In such cases, shape factors (Table 3.2) are computed twice by interchanging B' and L'. Also, i factors (Table 3.3) are also computed twice by replacing H_i once with H_L and then with H_B. Finally, the B' term in the q_{ult} expression also gets replaced by L'. Thus, in such cases, one would obtain two distinct q_{ult} values. The lesser of these values is compared to P/A for the footing design.

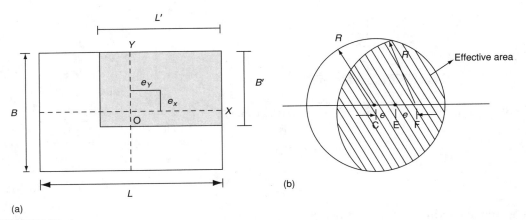

FIGURE 3.8
(a) Rectangular footings with eccentricity. (b) Circular footings with eccentricity.

In the case of circular footings having load eccentricity e and radius R, one must first locate the diameter corresponding to the eccentricity (point E in Figure 3.8b) and then construct a circular arc centered at F (EF = CE) with a radius equal to that of the footing. Then, the shaded area represents the effective footing area. Since the effective footing area is not of a geometrically regular shape, typically this is transformed into an equivalent rectangular footing of dimensions B' and L'. The effective dimensions can be found from the following expression:

$$B'L' = 2R\left[R\ \cos^{-1}\left(\frac{e}{R}\right) - e\right] \tag{3.19}$$

However, it must be noted that the unmodified B and L must be used when determining the depth factors (d) in the bearing capacity equations.

When footings are to be designed for a column that carries an unbalanced moment, M, and an axial force, P, which are fixed in magnitude, the resulting eccentricity ($e = M/P$) induced on the footings can be avoided by offsetting the column by a distance of $x = -e$, as shown in Figure 3.9. It is seen how the axial force in the column creates an equal and opposite moment to counteract the moment in the column. However, this technique cannot be employed to prevent footing eccentricities when eccentricities are introduced by variable moments due to wind and wave loading.

3.2.7 Bearing Capacity Using *In Situ* Test Data

3.2.7.1 *Cone Penetration Test Data*

Cone penetration data can be used to obtain the undrained strength of saturated fine-grained soils using the following expression:

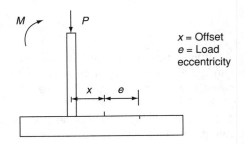

FIGURE 3.9
Designing footings to avoid eccentricity.

$$S_u = \frac{q_c - \sigma'_{v0}}{N_k} \tag{3.20}$$

where N_k is the cone factor that ranges between 15 and 19 for normally consolidated clay and between 27 and 30 for over-consolidated clays. Bowles (1995) suggests the following expression for N_k:

$$N_k = 13 + \frac{5.5}{50} PI \tag{3.21}$$

where PI is the plasticity index.

To determine an average q_c for a footing design, one would consider a footing influence zone that extends $2B$ below the footing and $\frac{1}{2}D$ above the footing.

3.2.7.2   Standard Penetration Test Data

Parry (1977) provided the following expression for the allowable bearing capacity (in kPa) of spread footings on cohesionless soils. For $D_f < B$:

$$q_{n,all} = 30N_{55}\left(\frac{s}{25.4}\right) \tag{3.22}$$

where N_{55} is the corrected SPT blow count corresponding to a 55% hammer efficiency and s is the settlement in millimeters. A modified and more versatile form of this expression is provided in Section 4.3.1.

Typically, when SPT data are provided, one can use the following correlation to estimate an equivalent angle of friction ϕ for the soil and determine the bearing capacity using the methods presented in Section 3.2:

$$\phi = 25 + 28\left(\frac{N_{55}}{\bar{q}}\right)^{1/2} \text{degrees} \tag{3.23}$$

where $\bar{q}$ = effective overburden stress at the evaluated location.

The footing influence zone suggested in Section 3.2.7.1 can be employed for computations involving Equations (3.22) and (3.23) as well.

3.2.7.3   Plate Load Test Data

Figure 3.10 shows a typical plot of plate-load test results on a sand deposit. When one scrutinizes Figure 3.10, it is seen that the ultimate bearing capacity of the plate can be estimated from the eventual flattening of the load–deflection curve. Knowing the ultimate bearing capacity of the plate, one can predict the expected bearing capacity of a footing to be placed on the same location using the following expressions:

Clayey soils

$$q_{u(f)} = q_{u(p)} \tag{3.24}$$

Sandy soils

$$q_{u(f)} = q_{u(p)}\left(\frac{B_f}{B_p}\right) \tag{3.25}$$

where B_p is the plate diameter and B_f is the equivalent foundation diameter, which can be determined as the diameter of a circle having an area equal to that of the footing.

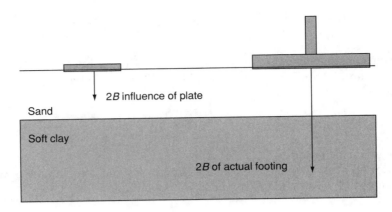

FIGURE 3.10
Illustration of influence zones.

It must be noted that the above expressions can be applied if it is known that the influence zone (Figure 3.10) of both the plate and when the footing is confined to the same type of soil and the effects of the groundwater table would be similar in both cases.

3.2.8 Presumptive Load-Bearing Capacity

The building codes of some cities suggest bearing capacities for certain building sites based on the classification of the predominant soil type at that site. Table 3.5–Table 3.7 present a comprehensive list of presumptive bearing capacities for various soil types. However, it must be noted that these values do not reflect the foundation shape, depth, load inclination, location of the water table, and the settlements that are associated with the sites. Hence, the use of these bearing capacity factors are advocated primarily in situations where a preliminary idea of the potential foundation size is needed for the subsequent site investigation followed by detailed design.

3.3 Settlement Analysis

Methodologies used for computation of ground settlement under building foundations have been discussed in detail in Section 1.5. Therefore, in this section, a number of techniques that are commonly employed to evaluate the ground stress increase due to footings will be reviewed. Then a number of examples will be provided to illustrate the application of the above techniques.

3.3.1 Stress Distribution in Subsurface Soils due to Foundation Loading

3.3.1.1 Analytical Methods

The vertical stress induced in the subsurface by a concentrated vertical load, such as the load on a relatively small footing founded on an extensive soil mass, can be approximately estimated by Boussinesq's elastic theory as follows:

$$\Delta\sigma_z = \frac{3P}{2\pi} \frac{z^3}{(r^2 + z^2)^{5/2}} \tag{3.26}$$

where r and z are indicated in Figure 3.11.

Equation (3.26) can be used to derive the magnitude of vertical stress imposed at any depth z vertically below the center of a circular foundation (of radius R) carrying a uniformly distributed load of q as (Figure 3.12)

$$\Delta\sigma_z = q\left[1 - \frac{1}{[1 + (R/z)^2]^{3/2}}\right] \tag{3.27a}$$

Stress increments in the horizontal (x and y) and vertical (z) directions due to other shapes of uniformly loaded footings (e.g., rectangular, strip, etc.) can be estimated based on analytical expressions presented in Harr (1966). Equation (3.26) can also be used to derive the vertical stress imposed at any depth z vertically below the corner of a rectangular foundation carrying a distributed load of q as (Figure 3.12b) expressed below:

$$\Delta\sigma_z = qK(m, n) \tag{3.27b}$$

where m = length/width of the foundation and n = z/foundation width. Values of $K(m, n)$ are tabulated in Table 3.8. Equation (3.27b) can also be applied to determine the stress increase at any point under the loaded area using partitions of the loaded area in which the corners coincide in plan with the point of interest. This technique is illustrated in a settlement estimation problem in Example 4.3.

3.3.1.2 Approximate Stress Distribution Method

At times it is more convenient to estimate the subsurface stress increments due to footings using approximate distributions. A commonly used distribution is the 2:1 distribution shown in Figure 3.13. Based on Figure 3.13, it can be seen that the stress increment caused by a uniformly loaded rectangular footing ($B \times L$) at a depth of z is

$$\Delta\sigma_z = q\left[\frac{BL}{(B + z)(L + z)}\right] \tag{3.28}$$

Example 3.3

Assume that it is necessary to compute the ultimate consolidation settlement and the 10-year settlement of the 1.5 m × 1.5 m footing carrying a 200-kN load as shown in Figure 3.14. Soil properties are provided in Table 3.8. Also assume the laboratory consolidation characteristics of a representative sample (from the mid-plane area of the clay layer) are represented by Figure 3.15 and the coefficient of consolidation (C_v) of the clay was determined to be $1.0 \times 10^{-8}\,\mathrm{m^2/sec}$ based on the methodology presented in Section 1.5.

From Figure 3.15, preconsolidation pressure = p_c = 60 kPa

Contact pressure = q = $200/(1.5)^2$ = 88.89 kPa

Overburden pressure at the footing depth = 16.5×1.0 = 16.5 kPa

The average stress increase in the clay layer can be obtained using Newmark's influence chart (reproduced in Figure 3.16) by considering the mid-plane depth of clay. This can be determined from Figure 3.16 by mapping the footing to the scale indicated at the bottom of the figure, i.e., d_c (the depth from the footing to the location where the stress increase is needed) = the distance indicated as OQ. In this example, one can see that d_c = 3.75 m.

TABLE 3.4A

Factors of Safety on Ultimate Geotechnical Capacity of Spread Footings for
Bearing Capacity and Sliding Failure (AASHTO, 1996)

Failure Condition	Required Minimum Factor of Safety (FS)
Bearing capacity of footing on soil or rock	3.0
Sliding resistance of footing on soil or rock	1.5

Source: From AASHTO, 1996, *Standard Specifications for Highway Bridges*, American
Association for State Highway and Transportation Officials, Washington, DC. With
permission.

TABLE 3.4B

Factors of Safety on Ultimate Bearing Capacity of Spread Footings on Soils

Basis for Soil Strength Estimate	Suggested Minimum Factor of Safety (FS)
Standard penetration tests	3.0
Laboratory/field strength tests	2.5

Source: From Federal Highway Administration, 1998, *Load and Resistance Factor Design
(LRFD) for Highway Bridge Superstructures*, Washington, DC. With permission.

TABLE 3.4C

Variable Factors of Safety on Ultimate Bearing Capacity of Spread Footings

			Required Minimum Factor of Safety (FS)			
			Permanent Structures		Temporary Structures	
Category	Typical Structures	Category Characteristics	Complete Soil Exploration	Limited Soil Exploration	Complete Soil Exploration	Limited Soil Exploration
A	Railway bridges Warehouses Blast furnaces Hydraulic Retaining walls Silos Highway bridges	Maximum design load likely to occur often; consequences of failure disastrous Maximum design load may occur occasionally; consequences of failure serious	3.0	4.0	2.3	3.0
B	Light industrial and public buildings		2.5	3.5	2.0	2.6
C	Apartment and office buildings	Maximum design load unlikely to occur	2.0	3.0	2.0	2.3

Source: From Federal Highway Administration, 1998, *Load and Resistance Factor Design (LRFD) for Highway Bridge
Superstructures*, Washington, DC. With permission.

TABLE 3.5

Presumptive Bearing Capacities from Indicated Building Codes, kPa

Soil Description	Chicago, 1995	Natl. Board of Fire Underwriters, 1976	BOCA,[a] 1993	Uniform Building Code, 1991[b]
Clay, very soft	25			
Clay, soft	75	100	100	100
Clay, ordinary	125			
Clay, medium stiff	175	100		100
Clay, stiff	210		140	
Clay, hard	300			
Sand, compact and clean	240	—	140	200
Sand, compact and silty	100	}		
Inorganic silt, compact	125	}		
Sand, loose and fine		}	140	210
Sand, loose and coarse, or sand–gravel mixture, or compact and fine		140–400	240	300
Gravel, loose and compact coarse sand	300	}	240	300
		}		
Sand–gravel, compact		—	240	300
Hardpan, cemented sand, cemented gravel	600	950	340	
Soft rock				
Sedimentary layered rock (hard shale, sandstone, siltstone)			6000	1400
Bedrock	9600	9600	6000	9600

Note: Values converted from pounds per square foot to kilopascals and rounded. Soil descriptions vary widely between codes. The following represents author's interpretations.
[a]Building Officials and Code Administrators International, Inc.
[b]Bowles (1995) interpretation.
Source: From Bowles, J.E., 2002, *Foundation Analysis and Design*, McGraw-Hill, New York. With permission.

The stress increase at a depth d_c can be found using Equation (1.19):

$$\Delta p = NqI \qquad\qquad (1.19)$$

where N and I are the number of elements of Newmark's chart covered by the scaled footing and the influence factor of the diagram respectively. For the chart shown in Figure 3.16, $I = 0.001$. If the footing were to behave as a flexible footing, the center settlement would be the maximum while the corner settlement would be the minimum within the footing. Thus,

$$\Delta p_{center} = (4 \times 19) \times 88.89 \times 0.001 = 6.75 \text{ kPa}$$
$$\Delta p_{corner} = (58) \times 88.89 \times 0.001 = 5.2 \text{ kPa}$$

On the other hand, if the footing were to behave as a rigid footing, then the average stress increase at the mid-plane level of the clay layer within the footing can be determined by using appropriate stress attenuation (Figure 3.13). Using the commonplace 2:1 stress attenuation (Equation (3.28)), one can estimate the stress increase as

$$\Delta p = q \left[\frac{BL}{(B + z_c)(L + z_c)} \right]$$

where B and L are footing dimensions.

Thus, $\Delta p_{average} = 88.89[1.5/(1.5 + 3.75)]^2 = 7.256 \text{ kPa}$

TABLE 3.6A

Presumptive Bearing Capacities for Foundations in Granular Soils Based on SPT Data
(at a Minimum Depth of 0.75 m Below Ground Level)

Description of Soil	N-Value in SPT	Presumed Bearing Value (kN/m²) for Foundation of Width		
		1 m	2 m	4 m
Very dense sands and gravels	>50	800	600	500
Dense sands and gravels	30–50	500–800	400–600	300–500
Medium-dense sands and gravels	10–30	150–500	100–400	100–300
Loose sands and gravels	5–10	50–150	50–100	30–100

Note: The water table is assumed not to be above the base of foundation. Presumed bearing values for pad foundations up to 3 m wide are approximately twice the above values.
Source: From Tomlinson, M.J. and Boorman, R., 1995, *Foundation Design and Construction*, Longman Scientific and Technical, Brunthill, Harlow, England. With permission.

TABLE 3.6B

Presumptive Bearing Capacities for Foundations in Clayey Soils Based on Undrained Shear Strength (at a Minimum Depth of 1 m Below Ground Level)

Description	Undrained Shear Strength (kN/m²)	Presumed Bearing Value (kN/m²) for Foundation of Width		
		1 m	2 m	4 m
Hard boulder clays, hard-fissured clays (e.g., deeper London and Gault Clays)	>300	800	600	400
Very stiff boulder clay, very stiff "blue" London Clay	150–300	400–800	300–500	150–250
Stiff-fissured clays (e.g., stiff "blue" and brown London Clay), stiff weathered boulder clay	75–150	200–400	150–250	75–125
Firm normally consolidated clays (at depth), fluvio-glacial and lake clays, upper weathered "brown" London Clay	40–75	100–200	75–100	50–75
Soft normally consolidated alluvial clays (e.g., marine, river and estuarine clays)	20–40	50–100	25–50	Negligible

Source: From Tomlinson, M.J. and Boorman, R., 1995, *Foundation Design and Construction*, Longman Scientific and Technical, Brunthill, Harlow, England. With permission.

It must be noted that if one were to have averaged the above stress estimates for the center and corner of the footing, one would have obtained

$$\Delta p_{average} = (1/2)(6.75 + 5.2) = 5.975 \text{ kPa}$$

Since the estimates are significantly different, the author suggests using the averages of the estimates from Figure 3.15 as opposed to the approximate estimate obtained from Figure 3.13. The average effective overburden pressure at the mid-plane of the clay layer is found from Equation (1.4b) as

$$\sigma'_{v0} = 16.5(2) + 17.5(1.5) + 18.0(1.25) - 9.8(2.75) = 54.8 \text{ kPa}$$

Since $\sigma'_{v0} < p_c$, one can assume that the overall clay layer is in an over-consolidated state.

TABLE 3.7

Presumptive Bearing Capacities for Foundations on Rock Surface (Settlement Not Exceeding 50 mm)

Rock Group	Strength Grade	Discontinuity Spacing (mm)	Presumed Allowable Bearing Value (kN/m²)
Pure limestones and dolomites, carbonate sandstones of low porosity	Strong	60 to >1,000	>12,500[a]
	Moderately strong	>600	>10,000[b]
		200–600	7,500–10,000
		60–200	3,000–7,500
	Moderately weak	600 to >1,000	>5,000[a]
		200–600	3,000–5,000
		60–200	1,000–3,000
	Weak	>600	>1,000[a]
		200–600	750–1,000
		60–200	250–750
	Very weak		See note[b]
Igneous, oolitic, and marly limestones; well-cemented sandstones; indurated carbonate mudstones; metamorphic rocks (including slates and schists with flat cleavage/foliation)	Strong	200 to >1,000	10,000 to >12,500[a]
		60–200	5,000–10,000
	Moderately strong	600 to >1,000	8,000 to >100,000[a]
		200–600	4,000–8,000
		60–200	1,500–4,000
	Moderately weak	600 to >1,000	3,000 to >5,000[a]
		200–600	1,500–3,000
		60–200	500–1,500
	Weak	600 to >1,000	750 to >1,000[a]
		>200	See note[b]
	Very weak	All	See note[b]
Very marly limestones: poorly cemented sandstones; cemented mudstones and shales; slates and schists with steep cleavage/foliation	Strong	600 to >1,000	10,000 to >12,500[b]
		200–600	5,000–10,000
		60–200	2,500–5,000
	Moderately strong	600 to >1,000	4,000 to >6,000[b]
		200–600	2,000 to >4,000
		60–200	750–2,000
	Moderately weak	600 to >1,000	2,000 to >3,000[b]
		200–600	750–2,000
		60–200	250–750
	Weak	600 to >1,000	500–750
		200–600	250–500
		<200	See note[b]
	Very weak	All	See note[b]
Uncemented mudstones and shales	Strong	200–600	250–5,000
		60–200	1,250–2,500
	Moderately strong	200–600	1,000–2,000
		60–200	1,300–1,000
	Moderately weak	200–600	400–1,000
		60–200	125–400
	Weak	200–600	150–250
		60–200	See note[b]
	Very weak	All	See note[b]

Notes: Presumed bearing values for square foundations up to 3 m wide are approximately twice the above values, or equal to the above values if settlements are to be limited to 25 mm.

[a] Bearing pressures must not exceed the unconfined compression strength of the rock if the joints are tight. Where the joints open the bearing pressure must not exceed half the unconfined compression strength of the rock.

[b] Bearing pressures for these weak or closely jointed rocks should be assessed after visual inspection, supplemented as necessary by field or laboratory tests to determine their strength and compressibility.

Source: From Tomlinson, M.J. and Boorman, R., 1995, *Foundation Design and Construction*, Longman Scientific and Technical, Brunthill, Harlow, England. With permission.

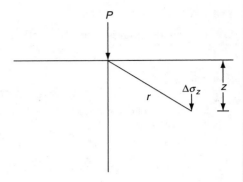

FIGURE 3.11
Stress increase due to a concentrated load.

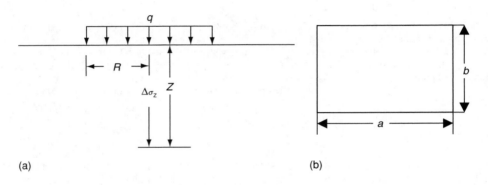

(a)   (b)

FIGURE 3.12
(a) Stress increase due to a distributed circular footing. (b) Stress increase to a distributed rectangular footing.

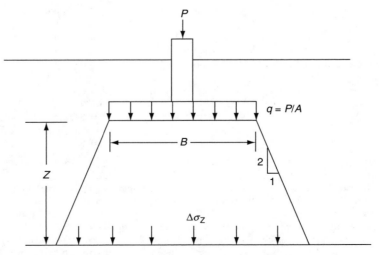

FIGURE 3.13
Approximate estimation of subsurface vertical stress increment.

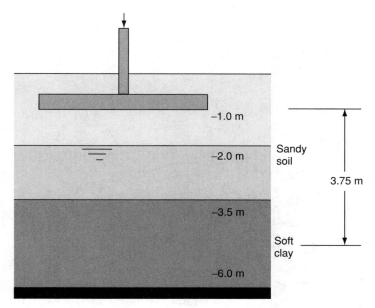

FIGURE 3.14
Illustration for Example 3.3.

TABLE 3.8

"*K*" Values for Equation (3.27b)

n	M	1	1.4	2	3	4	6	10
1		0.175	0.187	0.195	0.2	0.201	0.202	0.203
2		0.084	0.105	0.12	0.128	0.132	0.136	0.138
3		0.046	0.058	0.072	0.088	0.093	0.097	0.099
4		0.027	0.035	0.048	0.06	0.068	0.075	0.077
5		0.018	0.025	0.032	0.045	0.05	0.057	0.062
6		0.012	0.017	0.025	0.032	0.039	0.047	0.052
7		0.01	0.013	0.018	0.025	0.03	0.036	0.043
8		0.008	0.011	0.014	0.02	0.025	0.032	0.036
9		0.006	0.008	0.012	0.015	0.02	0.025	0.032
10		0.005	0.007	0.01	0.012	0.018	0.022	0.028

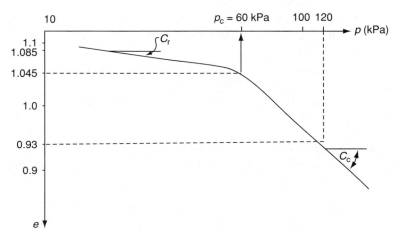

FIGURE 3.15
Laboratory consolidation curve.

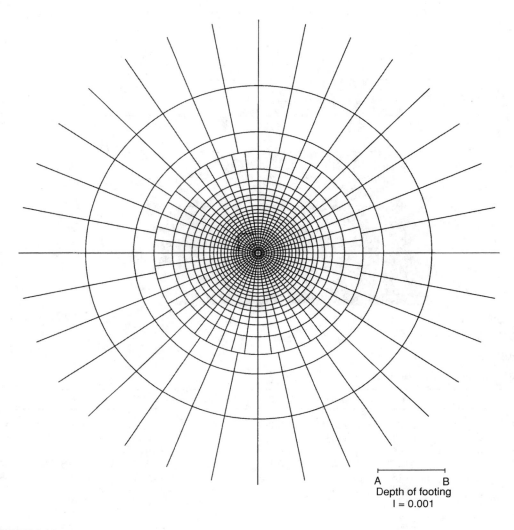

A   B
Depth of footing
I = 0.001

FIGURE 3.16
Use of Newmark's chart in Example 3.3.

Ultimate settlement beneath the center of the (flexible) footing
The following expression can be used to estimate the ultimate consolidation settlement
since $\sigma'_{v0} + \Delta p_{center} > p_c$ (Figure 1.20c and Figure 3.15):

$$s_{center} = \left[\frac{H}{1 + e_0}\right]\left[C_r \log\frac{p_c}{\sigma'_{v0}} + C_c \log\frac{\sigma'_{v0} + \Delta p}{p_c}\right] \tag{1.18a}$$

$$s_{center} = \left[\frac{2.5}{1 + 1.06}\right]\left[0.064 \log\frac{60}{54.8} + 0.382 \log\frac{54 + 6.75}{60}\right] = 8.19 \text{ mm}$$

Ultimate settlement beneath the corner of the (flexible) footing
The following expression can be used to estimate the ultimate consolidation settlement
since $\sigma'_{v0} + \Delta p_{corner}$ (Figure 1.20b and Figure 3.15):

$$s_{center} = \left[\frac{H}{1 + e_0}\right]\left[C_r \log\frac{\sigma'_{v0} + \Delta p}{\sigma'_{v0}}\right] \tag{1.18b}$$

$$S_{center} = \left[\frac{2.5}{1 + 1.06}\right]\left[0.064 \log\frac{54.8 + 5.2}{54.8}\right] = 3.06 \text{ mm}$$

Average ultimate settlement of the footing (rigid)
The following expression can be used to estimate the average ultimate consolidation settlement since $\sigma'_{v0} + \Delta p_{average} > p_c$ (Figure 1.20c and Figure 3.15):

$$S_{average} = \left[\frac{H}{1 + e_0}\right]\left[C_r \log\frac{p_c}{\sigma'_{v0}} + C_c \log\frac{\sigma'_{v0} + \Delta p}{p_c}\right] \tag{1.18c}$$

$$S_{average} = \left[\frac{2.5}{1 + 1.06}\right]\left[0.064 \log\frac{60}{54.8} + 0.382 \log\frac{54.8 + 5.975}{60}\right] = 5.64 \text{ mm}$$

Estimation of the 10-year settlement
The settlement of the footing at any intermediate time (t) can be estimated by using the average degree of consolidation, U_{ave}, of the clay layer corresponding to the particular time t in combination with any one of the above ultimate settlement estimates

$$s_t = U_{avg}s_{ult} \tag{3.29}$$

Using Terzaghi's theory of 1D consolidation (Terzaghi, 1943), the average degree of consolidation at time t, U_{ave}, can be determined from Table 1.8 knowing the time factor (T) corresponding to the time t. T can be determined using the following expression:

$$T = \frac{C_v t}{H_{dr}^2} \tag{1.16}$$

where H_{dr} is the longest path accessible to draining pore water in the clay layer. From Figure 3.14, one can see that, for this example, $H_{dr} = 2.5$ m. Then,

$$T = \frac{10^{-8}(10 \times 365 \times 24 \times 60 \times 60)}{2.5^2} = 0.504$$

From Table 1.8, $U_{ave} = 0.77$.

Example 3.4
Assume that it is necessary to compute the ultimate total differential settlement of the foundation shown in Figure 3.14, for which the strain-influence factor plot is shown in Figure 3.17. The average CPT values for the three layers are given in Table 3.9.

Solution
For the above data,

Contact pressure $\Delta\sigma = 200/(1.5)^2$ kPa $= 88.89$ kPa
Overburden pressure at footing depth (q) $= 16.5 \times 1.0$ kPa $= 16.5$ kPa

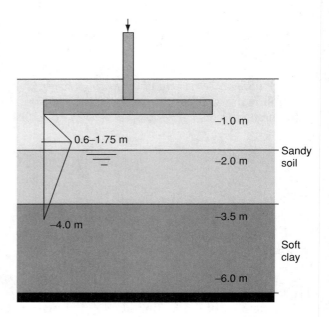

FIGURE 3.17
Immediate settlement computation.

Immediate settlement. Areas of the strain-influence diagram covered by different elastic moduli are

$$A_1 = 0.5(0.75 \times 0.6) + 0.5(0.25)(0.533 + 0.6) = 0.367 \text{ m}$$
$$A_2 = 0.5(1.5)(0.533 + 0.133) = 0.5 \text{ m}$$
$$A_3 = 0.5(0.5)(0.133) = 0.033 \text{ m}$$

Then, by applying Equation (1.13), one obtains the immediate settlement as

$$S_{center} = \{1 - 0.5[16.5/(88.89 - 16.5)]\}[1.0][88.89 - 16.5][0.367(1.0)/(11.5 \times 10^3)$$
$$+ 0.5/(10.7 \times 103) + 0.033/(2.57 \times 10^3)] = 5.87 \text{ mm}$$

From Equation (1.12), S_{corner} can be deduced as $0.5(5.87) = 2.94$ mm.

Therefore, the total settlement at the center of the footing will be 14.06 ($= 8.19 + 5.87$) mm or 0.55 in., while that at the corner will be 6.0 ($3.06 + 2.94$) mm or 0.24 in.

$$S_{center} = \{1 - 0.5[16.5/(88.89 - 16.5)]\}[1.0][88.89 - 16.5][0.367(1.0)/(11.5 \times 10^3)$$
$$+ 0.5/(10.7 \times 10^3) + 0.033/(2.57 \times 10^3)] = 5.87 \text{ mm}$$

Total settlement check. Most building codes stipulate the maximum allowable total settlement to be 1.0 in. Hence, the above value is acceptable.

TABLE 3.9

Soil Properties Used in Example 3.4

Soil Type	q_c	E_s
Dry sand	2.875 MPa	11.5 MPa ($E_s = 4q_c$ from Table 1.7)
Wet sand	2.675 MPa	10.7 MPa
Clay	5 MPa	10 MPa ($E_s = 2q_c$ from Table 1.7)

Differential settlement check. The differential settlement is equal to

$(s_{center} - s_{corner})$/distance from center to corner

or $(14.00 - 6.00)/(1.06)/1000 = 0.007$

According to most building codes, the maximum allowable differential settlement to prevent structural cracks in concrete is 0.013 (1 in 75). Hence, the differential settlement criterion is also satisfied.

3.3.2 Settlement Computation Based on Plate Load Test Data

The immediate settlement of a shallow footing can be determined from a plate load test that is performed at the same location and the depth at which the footing would be constructed. For the same magnitude in the contact stress level, settlement of the foundation can be estimated based on the settlement of the plate and the following expressions:

Clayey soils

$$s_f = s_p \frac{B_f}{B_p} \tag{3.30}$$

Sandy soils

$$s_f = s_p \left(\frac{2B_f}{B_f + B_p} \right)^2 \tag{3.31}$$

where B_p is the plate diameter and B_f is the equivalent foundation diameter, which can be determined as the diameter of a circle having an area equal to that of the footing.

3.3.3 Computation of Settlement in Organic Soils

Foundations constructed in organic soils exhibit prolonged settlement due to secondary compression, which is relatively larger in magnitude than the primary consolidation. This is particularly the case when the organic content of the soil deposit is significant. Therefore, foundation designers, who do not recommend the removal of organic soils from potential building sites, must alternatively employ specific analytical techniques to estimate the expected secondary compression component that predominates the total settlement of the foundation. The following analytical treatise is presented to address this need.

The organic content of a soil (OC) is defined as

$$OC = \frac{W_o}{W_s} (100)\% \tag{3.32}$$

where W_o is the weight of organic matter in the soil sample (usually determined based on the loss of weight of the sample on combustion) and W_s is the total weight of the solids in the soil sample.

Many researchers (Andersland et al., 1980; Gunaratne et al., 1997) have discovered linear relationships between the organic content of organic soils and their initial void ratios and water contents. Gunaratne et al. (1997) determined the following specific relationships for Florida organic soils, based on an extensive laboratory testing program:

$$e_\infty = 0.46 + 1.55(\text{OC}) \tag{3.33}$$

$$\text{OC} = w * 0.136 + 2.031 \tag{3.34}$$

where e_∞ and w are the ultimate void ratio and the water content, respectively. The ultimate 1D compressibility of organic soils (vertical strain per unit load increment) constitutes a primary compressibility component, a, and a secondary compressibility component, b, as expressed below

$$\varepsilon_{\text{ult}} = \Delta\sigma[a + b] \tag{3.35}$$

Parameters a and b, specific to any organic soil, can be expressed in terms of the primary and secondary void ratio components (e_p and e_s, respectively) of the initial void ratio, e_0, as illustrated in Equations (3.36)

$$a = -\frac{1}{(1+e)}\frac{\partial e_p}{\partial\sigma} \tag{3.36a}$$

$$b = -\frac{1}{(1+e)}\frac{\partial e_s}{\partial\sigma} \tag{3.36b}$$

Based on observed linear relationships such as that in Equation (3.33), Gunaratne et al. (1997) also determined that

$$a = -\frac{1}{(1+e)}\left[\frac{d}{d\sigma}\overline{I_p(\sigma)} + \text{OC}\frac{d}{d\sigma}\overline{M_p(\sigma)}\right] \tag{3.37a}$$

$$b = -\frac{1}{(1+e)}\left[\frac{d}{d\sigma}\overline{I_s(\sigma)} + \text{OC}\frac{d}{d\sigma}\overline{M_s(\sigma)}\right] \tag{3.37b}$$

where $[\overline{I_p(\sigma)}, \overline{M_p(\sigma)}]$ and $[\overline{I_s(\sigma)}, \overline{M_s(\sigma)}]$ are stress-dependent functions associated with primary and secondary compressibilities, respectively. Finally, by employing Equations (3.37), Gunaratne et al. (1997) derived the following specific relationships for Florida organic soils:

$$a = \frac{\dfrac{97.79}{(1.27\sigma + 97.79)^2} + \dfrac{23.13 \cdot \text{OC}}{(0.16\sigma + 23.13)^2}}{F(\text{OC},\sigma)} \tag{3.38a}$$

$$b = \frac{\dfrac{360.17}{(1.86\sigma + 360.17)^2} + \dfrac{40.61 \cdot \text{OC}}{(0.52\sigma + 40.61)^2}}{F(\text{OC},\sigma)} \tag{3.38b}$$

where

$$F(\text{OC},\sigma) = \left[2.79 - \frac{\sigma}{(0.78\sigma + 74.28)}\right] + \text{OC}\left[9.72 - \frac{\sigma}{(0.12\sigma + 15.33)}\right]$$

Parameters a and b, specific for a field organic soil deposit, would be dependent on the depth of location, z, due to their strong stress dependency.

The vertical strain in a layer of thickness, Δz, can be expressed in terms of its total (primary and secondary) 1D settlement, Δs_{p+s}, as in following equation:

$$\varepsilon_{ult} = \frac{\Delta s_{p+s}}{\Delta z} \tag{3.39}$$

Hence, the total 1D settlement can be determined as

$$s_{p+s} = \int_0^H [a(z) + b(z)](\Delta\sigma_z)dz \tag{3.40}$$

where $a(z)$ and $b(z)$ are a and b parameters in Equations (3.38) expressed in terms of the average current stress (initial overburden stress, $\sigma'_{v0} + \frac{1}{2}$, stress increment, ($\Delta\sigma_z$, produced due to the footing at the depth z).) $\Delta\sigma_z$ can be determined using Bousinesq's distribution (Equation (3.27)) or any other appropriate stress attenuation such as the 2:1 distribution (Equation (3.28)) commonly employed in foundation design.

Due to the complex nature of functions a and b (Equations (3.38)), one can numerically integrate Equation (3.40) to estimate the total settlement of an organic soil layer due to a finite stress increment imposed by a foundation.

Example 3.5
Assume that, based on laboratory consolidation tests, one wishes to predict the ultimate 1D settlement expected in a 1-m thick organic soil layer (OC = 50%) and the current overburden pressure of 50 kPa due to an extensively placed surcharge of 50 kPa.

Solution
Since there is no significant stress attenuation within 1 m due to an extensive surcharge, the final pressure would be

$$\sigma_v + \Delta\sigma_z = 50 + 50 = 100 \text{ kPa}$$

throughout the organic layer. Then, by applying Equation (3.35)

$$\varepsilon_{ult} = \int_{50}^{100} [a(\sigma) + b(\sigma)](d\sigma)$$

where $a(\sigma)$ and $b(\sigma)$ are obtained from Equation (3.38) using OC and σ values of 0.5 and 50 kPa, respectively.

Finally, on performing the integration numerically, one obtains primary and secondary compressions of 0.107 and 0.041 m. which produces a total settlement of 0.148 m.

Fox et al. (1992) used the C_α/C_c concept to predict the secondary settlement of organic soils.

3.4 Load and Resistance Factor Design Criteria

Two design philosophies are commonly used in design of foundations:

1. Allowable stress design (ASD)
2. Load and resistance factor design (LRFD)

Of the two, the more popular and historically successful design philosophy is the ASD, which has also been adopted in this chapter so far. ASD can be summarized by the following generalized expressions:

$$R_n/F \geq \sum Q_i \tag{3.41}$$

where R_n is the nominal resistance, Q_i is the load effect, and F is the factor of safety.

The main disadvantages of the ASD methods are: (1) F is applied only to the resistance part of the equation without heeding the fact that varying types of loads have different levels of uncertainty, (2) F is only based on judgment and experience, and (3) no quantitative measure of risk is incorporated in F.

The design of spread footings using LRFD requires evaluation of the footing performance at various limit states. The primary limit states for spread footing design include strength limits such as bearing capacity failure or sliding failure and service limits such as excessive settlements or vibration.

The goal of LRFD is to design, without being conservative as to be wasteful of resources, a foundation that serves its function without reaching the limit states.

3.4.1   Load and Resistance Factor Design Philosophy

LRFD-based evaluation of strength limit state can be summarized as

$$\phi R_n \geq \eta \sum \gamma_i Q_i \tag{3.42}$$

where ϕ is the resistance factor, γ_i are the load factors, and η is the load modifier.

Load factors account for the uncertainties in magnitude and direction of loads, location of application of loads, and combinations of loads.

On the other hand, resistance factors can be made to incorporate variability of soil properties, reliability of predictive equations, quality control of construction, extent of soil exploration, and even the consequences of failure. The main advantages of LRFD are that it accounts for variability in both resistance and loads and provides a qualitative measure of risk related to the probability of failure. However, LRFD also has the limitation of not facilitating the selection of appropriate resistance factors to suit the design of different foundation types. The LRFD-based evaluation of service limit state can be described by Equation (3.1b).

Three different methods are adopted to select the resistance and load factors (FHWA, 1998):

1. Calibration by judgment (requires extensive experience)
2. Calibration by fitting to ASD
3. Calibration by the theory of reliability

The procedure used for the selection is known as the calibration of LRFD. The two latter procedures will be discussed in this chapter.

3.4.2   Calibration by Fitting to ASD

Using Equations (3.41) and (3.44) and assuming $\eta = 1.0$

$$\phi = \frac{\gamma_D Q_D + \gamma_L Q_L}{F(Q_D + Q_L)} \tag{3.43}$$

where Q_D is the dead load and Q_L is the live load.

If one assumes a dead load–live load ratio (Q_D/Q_L) of 3.0, F $= 2.5$, and load factors of $\gamma_D = 1.25$ and $\gamma_L = 1.75$, then

$$\phi = \frac{1.25(3.0) + 1.75(1.0)}{2.5(3.0 + 1.0)} = 0.55$$

Hence, the resistance factor, ϕ, corresponding to an ASD safety factor of 2.5 and a dead–live load ratio of 3 is 0.55. Similarly, one can estimate the ϕ values corresponding to other FS and Q_D/Q_L values as well.

3.4.3 Calibration by Reliability Theory

In the LRFD calibration using the theory of reliability, the foundation resistance and the loads are considered as random variables. Therefore, the resistance and the loads are incorporated in the design using their statistical distributions. Today, these concepts have been included in the bridge design guidelines of the United States Federal Highway Administration (FHWA, 1998). Based on these guidelines, the statistical concepts relevant to the calibration procedure are discussed in the next section.

3.4.3.1 *Variability of Soil Data*

A quantitative measure of the variability of site soil can be provided by the coefficient of variation (COV) of a given soil property, X, defined as follows:

$$\text{COV}(X) = \frac{\sigma}{\mu} \tag{3.44}$$

where μ is the mean of the entire population of X at the site and σ is the standard deviation of the entire population of X at the site.

However, both μ and σ can be estimated by their respective sample counterparts and s obtained from an unbiased finite sample of data (on X) of size n, obtained at the same site using the following expressions:

$$\bar{x} = \frac{\sum_{i=1}^{n} x_i}{n} \tag{3.45}$$

$$s_x = \frac{\sum_{i=1}^{n} (x_i - \bar{x})^2}{n - 1} \tag{3.46}$$

Using data from Teng et al, (1992) (Figure 3.18), it can be illustrated how the sample standard deviation is related to the population standard deviation. Figure 3.18 shows the estimation of the undrained shear strength (S_u) of clay at a particular site using three different methods: (1) cone penetration tests (CPT) (2) vane shear test (VST), and (3) laboratory consolidation tests based on the preconsolidation pressure (σ'_p). It is seen from Figure 3.18 that in each case the estimation can be improved by increasing the sample size up to an optimum size of about 7. The corresponding standard deviation estimate can possibly be interpreted as the population standard deviation. However, the best estimate of the standard deviation that one can make varies with the specific technique used in the estimation. Moreover, Figure 3.18 also shows that, based on the laboratory prediction method, VST provides a much more accurate estimate of the "true" standard deviation of

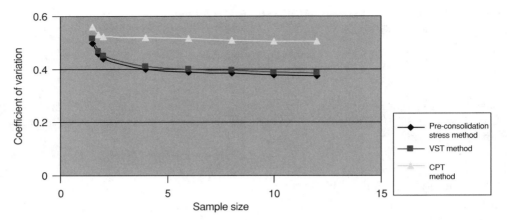

FIGURE 3.18
Reliability variation of undrained strength prediction with sample size. (From Teng, W.H., Mesri, G., and Halim, I., 1992, *Soils and Foundations*, 32(4): 107–116. With permission.)

the undrained shear strength (S_u) of a clayey site soil. Alternatively, the information contained in research findings such as in Figure 3.18 can be utilized in planning subsurface investigations. Intuitively, one also realizes that the standard deviation estimates obtained from a given evaluation method correlate well with reliability of the evaluation method, i.e., a relatively higher standard deviation indicates a less reliable evaluation method.

The typical variability associated with soil index tests and strength tests as reported by Phoon et al. (1995) are shown in Table 3.10 and Table 3.11, respectively. For analytical purposes, one can completely describe a random variable using an appropriate probability density function (in the case of a continuous random variable) or probability mass function (in the case of a discrete random variable) that satisfies the statistics of that particular random variable. The distribution that satisfies all the statistical properties of the random variable would obviously be its own histogram. However, what is assumed in many instances is a mathematical function that would closely "model" the statistical properties of the considered random variable. When selecting an appropriate mathematical distribution for a given variable, it is most common to match only the mean and the standard deviation of that variable with the corresponding quantities that are computed using the mathematical equation of the considered distribution as follows:

$$\mu = \int_{-\infty}^{\infty} x f(x) \, dx$$

= moment of area of the distribution about the origin of the x-axis
= centroidal location   (3.47)

$$\sigma^2 = \int_{-\infty}^{\infty} (x - \mu)^2 f(x) \, dx$$

= second moment of area of the distribution about the centroidal location (mean)

(3.48)

TABLE 3.10

Soil Variability in Index Tests (Phoon et al., 1995)

Property	Soil Type	Inherent Soil Variability (COV)	Measurement Variability (COV)
Natural water content	Fine grained	0.18	0.08
Liquid limit	Fine grained	0.18	0.07
Plastic limit	Fine grained	0.16	0.1
Plasticity index	Fine grained	0.29	0.24
Bulk density	Fine grained	0.09	0.01
Dry density	Fine grained	0.07	—
Relative density — direct	Sand	0.19	—
Relative density — indirect	Sand	0.61	—

Source: From Phoon, K., Kulhawy, F.H., and Grigoriu, M.D., 1995, Reliability-Based Design of Foundations for Transmission Line Structures, TR 105000 Final report, report prepared by Cornell University for the Electric Power Research Institute, Palo Alto, CA. With permission.

Two very commonly employed distributions that merely satisfy the above-mentioned mean and the standard deviation criteria (Equations (3.47) and (3.48)) only are the normal and the lognormal distributions. However, in the case of a given variable, if the analyst is forced to select a probability distribution that would represent the random variation of that variable more accurately, then in addition to the mean and the standard deviation estimates, one could also compute the coefficients of skewness and kurtosis (flatness) computed from the sample data (Harr, 1977). It must be noted that the coefficients of skewness and kurtosis for the population can be related to the third and the fourth moments of area of the probability distribution about the mean, respectively.

3.4.3.2 Normal Distribution

If a continuous random variable X is normally distributed, its probability density function is given by

$$f(x) = \frac{1}{\sigma\sqrt{2\pi}} \exp\left[-\frac{1}{2}\left\{\frac{x-\mu}{\sigma}\right\}^2\right] \qquad (3.49)$$

TABLE 3.11

Soil Variability in Strength Tests (Phoon et al., 1995)

Property	Soil Type	Inherent Soil Variability (COV)	Measurement Variability (COV)
Undrained strength (unconfined compression testing)	Fine grained	0.33	—
Undrained strength (unconsolidated undrained triaxial testing)	Clay, silt	0.22	—
Undrained strength (preconsolidation pressure from consolidated undrained triaxial testing)	Clay	0.32	0.19
Tan Φ (triaxial compression)	Clay, silt	0.20	—
	Sand, silt	—	0.08
Tan Φ (direct shear)	Clay, silt	0.23	—
	Clay	—	0.14

Source: From Phoon, K., Kulhawy, F.H., and Grigoriu, M.D., 1995, Reliability-Based Design of Foundations for Transmission Line Structures, TR 105000 Final report, report prepared by Cornell University for the Electric Power Research Institute, Palo Alto, CA. With permission.

It can be shown that Equation (3.49) automatically satisfies the conditions imposed by Equations (3.47) and (3.48).

3.4.3.3 Lognormal Distribution

If a continuous random variable X is log-normally distributed, then the natural logarithm of x, $\ln(x)$, is normally distributed and its probability density function is given by Equation (3.50a)

$$f(x) = \frac{1}{\xi\sqrt{2\pi}} \exp\left[-\frac{1}{2}\left\{\frac{\ln(x) - \lambda}{\xi}\right\}^2\right] \tag{3.50a}$$

where λ and ξ are the mean and the standard deviation of $\ln(x)$, respectively. The statistics of $\ln(x)$ can be expressed by those of X as

$$\lambda = \ln\left[\frac{\mu}{\sqrt{(1 + \text{COV}^2(X))}}\right] \tag{3.50b}$$

and

$$\xi = \sqrt{\ln(1 + \text{COV}^2(X))} \tag{3.50c}$$

Furthermore, it can be shown that when the random variable X exhibits a variation within a relatively minor range, i.e., when the COV(X) is relatively small (<0.2), the above expressions simplify to

$$\lambda = \ln(\mu) \tag{3.50d}$$

and

$$\xi = \text{COV}(X) \tag{3.50e}$$

3.4.3.4 Estimation of Probabilities

A primary use of mathematically expressed probability distributions like the normal or the lognormal distribution is the convenience that such a distribution provides in the computation of probability estimates. Similar computations also greatly enhance the assessment of reliability estimates in the design procedures that incorporate random characteristics of loads applied on earthen structures and the relevant geotechnical parameters of the foundation soil. Accordingly, if X is a random variable that assumes values in the range of $[a, b]$, then the probability of finding values of X less than c can be expressed in terms of its probability distribution as

$$P(X < c) = \int_a^c f(x)\,\mathrm{d}x \tag{3.51}$$

3.4.3.5 Reliability of Design

If the effect of a load applied on a substructure such as a foundation and the resistance provided by the shear strength of the foundation soil are expressed in terms of random variables Q and R, respectively, then the reliability of the design can be expressed as

$$\text{Re} = P(R \geq Q) \tag{3.52}$$

In order to compute the reliability of a design that involves randomly distributed load effects, Q, and soil resistance, R, it is convenient to express the interaction between R and Q in terms of the combined random variable $g(R, Q) = (R - Q)$.

The central axis theorem of statistics (Harr, 1977) states that, if both R and Q are normally distributed, i.e., normal variates, then $g(R, Q)$ would be normally distributed as well. Therefore, it follows that, if both R and Q are log-normally distributed, i.e., log-normal variates, then $g'(R, Q) = \ln R - \ln Q$ would be normally distributed as well.

The following statistical relations can be derived between two different random variables a and b:

$$E(a + b) = E(a) + E(b) \tag{3.53a}$$

where E indicates the expected value or the mean

$$\sigma^2(a + b) = \sigma^2(a) + \sigma^2(b) \tag{3.53b}$$

Based on Equation (3.53), $g'(R, Q)$ would have the following characteristics:

$$\text{Mean}[g'(R,Q)] = \text{Mean}(\ln R) - \text{Mean}(\ln Q) \tag{3.54a}$$

$$\text{Standard deviation}[g'(R,Q)] = \sqrt{\sigma^2_{\ln (R)} + \sigma^2_{\ln (Q)}} \tag{3.54b}$$

Using Equation (3.50b),

$$\begin{aligned}
\text{Mean}[g'(R,Q)] &= \ln \left[\frac{\bar{R}}{\sqrt{1 + \text{COV}^2_R}} \right] - \ln \left[\frac{\bar{Q}}{\sqrt{1 + \text{COV}^2_Q}} \right] \\
&= \ln \left[\frac{\bar{R}}{\bar{Q}} \frac{\sqrt{1 + \text{COV}^2_Q}}{\sqrt{1 + \text{COV}^2_R}} \right]
\end{aligned} \tag{3.55a}$$

Similarly, using Equation (3.50c)

$$\begin{aligned}
\text{Standard deviation}[g'(R, Q)] &= \sqrt{\left[\ln (1 + \text{COV}^2_R) \right] + \left[\ln (1 + \text{COV}^2_Q) \right]} \\
&= \sqrt{\ln \left[(1 + \text{COV}^2_R)(1 + \text{COV}^2_Q) \right]}
\end{aligned} \tag{3.55b}$$

Then, the reciprocal of the coefficient of variation of $g'(R, Q)$ can be expressed as

$$\beta = \frac{\mu}{\sigma} = \frac{\ln\left[\dfrac{\bar{R}}{\bar{Q}}\dfrac{\sqrt{(1+COV_Q^2)}}{\sqrt{(1+COV_R^2)}}\right]}{\sqrt{\ln\left[(1+COV_R^2)(1+COV_Q^2)\right]}} \qquad (3.55c)$$

If one expresses the mathematical expression for the normal distribution (Equation (3.49)) in terms of the standard normal variate z, where

$$R - Q = X \qquad (3.56a)$$

$$z = \frac{x - \mu}{\sigma} \qquad (3.56b)$$

then Equation (3.49) simplifies to

$$f(z) = \frac{1}{\sigma\sqrt{2\pi}}\exp\left[-\frac{1}{2}z^2\right] \qquad (3.57)$$

Then, from the differential form of Equation (3.56),

$$\sigma(dz) = dx \qquad (3.58)$$

Therefore, the estimation of probability in Equation (3.51) would be simplified as follows:

$$P(X < c) = \int_{-\infty}^{c} f(x)\,dx \qquad (3.59)$$

$$P(X < c) = P\left(Z < \frac{c - \mu}{\sigma}\right) = \int_{-\infty}^{(c-\mu)/\sigma} f(z)\,dx$$

Substituting from Equations (3.57) and (3.38),

$$P(X < c) = \int_{-\infty}^{(c-\mu)/\sigma} \frac{1}{\sigma\sqrt{2\pi}}\exp\left[-\frac{1}{2}z^2\right]\sigma(dz)$$

$$= \int_{-\infty}^{(c-\mu)/\sigma} \frac{1}{\sqrt{2\pi}}\exp\left[-\frac{1}{2}z^2\right]dz \qquad (3.60)$$

$$= \frac{1}{\sqrt{2\pi}}\int_{-\infty}^{(c-\mu)/\sigma} \exp\left[-\frac{1}{2}z^2\right]dz$$

3.4.3.6 Reliability Index

The reliability of the design can be computed using Equations (3.56a) and (3.52) as

$$\text{Re} = P(R \geq Q) = P(R - Q \geq 0) = 1 - P[(R - Q) \prec 0] = 1 - P(X \prec 0)$$

Then, setting $c = 0$ in Equation (3.60),

$$-\text{Re} = \frac{1}{\sqrt{2\pi}} \int_{-\infty}^{-\mu/\sigma} \exp\left[-\frac{1}{2}z^2\right] dz$$

since $\beta = \mu/\sigma$

$$-\text{Re} = \frac{1}{\sqrt{2\pi}} \int_{-\infty}^{-\beta} \exp\left[-\frac{1}{2}z^2\right] dz$$

If one defines the above-mentioned integrals in terms of the error function (erf), which is conveniently tabulated in standard normal distribution tables, as follows:

$$F(-\beta) = \frac{1}{\sqrt{2\pi}} \text{erf}(-\beta) = \frac{1}{\sqrt{2\pi}} \int_{-\infty}^{-\beta} \exp\left[-\frac{1}{2}z^2\right] dz \qquad (3.61)$$

Then, the reliability of the design is $F(-\beta)$, where the reliability index β is defined in terms of the load and resistance statistics in Equation (3.55a) and Equation (3.55b) as

$$\beta = \frac{\ln\left[\dfrac{\bar{R}}{\bar{Q}} \dfrac{\sqrt{1 + \text{COV}_Q^2}}{\sqrt{1 + \text{COV}_R^2}}\right]}{\sqrt{\ln\left[(1 + \text{COV}_R^2)(1 + \text{COV}_Q^2)\right]}} \qquad (3.55c)$$

3.4.3.7 Resistance Statistics

The measured resistance R_m can be expressed in terms of the predicted resistance, R_n, as

$$R_m = \lambda_R R_n \qquad (3.62)$$

where λ_R represents the bias factor for resistance. The bias factor includes the net effect of various sources of error such as the tendency of a particular method (e.g., Hansen's bearing capacity) to under-predict foundation resistance, energy losses in the equipment in obtaining SPT blow counts, soil borings in strata not being representative of the site, etc. For n number of sources of error with individual factors affecting the strength of resistance prediction procedure, the mean bias factor can be expressed as follows:

$$\lambda_R = \lambda_1 \lambda_2 \ldots \lambda_n \qquad (3.63a)$$

TABLE 3.12

Resistance Statistics

| Correction | Statistics for Correction Factors | |
	λR	COV_R
Model error	1.3	0.5
Equipment/procedure used in SPT	1.0	0.15–0.45 (use 0.3)
Inherent spatial variability	1.0	$(0.44/L)^{0.5}$

L=length of pile.
Source: From Federal Highway Administration, 1998, *Load and Resistance Factor Design (LRFD) for Highway Bridge Superstructures*, Washington, DC. With permission.

Then, based on the principles of statistics, the coefficient of variation of λ_R is given by

$$COV_R^2 = COV_1^2 + COV_2^2 + \cdots + COV_n^2 \tag{3.63b}$$

Table 3.12 indicates the values recommended by FHWA (1998) for λ_R and COV_R.

3.4.3.8   Load Statistics

Similarly for the measured load, one can write

$$Q_m = \lambda_{QD}Q_D + \lambda_{QL}Q_L \tag{3.64}$$

where the load bias factor includes various uncertainties associated with dead and live loads. λ_{QD} values for commonplace materials are found in Table 3.13. On the other hand, the AASHTO LRFD live load model specifies $\lambda_{QL} = 1.15$ and $COV_{QD} = 0.18$. As an example, if there are m significant sources of bias for dead loads in a given design situation, then from Equations (3.63)

$$\lambda_{QD} = \lambda_1 \lambda_2 \ldots \lambda_m \tag{3.65a}$$

$$COV_{QD}^2 = COV_1^2 + COV_2^2 + \cdots + COV_m^2 \tag{3.65b}$$

3.4.3.9   Determination of Resistance Factors

By rearranging Equation (3.55c), one obtains

$$\bar{R} = R_m = \lambda_R R_n = Q_m \exp\left[\beta_T \sqrt{\ln\left[(1 + COV_R^2)(1 + COV_Q^2)\right]}\right] \frac{\sqrt{1 + COV_R^2}}{\sqrt{1 + COV_Q^2}} \tag{3.66a}$$

TABLE 3.13

Bias Factors and Coefficients of Variation for Bridge Foundation Dead Loads

Component	λ_{QD}	COV_{QD}
Factory made	1.03	0.08
Cast-in-place	1.05	0.10
Asphaltic wearing surface	1.00	0.25
Live load	$(1.3 = 8_{QL})$	$(0.7 = COV_{QL})$

Source: From Federal Highway Administration, 1998, *Load and Resistance Factor Design (LRFD) for Highway Bridge Superstructures*, Washington, DC. With permission.

TABLE 3.14

Relationship between Probability of Failure and Reliability
Index for Lognormal Distribution

Reliability Index	Probability of Failure
2.0	0.85×10^{-1}
2.5	0.99×10^{-2}
3.0	1.15×10^{-3}
3.5	1.34×10^{-4}
4.0	1.56×10^{-5}
4.5	1.82×10^{-6}
5.0	2.12×10^{-7}
5.5	2.46×10^{-8}

Source: From Federal Highway Administration, 1998, *Load and Resistance
Factor Design (LRFD) for Highway Bridge Superstructures*, Washington,
DC. With permission.

From Equation (3.42)

$$\phi_R R_n = \gamma Q_n = \gamma_D Q_D + \gamma_L Q_L \tag{3.66b}$$

By eliminating R_n from Equations (3.66a) and (3.66b), and using the relation

$$COV_Q^2 = COV_{QD}^2 + COV_{QL}^2 \tag{3.66c}$$

the resistance factor can be derived as

$$\phi_R = \frac{\lambda_R [\gamma_D Q_D + \gamma_L Q_L] \sqrt{\dfrac{(1 + COV_{QD}^2 + COV_{QL}^2)}{(1 + COV_R^2)}}}{Q_m \, \exp\left\{ \beta_T \sqrt{\ln\left[(1 + COV_{QD}^2 + COV_{QL}^2)(1 + COV_R^2)\right]} \right\}} \tag{3.67}$$

where β_T is the target reliability index evaluated from Table 3.14 corresponding to an
anticipated probability of failure.

Finally, the bias factors suggested by FHWA (1998) for SPT and CPT results based on
selected reliability indices are provided in Table 3.15.

Finally, Table 3.16 outlines the suggested resistance factors for a variety of foundation
strength prediction methods in common use.

TABLE 3.15

Bias Factors and Coefficients of Variation for Soil Strength
Measurements

Test	λ_R	COV_R
SPT	1.3	0.6–0.8
CPT	1.0	0.4
Angle of friction (N)	1.0	0.1
Cohesion	1.0	0.4
Wall friction (−)	1.0	0.2
Earth pressure coefficient (K)	1.0	0.15

Source: From Federal Highway Administration, 1998, *Load and Resistance
Factor Design (LRFD) for Highway Bridge Superstructures*, Washington,
DC. With permission.

TABLE 3.16

Resistance Factors for Geotechnical Strength Limit State for
Shallow Foundations

Method/Soil/Condition	Resistance Factor
Sand	
Semiempirical procedure using SPT data	0.45
Semiempirical procedure using CPT data	0.55
Rational method using shear strength (*N*) from	
SPT data	0.35
CPT data	0.45
Clay	
Semiempirical procedure using CPT data	0.5
Rational method using shear strength (S_u) from	
CPT data	0.5
Lab. test (UU triaxial)	0.6
Field vane shear tests	0.6
Rock	
Semiempirical procedure	0.6
Plate load test	0.55

Source: From Federal Highway Administration, 1998, *Load and Resistance Factor Design (LRFD) for Highway Bridge Superstructures*, Washington, DC. With permission.

3.4.3.10 Determination of the Simplified Resistance Factor

The denominator of Equation (3.55c) can be simplified as follows:

$$\ln\left[(1 + COV_R^2)(1 + COV_Q^2)\right] = \ln(1 + COV_R^2) + \ln(1 + COV_Q^2) \qquad (3.68)$$

Using the Taylor series expansion for relatively small values of COV (e.g., <0.3), Equation (3.68) can be written as

$$\ln\left[(1 + COV_R^2)(1 + COV_Q^2)\right] \approx COV_R^2 + COV_Q^2$$

Similarly, the numerator of Equation (3.55c) can be simplified as

$$\ln\left[\frac{\bar{R}}{\bar{Q}} \frac{\sqrt{1 + COV_Q^2}}{\sqrt{1 + COV_R^2}}\right] = \ln\frac{\bar{R}}{\bar{Q}} + \ln\sqrt{1 + COV_Q^2} - \ln\sqrt{1 + COV_R^2}$$

For relatively small values of COV (e.g., <0.3), the above expression can be simplified to

$$\ln\frac{\bar{R}}{\bar{Q}} + \ln\sqrt{1 + COV_Q^2} - \ln\sqrt{1 + COV_R^2} \approx \ln\frac{\bar{R}}{\bar{Q}}$$

Hence, the expression for β can be simplified to

$$\beta = \frac{\ln \bar{R}/\bar{Q}}{\sqrt{COV_R^2 + COV_Q^2}} \qquad (3.69)$$

By defining the α factor as follows:

$$\text{COV}_R^2 + \text{COV}_Q^2 = \alpha(\text{COV}_R + \text{COV}_Q)$$

and rearranging terms in Equation (3.69), one obtains

$$\frac{\bar{R}}{\bar{Q}} = e^{\alpha\beta(\text{COV}_R+\text{COV}_Q)} \tag{3.70a}$$

Separately combining R and Q terms, one obtains

$$\bar{R}e^{-\alpha\beta\text{COV}_R} = \bar{Q}e^{\alpha\beta\text{COV}_Q} \tag{3.70b}$$

Using the definition of the nominal resistance in Equation (3.62),

$$\lambda_R R_n e^{-\alpha\beta\text{COV}_R} = \left(\sum\lambda_n Q_n\right)e^{\alpha\beta\text{COV}_Q} \tag{3.70c}$$

Recalling Equation (3.42),

$$\phi R_n = \sum\gamma_n Q_n \tag{3.42}$$

From Equation (3.70c), it is seen that the load and resistance factors, γ and ϕ, respectively, depend on the statistics of each other (COV_R and COV_Q) as well. However, for convenience if one assumes that the resistance and load factors are independent of each other's statistics, then comparison of Equation (3.70c) with Equation (3.42) yields a convenient and approximate method to express the resistance factors, independent of the load factors, as follows:

$$\phi = \lambda_R e^{-\alpha\beta_T\text{COV}_R} \tag{3.70d}$$

where β_T is the target reliability.

Table 3.17 illustrates the selection of appropriate resistance factors for spread footing design based on SPT and CPT.

Example 3.6
Estimate a suitable resistance factor for a bridge footing that is to be designed based on SPT tests.

Solution
From Table 3.13 and Equation (3.65),

$$\lambda_{QD} = 1.03(1.05)(1.00) = 1.08$$

$$\text{COV}_{QD} = \sqrt{(0.08^2 + (0.1)^2 + (0.25)^2} = 0.0953$$

From Table 3.15, for SPT, $\lambda_R = 1.3$, $\text{COV}_R = 0.7$.

Also, since it is recommended that $\lambda_{QL} = 1.15$, $\text{COV}_{QL} = 0.18$ (FHWA, 1998), and assuming that $\gamma_L = 1.75$ and $\gamma_D = 1.25$ (Table 3.18) and applying Equation (3.67),

TABLE 3.17

Resistance Factors for Semiempirical Evaluation of Bearing Capacity for Spread Footings on Sand Using Reliability-Based Calibration

Estimation Method	Factor of Safety, FS	Average Reliability Index, β	Target Reliability Index, β_r	Span (m)	Resistance Factor Fitting with ASD	Resistance Factor Reliability Based	Selected Φ
SPT	4.0	4.2	3.5	10	0.37	0.49	0.45
				50	0.37	0.53	0.45
CPT	2.5	3.2	3.5	10	0.60	0.52	0.55
				50	0.60	0.57	0.55

Source: From Federal Highway Administration, 1998, *Load and Resistance Factor Design (LRFD) for Highway Bridge Superstructures*, Washington, DC. With permission.

$$\phi_R = \frac{1.3\left[1.25\dfrac{Q_D}{Q_L} + 1.75\right]\sqrt{\dfrac{(1 + (0.289)^2 + (0.18)^2)}{(1 + (0.7)^2)}}}{\left[(1.08)\dfrac{Q_D}{Q_L} + 1.15\right]\exp\left\{\beta_T\sqrt{\ln\left[(1 + (0.289)^2 + (0.18)^2)(1 + (0.7)^2)\right]}\right\}}$$

$$\phi_R = \frac{1.13\left[1.25\dfrac{Q_D}{Q_L} + 1.75\right]}{\left[(1.08)\dfrac{Q_D}{Q_L} + 1.15\right]\exp\{0.713\beta_T\}}$$

TABLE 3.18

Load Factors for Permanent Loads

Type of Load	Load Factor Maximum	Load Factor Minimum
Components and attachments	1.25	0.90
Downdrag	1.8	0.45
Wearing surfaces and utilities	1.5	0.65
Horizontal earth pressure		
Active	1.5	0.9
At rest	1.35	0.9
Vertical earth pressure		
Overall stability	1.35	N/A
Retaining structure	1.35	1.00
Rigid buried structure	1.30	0.90
Rigid frames	1.35	0.90
Flexible buried structure	1.95	0.90
Flexible metal box	1.50	0.90
Culverts		
Earth structures	1.50	0.75

Source: From Federal Highway Administration, 1998, *Load and Resistance Factor Design (LRFD) for Highway Bridge Superstructures*, Washington, DC. With permission.

Using Equation (3.67) and Table 3.14, the resistance factor can be expressed in terms of the probability of failure and the dead load–live load ratio (Table 3.19).

Example 3.7

For the column shown in Figure 3.19, use LRFD concepts to design a suitable footing to carry a column load of 400 kN. The subsoil can be considered as homogenous silty clay with the following properties: assume that the ground water table is not in the vicinity

$$\text{Unit weight} = \gamma = 17 \text{ kN/m}^3$$
$$\text{Internal friction} = \phi = 15°$$
$$\text{Unit cohesion} = c = 20 \text{ kPa}$$

Assume resistance factors ϕ_R of 0.6 and 0.6 (Table 3.16) for $\tan \phi$ and c, respectively,

$$\phi' = \tan^{-1}[0.6 \times \tan \phi] = 9°$$
$$c' = 20(0.6) = 12 \text{ kPa}$$

Table 3.1 indicates the following bearing capacity parameters:

Using Hansen's bearing capacity expression (Equation (3.5)),
   $N_c = 8, N_q = 2, N_\gamma = 0.4$
   $s_c = 1.359, s_q = 1.26, s_\gamma = 0.6$
   $d_c = 1.4, d_q = 1.294, d_\gamma = 1.0$
The vertical effective stress at the footing base level $= q = (17)(\text{depth}) = 17B$
Then, the following expressions can be written for the ultimate bearing capacity:

$$q_{ult} = (12)(8)(1.359)(1.4) + (17B)(2)(1.26)(1.294) + 0.5(17)(B)(0.4)(0.6)(1.0)$$
$$= 182.65 + 57.47B$$

Factored contact stress at the foundation level $= 1.25 \times 4 \times 400/(AB^2) + (1.0)17B$.
   The load factor for the dead load is obtained from Table 3.18. It must be noted that the recommended load factor for recompacted soil is 1.0.
   By applying $\phi R_n = \eta \sum \gamma_i Q_i$ with no load modifier ($\eta = 1.0$)

$$q_{ult} = 1.25 \times 4 \times 400/(AB^2) + 17B$$

TABLE 3.19

Variation of Resistance Factor ϕ_R with Q_D/Q_L and the Required Reliability

	Probability of Failure			
Q_D/Q_L	0.085	0.0099	0.00115	0.000134
1	0.363602	0.254558	0.178216	0.12477
2	0.347033	0.242958	0.170095	0.119084
3	0.338616	0.237066	0.16597	0.116196

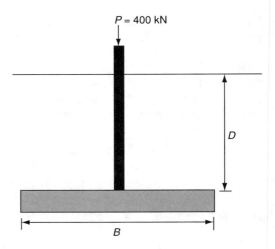

FIGURE 3.19
Illustration for Example 3.7.

From Hansen's expression

$$637/B^2 + 17B = 182.65 + 57.47B$$
$$637/B^2 = 182.65 + 40.47B$$

$B = 1.6\,\text{m}.$

When one compares the above footing width with $B = 1.55\,\text{m}$ obtained from the ASD method, the limit state design is seen to be slightly more conservative.

3.5 Design of Footings to Withstand Vibrations

Foundations subjected to dynamic loads such as that due to operating machines, wave loadings, etc. have to satisfy special criteria in addition to the regular bearing capacity and settlement criteria. Table 3.20 lists a number of criteria that may be considered during the design of a foundation that would be subjected to vibrations. However, the main design criteria are related to the limiting amplitude of vibration and the limiting acceleration for a given operating frequency. Figure 3.20 indicates the order of magnitudes of vibration corresponding to practically significant levels of severity based on Richart (1972) and the author's judgment.

For steady-state harmonic oscillations, the limiting accelerations can be deduced from the limiting amplitudes in terms of the frequency of oscillation (ω) as

$$\text{acceleration}_{\text{limit}} = \text{displacement}_{\text{limit}}\omega^2 \tag{3.71}$$

On the other hand, the maximum amplitudes (or accelerations) undergone by a given vibrating foundation can be determined by the principles of soil dynamics. Analytical formulations available from such analyses are provided in the ensuing sections for a number of different modes of vibration.

3.5.1 Vertical Steady-State Vibrations

The equation of motion for a rigid foundation of mass m subjected to a vertical steady-state constant amplitude simple harmonic force can be written as (Lysmer, 1966)

TABLE 3.20

List of Criteria for Design of Vibrating Footings

I. Functional considerations of installation
    A. Modes of failure and the design objectives
    B. Causes of failure
    C. Total operational environment
    D. Initial cost and its relation to item A
    E. Cost of maintenance
    F. Cost of replacement
II. Design considerations for installations in which the equipment produces exciting forces
    A. Static bearing capacity
    B. Static settlement
    C. Bearing capacity: static + dynamic loads
    D. Settlement: static + repeated dynamic loads
    E. Limiting dynamic conditions
        1. Vibration amplitude at operating frequency
        2. Velocity
        3. Acceleration
    F. Possible modes of vibration — coupling effects
    G. Fatigue failures
        1. Machine components
        2. Connections
        3. Supporting structure
    H. Environmental demands
        1. Physiological effect on persons
        2. Psychological effect on persons
        3. Sensitive equipment nearby
        4. Resonance of structural components
III. Design considerations for installation of sensitive equipment
    A. Limiting displacement, velocity, or acceleration amplitudes
    B. Ambient vibrations
    C. Possible changes in ambient vibrations
        1. By construction
        2. By new equipment
    D. Isolation of foundations
    E. Local isolation of individual machines

Source: From Richart, F.E., Hall, J.R., and Woods, R.D., 1970, *Vibrations of Soils and Foundations*, Prentice Hall, Englewood Cliffs, NJ. With permission.

$$m\ddot{z} + c_z\dot{z} + k_z z = Q_0 e^{i\omega t} \tag{3.72}$$

If the foundation is circular, the spring and the damping constants are given by

$$k_z = \frac{2G_s B}{1 - \nu_s} \tag{3.73a}$$

and

$$c_z = \frac{0.85 B^2}{1 - \nu_s}\sqrt{G_s \rho_s} \tag{3.73b}$$

respectively, where B is the equivalent footing diameter, G_s, ρ_s, and ν_s are the shear modulus, mass density, and Poisson's ratio of the foundation soil, respectively (Figure 3.21).

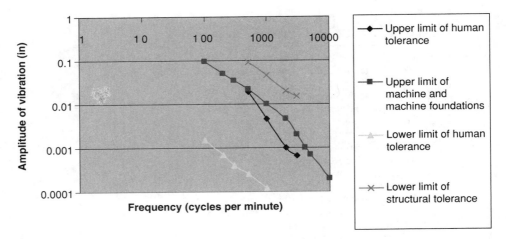

FIGURE 3.20
Limits of displacement amplitude.

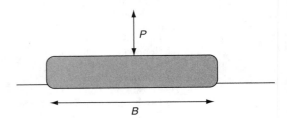

FIGURE 3.21
Footing subjected to vertical vibration.

Then, the following important parameters that relate to the vibratory motion can be derived using the elementary theory of vibrations:

1. *Natural frequency of vibration*

$$f_n = \frac{1}{2\pi} \sqrt{\frac{2G_s B}{1 - \nu_s} \frac{1}{m}} \tag{3.74a}$$

2. *Resonant frequency*
   For force-type excitation

$$f_m = \frac{1}{\pi B} \sqrt{\frac{G_s(B_z - 0.36)}{\rho_s B_z}} \quad \text{for } B_z > 0.3 \tag{3.74b}$$

   where the modified dimensionless mass ratio B_z is given by

$$B_z = 2(1 - \nu_s) \frac{m}{G_s B^3} \tag{3.74c}$$

3. *Damping ratio*
   Damping ratio $= D =$ (damping constant)/(critical damping constant)
   Critical damping constant $= 2(km)^{1/2}$

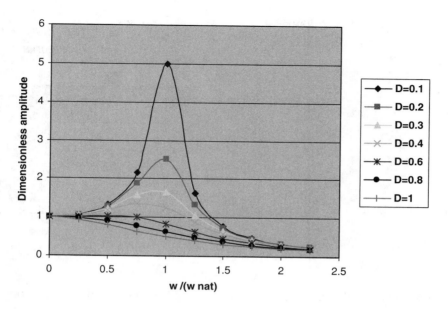

FIGURE 3.22
Plot of Magnification factors. (*Note:* Use amplitude $= \dfrac{A_z}{(Q_o/k_z)}$ for vertical vibration, $= \dfrac{\theta}{(M_y/k_\theta)}$ for rocking oscillation, $= \dfrac{A_x}{(Q_o/k_x)}$ for sliding oscillation).

$$D = D_z = \frac{0.425}{B_z} \tag{3.74d}$$

4. *Amplitude of vibration*

 The amplitude of vibration can be expressed as follows:

$$A_z = \frac{Q_0}{k_z} M = Q_0 \frac{1 - \nu_s}{2 G_s B} M \tag{3.74e}$$

where M is the magnification factor, $A_z/(Q_0/k_z)$, which is plotted in Figure 3.22 against the nondimensional frequency, ω/ω_n, and the damping ratio, D, where

$$\omega_n = 2\pi f_n = \sqrt{\frac{2 G_s B}{1 - \nu_s} \frac{1}{m}} \tag{3.74a}$$

Example 3.8

A rigid circular concrete foundation supporting a machine is 4 m in diameter (Figure 3.23). The total weight of the machine and foundation is 700 kN. The machine imparts a vertical vibration of $25 \sin 20t$ kN on the footing. If the foundation soil is dense sand having the following properties: unit weight $= 17 \, \text{kN/m}^3$ and elastic modulus $= 55 \, \text{MPa}$. Determine (1) the resonant frequency, (2) the amplitude of the vibration at the resonant frequency, and (3) the amplitude of the vibration at the operating frequency.

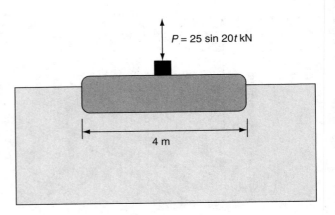

FIGURE 3.23
Illustration for Example 3.8.

Solution
For sandy soil, Poisson's ratio can be assumed to be 0.33. Hence, the shear modulus and the mass ratio can be computed as

$$G_s = \frac{E_s}{2(1 + \nu_s)} = (55)/2/1.33 = 20.7 \, \text{MPa}$$

$$B_z = 2(1 - \nu_s)\frac{m}{\rho_s B^3} = 2(1 - 0.33)(700)/(17)(4)^3 = 0.86$$

(1) *Resonant frequency*

$$f_m = \frac{1}{\pi B}\sqrt{\frac{G_s(B_z - 0.36)}{\rho_s B_z}} = \frac{1}{\pi(4)}\sqrt{\frac{(20.7 \times 1,000,000)(0.86 - 0.36)}{(17 \times 1,000/9.8)(0.86)}} = 6.63 \, \text{cps}$$

(2) *Natural frequency*

$$f_n = \frac{1}{2\pi}\sqrt{\frac{2G_s B}{1 - \nu_s}\frac{1}{m}} = \frac{1}{2\pi}\sqrt{\frac{2(20.7)(1,000,000)(4)}{1 - 0.33}\frac{1}{700(1,000)/(9.8)}} = 9.36 \, \text{cps}$$

(3) *Operating frequency*
$f_o = 20/(2\pi) = 3.18 \, \text{cps}$
Hence, $f_m/f_o = 6.63/3.18 = 2.08 > 2$
Thus, the operating frequency range is considered safe.

(4) *Amplitude of vibration*
$T_m/T_n = 6.63/9.36 = 0.71$
$D = 0.425/B_z = 0.425/0.86 = 0.491$

From Figure 3.21, the magnification factor, $M = 1.2$, is

$$A_z = \frac{Q_0}{k_z}M = Q_0\frac{1 - \nu_s}{2G_s B}M = 25(1 - 0.33)(1.2)/(2 \times 20.7 \times 1000 \times 4) = 0.27 \, \text{mm}$$

Based on an operating frequency of 3.18 cps or 191 cpm, the above amplitude of 0.27 mm or 0.011 in. would fall in the "troublesome range" in Figure 3.20.

3.5.2 Rocking Oscillations

The motion of a rigid foundation subjected to a steady-state constant amplitude harmonic rocking moment about the y-axis can be written as (Hall, 1967) (Figure 3.24)

$$I_y \ddot{\theta} + c_\theta \dot{\theta} + k_z \theta = M_y e^{i\omega t} \tag{3.75a}$$

where

$$I_y = m \left(\frac{B^2}{16} + \frac{h^2}{3} \right) \tag{3.75b}$$

If the foundation is circular, the spring and the damping constants are given by

$$k_\theta = \frac{G_s B^3}{3(1 - \nu_s)} \tag{3.76a}$$

and

$$c_\theta = \frac{0.05 B^4}{(1 - \nu_s)(1 + B_\theta)} \sqrt{G_s} \tag{3.76b}$$

respectively, where B_θ, the inertia ratio, is given by

$$B_\theta = 12(1 - \nu_s) \frac{I}{\rho B^5} \tag{3.76c}$$

Then, the following parameters relevant to the vibratory motion can be derived using the elementary theory of vibrations:

1. *Natural frequency of vibration*

$$f_n = \frac{1}{2\pi} \sqrt{\frac{k_\theta}{I_0}} \tag{3.77a}$$

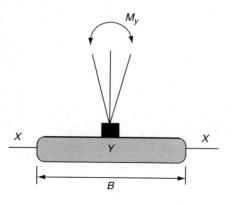

FIGURE 3.24
Footing subjected to rocking.

2. *Resonant frequency*

   Moment-type excitation

$$f_m = f_n \sqrt{1 - \frac{0.45}{B_\theta (1 + B_\theta)^2}} \qquad (3.77b)$$

3. *Damping ratio*

$$D = D_\theta = \frac{0.15}{\sqrt{B_\theta}(1 + B_\theta)} \qquad (3.77c)$$

4. *Amplitude of vibration*

   The amplitude of vibration can be expressed as follows:

$$\theta = \frac{M_y}{k_\theta} M \qquad (3.77d)$$

where M is the magnification factor, $\theta/(M_y/k_\theta)$, which is plotted in Figure 3.22 against the nondimensional frequency ω/ω_n and thus

$$\omega_n = 2\pi f_n = \sqrt{\frac{k_\theta}{I}} \qquad (3.77e)$$

The above relations can be applied to a rectangular footing (of the same height, h) using an equivalent B_e that is determined by equating the moment of area of the surface of the footing about the y-axis (I_y) to that of the equivalent circular footing. Thus,

$$\frac{1}{64} \pi B_e^4 = \frac{1}{12} BL^3$$

3.5.3 Sliding Oscillations

A mass-spring-dashpot analog was developed by Hall (1967) to simulate the horizontal sliding oscillations of a rigid circular footing of mass m (Figure 3.25). This can be expressed by

$$m\ddot{x} + c_x \dot{x} + k_x x = Q_0 e^{i\omega t} \qquad (3.78)$$

FIGURE 3.25
Footing subjected to sliding oscillation.

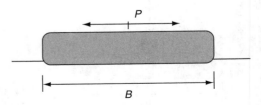

If the foundation is circular, the spring and the damping constants are given by

$$k_x = \frac{16(1 - \nu_s)G_sB}{7 - 8\nu_s} \tag{3.79a}$$

and

$$c_z = \frac{4.6(1 - \nu_s)B^2}{7 - 8\nu_s}\sqrt{G_s\rho_s} \tag{3.79b}$$

respectively.

Then, the following important parameters with respect to the above motion can be derived:

1. *Natural frequency of vibration*

$$f_n = \frac{1}{2\pi}\sqrt{\frac{16(1 - \nu_s)G_sB}{7 - 8\nu_s}\frac{1}{m}} \tag{3.80a}$$

2. *Resonant frequency*
   Moment-type excitation

$$f_m = f_n\sqrt{1 - \frac{0.166}{B_x}} \tag{3.80b}$$

where the modified dimensionless mass ratio B_x is given by

$$B_x = \frac{7 - 8\nu_s}{8(1 - \nu_s)}\frac{m}{\rho B^3} \tag{3.80c}$$

3. *Damping ratio*

$$D = D_x = \frac{0.288}{\sqrt{B_x}} \tag{3.80d}$$

4. *Amplitude of vibration*
   The amplitude of vibration can be expressed as follows:

$$A_x = \frac{Q_0}{k_x}M \tag{3.80e}$$

where M is the magnification factor, $A_x/(Q_0/k_x)$, which is also plotted in Figure 3.22 against the nondimensional frequency ω/ω_n and thus

$$\omega_n = 2\pi f_n = \sqrt{\frac{16(1 - \nu_s)G_sB}{7 - 8\nu_s}\frac{1}{m}} \tag{3.80a}$$

3.5.4   Foundation Vibrations due to Rotating Masses

If the foundation vibrations described in Sections 3.5.1 to 3.5.3 are created by unbalanced masses (m_1 with an eccentricity of e) rotating at an angular frequency of ω, then the following modifications must be made to Equations (3.72), (3.75a), and (3.78):

1. *Translational oscillations*
   $Q_0 = m_1 e \omega^2$ must be substituted in Equations (3.72) and (3.78) for Q_0.
2. *Rotational oscillations*
   $M_y = m_1 e z \omega^2$ must be substituted in Equation (3.75a) for M_y, where z is the moment arm of the unbalanced force.

   In all of the above cases, the new equations of motion corresponding to Equations (3.72), (3.75a), and (3.78) have to be solved to determine the resonance frequencies and the amplitudes of vibrations.

3.6   Additional Examples

Example 3.9

Predict the following settlement components for a circular footing of 2 m in diameter that carries a load of 200 kN as shown in Figure 3.26.

(a) Average consolidation settlement of the footing in 5 years (use the 2:1 distribution method)
(b) Maximum ultimate differential settlement
(c) Elastic settlement from Schertmann's method
(d) Total ultimate settlement of the center of the footing

   Consolidation properties of the clay layer can be obtained from Figure 3.15. Assume its coefficient of consolidation to be 1×10^{-8} m^2/sec.
   Suitable elastic parameters of the sandy soil can be obtained from Chapter 1 (Table 1.5 and Table 1.6).

(a) Increase stress at the center of the soft clay

2 m diameter footing

$$\Delta \sigma = \frac{Q}{\pi/4(D + Z)^2}$$

At mid-plane

$$\Delta \sigma = \frac{200}{\pi/4\,(2 + 2.8)^2}$$

$$\Delta \sigma = 11.05 \text{ kPa}$$

From Figure 3.15, preconsolidation pressure $P_c = 60$ kPa.

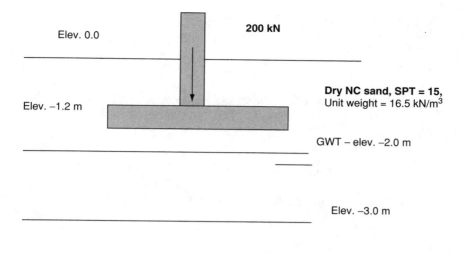

Elev. 0.0

200 kN

Elev. −1.2 m

Dry NC sand, SPT = 15,
Unit weight = 16.5 kN/m³

GWT − elev. −2.0 m

Elev. −3.0 m

Soft clay
Unit weight = 18 kN/m³

Elev. −5.0 m

FIGURE 3.26
Illustration for Example 3.9.

The average effective overburden pressure at the mid-plane of the soft clay,

$$\sigma'_{v0} = (16.5)(2) + (17.5)(1) + (18)(1) - (9.8)(2)$$
$$= 48.9 \text{ kPa} < P_c \quad \text{So, it is OC}$$

$$\sigma'_{v0} + \Delta\sigma = 48.9 + 11.05 = 59.95 \text{ kPa} < P_c \text{ (60 kPa)}$$

$$\therefore S_{avg} = \left[\frac{H}{1 + e_0}\right]\left[C_r \log\frac{\sigma'_{v0} + \Delta\sigma}{\sigma'_{v0}}\right]$$

(1.18b)

$$= \left[\frac{2}{1 + 1.06}\right]\left[0.064 \log\frac{59.95}{48.9}\right] = 5.5 \text{ m}$$

$T = 5$ years, $H_{dr} = 2$ m, $C_u = 1 \times 10^{-8}$ m²/sec

$$T = \frac{C_v\, t}{H_{dr}^2} = \frac{1 \times 10^{-8}\,(5 \times 365 \times 24 \times 60 \times 60)}{2^2} = 0.394$$

For $T = 0.394$, $U_{avg} = 0.69$. Therefore,

$$S_{5 \text{ years}} = (5.5)(0.69) = 3.795 \text{ mm}$$

(b) Using Newmark's chart, $d_c = 2.8\,\text{m} = OQ$ (Figure 3.16)

Footing radius $= 1\,\text{m} = \frac{1}{2.8}\,OQ = 0.36OQ$

Placing at the center of chart

$N_{\text{center}} = 48 \times 4 = 192$:

$$\Delta\sigma_{\text{center}} = 192 \times 0.001 \times \frac{200}{(\pi/4)(2)^2} = 12.2\,\text{kPa}$$

$N_{\text{edge}} = 150$:

$$\Delta\sigma_{\text{edge}} = 150 \times 0.00\,1 \times \frac{200}{(\pi/4)(2)^2} = 9.5\,\text{kPa}$$

$$\sigma'_{v0} + \Delta\sigma_{\text{center}} = 48.9 + 12.2 = 61.1\,\text{kPa} > p_c$$

$$
\begin{aligned}
S_{\text{center}} &= \left[\frac{H}{1+e_0}\right]\left[C_r \log\frac{p_c}{\sigma_{v0}} + C_c \log\frac{p_c + \Delta\sigma}{p_c}\right] \\[6pt]
&= \left[\frac{1}{1+1.06}\right]\left[0.064 \log\frac{60}{48.9} + 0.382 \log\frac{60 + 12.2}{60}\right] \quad (1.18c) \\[6pt]
&= 35.3\,\text{mm}
\end{aligned}
$$

$$\sigma'_{v0+} + \Delta\sigma_{\text{edge}} = 48.9 + 9.5 = 58.4 < p_c$$

$$
\begin{aligned}
S_{\text{edge}} &= \left[\frac{H}{1+e_0}\right]\left[C_r \log\frac{\sigma'_{v0} + \Delta\sigma}{\sigma'_{v0}}\right] \\[6pt]
&= \left[\frac{2}{1+1.06}\right]\left[0.064 \log\frac{48.9 + 9.5}{48.9}\right] \quad (1.18b) \\[6pt]
&= 4.79\,\text{mm}
\end{aligned}
$$

$$\Delta S = 35.3 - 4.79 = 30.5\,\text{mm}$$

(c)

$$
\begin{aligned}
S_c &= C_1 C_2 (\Delta\sigma - q) \sum_0^Z \frac{I_z}{E_s}\Delta z \\[6pt]
&= \left[1 - 0.5\,\frac{19.8}{63.66 - 19.8}\right]\left[1 + 0.2 \log\frac{0.1}{0.1}\right]\left[\frac{0.192}{15,000} + \frac{0.108 + 0.416}{6,750} + \frac{0.48}{10,000}\right](63.66 - 19.8) \\[6pt]
&= 4.7\,\text{mm}
\end{aligned}
$$

$$(1.13)$$

(d) Total ultimate settlement

$\qquad = $ Center consolidation $+$ elastic settlement

$\qquad = 35.3 + 4.7 = 40.0\,\text{mm}$

Example 3.10

Assuming that the depth of the embedment is 1.2 m, design a suitable *strip* footing for the wall that carries a load of 15 kN/m as shown in Figure 3.27. Suitable soil parameters for the site can be obtained from Chapter 2.

$SPT\bar{N} = 12$

From Table 2.2, medium stiff clay

$\gamma_{moist} = 18.9 \text{ kN/m}^3$
$\phi = 3° \text{ (Take } \phi = 0°)$

From Table 2.3

$\sigma_{sat} = 18.9 \text{ kN/m}^3$
$\sigma_{sub} = 9.1 \text{ kN/m}^3$
$\phi = 0°$
$C = \dfrac{N}{T} = \dfrac{12}{8}$
$= 1.5 \text{ psf} = 0.0718 \text{ kPa}$

Using Meyerhoff's bearing capacity: Equation (3.3),

$$q_{ult} = cN_cS_cd_c + qN_qS_qd_q + 0.5\,B_\gamma N_\gamma S_\gamma d_\gamma \tag{3.4}$$

From Table 3.3 (for $\phi = 0$ Meyerhoff),

$$N_c = 5.14\,, \quad N_q = 1.0\,, \quad N_\gamma = 0.0$$

From Table 3.2(b) (for $\phi = 0$)

$$K_p = \tan^2\left(45 + \frac{\phi}{2}\right) = 1.0 \quad \left(\frac{\phi}{2} \to 0\right)$$

$$S_c = 1 + 0.2\,(1)\,\frac{B}{1} = 1 + 0.2B$$

$$S_q = 1.0$$

$$d_c = 1 + 0.2\,\sqrt{1}\left(\frac{1.5}{B}\right) = 1 + \frac{0.3}{B}$$

$$d_q = 1.0$$

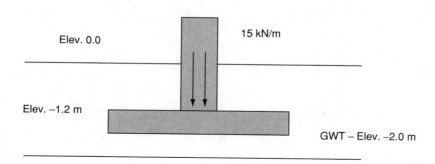

Elev. 0.0

15 kN/m

Elev. −1.2 m

GWT − Elev. −2.0 m

FIGURE 3.27
Illustration for Example 3.10.

Therefore,

$$q_{ult} = (0.0718)(5.14)(1 + 0.2B)\left(1 + \frac{0.3}{B}\right) + (18.9)(1.5)(1.0)(1.0)(1.0) + 0$$

$$\frac{P}{A} \leq \frac{q_{ult}}{F}$$

$$\frac{15}{(B \times 1)} \leq \frac{\left((0.37 + 0.0738B)\left(1 + \frac{0.3}{B}\right) + 28.35\right)}{2.5}$$

Therefore, $B > 1.3\,\text{m}$.

Example 3.11

A 5-kN horizontal load acts on the column shown in Figure 3.28 at a location of 1.5 m above the ground level. If the site soil is granular with an angle of friction 20° and a unit weight of 16.5 kN/m³, determine a suitable footing size. If the ground water table subsides to a depth outside the foundation influence zone, what would be the factor of safety of the footing you designed?

$$K_p = 2.04; \quad \phi = 20°; \quad \gamma = 16.5 \text{ kN/m}^3$$

Meyerhoff's bearing capacity expression:

$$q_{ult} = cN_cS_cd_ci_c + qN_qS_qd_qi_q + 0.5\,B\,\gamma'N_\gamma S_\gamma d_\gamma i_\gamma \tag{3.4}$$
$$(c = 0)$$

From Table 3.1, for $\phi = 20°$, $N_q = 6.4$, $N_\gamma = 2.9$.
    From Table 3.2(b),

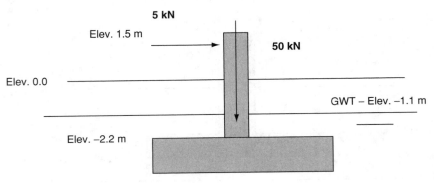

FIGURE 3.28
Illustration for Example 3.11.

$$S_q = S_\gamma = 1 + 0.1(2.04)\frac{B}{B} = 1.204 \text{ (assume circular or square footing)}$$

$$d_q = d_\gamma = 1 + 0.2\sqrt{2.04}\frac{D}{B} = 1 + \frac{0.314}{B}$$

$$i_q = \left(1 - \frac{5.71}{90}\right)^2 = 0.877, \ i_\gamma = \left(1 - \frac{5.71}{20}\right)^2 = 0.51$$

$$q = (16.5)(1.1) + (16.5 - 9.8)(1.1) = 25.52 \text{ kPa}$$

$$q_{ult} = (25.52)(6.4)\left(1 + \frac{0.314}{B}\right)(1.204)(0.877) + (0.5)(B)(16.5 - 9.8)(2.9)(1.204)\left(1 + \frac{0.314}{B}\right)(0.51)$$

$$= 172.46 + \frac{54.15}{B} + 5.97B + 1.87$$

or

$$q_{ult} = 174.33 + \frac{54.15}{B} + 5.97B$$

$$q_d \leq \frac{q_{ult}}{F}$$

$$\frac{50}{B \times B} + 25.52 + \frac{18.5 \times (B/2)}{(B \times B^2)/12} \leq \frac{174.33 + (54.15/B) \times 5.97 B}{2.5}$$

Therefore, $B = 1.43$ m.
  With no water within influence zone and
  $B = 1.43$ m
  $Q = (16.5)(2.2) = 36.3$ kPa

$$q_d = \frac{P}{A} + \gamma D + \frac{Mc}{I} = \frac{50}{B^2} + 36.3 + \frac{18.5 \times \frac{B}{2}}{\frac{B \times B^3}{12}} = 98.7 \text{ kPa}$$

$$q_{ult} = (36.3)(64.)\left(1 + \frac{0.314}{B}\right)(1.204)(0.877) + (0.5)(B)(16.5)(2.9)(1.204)\left(1 + \frac{0.314}{B}\right)(0.51)$$

$$= 324.8 \text{ kPa}$$

$$\text{FS} = \frac{324.8}{98.7} = 3.29$$

So, FS increases.

References

Andersland, O.B. and Al-Khafaji, A.A.W.N, 1980, Organic material and soil compressibility, *Journal of the Geotechnical Engineering Division*, ASCE, GT7: 749.

AASHTO, 1996, *Standard Specifications for Highway Bridges*, American Association for State Highway and Transportation Officials, Washington, DC.

Bowles, J.E., 2002, *Foundation Analysis and Design*, McGraw-Hill, New York.

Bowles, J.E., 1995, *Foundation Analysis and Design*, McGraw-Hill, New York.

Das, B.M., 1993, *Principles of Soil Dynamics*, PWS-Kent Publishers, Boston.

Federal Highway Administration, 1998, *Load and Resistance Factor Design (LRFD) for Highway Bridge Substructures*, NCHRP Course No. 13068, HI 98-032, July, Washington, DC.

Fox, P.J. and Edil, T.B., 1992, C_α/C_c concept applied to compression of peat, *Journal of the Geotechnical Engineering Division*, ASCE, 118(GT8): 1256–1263.

Gunaratne, M., Stinnette, P., Mullins, G., Kuo, C., and Echelberger, W., 1998, Compressibility relations for Florida organic material, *ASTM Journal of Testing and Evaluation*, 26(1): 1–9.

Hall, J.R., Jr., 1967, Coupled rocking and sliding oscillations of rigid circular footings, Proceedings of the International Symposium on Wave Propagation and Dynamic Properties of Earth Materials, Albuquerque, NM, August.

Hansen, J.B., 1970, A Revised and Extended Formula for Bearing Capacity, Danish Geotechnical Institute, Copenhagen, Bulletin No. 28.

Harr, M.E., 1966, *Foundations of Theoretical Soil Mechanics*, Mc-Graw-Hill, New York.

Harr, M.E., 1977, *Mechanics of Particulate Media*, McGraw-Hill, New York.

Lysmer, J. and Richart, F.E., Jr., 1966, Dynamic response of footings to vertical loading, *Journal of the Soil Mechanics and Foundations Division*, ASCE, 92(SM 1): 65–91.

Meyerhoff, G.G., 1951, The ultimate bearing capacity of foundations, *Geotechnique*, 2(4): 30–331.

Meyerhoff, G.G., 1963, Some recent research on the bearing capacity of foundations, *Canadian Geotechnical Journal*, 1(1): 16–26.

Meyerhoff, G.G., 1956, Penetration tests and bearing capacity of cohesionless soils, *Journal of the Soil Mechanics and Foundations Division*, ASCE, 82(SM 1): 1–19.

Parry, R.H., 1977, Estimating bearing capacity of sand from SPT values, *Journal of the Geotechnical Engineering Division*, ASCE, 103(GT9): 1014–1019.

Phoon, K., Kulhawy, F.H., and Grigoriu, M.D., 1995, Reliability-Based Design of Foundations for Transmission Line Structures, TR 105000 Final report, report prepared by Cornell University for the Electric Power Research Institute, Palo Alto, CA, 340 pp.

Richart, F.E., 1962, *Forced Vibrations*, Transactions, ASCE, Vol. 127, Part I, pp. 863–898.

Richart, F.E., Hall, J.R., and Woods, R.D., 1970, *Vibrations of Soils and Foundations*, Prentice Hall, Englewood Cliffs, NJ.

Teng, W.H., Mesri, G., and Halim, I., 1992, Uncertainty of mobilized undrained shear strength, *Soils and Foundations*, 32(4): 107–116.

Terzaghi, K., 1943, *Theoretical Soil Mechanics*, John Wiley, New York.

Tomlinson, M.J. and Boorman, R, 1995, *Foundation Design and Construction*, Longman Scientific and Technical, Brunthill, Harlow, England.

Vesic, A.S., 1973, Analysis of ultimate loads of shallow foundations, *Journal of the Soil Mechanics and Foundation Engineering Division*, ASCE, 99(SM 1): 45–763.

Vesic, A.S., 1975, *Foundation Engineering Handbook*, 1st ed., Chapter 3, Winterkorn H.F. and Fang H. (eds.), Van Nostrand Reinhold, New York.

4

Geotechnical Design of Combined Spread Footings

Manjriker Gunaratne

CONTENTS

4.1 Introduction

Combined spread footings can be employed as viable alternatives to isolated spread footings under many circumstances. Some of them are listed below:

1. When the ground bearing capacity is relatively low, the designers have to seek methods of lowering the bearing stress. A larger footing area that provides a common foundation to many columns would distribute the load and reduce the bearing stress. In addition, this modification will also reduce the footing settlement.

2. When the exterior columns or walls of a heavy structure are in close proximity to the property line or other structures, the designer would not have the

freedom to utilize the area required to design an isolated spread footing. In such cases, any adjoining interior columns can be incorporated to design a combined footing.

3. Isolated spread footings can become unstable in the presence of unexpected lateral forces. The stability of such footings can be increased by tying them to other footings in the vicinity.

4. When adjoining columns are founded on soils with significantly different compressibility properties, one would anticipate undesirable differential settlements. These settlements can be minimized by a common combined footing.

5. If the superstructure consists of a multitude of column loads, designing a single monolithic *mat* or *raft* footing that supports the entire system of structural columns would be more economical. This is especially the case when the total area required by isolated spread footings for the individual columns is greater than 50% of the entire area of the column plan (blueprint).

4.2 Design Criteria

Two distinct design philosophies are found in the current practice with respect to design of combined footings. They are: (1) conventional or the rigid method and (2) beams or slabs on elastic foundation or the flexible method. Of these, in the conventional design method, one assumes that the footing is infinitely rigid compared to the foundation soils and that the contact pressure distribution at the foundation–soil interface is uniform (in the absence of any eccentricity) or planar (with eccentricity). In other words, the deflection undergone by the footing is considered to be unrelated to the contact pressure distribution. This assumption can be justified in the case of a spread footing with limited dimensions or a stiff footing founded on a compressible soil. Therefore, the conventional method appears to be inadequate for footings with larger dimensions relative to their thickness and in the case of footings that are flexible when compared to the foundation medium. Although these drawbacks are addressed in such cases by the flexible footing

TABLE 4.1

From English	To SI	Multiply by	Quantity	From SI	To English	Multiply by
Ft	m	0.3048	Lengths	m	ft	3.28
In	m	0.0254		m	in	39.37
Tons	kN	8.9	Loads	kN	tons	0.114
Tsf	kPa	95.76	Stress/strength	kPa	tsf	0.0104
(pcf) lbs/ft^3	N/m^3	157.1	Force/Unit-volume	N/m^3	lbs/ft^3	0.0064
kips/ft^3	kN/m^3	157.1		kN/m^3	Kips/ft^3	0.0064
Lb-inch	N-mm	112.98	Moment or energy	N-mm	lb-inch	0.0089
kip-inch	kN-mm	112.98		kN-mm	kip-inch	0.0089
lb-ft	N-m	1.356		N-m	lb-ft	0.7375
kip-ft	kN-m	1.356		kN-m	Kip-ft	0.7375
ft-lb	Joule	1.356		Joule	ft-lb	0.7375
ft-kip	kJoule	1.356		kJoule	ft-kip	0.7375
s/ft	s/m	3.2808	Damping	s/m	s/ft	0.3048
Blows/ft	Blows/m	3.2808	Blow count	blows/m	blows/ft	0.3048

design method to some extent, one has to still assume that the soil behaves as an *elastic* foundation under the flexible footing.

According to ACI (1966), for relatively uniform column loads which do not vary more than 20% between adjacent columns and relatively uniform column spacing, mat footings may be considered as rigid footings if the column spacing is less than $1.75/\beta$ or when the mat is supporting a rigid superstructure. The characteristic coefficient of the elastic foundation, β, is defined by Equation (4.14).

4.2.1 Conventional Design Method

Three distinct design criteria are used in this approach.

4.2.1.1 Eccentricity Criterion

An effort must be made to prevent the combined footing from having an eccentricity, which could cause tilting and the need for a relatively high structural footing rigidity. In order to assure this condition, the footing must be dimensioned so that its centroid coincides with the resultant of the structural loads. Thus, if the coordinates of the centroid of the footing are $(\bar{X}, \bar{Y})$ and the locations of the column loads P_i are (x_i, y_i) (with respect to a local Cartesian coordinate system) (Figure 4.1), then the following conditions must be ensured:

$$\bar{X} = \frac{\sum x_i P_i}{\sum P_i} \tag{4.1}$$

$$\bar{Y} = \frac{\sum y_i P_i}{\sum P_i} \tag{4.2}$$

4.2.1.2 Bearing Capacity Criterion

The allowable stress design (ASD) can be stated as follows:

$$P/A \le \frac{q_{ult}}{F} \tag{4.3}$$

where q_{ult} is the ultimate bearing capacity of the foundation (kN/m^2, kPa, or ksf), P is the structural load (kN or kips), A is the footing area (m^2 or ft^2), and F is an appropriate safety

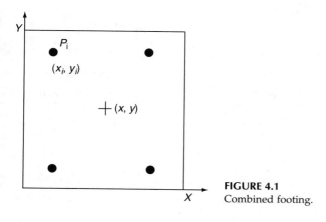

FIGURE 4.1
Combined footing.

factor that accounts for the uncertainties involved in the determination of structural loads (P) and the ultimate bearing capacity (q_{ult}).

One can typically use any one of the bearing capacity equations found in Section 3.2 to evaluate the bearing capacity of the foundation. For conversion of units see Table 4.1.

4.2.1.3 Settlement Criterion

The designer has to also ensure that the combined footing does not undergo either excessive total settlement or differential settlement within the footing. Excessive settlement of the foundation generally occurs due to irreversible compressive deformation taking place immediately or with time. Excessive time-dependent settlement occurs in saturated compressible clays where one will receive advanced warning through cracking, tilting, and other signs of building distress. Significant immediate settlement can also occur in loose sands or compressible clays and silts. Settlements can be determined based on the methods described in Section 3.3.

4.3 Conventional Design of Rectangular Combined Footings

When it is practical or economical to design a single footing to carry two column loads, a rectangular combined footing (Figure 4.2) can be considered.

Example 4.1

Use the conventional or the "rigid" method to design a combined footing for the two columns shown in Figure 4.2, if the average SPT value of the cohesionless foundation soil assumed to be reasonably homogeneous is about 22.

Step 1: Compute ultimate loads A
Column A

$$P_{ult} = 1.4(250) + 1.7(300) \text{ kN} = 860 \text{ kN}$$
$$M_{ult} = 1.4(100) + 1.7(20) \text{ kN m} = 174 \text{ kN m}$$

(This moment could be the result of external wind loads or some other nonvertical load due to bracing force.)

Column B
$$P_{ult} = 1.4(500) + 1.7(400) \text{ kN} = 1380 \text{ kN}$$
$$M_{ult} = 0$$

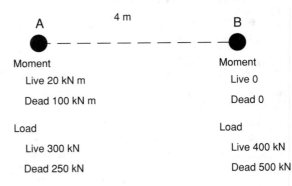

FIGURE 4.2

Footing configuration for Example 4.1.

(one could follow the same computational procedure even with a nonzero moment on the second column)

Step 2: Determine the footing length
The resultant force on the footing $= 860 + 1380$ kN $= 2240$ kN (at C)
The resultant moment $= 174$ kN m (at A)

Option 1
Let us design a rectangular footing that is adequate to compensate the eccentricity in the first column (eccentricity criterion, Equation 4.1):

$$\sum M_A = 0$$

$2240\,\bar{x} = 1380(4) + 174$
therefore $\bar{x} = 2.54$ m
Thus, a single upward reaction of 2240 kN acting at a distance of 2.54 m will statically balance the entire footing. One way to produce such a reaction force is to have a rigid footing that would have its centroid at C, inducing a uniform soil pressure distribution.
Hence the required length of the rigid footing would be $2(2.54 + 0.5)$ m or 6.08 m.
This requires an additional end-section next to column B of 1.58 m (Figure 4.3a).

Option 2
The other option is to curtail the end-section next to column B to about 0.5 m and design an eccentric footing (Figure 4.3b).
This footing would have a base eccentricity of $2.54 + 0.5 - \frac{1}{2}(2.54 + 0.5 + 1.96)$ or 0.9 m.

Step 3: Determine the footing width
One can employ the bearing capacity criterion (Equation 4.3) and SPT data to determine the allowable bearing capacity and then footing width as follows. It must be noticed that q_a is considered as q_{ult}/F:

$$q_a = 30N_{55} \text{ (kPa)} \tag{4.4}$$

where N_{55} is the average SPT value of the foundation at a depth of $0.75B$ below the base of the footing and B is the minimum width of the footing. Equation (4.4) is especially suitable for cohesionless soils.
Alternatively, if the foundation soil is not necessarily cohesionless, one could use the SPT value and Equation (4.5) to determine the allowable bearing capacity for an allowable settlement of s inches.
On the other hand, if soil investigation data is available in terms of CPT (cone penetration data), one could use the correlations presented in Chapter 2.
Using Equation (4.4), $q_a = 30\,(0.45)(22) = 300$ kPa
Then, one can determine the width of the footing (B) using Equation (4.3) as
$2240/(BL) < 300$
or
$B > 2240/(300L)$
therefore $B > 1.22$ m.
Thus, a rectangular footing of 6.08 m $\times$ 1.22 m is adequate to carry both column loads without any eccentricity.

Step 4: Plot shear and moment diagrams
Sign convention:
Shear: clockwise positive. Moment: sagging moments positive

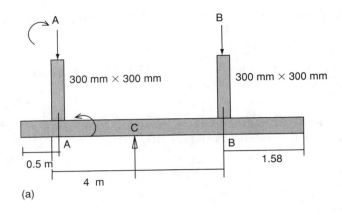

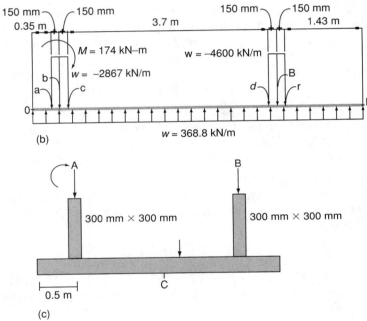

FIGURE 4.3
Design configurations for Example 4.1: (a) design option 1, (b) loading configuration for design option 1, and (c) design option 2.

The distributed reaction per unit length (1 m) on the footing can be computed as

$$w = 2240/6.08 = 368.42 \text{ kN/m}$$

When x is measured from the left edge of the footing, the shear and the moments of each segment of the footing can be found as follows:

In the segment to the left of column A ($0 < x < 0.35$ m)

$$S_1 = 368.42x \text{ kN}$$
$$M_1 = 368.42(x^2/2) \text{ kN m}$$

Within column A ($0.35 < x < 0.65$ m)

In this segment, the column load of 860 kN is distributed at an intensity of 2866.67 kN/m. Then,

$$S_2 = 368.42x - (2866.67)(x - 0.35) \text{ kN}$$

Similarly if one assumes that within column A, the column moment of 174 kN m is also distributed with an intensity of 580 kN m/m (174/0.3),

$$M_2 = 368.42(x^2/2) - (2866.67)(x - 0.35)^2/2 + 580(x - 0.35) \text{ kN m}$$

(with an inflexion point at $x = 0.63$ m).

In between the two columns ($0.65 < x < 4.35$ m)

$$S_3 = 368.42x - 860 \text{ kN}$$
$$M_3 = 368.42(x^2/2) - 860(x - 0.5) + 174 \text{ kN m}$$

(with an inflexion point at $x = 2.33$ m).

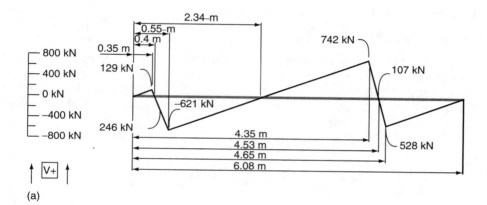

(a)

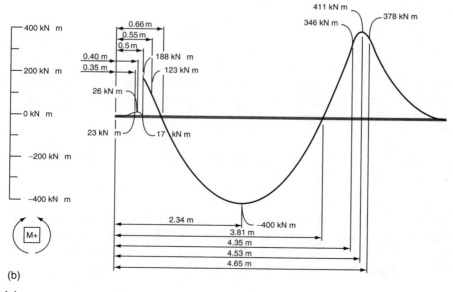

(b)

FIGURE 4.4
(a) Shear force and (b) bending moment diagrams for Example 4.1.

Within column B (4.35 < x < 4.65 m)

The distributed column load would be of intensity 1380/0.3 kN/m. Then,

$$S_4 = 368.42(x) - 860 - (4600)(x - 4.35)$$
$$M_4 = 368.42(x^2/2) - 860(x - 0.5) + 174 - (4600)(x - 4.35)^2/2 \text{ kN m}$$

(with an inflexion point at $x = 4.53$ m).

Within the end-section right of column B (4.65 < x < 6.08 m)

$$S_5 = 368.42(x) - 860 - 1380$$
$$M_4 = 368.42(x^2/2) - 860(x - 0.5) + 174 - 1380(x - 4.5) \text{ kN m}$$

The corresponding shear force and bending moment diagrams are plotted in Figure 4.4.

4.4 Conventional Design of Mat Footings

4.4.1 Bearing Capacity of a Mat Footing

One can use Equations (3.1)–(3.6) to proportion a mat footing if the strength parameters of the ground are known. However, since the most easily obtained empirical strength parameter is the standard penetration blow count, N, an expression is available that uses N to obtain the bearing capacity of a mat footing on a granular subgrade (Bowles, 2002). This is expressed as follows:

For $0 \leq D_f \leq B$ and $B > 1.2$ m

$$q_{n,\text{all}} = \frac{N}{0.08}\left(1 + \frac{1}{3.28B}\right)^2\left(1 + \frac{0.33D_f}{B}\right)\left(\frac{s}{25.4}\right) \tag{4.5}$$

For $B < 1.2$ m

$$q_{n,\text{all}} = \frac{N}{0.08}\left(\frac{s}{25.4}\right) \tag{4.6}$$

where $q_{n,\text{all}}$ is the net allowable bearing capacity in kilopascals, B is the width of the footing, s is the settlement in millimeters, and D_f is the depth of the footing in meters. Then a modified form of Equation (4.3) has to be used to avoid bearing failure:

$$P/A \leq q_{n,\text{all}} \tag{4.7}$$

It is again seen in Equation (4.6) that the use of a safety factor is precluded by employing an allowable bearing capacity.

Example 4.2

Figure 4.5 shows the plan of a column setup where each column is 0.5 m × 0.5 m in section. Design an adequate footing if the corrected average SPT blow count of the subsurface is 10 and the allowable settlement is 25.4 mm (1 in.). Assume a foundation depth of 0.5 m. Then the bearing capacity can be computed from Equation (4.5) as

$$q_{n,\text{all}} = \frac{10}{0.08}\left(1 + \frac{1}{3.28(5.0)}\right)^2\left(1 + \frac{0.33(0.5)}{5.0}\right)\left(\frac{25.4}{25.4}\right) = 136.87 \text{ kPa}$$

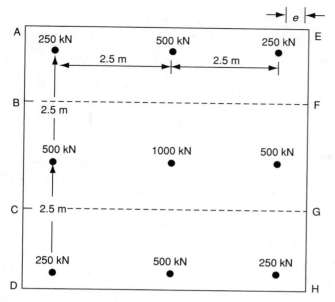

FIGURE 4.5
Illustration for Example 4.2.

By applying Equation (4.6),

$$4000/(5 + 2e)^2 \leq 136.87$$

therefore

$$e > 0.2029 \, \text{m}$$

Hence the mat can be designed with 0.25 m edge space as shown in Figure 4.5.

For the reinforcement design, one can follow the simple procedure of separating the slab into a number of strips as shown in Figure 4.5. Then each strip (BCGF in Figure 4.5) can be considered as an individual beam. The uniform soil reaction per unit length (ω) can be computed as $4000(2.5)/[(5.5)(5.5)] = 330.5 \, \text{kN/m}$. Figure 4.6 indicates the free-body diagram of the strip BCGF in Figure 4.5.

It can be seen from the free-body diagram that the vertical equilibrium of each strip is not satisfied because the resultant downward load is 2000 kN, as opposed to the resultant upward load of 1815 kN. This discrepancy results from the arbitrary separation of strips at the midplane between the loads where nonzero shear forces and moments exist. In fact,

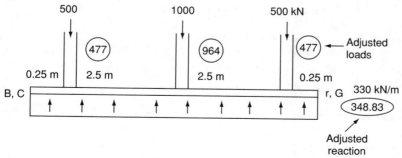

FIGURE 4.6
Free-body diagram for the strip BCGF in Figure 4.5.

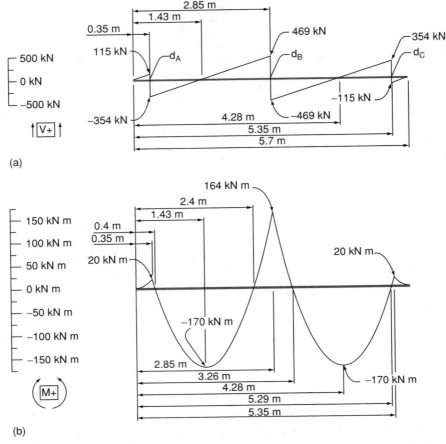

FIGURE 4.7
(a) Shear force and (b) bending moment diagrams for Example 4.2.

one realizes that the resultant upward shear at the boundaries BF and CG (Figure 4.5) accounts for the difference, that is, 185 kN. However, to obtain shear and moment diagrams of the strip BCGF, one can simly modify them as indicated in the figure. This has been achieved by reducing the loads by a factor of 0.954 and increasing the reaction by a factor of 1.051. The two factors were determined as follows:

For the loads, $[(2000 + 1815)/2]/2000 = 0.954$

For the reaction, $1815/[(2000 + 1815)/2] = 1.051$

The resulting shear and moment diagrams are indicated in Figure 4.7.

Then, using Figure 4.7, one can determine the steel reinforcements as well as the mat thickness. This effort is not repeated here since it is discussed in detail in Chapter 5.

4.5 Settlement of Mat Footings

The settlement of mat footings can also be estimated using the methods that were outlined in Section 3.3 and, assuming that they impart stresses on the ground in a manner similar

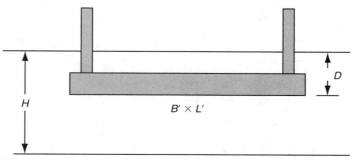

FIGURE 4.8
Immediate settlement computation for mat footings.

to that of spread footings. An example of the estimation of the immediate settlement under a mat footing is provided below (Figure 4.8).

4.5.1 Immediate Settlement

The following expression (Timoshenko and Goodier, 1951) based on the theory of elasticity can be used to estimate the *corner settlement* of a rectangular footing with dimensions of L' and B',

$$s_i = qB' \frac{1 - \nu_s^2}{E_s} \left[I_1 + \frac{1 - 2\nu_s}{1 - \nu_s} I_2 \right] I_F \tag{4.8}$$

where q is the contact stress, B' is the least dimension of the footing, ν_s is the Poisson ratio of the soil, and E_s is the elastic modulus of the soil. Factors I_1, I_2, and I_F are obtained from Table 4.2 and Figure 4.10, respectively, in terms of the ratios $N = H/B'$ (H = layer thickness), $M = L'/B'$ (L' = other dimension of the footing), and D/B.

The same expression (Equation 4.8) can be used to estimate the settlement of the footing at any point other than the corner by approximate partitioning of the footing as illustrated in this example. It must be noted that even if the footing is considered as a combination of several partitions (B' and L'), for determining the settlement of an intermediate (noncorner) location, the depth factor, I_F, is applied for the entire footing based on the ratio D/B.

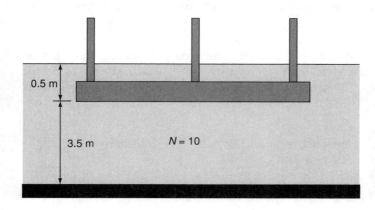

FIGURE 4.9
Illustration for Example 4.3.

TABLE 4.2

I_1 and I_2 for Equation (4.8).

M	1	1.1	1.2	1.3	1.4	1.5	1.6	1.7	1.8	1.9	2
N											
0.1	$I_1 = 0.002$	0.002	0.002	0.002	0.002	0.002	0.002	0.002	0.002	0.002	0.002
	$I_2 = 0.023$	0.023	0.023	0.023	0.023	0.023	0.023	0.023	0.023	0.023	0.023
0.2	0.009	0.008	0.008	0.008	0.008	0.008	0.007	0.007	0.007	0.007	0.007
	0.041	0.042	0.042	0.042	0.042	0.042	0.043	0.043	0.043	0.043	0.043
0.3	0.019	0.018	0.018	0.017	0.017	0.016	0.016	0.016	0.016	0.016	0.015
	0.055	0.056	0.057	0.057	0.058	0.058	0.059	0.059	0.059	0.059	0.059
0.5	0.049	0.047	0.046	0.045	0.044	0.043	0.042	0.041	0.041	0.040	0.040
	0.074	0.076	0.077	0.079	0.080	0.081	0.081	0.082	0.083	0.083	0.084
0.7	0.085	0.083	0.081	0.079	0.078	0.076	0.075	0.074	0.073	0.072	0.072
	0.082	0.085	0.088	0.090	0.092	0.093	0.095	0.096	0.097	0.098	0.098
0.9	0.123	0.121	0.119	0.117	0.115	0.113	0.112	0.110	0.109	0.108	0.107
	0.084	0.088	0.091	0.094	0.097	0.099	0.101	0.103	0.104	0.106	0.107
1	0.142	0.140	0.138	0.136	0.134	0.132	0.130	0.129	0.127	0.126	0.125
	0.083	0.088	0.091	0.095	0.098	0.100	0.102	0.104	0.106	0.108	0.109
3	0.363	0.372	0.379	0.384	0.389	0.393	0.396	0.398	0.400	0.401	0.402
	0.048	0.052	0.056	0.060	0.064	0.068	0.071	0.075	0.078	0.081	0.084
5	0.437	0.452	0.465	0.477	0.487	0.496	0.503	0.510	0.516	0.522	0.526
	0.031	0.034	0.036	0.039	0.042	0.045	0.048	0.050	0.053	0.055	0.058
7	0.471	0.490	0.506	0.520	0.533	0.545	0.556	0.566	0.575	0.583	0.590
	0.022	0.024	0.027	0.029	0.031	0.033	0.035	0.037	0.039	0.041	0.043
9	0.491	0.511	0.529	0.545	0.560	0.574	0.587	0.598	0.609	0.618	0.627
	0.017	0.019	0.021	0.023	0.024	0.026	0.028	0.029	0.031	0.033	0.034
10	0.498	0.519	0.537	0.554	0.570	0.584	0.597	0.610	0.621	0.631	0.641
	0.016	0.017	0.019	0.020	0.022	0.023	0.025	0.027	0.028	0.030	0.031
50	0.548	0.574	0.598	0.620	0.640	0.660	0.678	0.695	0.711	0.726	0.740
	0.003	0.003	0.004	0.004	0.004	0.005	0.005	0.005	0.006	0.006	0.006
100	0.555	0.581	0.605	0.628	0.649	0.669	0.688	0.706	0.722	0.738	0.753
	0.002	0.002	0.002	0.002	0.002	0.002	0.003	0.003	0.003	0.003	0.003
500	0.560	0.587	0.612	0.635	0.656	0.677	0.696	0.714	0.731	0.748	0.763
	0.000	0.000	0.000	0.000	0.000	0.000	0.001	0.001	0.001	0.001	0.001

Example 4.3

If the medium dense sandy soil layer underlying the footing in Example 4.2 is underlain by sound bedrock at a depth of 4.0 m below the surface (Figure 4.9), estimate the average immediate settlement and the maximum differential settlement of the mat footing.

Solution

Let us assume that in this case the sand is normally consolidated. Then, for an average SPT value of 10, E_s is approximately given by $500(N + 15)$ kPa or 12.5 MPa (Table 1.6). A Poisson's ratio of 0.33 can also be assumed in normally consolidated sand (Table 1.4).

Then the uniformly distributed contact stress $= 4000/(5.5)^2 = 132.23$ kPa
D/B for the entire footing $= 0.5/5.5 = 0.09$
From Figure 4.10, for $L/B = 1$, $I_F = 0.85$.

TABLE 4.2

I_1 and I_2 for Equation (4.8). — *Continued*

M	2.5	3.5	5	6	7	8	9	10	15	25	50	100
N												
0.1 $I_1=$	0.002	0.002	0.002	0.002	0.002	0.002	0.002	0.002	0.002	0.002	0.002	0.002
$I_2=$	0.023	0.023	0.023	0.023	0.023	0.023	0.023	0.023	0.023	0.023	0.023	0.023
0.2	0.007	0.006	0.006	0.006	0.006	0.006	0.006	0.006	0.006	0.006	0.007	0.006
	0.043	0.043	0.044	0.044	0.044	0.044	0.044	0.044	0.044	0.044	0.044	0.044
0.3	0.015	0.014	0.014	0.014	0.014	0.014	0.014	0.014	0.014	0.014	0.014	0.014
	0.060	0.061	0.061	0.061	0.061	0.061	0.061	0.061	0.061	0.061	0.061	0.061
0.5	0.038	0.037	0.036	0.036	0.036	0.036	0.036	0.036	0.036	0.036	0.036	0.036
	0.085	0.087	0.087	0.088	0.088	0.088	0.088	0.088	0.088	0.088	0.088	0.088
0.7	0.069	0.066	0.065	0.065	0.064	0.064	0.064	0.064	0.064	0.064	0.063	0.063
	0.101	0.104	0.105	0.106	0.106	0.106	0.106	0.107	0.107	0.107	0.107	0.107
0.9	0.103	0.099	0.097	0.096	0.096	0.095	0.095	0.095	0.095	0.095	0.094	0.094
	0.111	0.115	0.118	0.118	0.119	0.119	0.119	0.119	0.120	0.120	0.120	0.120
1	0.121	0.116	0.113	0.112	0.112	0.112	0.111	0.111	0.111	0.110	0.110	0.110
	0.114	0.119	0.122	0.123	0.123	0.124	0.124	0.124	0.125	0.125	0.125	0.125
3	0.402	0.396	0.386	0.382	0.378	0.376	0.374	0.373	0.370	0.368	0.367	0.367
	0.097	0.116	0.131	0.137	0.141	0.144	0.145	0.147	0.151	0.152	0.153	0.154
5	0.543	0.554	0.552	0.548	0.543	0.540	0.536	0.534	0.526	0.522	0.519	0.519
	0.070	0.090	0.111	0.120	0.128	0.133	0.137	0.140	0.149	0.154	0.156	0.157
7	0.618	0.646	0.658	0.658	0.656	0.653	0.650	0.647	0.636	0.628	0.624	0.623
	0.053	0.071	0.092	0.103	0.112	0.119	0.125	0.129	0.143	0.152	0.157	0.158
9	0.663	0.705	0.730	0.736	0.737	0.736	0.735	0.732	0.721	0.710	0.704	0.702
	0.042	0.057	0.077	0.088	0.097	0.105	0.112	0.118	0.136	0.149	0.156	0.158
10	0.679	0.726	0.758	0.766	0.770	0.770	0.597	0.768	0.753	0.745	0.738	0.735
	0.038	0.052	0.071	0.082	0.091	0.099	0.106	0.112	0.132	0.147	0.156	0.158
50	0.803	0.895	0.989	1.034	1.070	1.100	1.125	1.146	1.216	1.268	1.279	1.261
	0.008	0.011	0.016	0.019	0.022	0.025	0.028	0.031	0.046	0.071	0.113	0.142
100	0.819	0.918	1.020	1.072	1.114	1.150	1.182	1.209	1.306	1.408	1.489	1.499
	0.004	0.006	0.008	0.010	0.011	0.013	0.014	0.016	0.024	0.039	0.071	0.113
500	0.832	0.935	1.046	1.102	1.150	1.191	1.227	1.259	1.382	1.532	1.721	1.879
	0.001	0.001	0.002	0.002	0.002	0.003	0.003	0.003	0.005	0.008	0.016	0.031

Values of I_1 and I_2 to compute the Steinbrenner influence factor I_s for use in Equation (5.16a) for several $N = H/B'$ and $M = L/B$ ratios.

Source: From Bowles, J.E. (2002). *Foundation Analysis and Design.* McGraw-Hill, New York. With permission.

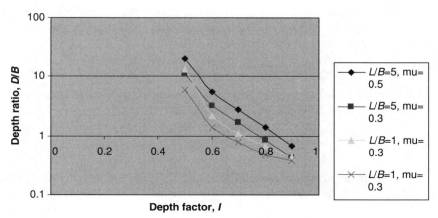

FIGURE 4.10
Plot of the depth influence factor I_F for Equation (4.8).

Therefore, the immediate settlement expression (Equation 4.8) can be simplified to:

$$s_i = 132.23 B' \frac{1 - 0.33^2}{12{,}500} [I_1 + 0.5 I_2] = 9.43 [I_1 + 0.5 I_2] B' \text{ mm}$$

For the corner settlement
$M = L/B = 1.0, N = H/B = 3.5/5.5 = 0.636$
From Table 4.2, $I_1 = 0.066, I_2 = 0.079$

$$s_i = 9.43 [0.066 + 0.5(0.079)](5.5) = 5.471 \text{ mm}$$

For the center settlement
$M = L/B = 2.85/2.85 = 1.0, N = H/B = 3.5/2.85 = 1.23$
From Table 4.2, $I_1 = 0.18, I_2 = 0.079$

$$s_i = 9.43 [0.18 + 0.5(0.079)](2.85)(4) = 23.596 \text{ mm}$$

"4" indicates the four equal partitions required to model the center by superposition of four corners of the partitions.

Maximum angular distortion within the footing $= (23.596 - 5.671)/(2)^{1/2}(2.85)(1000)$
$< 1/200$
It would be safe from any architectural damage.

4.6 Design of Flexible Combined Footings

Flexible rectangular combined footings or mat footings are designed based on the principles of beams and slabs on elastic foundations, respectively. In this approach, the foundation medium is modeled by a series of "elastic" springs characterized by the modulus of vertical subgrade reaction, k_s, and spread in two dimensions. First, it is essential to identify this important empirical soil parameter that is used in a wide variety of designs involving earthen material.

4.6.1 Coefficient of Vertical Subgrade Reaction

The coefficient of subgrade reaction is an empirical ratio between the distributed pressure induced at a point of an elastic medium by a beam or slab and the deflection (w_0) undergone by that point due to the applied pressure, q

$$k_s = q/w_0 \tag{4.9}$$

Egorov (1958) showed that the elastic deformation under a circular area of diameter B carrying a uniformly distributed load of q is given by

$$w_0 = \frac{Bq(1 - \mu_s^2)}{E_s} \tag{4.10}$$

where E_s is the elastic modulus of the medium and μ_s is the Poisson ratio of the medium.
By combining Equations (4.9) and (4.10), one obtains

$$k_s = \frac{E_s}{B(1 - \mu_s^2)} \tag{4.11}$$

Equation (4.11) clearly demonstrates that the subgrade modulus is not a soil parameter and it depends on the size of the loaded area. Thus, if designers use elastic properties of a foundation to determine k_s, then a suitable "B" would have to be used in Equation (4.11) along with E_s, the elastic modulus of foundation soil and μ_s, the Poisson ratio of the foundation soil. Typically, the following B values are used in different cases:

(i) Design of a combined footing — the footing width
(ii) Design of a pile — the pile diameter
(iii) Design of a sheet pile or laterally loaded pile — the pile width or pile diameter.

Similarly, if one uses plate load test results for evaluating k_s, then from a theoretical point of view, it is appropriate to adjust the k_s obtained from plate load tests as:

Clayey soils

$$k_{sf} = k_{sp} \frac{B_p}{B_f} \tag{4.12a}$$

Sandy soils

$$k_{sf} = k_{sp} \left(\frac{B_f + B_p}{2B_f} \right)^2 \tag{4.12b}$$

where B_p is the plate diameter and B_f is the equivalent foundation diameter, which can be determined as the diameter of a circle having an area equal to that of the footing.

Equation (4.12a) is typically used for clayey soil. A slightly modified version of Equation (4.12a) is used for granular soils (Equation 4.12b).

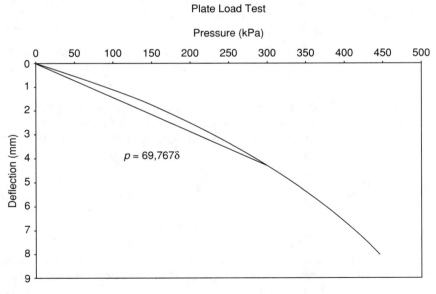

FIGURE 4.11
Plate load test data for Example 4.4.

Example 4.4

Estimate the coefficient of vertical subgrade reaction, k_s, for a 1 m × 1 m footing carrying a 300 kN load, using the data from a plate load test (Figure 4.11) conducted on a sandy soil using 0.45 m × 0.45 m plate.

The contact stress on footing $= 300 \, \text{kN}/(1 \times 1) = 300 \, \text{kPa}$.

The modulus of subgrade reaction for the plate $= 300 \, \text{kPa}/4.3 \, \text{mm} = 69{,}767 \, \text{kN/m}^3$ or 69.77 MN/m^3.

Applying Equation (4.12b)

$$k_{sf} = (69.77) \left(\frac{1 + 0.45}{2} \right)^2 = 36.67 \, \text{MN/m}^3$$

Alternatively, Equation (3.31) can be used to estimate the settlement of the footing at the same stress level of 300 kPa.

$$s_f = s_p \left(\frac{2B_f}{B_f + B_p} \right)^2 \tag{3.31}$$

$$s_f = (4.3) \left(\frac{2}{1 + 0.45} \right)^2 = 8.18 \, \text{mm}$$

Then, the modulus of subgrade reaction for the footing $= 300 \, \text{kPa}/8.18 \, \text{mm} = 36.67$ MN/m^3.

4.6.2 Analysis and Design of Rectangular Combined Footings

Based on the definition of k_s and the common relationship between the distributed load, shear and moment, the following differential equation that governs the equilibrium of a beam on an elastic foundation can be derived:

$$EI \frac{d^4 w}{dx^4} + k_s w = 0 \tag{4.13}$$

In solving the above equation, the most significant parameter associated with the design of beams or slabs on elastic foundations turns out to be the characteristic coefficient of the elastic foundation or the relative stiffness, β, given by the following expression:

For a rectangular beam

$$\beta = \left(\frac{3k_s}{Eh^3} \right)^{1/4} \tag{4.14}$$

where E is the elastic modulus of concrete, k_s is the coefficient of subgrade reaction of the foundation soil usually determined from a plate load test (Section 4.6.1) or Equation (4.10) (Egorov, 1958), and h is the beam thickness.

One can use the following criteria to determine whether a rectangular combined footing must be designed based on the rigid method or the flexible beam method:

For rigid behavior

$$\beta L < \frac{\pi}{4} \tag{4.15a}$$

For flexible behavior

$$\beta L > \pi \tag{4.15b}$$

Owing to their finite size and relatively large thickness, one can expect building foundation mats to generally exhibit rigid footing behavior. Therefore, applications of the flexible footing method are generally limited to concrete slabs used for highway or runway construction. Once β has been evaluated for a particular rectangular combined footing or a mat, the shear, moment, and reinforcing requirements can be determined from nondimensional charts that are based on the solution for a concentrated load (P) applied on a beam on an elastic foundation.

4.6.3 Design of Rectangular Combined Footings Based on Beams on Elastic Foundations

The following expressions can be used, along with Figure 4.12, for the evaluation of moments and shear due to concentrated loads in infinite beams (Table 4.3 and Table 4.4). If the loads/moments are applied at the end, the footing has to be semi-infinite and if the loads/moments are applied at the center, it has to be infinite in order to apply the following equations. The following criteria can be used to verify the semi-infinite or infinite state of a given footing:

For semi-infinite beams, $\beta L > 4$
For infinite beams, $\beta L > 8$

Case (1): Concentrated load at the center (inner column)

$$M = -\frac{P_0}{4\beta} C \tag{4.16a}$$

$$V = \frac{P_0}{2} D \tag{4.16b}$$

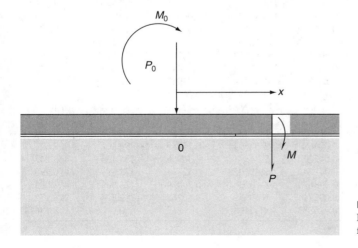

FIGURE 4.12
Moment and shear effects of external forces.

TABLE 4.3

Flexible Footing Shear Computation (Example 4.5)

Distance from A (m)	C Coefficient for Load at A (Figure 4.13a)	Equation (4.17b) for P = 860 kN	B Coefficient for Moment at A (Figure 4.14b)	Equation (4.19b) for M = +174 kN m	C Coefficient for Load at B (Figure 4.13a)	Equation (4.17b) for P = 1380 kN	Shear (kN)
0	−1	−860	0	0	0.00189	2.6082	−857.392
0.25	−0.56191	−483.244	−0.19268	−63.0058	−0.00586	−8.08128	−554.331
0.5	−0.24149	−207.685	−0.29079	−95.0871	−0.01769	−24.4067	−327.179
0.75	−0.02364	−20.3321	−0.32198	−105.289	−0.03435	−47.4044	−173.025
1.0	0.110794	95.28264	−0.30956	−101.226	−0.05632	−77.7147	−83.6581
1.25	0.181547	156.1307	−0.27189	−88.9076	−0.08349	−115.212	−47.989
1.5	0.206788	177.8373	−0.22257	−72.7808	−0.11489	−158.544	−53.4875
1.75	0.201966	173.6904	−0.17099	−55.9141	−0.14822	−204.539	−86.7631
2.0	0.179379	154.2663	−0.12306	−40.2406	−0.17938	−247.543	−133.517
2.25	0.148217	127.4669	−0.08201	−26.8167	−0.202	−278.754	−178.104
2.5	0.114887	98.80321	−0.04913	−16.0641	−0.20679	−285.367	−202.628
2.75	0.083487	71.7992	−0.0244	−7.9784	−0.18155	−250.535	−186.714
3.0	0.056315	48.43071	−0.00703	−2.29749	−0.11079	−152.896	−106.762
3.25	0.034351	29.54221	0.004195	1.371824	0.02364	32.6232	63.53724
3.5	0.017686	15.20979	0.010593	3.463825	0.24149	333.2562	351.9298
3.75	0.005856	5.03605	0.013442	4.39548	0.56191	775.4358	784.8673
4.0	−0.00189	−1.6249	0.013861	4.532652	1	1380	1382.908

TABLE 4.4

Flexible Footing Moment Computation (Example 4.5)

Distance from A (m)	B Coefficient for Load at A (Figure 4.14b)	Equation (4.17a) for P = 860 kN	A Coefficient for Moment at A (Figure 4.14a)	Equation (4.19a) for M = −174 kN m	B Coefficient for Load at B (Figure 4.14b)	Equation (4.17a) for P = 1380 kN	Moment (kN m)
0	0	0	−1	174	0.013861	20.34933	194.3493
0.25	−0.19268	−176.281	−0.94727	164.8247	0.013442	19.7342	8.277393
0.5	−0.29079	−266.04	−0.82307	143.2137	0.010593	15.55158	−107.275
0.75	−0.32198	−294.583	−0.66761	116.1639	0.004195	6.15868	−172.26
1.0	−0.30956	−283.216	−0.50833	88.44872	−0.00703	−10.3207	−205.088
1.25	−0.27189	−248.751	−0.36223	63.02802	−0.0244	−35.8216	−221.545
1.5	−0.22257	−203.63	−0.23835	41.47374	−0.04913	−72.1278	−234.284
1.75	−0.17099	−156.44	−0.14002	24.36289	−0.08201	−120.399	−252.476
2.0	−0.12306	−112.588	−0.06674	11.61288	−0.12306	−180.664	−281.639
2.25	−0.08201	−75.0294	−0.0158	2.749078	−0.17099	−251.03	−323.311
2.5	−0.04913	−44.945	0.016636	−2.89471	−0.22257	−326.755	−374.595
2.75	−0.0244	−22.3224	0.03469	−6.03604	−0.27189	−399.162	−427.52
3.0	−0.00703	−6.42804	0.042263	−7.35374	−0.30956	−454.465	−468.247
3.25	0.004195	3.838171	0.042742	−7.43707	−0.32198	−472.699	−476.298
3.5	0.010593	9.691294	0.038871	−6.7636	−0.29079	−426.909	−423.981
3.75	0.013442	12.29793	0.03274	−5.69668	−0.19268	−282.874	−276.272
4.0	0.013861	12.68172	0.025833	−4.49498	0	0	8.186742

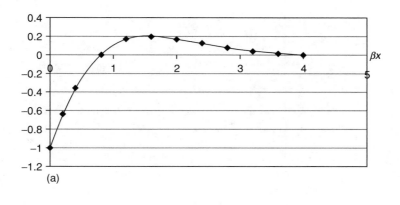

(a)

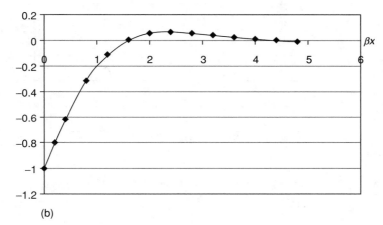

(b)

FIGURE 4.13
Plots of coefficients C and D for infinite beams. (a) Coefficient C, (b) coefficient D.

where the coefficients C and D are defined as

$$C = -(\cos \beta x - \sin \beta x)e^{-\beta x} \qquad (4.16c)$$

$$D = -(\cos \beta x)e^{-\beta x} \qquad (4.16d)$$

and are plotted in Figure 4.13.

Case (2): Concentrated load at the end (outer column) (Figure 4.14)

$$M = \frac{P_0}{\beta}B \qquad (4.17a)$$

$$V = P_0 C \qquad (4.17b)$$

where

$$B = -(\sin \beta x)e^{-\beta x} \qquad (4.17c)$$

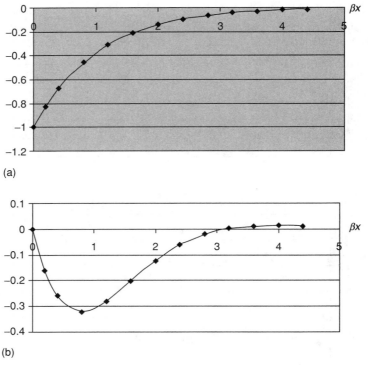

(a)

(b)

FIGURE 4.14
Plots of coefficients A and B for an infinite beam. (a) Coefficient A, (b) coefficient B.

Case (3): External moment at the center (inner column)
In order to determine the moments and shear due to externally applied moments in elastic beams, the Equations(4.16) can be modified in the following manner (Figure 4.15):

$$M = \frac{P_0}{\beta} C(\beta, x) - \frac{P_0}{\beta} C[\beta, (x - dx)]$$

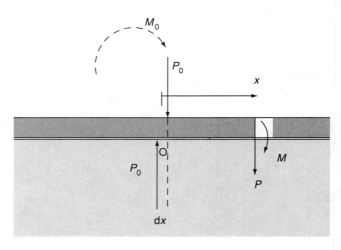

FIGURE 4.15
Transformation of external moment to an external force.

$$M = \frac{P_0}{\beta}[C(\beta,x) - C[\beta,(x - dx)]]$$

$$M = \frac{(P_0 \cdot dx)}{\beta} \frac{[C(\beta,x) - C[\beta,(x - dx)]]}{dx}$$

$$M = \frac{M_0}{\beta} \frac{dC}{dx}$$

$$M = -\frac{M_0}{\beta} D \tag{4.18a}$$

$$V = M_0 A \tag{4.18b}$$

where the coefficient A is defined as

$$A = -(\cos \beta x + \sin \beta x)e^{-\beta x} \tag{4.18c}$$

Case (4): External moment at the end (outer column)
Similarly, Equations (4.17) can be modified to obtain

$$M = -M_0 A \tag{4.19a}$$

$$V = 2\beta M_0 B \tag{4.19b}$$

Example 4.5
Plot the shear and moment diagram in the combined footing in Example 4.1, if the thickness of the concrete slab is 0.1 m. Assume that the soil type is clayey sand and average corrected SPT blow count is 9.

Solution
Since $N = 9$, one can obtain $E_s = 80$ tsf $= 7.6$ MPa (Table 1.7)
h < $[3(7.0)/(27,600)/(B/6.08)^4]^{1/3}$ m or h < 226 mm, and definitive rigid behavior only for
h > $[3(7.0)/(27,600)/(B/4 \times 6.08)^4]^{1/3}$ m or h > 1.437 m.
For cohesionless soils, assume a Poisson's ratio of 0.33.
From, Equation (4.11),

$$k_s = \frac{E}{B(1 - \mu^2)} = 7.6/1.22(1 - 0.33^2) = 7.0 \, MN/m^3$$

Then, by assuming a concrete elastic modulus of 27,600 MPa, β can be determined from Equation (4.14) as

$$\beta = [3(7.0)/(27,600)(0.1)^3]^{1/4} = 0.94 \, m^{-1}$$

But, $L = 6.08$ m
Hence, from Equation (4.15b), $\beta L = 5.7$, which confirms flexible behavior.
Inspection of Equations (4.15) indicates that under the given soil conditions, the footing in Example 4.2 would exhibit definitive flexible behavior for $h < [3(7.0)/(27,600)/(B/6.08)^4]^{1/3}$ m or h < 226 mm, and definitive rigid behavior only for $h > [3(7.0)/(27,600)/(B/4 \times 6.08)^4]^{1/3}$ m or h > 1.437 m.
Figure 4.16(a) and (b) show the shear and moment diagrams.
Analytical expressions for estimating the shear and moments in finite beams on elastic foundations are provided in Scott (1981) and Bowles (2002). When Example 4.1 is solved

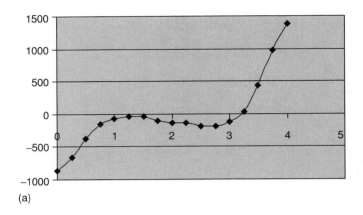

(a)

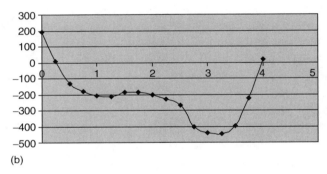

(b)

FIGURE 4.16
(a) Shear distribution (in kN) in the flexible footing. (b) Moment distribution (in kN m) in the flexible footing.

using the corresponding expressions for finite beams, more accurate shear and moment diagrams can be obtained (Figure 4.17).

4.6.4 Analysis of Mat Footings Based on Slabs on Elastic Foundations

The governing differential equation for a free (unloaded) an axisymmetrical structural plate or slab on an elastic foundation is given in polar coordinates (Figure 4.18) by

$$D\left(\frac{\mathrm{d}^4 w}{\mathrm{d}r^4} + \frac{2}{r}\frac{\mathrm{d}^3 w}{\mathrm{d}r^3} - \frac{1}{r^2}\frac{\mathrm{d}^2 w}{\mathrm{d}r^2} + \frac{1}{r^3}\frac{\mathrm{d}w}{\mathrm{d}r}\right) + k_s w = 0 \tag{4.20a}$$

where

$$D = \frac{Eh^3}{12(1-\mu^2)} \tag{4.20b}$$

It is noted that the axisymmetric conditions preclude the need for any θ terms in Equations (4.20). Then, as in the case of beams, the following characteristic coefficient expressing the relative stiffness of the footing can be defined to obtain the solution to Equations (4.20):

$$\beta = \left(\frac{12(1-\mu^2)k_s}{Eh^3}\right)^{1/4} \tag{4.21}$$

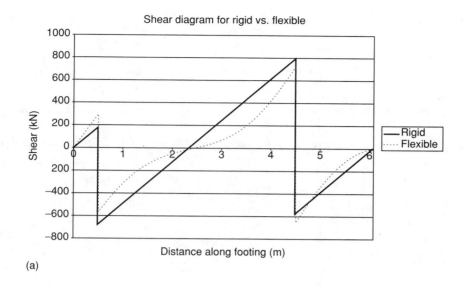

(a)

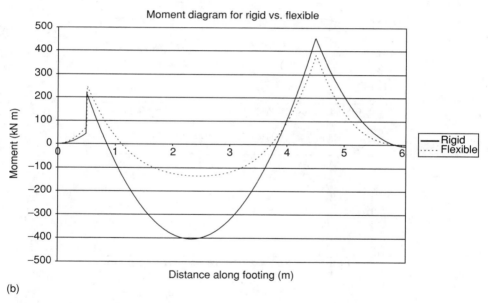

(b)

FIGURE 4.17
(a) Comparison of shear distribution (rigid vs. flexible methods). (b) Comparison of moment distribution (rigid vs. flexible methods).

Then, the radial and tangential moments and shear force at any radial distance r can be obtained from the following expressions:

$$M_r = -\frac{P}{4}C \tag{4.22a}$$

$$M_\theta = -\frac{P}{4}D \tag{4.22b}$$

$$V = -\frac{P\beta}{4}E \tag{4.22c}$$

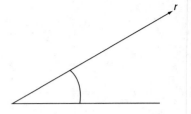

FIGURE 4.18
Illustration of the coordinate system for slabs.

The functions C, D, and E given by Equations (4.23) are plotted in Figure 4.19(a)–(c), respectively.

$$C = Z_3''(\beta r) + \frac{\nu_s}{\beta r} Z_3'(\beta r) \tag{4.23a}$$

$$D = \nu_s Z_3''(\beta r) + \frac{1}{\beta r} Z_3'(\beta r) \tag{4.23b}$$

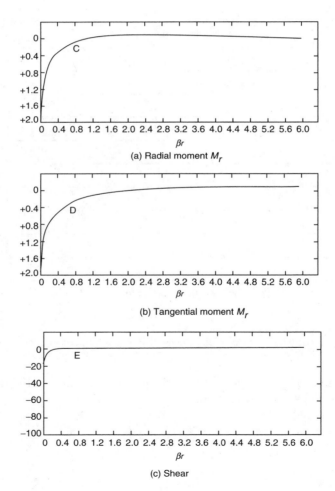

FIGURE 4.19
Radial and tangential moments and shear coefficients in a slab under point load. (From Scott, R.F. (1981). *Foundation Analysis*. Prentice Hall, Englewood Cliffs, NJ. With permission.)

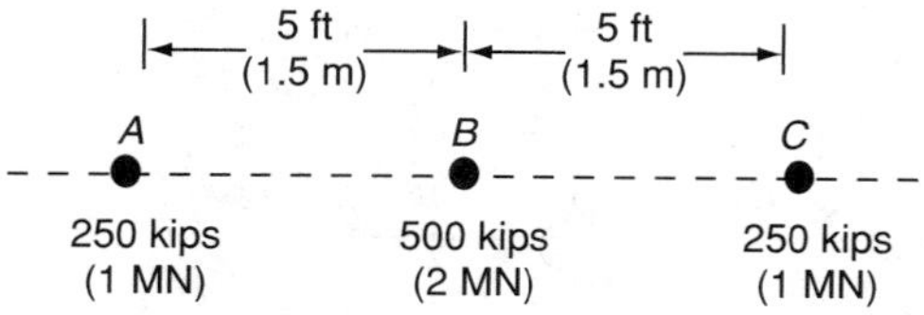

FIGURE 4.20
Illustration for Example 4.6.

$$E = Z_3'''(\beta r) + \frac{1}{\beta r} Z_3''(\beta r) - \frac{1}{(\beta r)^2} Z_3'(\beta r) \qquad (4.23c)$$

where $Z_3(\beta r)$ is the real part of Bessel functions of the third kind and zeroth order.

Example 4.6

(This example is solved in British/US units. Hence the reader is referred to Table 4.1 for conversion of the following units to SI.)

Plot the shear and moment distribution along the columns A, B, and C of the infinite slab of 8 in. thickness shown in Figure 4.20, considering it to be a flexible footing. Assume a coefficient of subgrade reaction of 2600 lb/ft^3. Since $E_c = 5.76 \times 10^8$ psf and $\mu_c = 0.15$, one can apply Equation (4.21) to obtain $\beta = 0.1156$ ft^{-1}.

Using the above results, Figure 4.19(a), and Equation (4.22a), Table 4.5 can be developed for the radial moment (the moment on a cross-section perpendicular to the line ABC in Figure 4.10). These moment values are plotted in Figure 4.21 and the moment coefficients are given in Table 4.5.

4.7 Structural Matrix Analysis Method for Design of Flexible Combined Footings

The stiffness matrix analysis method is also known as the finite element method of foundation analysis due to the similarity in the basic formulation of the conventional finite element method and this method. First, the footing has to be discretized into a number of one-dimensional (beam) elements. Figure 4.22 shows the typical discretization

TABLE 4.5

Moment Coefficients (C) for Flexible Mat

Distance from A (ft)	C Coefficient for Load at A	C Coefficient for Load at B	C Coefficient for Load at C	Radial Moment (kip ft)
0	1.6	0.18	0.0	−122.5
1.0	0.8	0.25	0.02	−32.25
2.0	0.5	0.4	0.05	−84.38
3.0	0.4	0.5	0.08	−92.5
4.0	0.25	0.8	0.1	−121.8
5.0	0.18	1.6	0.18	−222.5
6.0	0.1	0.8	0.25	−12.88
7.0	0.08	0.5	0.4	−92.5
6.0	0.05	0.4	0.5	−84.38
9.0	0.02	0.25	0.8	−32.25
10.0	0.0	0.18	1.6	−122.5

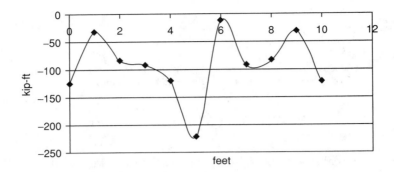

FIGURE 4.21
Moment distribution in the flexible mat.

of a footing in preparation for load–deflection analysis. Based on slope–deflection relations in structural analysis, the following stiffness relation can be written for a free pile element (i.e., 1, 2).

In Figure 4.22 the nodes are indicated in bold numbers while the degrees of freedom are indicated in regular numbers with arrows.

$$
\begin{bmatrix} P_1 \\ P_2 \\ P_3 \\ P_4 \end{bmatrix} = \begin{bmatrix} 4EI/L & 6EI/L^2 & 2EI/L & -6EI/L^2 \\ 6EI/L^2 & 12EI/L^3 & 6EI/L^2 & -12EI/L^3 \\ 2EI/L & 6EI/L^2 & 4EI/L & -6EI/L^2 \\ -6EI/L^2 & -12EI/L^3 & -6EI/L^2 & 12EI/L^3 \end{bmatrix} \begin{bmatrix} w_1 \\ w_2 \\ w_3 \\ w_4 \end{bmatrix}^{\mathrm{T}}
\tag{4.24}
$$

where P_i (*i* even) are the internal loads on beam elements concentrated (lumped) at the nodes; P_i (*i* odd) are the internal moments on beam elements concentrated (lumped) at the nodes; w_i (*i* even) is the nodal deflection of each beam element; w_i (*i* odd) is the nodal rotations of each beam element; and EI is the stiffness of the footing $[=(1/12)Bh^3]$; L is the length of each pile element and h is the thickness of the footing.

Since the beam is considered to be on an elastic foundation, the modulus of vertical subgrade reaction k_s at any point (p) can be related to the beam deflection (w) at that point by the following expression:

$$
p = k_s w
\tag{4.25}
$$

Hence, the spring stiffness K_j (force/deflection) can be expressed conveniently in terms of the modulus of vertical subgrade reaction k_s as follows:

$$
K_j = LBk_s
\tag{4.26}
$$

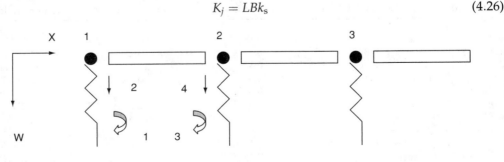

FIGURE 4.22
Discretization of footing for stiffness analysis method.

The element stiffness matrices expressed by equations such as Equation (4.24) can be assembled to produce the global stiffness matrix $[K]$ using basic principles of structural matrix analysis. During the assembling process, the spring stiffness K_j of each node can be added to the corresponding diagonal element of $[K]$. Hence,

$$[P] = [K][w] \qquad (4.27)$$

where $[P]$ and $[w]$ are the load and deflection vectors, respectively.

Next, knowing the global force vector one can solve Equation (4.24) for the global deflection vector. Finally, the moments and the shear forces within the pile can be determined by substituting the nodal deflections in the individual element equations such as Equation (4.24).

Example 4.7

Determine the external load vector for the footing shown in Figure 4.23(a).

Solution

By observing the column layout and the total length of the footing, it is convenient to consider the footing as consisting of three beam elements, each of length 1.2 m. The resulting nodes are shaded in Figure 4.23(a) and the nodal numbers are indicated in bold. In Figure 4.23(b), the eight degrees of freedom are also indicated with arrows next to them. Then the external force vector can be expressed as

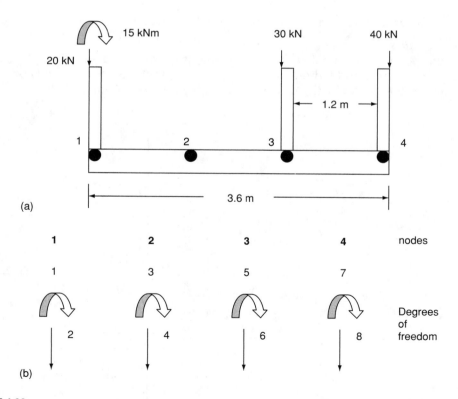

(a)

(b)

FIGURE 4.23
(a) Illustration for Example 4.7. (b) Illustration of the degrees of freedom.

$$[P] = [15 \ 20 \ 0 \ 0 \ 0 \ 30 \ 0 \ 40]^{\mathrm{T}}$$

4.8 Finite Difference Method of Flexible Mat Footing Design

The governing equation for loaded slabs on elastic foundations (Equation 4.20) can be expressed in Cartesian coordinates as

$$D\left(\frac{\partial^4 w}{\partial x^4} + \frac{\partial^4 w}{\partial x^2 \partial y^2} + \frac{\partial^4 w}{\partial y^4}\right) + k_s w = \frac{\partial^2 P}{\partial x \partial y} \tag{4.28}$$

where P and w are the external concentrated load and the deflection at (x,y), respectively, and D is given by Equation (4.20b). By expressing the partial derivatives using their numerical counterparts (Chapra and Canale, 1988), the above equation can be expressed in the finite difference form as follows:

$$\begin{aligned}
20w(x,y) &- 8[w(x+h,y) + w(x-h,y) + w(x,y+h) + w(x,y-h)] \\
&+ 2[w(x+h,y+h) + w(x+h,y-h) + w(x-h,y+h) + w(x-h,y-h)] \\
&+ [w(x+2h,y) + w(x,y+2h) + w(x-2h,y) + w(x,y-2h)] \\
&= \frac{1}{D}[k_s h^4 w(x,y) + P(x,y)h^2]
\end{aligned} \tag{4.29}$$

where h is the grid spacing chosen on the mat in both X and Y directions (Figure 4.24).

Solution procedure
For simplicity, let us assume that the column layout is symmetric.
Step 1. Establish the x and y directional grid points ($n \times n$) on the mat. If n is odd, from basic algebra, the reader will realize that there would be $(1/8)(n + 1)(n + 3)$ unique

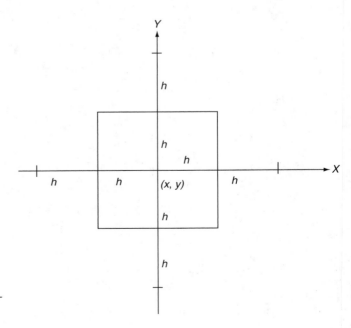

FIGURE 4.24
Finite difference formulation for deflection (w) of a given node (x,y).

columns and therefore, $(1/8)(n + 1)(n + 3)$ unknown nodal deflections on the mat. As an example, in the nine column layout (i.e., $n^2 = 9$, $n = 3$), shown in Figure 4.5, there are only $(1/8)(3+1)(3+3) = 3$ unique columns/nodes in the mat. All of the other columns would be similar to one of the unique columns. On the other hand, if n is even, there would be only $(1/8)(n)(n + 2)$ unknown nodal deflections. As an example, if the mat had a 16-column layout (i.e., $n^2 = 16$, $n = 4$), there would be only $(1/8)(4)(4+2) = 3$ unique columns/nodes (1, 2, and 3 in Figure 4.25) on that mat.

Step 2. Repeatedly apply Equation (4.29) to all of the unique nodal points, i.e., $(1/8)(n + 1)(n + 3)$ when n is odd or $(1/8)(n)(n + 2)$ when n is even, to produce as many equations in each case. However, one realizes that when Equation (4.29) is applied to edge nodes and then immediately inside the edge, a number of other unknown deflections from imaginary (or dummy) nodes outside the mat are also introduced. The unique ones out of them are illustrated as 4, 5, 6, 7, and 8 in Figure 4.25 for a 16-column layout. This introduces a further $(n + 2)$ unknowns when n is odd and $(n + 1)$ unknowns when n is even. As an example, in Figure 4.25, $n = 4$, since the additional number of unknowns is 5 (i.e., the deflections at 4, 5, 6, 7, and 8).

Step 3. Obtain additional equations by knowing that $M_x(x,y)$ (Equation 4.30a) along the two edges parallel to the X axis and $M_y(x,y)$ (Equation 4.30b and Figure 4.26) along the two edges parallel to the Y axis are zero. It is noted that this includes the condition of both above moments being zero at the four corners of the mat. The numerical form of the moments is given below:

$$M_x(x,y) = D\left[\frac{\partial^2 w}{\partial x^2} + v\frac{\partial^2 w}{\partial y^2}\right] = D\left[\frac{w(x+h,y) + w(x-h,y) - 2w(x,y)}{h^2}\right.$$
$$\left. + v\frac{w(x,y+h) + w(x,y-h) - 2w(x,y)}{h^2}\right] \qquad (4.30a)$$

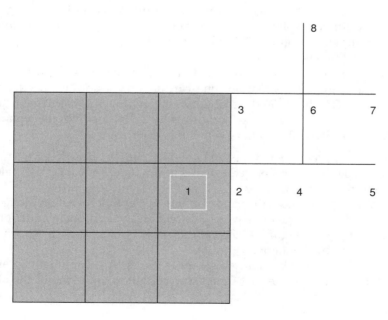

FIGURE 4.25
Illustration of internal and external nodes for an even number of columns.

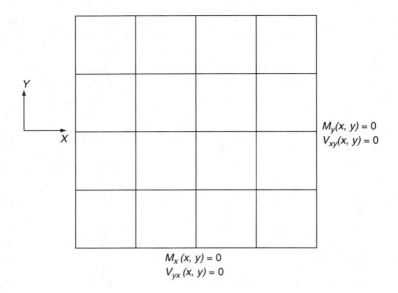

FIGURE 4.26
Layout of nodal points and boundary conditions.

$$M_y(x,y) = D\left[\nu\frac{\partial^2 w}{\partial x^2} + \frac{\partial^2 w}{\partial y^2}\right] = D\left[\frac{\nu w(x+h,y) + w(x-h,y) - 2w(x,y)}{h^2}\right.$$
$$\left. + \frac{w(x,y+h) + w(x,y-h) - 2w(x,y)}{h^2}\right] \qquad (4.30b)$$

In the case of a symmetric footing one can only apply this condition along one edge to obtain $\frac{1}{2}(n+1)$ additional equations when n is odd and $\frac{1}{2}n$ additional equations when n is even. As an example, in Figure 4.25, when $n = 4$, $M_y(x,y)$ is zero for nodes 2 and 3 producing two additional equations.

Furthermore, both of the above moments are zero at the four corners of the mat. Therefore, in the symmetric situation, one can obtain an additional equation for one corner. In Figure 4.25, this would be obtained by setting the moment in the diagonal direction to zero at node 3. Thus, step 3 produces $\frac{1}{2}(n+1) + 1$ additional equations when n is odd and $\frac{1}{2}n + 1$ additional equations when n is even.

Step 4. Obtain another set of equations by knowing that the shear force $\tau_{xy}(x,y)$ (Equation 4.31) along the two edges parallel to the Y axis and $\tau_{yx}(x,y)$ (Equation 4.31) along the two edges parallel to the X axis are zero.

In the case of a symmetric footing one can only apply this condition along one edge to obtain $\frac{1}{2}(n+1)$ additional equations when n is odd and $\frac{1}{2}n$ additional equations when n is even. As an example, in Figure 4.25 when $n = 4$, $\tau_{xy}(x,y)$ is zero for nodes 2 and 3 producing two additional equations.

It must be noted that since both the above shear forces are complimentary and therefore equal at the four corners of the mat under any condition, only one of the above can be applied for the corner node.

$$M_{xy}(x,y) = D(1-\nu)\frac{\partial^2 w}{\partial x \partial y} = M_{yx}(x,y)$$

$$= D(1-\nu)\frac{1}{4h^2}[w(x+h,y+h) - w(x-h,y+h) - w(x+h,y-h) + w(x-h,y-h)] \quad (4.31)$$

It is seen that step 4 also produces $\frac{1}{2}(n+1)$ additional equations when n is odd and $\frac{1}{2}n$ additional equations when n is even.

Step 5. From step 2, the total number of unknown deflections would be as follows:

n odd $(1/8)(n+1)(n+3) + (n+2)$

n even $(1/8)(n)(n+2) + n + 1$

From steps 3 and 4, the total number of equations available would be as follows:

n odd$(1/8)(n+1)(n+3) + \frac{1}{2}(n+1) + 1 + \frac{1}{2}(n+1) = (1/8)(n+1)(n+3) + (n+2)$

n even $(1/8)(n)(n+2) + \frac{1}{2}n + 1 + \frac{1}{2}n = (1/8)(n)(n+2) + n + 1$

It can be seen that the number of unknown deflections and the number of equations available would be the same.

Hence the deflections can be determined using a matrix based method or solution of simultaneous equations.

4.9 Additional Design Examples

Example 4.8
Design a combined footing to carry the loads shown in Figure 4.27

(a) Plot the shear and moment diagrams assuming that the columns are 0.3 m in diameter.
(b) Estimate the immediate settlement of the center and the corners of the footing using the elastic equation.

Suitable elastic parameters of the sandy soil can be obtained from Chapter 2. Also use the SPT–CPT correlations in Chapter 2

Total vertical load $= \bar{R} = 200 + 150 + 150 = 500$ KN

Location from left most column $= \bar{x} = \dfrac{150 \times 2.5 + 150 \times 4.5}{500}$
(by taking moment) $= 2.1$ m

Footing length without eccentricity:

Total length $= 2.9 + 2.9 = 5.8\,\text{m} = L$

$q_c = 4.5\,\text{MPa}, \quad R_f = f_s/q_c = \dfrac{4.5\,\text{kPa}}{4.5\,\text{MPa}} \times \dfrac{100}{100} = 0.1\%$ (friction ratio)

$q_c = 45$ bar

From Figure 2.20, for $q_c = 45$ bar, $R_f = 0.1\%$

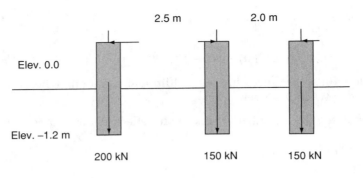

FIGURE 4.27
Illustration for Example 4.8.

$$\frac{q_c}{N} = 4 \text{ (sand to sandy silt, 4)}$$

$$\therefore N = \frac{q_c}{4} = \frac{45}{4} \approx 11$$

Allowable stress on soil, $q_a \le \frac{q_{ult}}{F} = 30N_{55}$ (kPa) $\hspace{3cm}$ (4.4)

$$\therefore q_a = 30(11) = 330 \text{ kPa}$$

Width of footing, $B = \dfrac{500}{B \times L} = 330$

$$\therefore B = \frac{500}{300 \times 5.8} = 0.29 \text{ m (use 1 m width)}$$

a) Plot the shear and moment diagrams

   The distributed reaction per unit length (1 m) on the footing can be computed as
   $$w = \frac{500}{5.8} \text{ kN/m} = 86.2 \text{ kN/m}$$

b) Immediate settlement

$$S_i = qB' \frac{1 - V_s^2}{E_s}\left[I_1 + \frac{1 - 2V_s}{1 - V_s} I_2\right] I_F \hspace{3cm} (4.8)$$

$q = \dfrac{500}{5.8 \times 1} = 86.2 \text{ kPa}, \ B' = 1 \text{ m}, \ <_s = 0.33$

$E_s = 500(N + 15) = 500(11 + 15) = 13000 \text{ kPa}$

$\dfrac{D}{B} = \dfrac{1.2}{1.0} = 1.2 \text{ (depth factor)}$

For
$$\left.\begin{array}{l}\dfrac{D}{B} = 1.2 \\[2mm] \dfrac{L}{B} = 5.8 \\[2mm] \mu = 0.33\end{array}\right\} \text{from Figure 4.10, } I_F = 0.74$$

For corner settlement:

$$\left.\begin{array}{l}N = \dfrac{H}{B'} = \dfrac{3}{1} = 3.0 \\[3mm] M = \dfrac{L}{B} = \dfrac{5.8}{1} = 5.8\end{array}\right\} \text{from Table 4.2, } I_1 = 0.383, I_2 = 0.136$$

$$S_{corner} = (86.2)(1)\frac{1 - 0.33^2}{13000}\left[0.383 + \frac{1 - 2(0.33)}{1 - 0.33}(0.136)\right](0.74)$$

$$= 1.976 \times 10^{-3} \text{ m} \approx 1.97 \text{ mm.}$$

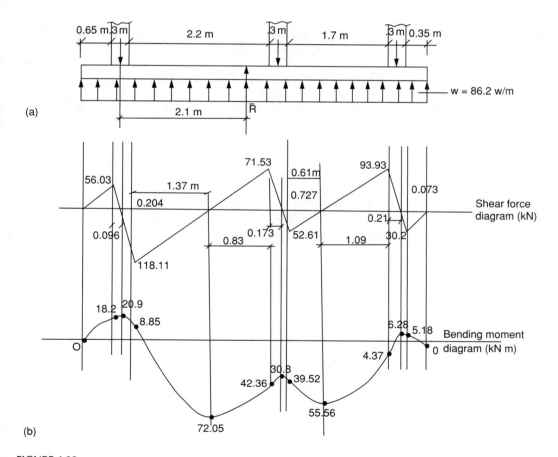

FIGURE 4.28
(a) Loads, (b) shear force, and bending moments for Example 4.8.

For center settlement:

$$N = \frac{3}{0.5} = 6$$
$$M = \frac{2.9}{0.5} = 5.8$$

$\left. \right\}$ from Table 4.2, $I_1 = 0.609, I_2 = 0.110$

$$S_{center} = (86.2)(0.5)\frac{1 - 0.33^2}{13000}\left[0.609 + \frac{1 - 2(0.33)}{1 - 0.33}(0.11)\right](0.74)$$

$$= 5.9 \times 10^{-3} \text{ m} \approx 5.9 \text{ mm}.$$

References

ACI Committee 436 Report (1966). Suggested design procedures for combined footings and mats, *ACI Journal*, October.

Bowles, J.E. (2002). *Foundation Analysis and Design*. McGraw-Hill, New York.

Chapra S.C. and Canale R.P. (1988). *Numerical Methods for Engineers*. McGraw-Hill, New York.

Das, B.M. (2002). *Principles of Foundation Engineering*. PWS Publishing, Boston, MA.

Egorov, E. (1958). Concerning the question of calculations for base under foundations with footings in the form of rings. *Mekhanica Gruntov*, Sb. Tr. No. 34, Gosstroiizdat, Moscow.

Scott, R.F. (1981). *Foundation Analysis*. Prentice Hall, Englewood Cliffs, NJ.

Timoshenko, S. and Goodier, J.M. (1951) *Theory of Elasticity*, McGraw-Hill, New York, 2nd edn.

5

Structural Design of Foundations

Panchy Arumugasaamy

CONTENTS

5.1 Introduction

Foundation substructures are structural members used to support walls and columns to transmit and distribute their loads to the ground. If these loads are to be properly transmitted, the substructure must be designed to prevent excessive settlement or rotation and to minimize differential settlement. In addition, it should be designed in such a way that the load bearing capacity of the soil is not exceeded and adequate safety against sliding and overturning is assured.

Cumulative floor loads of a building, a bridge, or a retaining wall are supported by the foundation substructure in direct contact with soil. The soil underneath the substructure becomes compressed and deformed during its interaction with the substructure. This deformation is the settlement that may be permanent due to dead loads or may be elastic due to transition live loads. The amount of settlement depends on many factors, such as the type of soil, the load intensity, the ground water conditions, and the depth of substructure below the ground level.

If the soil bearing capacity is different under different isolated substructures or footings of the same building a differential settlement will occur. Due to uneven settlement of supports the structural system becomes over stressed, particularly at column beam joints.

Excessive settlement may also cause additional bending and torsional moments in excess of the resisting capacity of the members, which could lead to excessive cracking and failures. If the total building undergoes even settlement, little or no overstressing occurs.

Therefore, it is preferred to have the structural foundation system designed to provide even or little settlement that causes little or no additional stresses on the superstructure. The layout of the structural supports varies widely depending upon the site conditions. The selection of the type of foundation is governed by the site-specific conditions and the optimal construction cost. In designing a foundation, it is advisable to consider different types of alternative substructures and arrive at an economically feasible solution. In the following sections, the design of a number of commonly used reinforced concrete foundation system types is presented. The reader is advised that, in keeping with the structural design practices in the United States, the English standard measurement units are adopted in the design procedures outlined in this chapter. However, the conversion facility in Table 5.1 is presented for the convenience of readers who are accustomed to the SI units.

5.2   Types of Foundations

Most of the structural foundations may be classified into one of the following types:

1. *Isolated spread footings:* These footings are used to carry individual columns. These may be square, rectangular, or occasionally circular in plan. The footings may be of uniform thickness, stepped, or even have sloped top (Figure 5.1) and reinforced in both directions. They are one of most economical types of foundation, when columns are spaced at a relatively long distance.

2. *Wall footings:* They are used to support partitions and structural masonry walls that carry loads from floors and beams. As shown in Figure 5.2, they have a limited width and continuous slab strip along the length of the wall. The critical section for bending is located at the face of the wall. The main reinforcement is placed perpendicular to the wall direction. Wall footings may have uniform thickness, be stepped, or have a sloped top.

TABLE 5.1

Unit Conversion Table

From English	To SI	Multiply by	Quantity	From SI	To English	Multiply by
lbs/ft^3	N/m^3	157.1	Force/unit-volume	N/m^3	lbs/ft^3	0.0064
kips/ft^3	kN/m^3	157.1		kN/m^3	kips/ft^3	0.0064
lb-in.	N mm	112.98	Moment; or energy	N mm	lb-in.	0.0089
kip-in.	kN mm	112.98		kN mm	kip-in.	0.0089
lb-ft	N m	1.356		N m	lb-ft	0.7375
kip-ft	kN m	1.356		kN m	kip-ft	0.7375
ft-lb	Joule	1.356		Joule	ft-lb	0.7375
ft-kip	kJ	1.356		kJ	ft-kip	0.7375
sec/ft	sec/m	3.2808	Damping	sec/m	sec/ft	0.3048
Blows/ft	Blows/m	3.2808	Blow count	Blows/m	Blows/ft	0.3048

Source: Courtesy of the New York Department of Transportation.

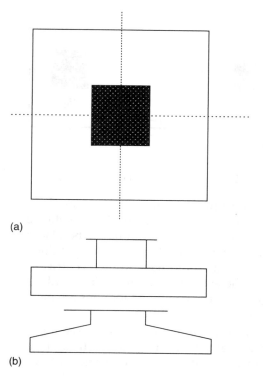

(a)

(b)

FIGURE 5.1
Isolated spread footing: (a) plan; (b) elevation.

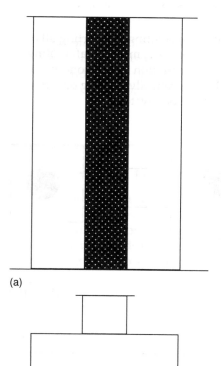

(a)

(b)

FIGURE 5.2
Wall footing: (a) plan; (b) elevation.

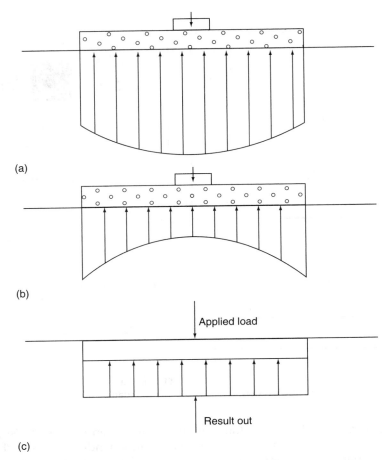

FIGURE 5.7
Pressure distribution under regular footings in different soil types: (a) pressure distribution in sandy soil; (b) pressure distribution in clayey soil; (c) simplified pressure distribution.

across the contact area between the footing and the soil. If a column footing is loaded with axial load P at or near the center of the footing, as shown in Figure 5.8, the contact pressure q under the footing is simply P/A. On the other hand, if the column is loaded with an axial load P and a moment of M, the stress under the footing is

$$q = P/A \pm MY/I \qquad (5.1)$$

where q is the soil pressure under the footing at any point, P is the applied load, A is the area of footing $= BD$ (B is the width of footing and D is the length of footing), M is the moment, Y is the distance from centroidal axis to point where the stress is computed, and I is the second moment of area of the footing ($I = BD^3/12$).

If e is the eccentricity of the load relative to the centroidal axis of the area A, the moment M can be expressed as Pe. The maximum eccentricity e for which Equation (5.1) applies is the one that produces $q = 0$ at some point. However, the larger eccentricities will cause a part of the footing to lift of the soil. Generally, it is not preferred to have the footing lifted since it may produce an uneconomical solution. In cases where a larger moment is involved, it is advisable to limit the eccentricity to cause the stress $q = 0$ condition at

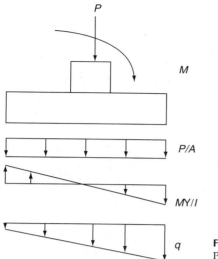

FIGURE 5.8
Pressure distribution under eccentric footings.

the edge of the footing. This will occur when the eccentricity e falls within the middle third of the footing or at a limit $B/6$ or $D/6$ from the centroidal point of footing. This is referred to as the *kern distance*. Therefore, the load applied within the kern distance will produce compression under the entire footing.

5.4 Determination of the Size of Footing

The footings are normally proportioned to sustain the applied factored loads and induced reactions that include axial loads, moments, and shear forces that must be resisted at the base of the footing or pile cap, in accordance with appropriate design requirements of the applicable codes. The base area of the footing or the number and the arrangement of piles are established after the permissible soil pressure or the permissible pile capacity has been determined by the principles of soil mechanics as discussed in Chapters 3, 4, and 6, on the basis of unfactored (service) loads such as dead, live, wind, and earthquake, whatever the combination that governs the specific design. In the case of footings on piles, the computation of moments and shear could be based on the assumption that the reaction from any pile is concentrated at the pile center.

5.4.1 Shear Strength of Footings

The strength of footing in the vicinity of the columns, concentrated loads, or reactions is governed by the more severe of two conditions: (a) wide beam action with each critical section that extends in a plane across the entire width needed to be investigated; and (b) footing subjected to two-way action where failure may occur by "punching" along a truncated cone around concentrated loads or reactions. The critical section for punching shear has a perimeter b_0 around the supported member with the shear strength computed in accordance with applicable provision of codes such as ACI 11.12.2. Tributary areas and corresponding critical sections for both wide-beam and two-way actions for isolated footing are shown in Figure 5.9.

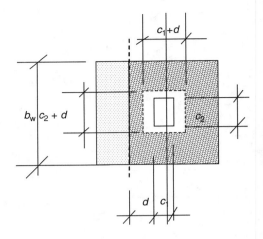

FIGURE 5.9
Tributary area and critical section for shear.

For footing design with no shear reinforcement, the shear strength of concrete V_c (i.e., $V_n = V_c$) is considered as the smallest of the following for two way action.

$$V_c = (2 + 4/\beta_c)b_0 d(f_c')^{0.5} \quad \text{(ACI formula 11 − 36)} \tag{5.2}$$

$$V_c = (2 + \alpha_s/(b_0/d))b_0 d(f_c')^{0.5} \quad \text{(ACI formula 11 − 37)} \tag{5.3}$$

$$V_c = 4b_0 d(f_c')^{0.5} \quad \text{(ACI formula 11 − 38)} \tag{5.4}$$

where b_0 is the perimeter of critical section taken at $d/2$ from the loaded area

$$b_0 = 2(c_1 + c_2) + 4d \tag{5.5}$$

d is the effective depth of the footing, β_c is the ratio of the long side to the short side of the loaded area, and $\alpha_s = 40$ for interior columns, 30 for edge columns, and 20 for corner columns.

In the application of above ACI Equation 11-37, an "interior column" is applicable when the perimeter is four-sided, an "edge column" is applicable when the perimeter is three-sided, and finally a "corner column" is applicable when the perimeter is two-sided.

Design Example 5.1
Design for base area of footing (Figure 5.10).

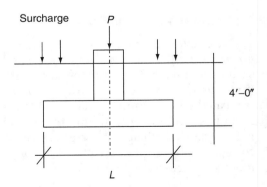

FIGURE 5.10
Illustration for Example 5.1.

Problem Statement

Determine the base area A_f required for a square footing of a three-storey building interior column with the following loading conditions:

Service dead load $= 400$ kips
Service live load $= 280$ kips
Service surcharge (fill) $= 200$ psf
Permissible soil pressure $= 4.5$ ksf
Column dimensions $= 24 \times 15$ in.

Solution

The base area of the footing is determined using service (unfactored) loads with the net permissible soil pressure.

1. *Determination of base area:*

Let us assume that the bottom of the footing is 4 ft below the ground level:

Average weight of soil $= 125.00$ pcf
Total weight of surcharge $= (0.125 \times 4) + 0.2 = 0.70$ ksf
Permissible soil pressure $= 4.50$ ksf
Net permissible soil pressure $= 4.5 - 0.7 = 3.80$ ksf

Given

Service DL $= 400.00$ kips
Service LL $= 280.00$ kips

Required base area of footing:

$$A_f = L^2 = \frac{400 + 280}{3.80} = 178.95 \text{ ft}^2$$

$$L = 13.38 \text{ ft}$$

Use a $13'-6'' \times 13'-6''$ square footing, $A_f = 182.25$ ft^2.

2. *Factored loads and soil reaction:*

To proportion the footing for strength (depth and area of steel rebar) factored loads are used:

Safety factor for DL $= 1.40$
Safety factor for LL $= 1.70$

$$P_u = 1.4(400) + 1.7(280) = 1036 \text{ kips}$$
$$q_s = 5.68 \text{ ksf}$$

Example 5.2

For the design conditions of Example 5.1, determine the overall thickness of footing and the required steel reinforcement given that $f'_c = 3,000$ psi and $f_y = 60,000$ psi (Figure 5.11).

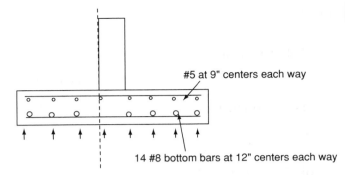

FIGURE 5.12
Determination of reinforcement for
Example 5.2.

#5 at 9" centers each way

14 #8 bottom bars at 12" centers each way

the moment at a foot distance from the edge of the footing. Hence, it is good practice to have rebars bent up at the end so that it provides a mechanical means of locking the bar in place.

The basic development length

$$l_d = 0.04 A_b f_y / (f_c')^{0.5} = 0.04 \times 0.79 \times 60/(3)^{0.5} = 34.62 \text{ in.}$$

Clear spacing of bars $= (13.5 \times 12 - 2 \times 3 - 1.0)/(14 - 1) = 11.92$ in. $> 3D_b$, OK
Hence multiplier, ACI 12.2.3.1-3 $= 1$
Since cover is not less than $2.5D_b$ a reduction factor of 0.8 may be used. Hence, $l_d = 27.69$ in.
In any case l_d should not be less than 75% of the basic development length $= 25.96$ in.
Hence, provide a development length $= 28$ in.
But in reality the bar has a hook at the end. Hence it is satisfactory.

(iv) Temperature reinforcement (Figure 5.12):
It is good practice to provide a top layer of minimum distribution reinforcement to avoid cracking due to any rise in temperature caused by heat of hydration of cement or premature shrinkage of concrete.
It is advised to provide at least the minimum area of steel required in both directions.

$$A_s \text{ min} = 0.11 A_g / 2 f_y = 0.11 \times 12 \times 36/(2 \times 60) \text{ (AASHTO LRFD provision)}$$
$$= 0.40 \text{ in.}^2/\text{ft}$$

Area of #5 $= 0.31$ in.^{2}

Provide #5 at 9 in. centers, $A_s = 0.41$ in.^{2}/ft

Use #5 bars at 9 in. centers in both directions.

5.5   Strip or Wall Footings

A wall footing generally has cantilevers out on both sides of the wall as shown in Figure 5.13. The soil pressure causes the cantilever to bend upward and, as a result, reinforcement is required at the bottom of the footing, as shown in Figure 5.13.

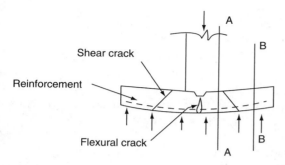

FIGURE 5.13
Structural action in a wall footing.

The critical sections for design for flexure and anchorage are at the face of the wall (section A–A in Figure 5.13). One-way shear is critical at the section at a distance d from the face of the wall (section B–B in Figure 5.13).

Example 5.3
A 8-in. thick wall is a part of a vertical load carrying member of an eight-storey condominium and hence carries seven floors and the roof. The wall carries a service (unfactored) dead load of 1.5 kips per foot per floor including the roof and a service live load of 1.25 kips per foot per floor. The allowable soil net bearing pressure is 5.0 ksf at the level of the base of the footing, which is 5 ft below the ground surface. The floor-to-floor height is 10 ft including the roof. Design the wall footing assuming $f_c' = 3,000$ psi and $f_y = 60,000$ psi.

Solution
(1) Estimate the total service load. Consider 1-ft width of the wall
    Dead load from self weight of the wall $W_{d1} = (8 \times 10 + 5$: height$) \times (8/12$: thickness wall$) \times (0.15$ kips/ft$)$

$$W_{d1} = 8.50 \text{ kips/ft}$$

Dead load from floors

$$W_{d2} = 8 \times 1.5 = 12.00 \text{ kips/ft}$$
$$\text{Total DL} = W_{d1} + W_{d2} = 8.5 + 12.0$$
$$= 20.50 \text{ kips/ft}$$
$$\text{Liveload} = 8 \times 1.25$$
$$= 10.00 \text{ kips/ft}$$

Note that the net bearing pressure at the footing level is given, and hence the self-weight of the footing does not need to be considered.

(2) Compute the width of the wall

$$\text{Width required} = \frac{20.5 \text{ kips} + 10.0 \text{ kips}}{5 \text{ ksf}} = 6.1 \text{ ft}$$

Try a footing 6 ft 4 in. wide; $w = 6.33$

Factored net pressure $q_u = \dfrac{1.4 \times 20.5 + 1.7 \times 10.0}{6.33} = 7.22$ ksf

In the design of the concrete and reinforcement, we will use $q_u = 7.22$ ksf.

(3) Check for shear

Shear usually governs the thickness of footing. Only one-way shear is significant for a wall footing. We need to check it at a distance d away from the face of the wall (section B–B in Figure 5.13).

Now let us assume a thickness of footing = 16 in.

$$d = 16 - 3 \text{ (cover)} - 0.5 \text{ (bar diameter)}$$
$$= 12.5 \text{ in.}$$

Clear cover (since it is in contact with soil) = 3 in.

$\phi = 0.85$

$f_c' = 3000\,\text{psi}$

$f_y = 60,000\,\text{psi}$

$V_u = 7.22\ \text{ksf} \times (21.50/12 \times 1)\ \text{ft}^2 = 12.9\ \text{kips/ft}$

$$\phi V_c = \phi(2f_c'^{0.5} b_w d) = 0.85 \times 2 \times 3,000^{0.5} \times 12 \times 12.5/1,000$$
$$= 13.97 \text{ kips/ft}$$

Since $\phi V_c > V_u$ the footing depth is satisfactory.

(4) Design of reinforcement

The critical section for moment is at the face of the wall section A–A in Figure 5.13. The tributary area for moment is shown shaded in Figure 5.14.

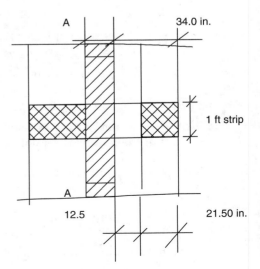

FIGURE 5.14
Plan view of footing (Example 5.3).

$$M_u = \frac{7.22(43.02/12)^2 \times 1}{2} = 29.01 \text{ ft-kips/ft}$$

$$M_u = \phi M_n = \phi A_s f_y jd$$

Let us assume $j = 0.9$, $jd = 11.25$

$$A_s = \frac{12,000 \, M_u}{\phi f_y jd}, \quad \text{where } \phi = 0.9$$

$$A_s = 0.57 \text{ in.}^2/\text{ft}$$

From ACI sections 10.5.3 and 7.12.2

Minimum $A_s = 0.0018bh = 0.0018 \times 12 \times 16 = 0.35 \text{ in.}^2/\text{ft}$
Spacing of #5 bars at 6-in. centers, $A_s = 0.62 \text{ in.}^2/\text{ft}$; provide #5 bars at 6-in. centers.
Maximum spacing allowed in the ACI section 7.6.5 $= 3h$ or 18 in.
Now compute

$$a = A_s f_y / 0.85 f_c' b$$
$$= \frac{0.62 \times 60,000}{0.85 \times 3,000 \times 12} = 1.22 \text{ in.}$$

$$\phi M_n = \frac{0.9 \times 0.62 \times 60,000(12.5 - 1.22/20)}{12000.00} = 33.18 \text{ ft-kips} > M_u$$

The design is satisfactory (Figure 5.15).

(5) Check the development length

Basic development length for #5 bars in 3,000 psi concrete $= l_{db}$
ACI code provision: furnish the following criterion:

$$l_{db} = 0.04A_b f_y / f_c'^{0.5} = 0.04 \times 0.31 \times 60,000/3,000^{0.5}$$
$$= 14 \text{ in.}$$

ACI 12.2.3. (a) No transverse steel (stirrups): does not apply

(b) and (c) Do not apply if flexural steel is in the bottom layer
(d) Cover $= 3$ in. and clear spacing $= 5.325$ in. $> 3d_b$ and therefore 12.3.3.1 (d)
   applies $\times 1.0$
ACI 12.2.3.4. Applies with a factor of 0.8

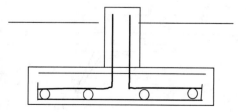

FIGURE 5.15
Configuration of reinforcement layout (Example 5.3).

ACI 12.2.4. Bottom bar, ×1.0; normal weight concrete, ×1.0; and standard deformed bar, ×1.0

$$l_{db} = 14 \times 1.0 \times 0.8 = 10.87 \text{ in.}$$

ACI 12.2.3.6. $l_{db} => 0.03 d_b f_y / f_c'^{0.5} = 0.03 \times 5/8 \times 60{,}000/3{,}000^{0.5} = 21$ in.

The length of the bar from the maximum stress point at the face of the wall is $34 - 3 = 31$ in., which is >21 in. and hence is satisfactory.

(6) Temperature and shrinkage ACI 7.12.2

$$A_s = 0.0018 bh = 0.0018 \times 12 \times 16$$
$$= 0.35 \text{ in.}^2$$

At least two thirds of this should be placed as top reinforcement in the transverse direction as the concrete exposed to the dry weather (low humidity and high temperature) until covered.

Provide #5 at 12-in. centers; $A_s = 0.31$ in.^{2}/ft and is thus satisfactory

$$A_s = 0.0018 bh = 0.0018 \times 76 \times 12$$
$$= 1.64 \text{ in.}^2/\text{ft}$$

This reinforcement should be divided between top and bottom layers in the longitudinal direction (Figure 5.16).

Provide 6 #4 at 14-in. centers both top and bottom

$$A_s = 6 \times 2 \times 0.2 = 2.40 \text{ in.}^2/\text{ft}$$
$$> 1.64 \text{ in.}^2/\text{ft and hence is satisfactory}$$

Example 5.4

You have been engaged as an engineer to design a foundation for a three-storey office building. It is required that the footings are designed for equal settlement under live loading. The footings are subjected to dead and live loads given below. However, statistics show that the usual load is about 50% for all footings. Determine the area of footing required for a balanced footing design. It is given that the allowable net soil bearing pressure is 5 ksf (Table 5.2–Table 5.4).

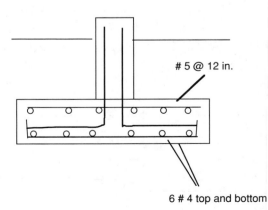

5 @ 12 in.

6 # 4 top and bottom

FIGURE 5.16
Details of reinforcement (Example 5.3).

Solution

TABLE 5.2

Details of Loads (Example 5.4)

Footing number	1	2	3	4	5	6
Dead load (kips)	130	170	150	190	140	200
Live load (kips)	160	210	200	180	210	250

TABLE 5.3

Computation of Load Ratios (Example 5.4)

Footing number	1	2	3	4	5	6
Ratio	1.231	1.235	1.333	0.947	1.500	1.250

TABLE 5.4

Computation of Factored Loads (Example 5.4)

Footing number	1	2	3	4	5	6
Usual load (DL + 0.5LL) (kips)	210	275	250	280	245	325

(a) Determine the footing that has the largest ratio of live load to dead load. Note that this ratio is 1.5 for footing #5.

(b) Calculate the usual load for all footings.

(c) Determine the area of footing that has the highest ratio of LL to DL (footing #5)

Area of footing # 5 $= (DL + LL)/(\text{allow soil pressure})$

$$= \frac{140 + 210}{5} = 70 \text{ ft}^2$$

Usual soil pressure under footing #5 $= (\text{usual load})/(\text{area of footing})$

$$= \frac{245}{70} = 3.5 \text{ ksf}$$

(e) Compute the area required for each footing by dividing its usual load by the usual soil pressure footing #5. For example, for footing #1,

$$\text{Required area} = 210/3.5 = 60 \text{ ft}^2$$

For other footings, the computations are shown below:

TABLE 5.5

Computation of Areas (Example 5.4)

Footing number	1	2	3	4	5	6
Usual load (DL $\pm$ 0.5LL) (kips)	210	275	250	280	245	325
Required area (ft^2)	60	78.57	71.43	80.00	70.00	92.86

TABLE 5.6

Computation of Soil Pressure (Example 5.4)

Footing number	1	2	3	4	5	6
Soil pressure (ksf)	4.83	4.84	4.90	4.63	5.00	4.85

(f) For verification, compute the soil pressure under each footing for the given loads.

Note that the soil pressure under footing #5 is 5 ksf, whereas under other footings it is less than 5 ksf.

5.6   Combined Footings

Combined footings are necessary to support two or more columns on one footing as shown in Figure 5.17. When an exterior column is relatively close to a property line (in an urban area) and a special spread footing cannot be used, a combined footing can be used to support the perimeter column and an interior column together.

The size and the shape of the footing are chosen such that the centroid of the footing coincides with the resultant of the column loads. By changing the length of the footing, the centroid can be adjusted to coincide with the resultant loads. The deflected shape and the reinforcement details are shown for a typical combined footing in Figure 5.17(b) and an example is given below to further illustrate the design procedure of such combined footings.

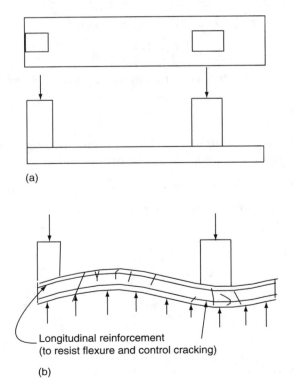

(a)

Longitudinal reinforcement
(to resist flexure and control cracking)

(b)

FIGURE 5.17
Typical combined footing: (a) under unloaded conditions; (b) under loaded conditions.

Example 5.5

In a three-storey building, an exterior column having a section of 24 in. × 18 in. carries a service dead load of 70 kips and service live load of 50 kips at each floor. At the same time the nearby 24 square interior column carries a service dead load of 100 kips and a live load of 80 kips at each floor. The architects have hired you as an engineer to design the footing for these columns. The specific site condition dictates that a combined footing be chosen as an economical solution. Both columns carry three floors above them and are located 18 ft apart. The geotechnical engineer has advised that the soil bearing pressure at about 4 ft below the ground is 5 ksf. The ground floor, which is going to be slab on grade with 6 in. concrete, supports a service live load of 120 psf. The soil below this floor is well compacted. The available concrete strength $f'_c = 3000$ psi and steel strength $f_y = 60,000$ psi. Design an economical footing.

Solution

Step 1. Determine the size and the factored soil pressure (Figure 5.18)

Allowable soil pressure = 5.00 ksf

Depth of soil above the base = 4.00 ft

Assume the unit weight of soil = 120.00 pcf

Allowable net soil pressure = $5 - 0.5 \times 0.15 - 3.5 \times 0.12 - 0.120$ ksf = 4.33 ksf

$$\text{Area required for the footing} = \frac{3 \times (70 + 50) + 3 \times (100 + 80)}{4.33} \text{ ft}^2 = 208.09 \text{ ft}^2$$

The resultant of the column load located at X from the center of external column

$$X = \frac{18 \times 3(100 + 80)}{3(70 + 50) + 3(100 + 80)} = 10.80 \text{ ft}$$
$$= 129.60 \text{ in.}$$

Distance from the external face of the exterior column = $129.6 + 9$ in. = 139 in. = 11.55 ft

Width of the footing = $208.09/(2 \times 11.55)$ = 9 ft

Factored external column load = $3 \times (1.4 \times 70 + 1.6 \times 50)$ kips = 534.00 kips

Factored internal column load = $3 \times (1.4 \times 100 + 1.6 \times 80)$ kips = 804.00 kips

Total = 1338.00 kips

Now we can compute the net factored soil pressure which is required to design the footing.

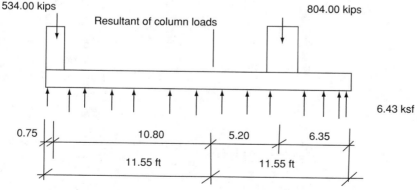

FIGURE 5.18

Factored load and factored net soil pressure (Example 5.5).

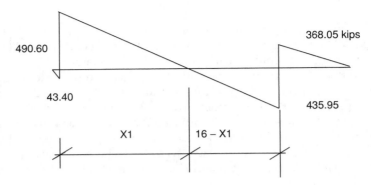

FIGURE 5.19
Shear force diagram (Example 5.5) (forces in kips).

$$q_{nu} = \frac{1338.0}{11.55 \times 9 \times 2.0} = 6.43 \text{ ksf}$$

Step 2. Draw the shear force and bending moment diagram (Figure 5.19)
If shear is zero at X1

$$X1 = \frac{16 \times 490.6}{490.6 + 435.95} = 8.47 \text{ ft}$$

Step 3. Determine the thickness of footing (Figure 5.20)

In this case, the footing acts as a wide (9 ft) heavy duty beam. It is better to determine the thickness based on the moment and check it for shear. We can start with minimum reinforcement of $200/f_y$ as per ACI Section 10.5.1.

$$\frac{200}{f_y} = \rho = 0.0033$$

$$f_c' = 3{,}000 \text{ psi}$$
$$f_y = 60{,}000 \text{ psi}$$
$$\phi = 0.9$$

$$\frac{M_n}{bd^2 f_c'} = \frac{\rho f_y}{f_c'} [1 - 0.59 \rho f_y / f_c']$$

and

$$M_n = \frac{M_u}{\phi} = \frac{2063.30}{0.90} = 2292.51 \text{ ft-kips}$$

$$\frac{\rho f_y}{f_c'} [1 - 0.59 \rho f_y / f_c'] = \frac{0.0033 \times 60{,}000}{3000} (1 - 0.59 \times 0.0033 \times 6{,}000/3{,}000) = 0.064$$

Therefore,

$$\frac{M_n}{bd^2 f_c'} = 0.064044 = \frac{12{,}000 \times 2292.51}{3{,}000 \times 9 \times 12 \times d^2}$$

$$d^2 = 1325.77 \text{ and hence } d = 36.41 \text{ in.}$$

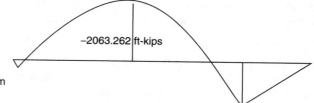

−2063.262 ft-kips

Maximum BM occurs at 8.47 ft from
the external column

Max BM = 6.43 × 9 × (8.47 + 0.75)²/2 − 534 × 8.47
         = 2063.262 ft-kips

At the face of the column
BM = 9 × 6.43 (6.35 −1)²
     ─────────────────────
              2
    = 828.192 ft-kips

FIGURE 5.20
Bending moment diagram (Example 5.5).

Now we will choose the total depth $h = 40$ in. and the area of steel will be more than
minimum steel required.

$h = 40.00$ in.

$d = 40 − 4$ (cover to reinforcement of 3 in. and 1 in. for reinforcement)

 $= 36.00$ in.

Step 4. Check two-way shear at the interior column

The critical perimeter is a square with sides 24 in. + 36 in. = 60.00 in.
Therefore,

$$b_0 = 4 \times 60 = 240.00 \text{ in.}$$

The shear, V_u, is the column load corrected for (minus) the force due to the soil pressure
on the area enclosed by the above perimeter, see Figure 5.21.

$$V_u = 804 − 6.43 \times (60/12)^2$$
$$= 643.25 \text{ kips}$$
$$\phi = 0.85$$

ϕV_c is the smallest of the following:

(a) $\phi V_c = \phi(2+4/\beta)b_0 df_c'^{0.5} = 0.85(2+4)240 \times 36 \times (3000)^{0.5} = 2413.5$ kips $(\beta = 1)$

(b) $\phi V_c = \phi(2+d\alpha/b_0)b_0 df_c'^{0.5} = 0.85(2+40 \times 36/240)240 \times 36 \times (3000)^{0.5}$ where
 $\alpha = 40$ $\phi V_c = 3217.98$ kips

(c) $\phi V_c = \phi 4 b_0 df_c'^{0.5} = 0.85 \times 4 \times 240 \times 36 \times (3000)^{0.5} = 1608.99$ kips

60

60

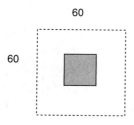

FIGURE 5.21
Shear perimeter (Example 5.5 Interior Column).

Since V_u is less than the smallest value of 1609 kips of all three above conditions, the depth of the footing is adequate to support the interior square column load.

Step 5. Check two-way shear at the exterior column

The critical perimeter is a rectangle with sides of $24 + 36 = 60.00$ in. and a width of $18 + 36/2 = 36.00$ in. The shear, V_u, is the column load minus the force due to the soil pressure on the area enclosed by the above perimeter

$$b_0 = 60 + 2 \times 36 = 132.00 \text{ in.}$$
$$b_1 = 36.00 \text{ in.}$$
$$b_2 = 60.00 \text{ in.}$$
$$V_u = 534.0 - 6.43 \times (60 \times 36/144)$$
$$= 437.55 \text{ kips}$$
$$\phi = 0.85$$

The shear perimeter around the column is three-sided, as shown in Figure 5.22. The distance from line B–C to the centroid of the shear perimeter is given by X_2

$$X_2 = 2(b_1 d^2/2)/(2b_1 d + b_2 d)$$
$$= 2\left(\frac{36 \times 36^2}{2}\right)/(2 \times 36 \times 36 + 60 \times 36)$$
$$= 9.82 \text{ in.}$$

The force due to the soil pressure on the area enclosed by the perimeter is

$$= 6.43 \times (60 \times 36/144) = 96.45 \text{ kips}$$

Then, summing up the moment about the centroid of the shear perimeter gives

$$M_u = 534 \times (36 - 9.82 - 9) - 96.45 \times (18 - 9.82)$$
$$= 8385.16 \text{ in.-kips}$$

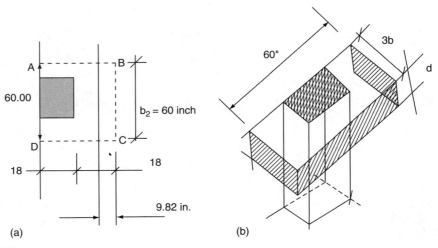

FIGURE 5.22
Illustration of the shear perimeter (Example 5.5 (a) Exterior column; (b) interior column).

This moment must be transferred to the footing through shear stress and flexure. The moment of inertia of the shear perimeter j_c is

$$j_c = 2[[36 \times 36^3/12] + [36 \times 36^3/12] + [36 \times 36][18 - 9.82^2]] + [(60 \times 36) \times 9.82^2]$$
$$j_c = 646590.5 + 208293.984 = 854884.5 \text{ in.}^4$$

The fraction of moment transferred by flexure is

$$\gamma_f = \frac{1}{1 + 2(b_1/b_2)^{0.5}/3} = \frac{1}{1 + 2(36/60)^{0.5}/3} = 0.659$$

The fraction transferred by shear $= (1 - \gamma_f) = (1 - 0.659) = 0.341$
The shear stress due to the direct shear and due to moment transfer will be additive at points A and D in Figure 5.22, giving the largest shear stresses on the critical shear perimeter

$$V_u = \frac{V_u}{b_0 d} + \frac{(1 - \gamma_f)M_u C}{j_c}$$

$$= \frac{437.55}{132 \times 36} + \frac{0.341 \times 8385.16 \times (36 - 9.82)}{854884.5}$$
$$= 0.092078 + 0.087564$$
$$= 0.180 \text{ kips}$$

Now we will compute ϕV_c from the following condition using ACI equations 11-36 to 11-38, which is the smallest of the following:

(a) $\phi V_c = \phi(2 + 4/\beta)f_c'^{0.5} = 0.85(2 + 4/(24/18)) \times (3000)^{0.5} = 0.2328 \text{ ksi}$
(b) $\phi V_c = \phi(2 + d\alpha/b_0)f_c'^{0.5} = 0.85(2 + 30 \times 36/132)(3000)^{0.5}$ where $\alpha = 30 \; \phi V_c$
    $= 0.474 \text{ ksi}$
(c) $\phi V_c = 4f_c'^{0.5} = 0.85 \times 4 \times (3000)^{0.5} = 0.186 \text{ ksi}$

Since V_u is less than the smallest value of 0.186 ksi of all the above three conditions, the depth of the footing is adequate to support the interior square column load.

Step 6. Check one-way shear

The shear force diagram shows that the maximum shear is near the exterior column and, hence, one-way shear is critical at a distance d from the face of the exterior column:

$$V_u = 490.60 - (36 + 9)/12 \times (6.43 \times 9) = 490.60 - 217.01 = 273.59 \text{ kips}$$

$$\phi V_c = 0.85 \times 2(3000)0.5 \times (9 \times 12) \times 36/1000 = 362.02 \text{ kips}$$

Since V_u is less than ϕV_c, the depth is adequate to support the required shear condition.

Step 7. Design the flexural reinforcement

(a) The mid-span (negative moment):
    $b = 108 \text{ in.}$
    $d = 36 \text{ in.}$
    $f_c' = 3 \text{ ksi}$
    $f_y = 60 \text{ ksi}$
    $M_u = 2063.3 \text{ ft-kips}$

Per ACI handbook,

$$\frac{12M_u}{0.9f'_c bd^2} = \frac{12 \times 2063.3}{0.9 \times 3 \times 108 \times 36^2} = Q = 0.0655$$

$$= q(1 - 0.59q) = 0.0671 \quad \text{for } q = 0.0700$$

$$A_s = bdqf'_c/f_y = 13.61 \text{ in.}^2$$

Provide #8 bars at 6-in. spacing at the top. Total area $A_s = 14.137$ in.^{2}

(b) At the face of interior column (positive moment):

$b = 108$ in.

$d = 36$ in.

$f'_c = 3$ ksi

$f_y = 60$ ksi

$M_u = 828.192$ ft-kips

Per ACI handbook,

$$\frac{12M_u}{0.9f'_c bd^2} = \frac{12 \times 2063.3}{0.9 \times 3^{0.5} \times 108 \times 36^2} = Q = 0.0455$$

$$= q(1 - 0.59q) = 0.0457 \quad \text{for } q = 0.0470$$

$$A_s = dqf'_c/f_y = 9.14 \text{ in.}^2$$

Provide #8 bars at 6-in. spacing at the bottom. Total area $A_s = 14.137$ in.^{2}, which is satisfactory.

Check for minimum area of steel required $= (200/f_y)bd = 12.96$ in.^{2}

Since the provided area of steel is greater than the minimum steel required, the flexural reinforcement provided is adequate.

Step 8. Check the development length

Basic development length l_{db} for #8 bars in 3000 psi concrete (ACI code provision) is given by

$$l_{db} = 0.04Ab/f_c^{0.5} = 0.04 \times 0.79 \times 60,000/3000^{0.5}$$

$$= 35 \text{ in.}$$

ACI 12.2.3.1:

(a) No transverse steel (stirrups): does not apply

(b) Does not apply

(c) Does not apply if flexural steel is in the bottom layer

(d) Cover $= 3$ in. and clear spacing $= 5.0$ in $> 3d_b$ and therefore 12.3.3.1 (d) applies with a factor of 1.0

ACI 12.2.3.4 Applies with a factor of 0.8

ACI 12.2.4 Bottom bar, $\times 1.0$; normal wt concrete, $\times 1.0$; and standard deformed bar, $\times 1.0$

ACI 12.2.4.1 Top bar with a factor of 1.3

$$l_{db} = 35 \times 1.0 \times 0.8 \times 1.3 = 36 \text{ in. (top bar)}$$

ACI 12.2.3.6 $l_{db} => 0.03d_b f_y / f_c' 0.5 = 0.03 \times 1 \times 60,000/3000^{0.5} = 33$ in.

The length of the bar from the maximum stress point at the face of the columns is $67 - 3 = 64$ in., which is >36 in. and hence is satisfactory.

Step 9. Temperature and shrinkage reinforcement (ACI 7.12.2)

$$A_s = 0.0018bh = 0.0018 \times 12 \times 40$$
$$= 0.86 \text{ in.}^2/\text{ft}$$

At least two thirds of this should be placed as top reinforcement in the transverse direction as the concrete exposed to the dry and hot weather until covered by earth (backfill).

Provide #7 at 12 in. on centers; $A_s = 0.6 \text{ in.}^2/\text{ft}$

$$A_s = 0.0018bh = 0.0018 \times 12 \times 40 = 0.86 \text{ in.}^2/\text{ft}$$

This reinforcement should be divided between the top and the bottom and should provide two thirds of it at the top since it will be exposed to temperature and half of it at the bottom layers in the longitudinal direction or provide 6 #7 at 12-in. centers at the top at the interior column and at the bottom at the exterior column.

$$A_s = 2 \times 0.61 = 1.22 \text{ in.}^2/\text{ft}$$
$$> 0.86 \text{ in.}^2/\text{ft}, \quad \text{satisfactory}$$

Step 10. Design of the transverse "beam"

The transverse strips under each column will be assumed to transmit the load evenly from the longitudinal beam strips into the column strip. The width of the column strip will be assumed to extend $d/2$ on either side of the interior column and one side of the exterior column (Figure 5.23).

(a) The maximum factored load for the interior column = 804 kips
   This load is carried by a 9-ft beam and, hence,

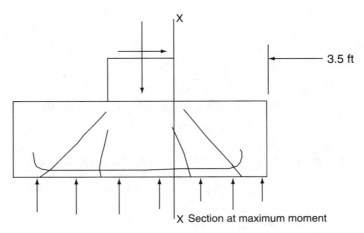

X Section at maximum moment

3.5 ft

FIGURE 5.23
Cross section of beam (Example 5.5).

$$\text{load per ft} = 804/9 = 89.33 \text{ kips}$$
$$\text{Maximum } M_u = 89.33 \times 3.5^2/2$$
$$= 547.15 \text{ ft-kips}$$
$$b = 60 \text{ in.}$$
$$d = 36 \text{ in.}$$
$$f_c' = 3 \text{ ksi}$$
$$f_y = 60 \text{ ksi}$$
$$M_u = 547.15 \text{ ft-kips}$$

Per ACI handbook,

$$\frac{12M_u}{0.9f_c'bd^2} = \frac{12 \times 547.15}{0.9 \times 3^{0.5} \times 60 \times 36^2} = Q = 0.0542$$

$$= q(1 - 0.59q) = 0.0544 \quad \text{for } q = 0.0563$$
$$A_s = bdqf_c'/f_y = 6.08 \text{ in.}^2$$

Provide #8 bars at 6-in. spacing at the bottom. This amounts to a total of ten bars

$$\text{Total } A_s = 7.85 \text{ in.}^2, \text{ Satisfactory}$$

(b) The maximum factored load for the interior column = 534 kips
This load is carried by a 9-ft beam and, hence, load per ft = 534/9

$$= 59.33 \text{ kip}$$
$$\text{Maximum } M_u = 59.33 \times 3.5^2/2$$
$$= 363.42 \text{ ft-kips}$$
$$b = 42 \text{ in.}$$
$$d = 36 \text{ in.}$$
$$f_c' = 3 \text{ ksi}$$
$$f_y = 60 \text{ ksi}$$
$$M_u = 363.42 \text{ ft-kips}$$

Per ACI handbook,

$$\frac{12M_u}{0.9f_c'bd^2} = \frac{12 \times 363.42}{0.9 \times 3^{0.5} \times 42 \times 36^2} = Q = 0.0514$$
$$= q(1 - 0.59q) = 0.0523 \quad \text{for } q = 0.0540$$
$$A_s = bdqf_c'/f_y = 4.08 \text{ in.}^2$$

Provide #8 bars at 6-in. spacing at the bottom. This amounts to a total of seven bars

$$A_s = 5.50 \text{ in.}^2 \text{ satisfactory}$$

Step 11. Details of reinforcement (Figure 5.24)

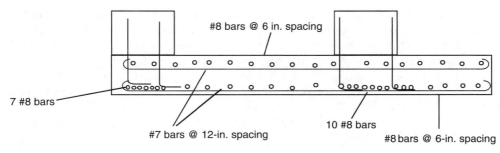

#8 bars @ 6 in. spacing

7 #8 bars

#7 bars @ 12-in. spacing

10 #8 bars

#8 bars @ 6-in. spacing

FIGURE 5.24
Details of reinforcement (Example 5.5).

5.7 Pile Foundations

A structure is founded on piles if the soil immediately below its base does not have adequate bearing capacity, or if the foundation cost estimate indicates that a pile foundation may be more economical and safer than any other type of foundation. In this discussion, we will consider only piles that are commonly available and driven into the ground by a mechanical driving devise known as a pile driver. Please note that the general principles are also applicable to other types of pile foundations, with minor modifications. Piles may be divided into three categories based on the method of transferring the load into the ground (Figure 5.25–Figure 5.28):

1. *Friction piles in coarse-grained very permeable soils:* These piles transfer most of their loads to the soil through skin friction. The process of driving such piles close to each other (in groups) greatly reduces the porosity and compressibility of the soil within and around the group.

2. *Friction piles in very fine-grained soils of low permeability:* These piles also transfer their loads to the soils through skin friction. However, they do not compact the soil during driving as in case 1. Foundations supported by piles of this type are commonly known as floating pile foundations.

3. *Point bearing piles:* These piles transfer their loads into a firm stratum or a soil layer. Depending on the geographical location, these piles have to be driven to a considerable depth below the base of the footing.

In practice, piles are used to transfer their loads into the ground using a combination of the above mechanisms.

5.7.1 Analysis of Pile Groups

The function of a pile cap, a relatively rigid body, is to distribute the loads to each pile in a group of piles. The loads could be vertical or horizontal loads or moments from the superstructure. The horizontal forces at the base are generally resisted by battered or raked piles. The batter can be as steep as 1 on 1, but it is economical to limit the batter to

1.0 horizontal to 2.5 vertical (approximately 22° of an inclination to vertical). The horizontal forces may be carried by vertical piles in the form of shear and moments. The shear capacity of piles is limited by the material property of the pile. However, it is advisable to resist by the horizontal component of the axial load in a battered pile.

When a footing consisting of N number of piles is subjected to a vertical load of P, moments of M_x and M_y, and a horizontal force of H, the following equation can be used to determine the force attributed to each pile. After determining the force in each pile, the horizontal resistance force may be provided by battering or raking the piles to develop adequate horizontal resistance:

$$\text{Load in a pile} = \frac{P}{N} \pm \frac{M_x d_x}{\sum d_x^2} \pm \frac{M_y d_y}{\sum d_y^2} \tag{5.6}$$

where P is the total vertical load in the pile cap, M_x is the moment at the pile cap about the x-axis, M_y is the moment of the pile cap about the y-axis, d_x is the x-directional distance of the pile from the center of the pile group, and d_y is the y-directional distance of the same pile from the center of the pile group.

The above principle is illustrated by the following example in an actual design situation.

Example 5.6

You have been engaged as the engineer to design the footing of a pier foundation for a major bridge. The bridge engineer has determined that the foundation needs to be designed for a factored load of 3650 kips, a transverse factored moment of 7050 ft-kips, and a longitudinal moment of 2400 ft-kips. The bridge pier is 8 ft (longitudinal direction) × 10 ft (transverse direction). The bridge engineer has proposed to use 18-in. square PC piles. The geotechnical engineer has recommended limiting the pile capacity to 325 kip (factored load). The group has to resist a lateral force of 125 kips in the transverse direction and 75 kips in the longitudinal direction. The bridge engineer has estimated that 17 to 18 piles would be adequate. The shear capacity of the 18-in. square pile is limited to 10 kips.

$$f_c' = 4,600 \text{ psi}$$

$$f_y = 60,000 \text{ psi}$$

Solution

This is a bridge foundation design example and hence AASHTO provisions apply:

1. Determine the number of piles and the spacing required to resist the given loading condition. It is given that the bridge engineer presumes that 18 piles would be required. The spacing between piles is more than three times the pile diameter = $3 \times 18 = 54$ in. Provide piles at spacing of 60 in. (5 ft) in both directions.

2. Determine the size of the pile cap or the footing.
   Careful study of the situation indicates that the pile cap should provide higher resistance in the transverse direction. An edge distance of 2 ft should be sufficient. If the pile group is arranged with five piles in the transverse direction and four piles in the longitudinal direction,

   Length of the pile cap = 4×5 ft + 2×2 ft = 24 ft

Width of the pile cap $= 3 \times 5\,\text{ft} + 2 \times 2 = 19\,\text{ft}$

$h = 5.00$ ft

$d = 3.75$ ft

$L = 24.00$ ft

$B = 19.00$ ft

3. Analysis of pile group.

   Analysis of the pile group can be carried out in a tabular form as given below:
   Pile load analysis — Transverse direction (Table 5.7)
   Pile load analysis — Longitudinal direction (Table 5.8)
   Combined loading effect: Load per pile (kips) (Table 5.9)

$$\text{Axial load due to moment } M_y = \frac{7050}{1000} \times 10 = 70.5 \text{ kips}$$

4. Consider one-way shear action:
   Critical section for shear $= d$ from the face of the piles

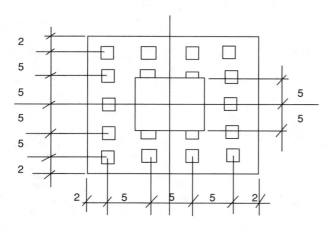

FIGURE 5.25
Arrangement of piles and the pile cap.

TABLE 5.7

Computations for Transverse Directional Analysis

N	X	d^2	$N \times d^2$	P_u kips 3650.00	M_y ft-kips 7050.00	$P_u \pm M_y$ kips	M_x ft-kips 2400.00
4	10	100	400	202.78	70.50	273.28	
4	5	25	100	202.78	35.25	238.03	
2	0	0	0	202.78	0.00	202.78	
4	−5	25	100	202.78	−35.25	167.53	
4	−10	100	400	202.78	−70.50	132.28	
Total 18			1000				

$$b = 228 \text{ in. } (19.0 \text{ ft})$$

Shear loads $= 1093.11$ kips

Shear area $= 10{,}260 \text{ in.}^2 \ (228 \times 45)$

Shearstress $= 106.54 \text{ psi} < 2f_c^{\prime 0.5} = 135.65 \text{ psi}$, Satisfactory

5. Consider two-way shear action:
   Critical section for shear $= d/2$ from the face of a single pile.

   $b_0 = 99$ in.
   Shear loads $= 302.67$ kips

   Shear area $= 4455 \text{ in.}^2 \ (45 \times 99)$

   Shearstress $= 67.938 \text{ psi} < 2f_c^{\prime 0.5} = 135.65 \text{ psi}$, Satisfactory

6. Flexural behavior — Determination of reinforcements in the longitudinal direction:

   Minimum cover to reinforcement $= 3$ in.

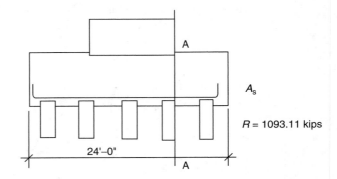

FIGURE 5.26
Pile cap layout — longitudinal direction.

TABLE 5.8

Computations for Longitudinal Directional Analysis

N	X	d^2	$N \times d^2$	M_x ft-kips
5	7.5	56.25	281.25	29.39
4	2.5	6.25	25	9.80
4	−2.5	6.25	25	−9.80
5	−7.5	56.25	281.25	−29.39
	Total		612.5	

TABLE 5.9

Computations of Combined Analysis

	Transverse					
Longitudinal	**T1**	**T2**	**T3**	**T4**	**T5**	
L1	161.67	196.92	232.17	267.42	302.67	1160.85
L2	142.07	177.32		247.82	283.07	
L3	122.48	157.73		228.23	263.48	
L4	102.89	138.14	173.39	208.64	243.89	

Notes: Horizontal force $= 8.10$ kips; high load on a pile $= 302.67$ kips. $\sum = 1093.11$ kips

$R = 1093.11$ kips

Max. moment at A $= 1093.11 \times 5$ kip-ft

$M_u = 5514.75$ kip-ft

$f'_c = 4.6$ ksi

$b = 216$ in.

$d = 45.00$ in.

$Q = 12M_u/\phi f'_c bd^2 = q(1 - 0.59q)$

$f_y = 60,000$ psi

$Q = 0.04348 < q_{min} = (200/f_y) \times x$

$A_s = qbd/x$, where $x = f_y/f'_c = 13.04$

$\quad = 32.40$ in.^{2}

Number of #10 bars $= 26$ bars

Spacing $= 8.15$ in. in the longitudinal direction

Check for minimum reinforcement (AASHTO Section 8.17.1):

$\phi M_n > 1.2 M_{cr}$

$M_{cr} = 7.5 f'^{0.5}_c \times l_g/y_t$

where

$l_g = bh^3/12 = 3,888,000$ in.^{4}

$y_t = 30$ in.

$M_{cr} = 5493.69$ kip-ft

$1.2 M_{cr} = 6592.42$ kip-ft

$\phi M_n = \phi A_s f_y j d$ kip-ft

$A_s = 27$ #10 bars $= 34.29$ in.^{2}

$T = A_s f_y = 2057.4$ kips

$C = 0.85 \beta b a f'_c$

where

$\beta = 0.85 - (4,600 - 4,000)/1,000 \times 0.05 = 0.82$

$C = 692.539a$

$a = T/C = 2.97$ in., $j_d = 43.5146$ in.

$\phi M_n = 6714.52$ ft-kips $> 1.2 M_{cr}$, Satisfactory

Provide 28 #10 bars at the bottom and provide #7 bars at the top at 8-in. spacing.

7. Determination of reinforcement in transverse direction:

Min. cover $= 3$ in.

$R = 1160.83$ kips

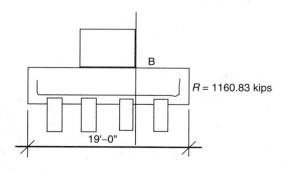

$R = 1160.83$ kips

19'–0"

FIGURE 5.27
Pile cap layout — transverse direction.

Max. moment at B $= 1120.3 \times 3.75$ kip-ft

$f'_c = 4{,}600$ psi

$M_u = 4201.1$ kip-ft

$f_y = 60{,}000$ psi

$f'_c = 4.6$ ksi $f_y = 60$ ksi

$b = 288$ in.

$d = 43.75$ in.

$Q = 12M_u/\phi f'_c bd^2 = q(1 - 0.59q)$

$f'_c = 4{,}600$ psi

$Q = 0.0221 \Rightarrow q = 0.027$

$q = 0.043478 < = q_{min} = \left(\frac{200}{f_y}\right)x$ where $x = \frac{f_y}{f'_c}$

$A_s = qbd/x$, where $x = f_y/f'_c = 13.04 = 42$ in.^{2}

Number of #10 bars $= 33.071$ bars

Spacing $= 8.5$ in.

Check for minimum reinforcement AASHTO Section 8.17.1:

$\phi M_n \geqslant 1.2 M_{cr}$

$M_{cr} = 7.5f'_c 0.5 \times l_g/y_t$

where $l_g = bh^3/12 = 518{,}4000$ in.^{4}

$Y_t = 30$ in.

$M_{cr} = 7324.9$ kip-ft

$1.2M_{cr} = 8789.9$ kip-ft

$\phi M_n = \phi A_s f_y j_d$ kip-ft

$A_s = 38$ #10 bars $= 47.752$ in.^{2}

$T = A_s f_y = 2865.1$ kips

$C = 0.85\beta baf'_c$, where $\beta = 0.85 - (4{,}600 - 4{,}000)/1{,}000 \times 0.05 = 0.82$

$C = 923.39a$

$a = T/C = 3.10$ in.

$jd = 42.19858$ in.

$\phi M_n = 9067.8$ kip-ft $> 1.2M_{cr}$, Satisfactory

Provide #10 bars at 7.5-in. spacing at the bottom.

8. Lateral force resistance of the pile cap:

   The lateral force could be due to a centrifugal force, wind force, or even due to earthquake motions. In this example, it is a combination of forces per section 3 of AASHTO specifications. The lateral forces were computed per AASHTO and found that the piles need to resist the following forces:

   $F_x = 75$ kips

   $F_y = 125$ kips

   $R = (F_x^2 + F_y^2)^{0.5} = 145.77$ kips

   Lateral force per pile $= 145.77/18.00 = 8.10$ kips

   Since the lateral force of 8.1 kips < 10 kips per pile, no pile requires any battering or raking.

9. Shrinkage and temperature reinforcement:

Average $R_H = 75\%$
Assumed shrinkage $= 150$ microstrains
Correction for $R_H = 1.4 - 0.01 \times R_H$
$S_H = 97.5$ microstrains
$E_c = 4E + 0.6$ psi
Concrete stress $= E_c \times S_H$ psi
$\quad = 376.93$ psi
Depth of shrinkage effect $= 5.00$ in. from the surface
The shrinkage induced force per ft $= 22.616$ kips
$\quad$ This force has to be resisted by steel reinforcement. Otherwise the concrete will develop cracking.
Required steel to prevent cracking $A_s = 22.616/0.85f_y$, $A_s = 0.443445$ in.^{2}/ft

10. Temperature effect:

Temperature rise during the initial stage of concrete curing does more damage to concrete than at latter stages

Temperature rise could be $= 25°C$
Temperature strains $= \alpha t$ and $\alpha = 6.5 \times 10^{-6}$
$\quad = 162.5$ microstrains
Concrete stress $= 162.5 \times E_c/3$
$\quad = 209.4$ psi
Assuming the depth of the temperature rising effect to be 6 in.
Temperature-induced force $= 15.077$ kips/ft
The required steel area $= 15.077/0.85f_y$
$\quad = 0.2956$ in.^{2}/ft
Total area $= 0.7391$ in.^{2}/ft
Spacing of #7 bars $= 9.9043$ in.
Provide #7 bars at 9-in. spacing at the top and the vertical face.
Some transportation agencies recognize shrinkage and temperature-related cracking of RC members and require that the minimum reinforcement is provided. For example, Florida Department of Transportation requires the following:

Two-way cage reinforcement must be provided on all faces of pier footings

(1) 5 bars at 12-in. centers as minimum

(2) When the minimum dimension exceeds 3.28 ft and volume–surface area ratio is greater than 12 in.

$$V/A = 20.39 > 12 \text{ in.}$$

The pile cap meets the mass concrete requirements

$$\Sigma A_b \Rightarrow S(2d_c + d_b)/100$$

where A_b = minimum area of bar (mm^2) $= 285$
S = spacing of bar (mm) $= 300$
d_c = concrete cover measured to the center of the bar (mm) $= 85.73$
d_b = diameter of the reinforcing bar $= 19.05$

$$2d_c + d_b = 190.5 \text{ mm}$$

But $(2d_c + d_b)$ need not be greater than 75 mm

$$\sum A_b = 0.75S = 225$$

Therefore, provide #6 bars at 12-in. centers.

11. Reinforcement development length:

   In this design, the reinforcement must be effective just outside of the piles within the pile caps. This is made possible by providing mechanically anchored bent-up bars (through a 90° bend). This is the most economical way of providing sufficient development. Otherwise the footing needs to be extended and may become uneconomical.

12. Reinforcement details

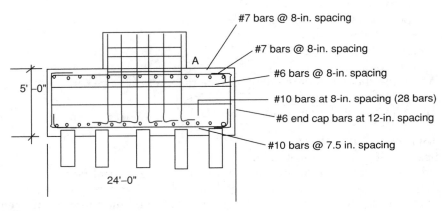

FIGURE 5.28
Typical pile cap details.

5.8 Design of Grade Beams

Example 5.7
One of your clients approaches you to design a foundation for a wood-framed (construction) building. The geotechnical engineer has advised you to use a grade beam supported by wooden timber piles. Twelve-inch diameter timber piles driven to a depth of 35 ft could carry a working load of 35 kips per pile. The grade beam has to carry the wall load of 2.5 kips per foot of dead load and 1.3 kips per foot of live load. The structural engineer advised you that the timber piles need to be staggered at least 1 ft 6-in. centers apart. If the building length is 85 ft, determine the pile spacing along the length of the building and design the grade beam given the following: $f'_c = 3,000$ psi, $f_y = 60,000$ psi. The frost depth is 2 ft 4 in (Figure 5.29 and Figure 5.30).

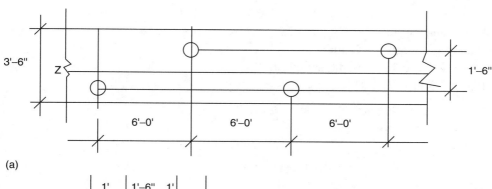

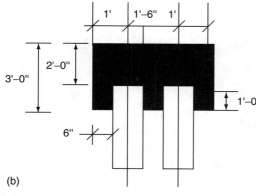

(a)

(b)

FIGURE 5.29
Illustration for Example 5.7: (a) front elevation; (b) side elevation.

Answer: Grade beam design
Data: Grade beam woodframe wall

$$\left.\begin{array}{l} DL = 2.5 \text{ kips/ft} \\ LL = 1.3 \text{ kips/ft} \end{array}\right\} = 3.8 \text{ kips}$$

Grade beam supported by timber piles driven to 35 ft
Pile capacity $= 35$ kips
Ultimate load $= 35 \times 2 = 70$ kips
Beam width $= 3.5$ ft
Beam depth >2 ft to 4 in. $= 3.0$ ft
Self weight of beam $= 1.575$ kips/ft

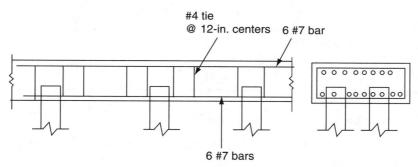

FIGURE 5.30
Reinforcement details for grade beam (Example 5.7).

When piles are spaced at 6–0

Load/pile $= (1.575 + 3.8)6 = 32.25$ kips < 35 kips

Thus, the design is adequate

$W_d = 2.5$ and $1.575 = 4.075$ kips/ft

$W_L = 1.3$ kips/ft

$W_u = 1.4 \times 4.075$ and 1.7×1.3

$\quad = 7.92$ kips/ft

$$M_u = \frac{W_u l^2}{9} = \frac{7.92 \times 6^2}{9} = 31.7 \text{ kips/ft}$$

$D = 36 \text{ in.} - 3 \text{ in.} - 0.5 \text{ in.} = 32.5 \text{ in.}$

$$Q = \frac{12 M_u}{0.9 f'_c b d^2} = \frac{12 \times 31.7}{0.9 \times 3 \times 42 \times 32.5^2} = 0.0032 = q \, (1 - 0.59q) \Rightarrow q = 0.0033$$

$$A_s = bdq\frac{f'_c}{f_y} = \frac{42 \times 32.5 \times 0.0033 \times 3}{60} = 0.3 \text{ m}^2$$

$$\text{Minimum reinforcement} = \rho = \frac{200}{f_y} = \frac{200}{60,000}$$

$$A_s = \frac{200}{60,000} \times 42 \times 32.5 = 4.55 \text{ m}^2$$

A_s required $< A_s$ min

Provide 6 #7 bars at the top and at the bottom

$$A_s = 6 \times 061 = 3.66 \text{ m}^2$$

A_s provided $> A_s$ required $\times 1.33$, OK.

Shear check:

$$\text{Shear force} = 0.55 \times 7.92 \times 6$$
$$Q = 26.14 \text{ kips}$$

$$\text{Shear stress} = \frac{Q}{bd} = \frac{26.14 \times 10^3}{42 \times 32.5}$$
$$= 19 \text{ psi}$$
$$< \sqrt{f'_c} = 54.8 \text{ psi}$$

provide #4 tie at 12 in. centers.

5.9   Structural Design of Drilled Shafts

The construction of high rise and heavier buildings in cities, where the subsurface conditions consist of relatively thick layers of soft to medium bearing strata overlying deep bedrock, led to the development of drilled shaft foundations. Therefore, the function of a drilled shaft (similar to pile foundations) is to enable structural loads to be taken down through deep layers of weak soil on to a hard stratum called for a very conservative value for bearing pressure for the hard strata around 8 to 10 kips per square foot.

However, the rapid advancement in the construction technology followed by the development of theories for design and analytical techniques, the use of computers, and full-scale testing led to the production of a better understanding of drilled shaft behavior. There are marked differences between the behavior of driven piles and drilled shaft. The drilled shaft is also known as caisson, drilled caisson, or drilled piers.

Drilled shafts have proved to be reliable foundations for transferring heavy loads from superstructure to be the suitable bearing strata beneath the surface of the ground. Economic advantages of a drilled shaft are often realized due to the fact that a very large drilled shaft can be installed to replace groups of driven piles, which in turn obviates the need for a pile cap. The drilled shaft is very often constructed to carry both vertical and horizontal loads.

5.9.1 Behavior of Drilled Shafts under Lateral Loads

Figure 5.31 shows views of two types of foundations used for column support in two buildings. Figure 5.31(a) shows two shaft foundations and Figure 5.31(b) shows a single-shaft support. The two-shaft system resists the wind moment by added tension and compression (a "push–pull" couple) in the shaft, although some bending is required to resist the wind shear, while the single-shaft foundation resists both the moment and shear produced by the wind load through bending.

5.9.2 Methodology for Design of Drilled Shafts

Drilled shafts are more often used to transfer both vertical and lateral loads. The design of a drilled shaft for lateral loading requires step-by-step procedures to be followed:

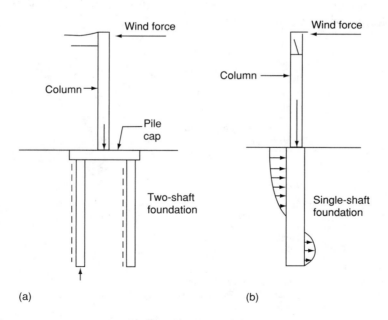

FIGURE 5.31
Elevation view of: (a) two-shaft foundation; (b) single-shaft foundation. (From *LRFD Bridge Design Specifications, Customary U.S. Units*, 2nd ed., American Association of State Highway and Transportation Officials, Washington, DC, 1998 (with 1999 interim revisions). With permission.)

1. Determine the depth of the drilled shaft to carry the computer-generated vertical load without undergoing excessive moment.

2. Determine the size (diameter) and mechanical properties of the concrete to resist the bending moment, shear force, and axial load that will be imposed on the drilled shaft by lateral loads in combination with axial loads.

3. Determine the deformation or stiffness of the drilled shaft in lateral translation and rotations to ensure that lateral deformation falls within acceptable limits.

There are three methods that can be used to analyze laterally loaded drill shafts. Brom's method can be used to estimate ultimate strength–state resistance. The other two methods include the "characteristic load method" and the "P–Y methods," which can deal better with the nonlinear aspects of the problem. In the following section Brom's method is presented.

5.9.2.1 Brom's Method of Design

Brom's method is a straightforward hand-calculation method for lateral load analysis of a single drilled shaft or pile. The method calculates the ultimate soil resistance to lateral load as well as the maximum moment induced in the pile. Brom's method can be used to evaluate fixed or free head condition in either purely cohesive or purely cohesionless soil profiles. The method is not conducive to lateral load analyses in mixed cohesive and cohesionless soil profiles. For long fixed head piles in sands, the method can also over-predict lateral load capacities (Long, 1996). Therefore, for mixed profiles and for long fixed head shaft in sands, the COM624P program should be used. A step-by-step procedure developed by the New York State Department of Transportation (1977) on the application of Brom's method is provided below:

Step 1. Determine the general soil type (i.e., cohesive or cohesionless) within the critical depth below the ground surface (about 4 or 5 shaft diameters).

Step 2. Determine the coefficient of horizontal subgrade reaction, K_h, within the critical depth for cohesive or cohesionless soils

TABLE 5.10

Values of Coefficients of n_1 and n_2 for Cohesive Soils

Unconfined compression strength, q_u (kPa)	n_1
<8	0.32
48–191	0.36
>191	0.40
Pile material	n_2
Steel	1.00
Concrete	1.15
Timber	1.30

Source: From *LRFD Bridge Design Specifications, Customary U.S. Units,* 2nd edn, American Association of State Highway and Transportation Officials, Washington, DC, 1998 (with 1999 interim revisions). With permission.

(a) Cohesive soils:

$$K_h = \frac{n_1 n_2 80 q_u}{b} \tag{5.7}$$

where q_u is the unconfined compressive strength (kPa), b is the width or diameter of the shaft (m), and n_1 and n_2 are the empirical coefficients taken from Table 5.10

(b) Cohesionless soils:

Choose K_h from the Table 5.11. (The values of K_h given in Table 5.11 were determined by Terzaghi.)

Step 3. Adjust K_h for loading and soil conditions
(a) Cyclic loading (or earthquake loading) in cohesionless soil:
 1. $K_h = \frac{1}{2} K_h$ from Step 2 for medium to dense soil.
 2. $K_h = \frac{1}{4} K_h$ from Step 2 for loose soil.
(b) Static loads resulting in soil creep (cohesive soils)

 1. Soft and very soft normally consolidated clays
  $K_h = (1/3 \text{ to } 1/6)K_h$ from Step 2
 2. Stiff to very stiff clays
  $K_h = (1/4 \text{ to } 1/2)K_h$ from Step 2

Step 4. Determine the pile parameters
(a) Modulus of elasticity, E (MPa)
(b) Moment of inertia, I (m^4)
(c) Section modulus, S (m^3), about an axis perpendicular to the load plane
(d) Yield stress of pile material, f_y (MPa), for steel or ultimate compression strength, f_c (MPa), for concrete
(e) Embedded pile length, D (m)
(f) Diameter or width, b (m)
(g) Eccentricity of applied load e_c for free-headed piles — i.e., vertical distance between ground surface and lateral load (m)
(h) Dimensionless shape factor C_s (for steel piles only):

TABLE 5.11

Values of K_h in Cohesionless Soils

	K_h (kN/m^3)	
Soil Density	**Above Groundwater**	**Below Groundwater**
Loose	1,900	1,086
Medium	8,143	5,429
Dense	17,644	10,857

Source: From *LRFD Bridge Design Specifications, Customary U.S. Units*, 2nd ed., American Association of State Highway and Transportation Officials, Washington, DC, 1998 (with 1999 interim revisions). With permission.

1. Use 1.3 for pile with circular section
2. Use 1.1 for H-section pile when the applied lateral load is in the direction of the pile's maximum resisting moment (normal to the pile flanges)
3. Use 1.5 for H-section pile when the applied lateral load is in the direction of the pile's minimum resisting moment (parallel to the pile flanges)

(i) M_y the resisting moment of the pile
1. $M_y = C_s f_y S$ (kN m) (for steel piles)
2. $M_y = f_c S$ (kN m) (for concrete piles)

Step 5. Determine β_h for cohesive soils or η for cohesionless soils
(a) $\beta_h = \sqrt[4]{K_h b/(4EI)}$ for cohesive soil, or
(b) $\eta = \sqrt[5]{K_h/EI}$ for cohesionless soil

Step 6. Determine the dimensionless length factor
(a) $\beta_h D$ for cohesive soil, or
(b) ηD for cohesionless soil

Step 7. Determine if the pile is long or short
(a) Cohesive soil:
1. $\beta_h D > 2.25$ (long pile)
2. $\beta_h D < 2.25$ (short pile)

   Note: It is suggested that for $\beta_h D$ values between 2.0 and 2.5, both long and short pile criteria should be considered in Step 9, and then the smaller value should be used.

(b) Cohesionless soil:
1. $\eta D > 4.0$ (long pile)
2. $\eta D < 2.0$ (short pile)
3. $2.0 < \eta D < 4.0$ (intermediate pile)

Step 8. Determine other soil parameters over the embedded length of pile
(a) The Rankine passive pressure coefficient for cohesionless soil, K_p
$K_p = \tan^2(45 + \phi/2)$ where ϕ is the angle of internal friction
(b) The average effective weight of soil, y (kN/m^3)
(c) The cohesion, c_u (kPa)
$c_u = \frac{1}{2}$ the unconfined compressive strength, q_u

Step 9. Determine the ultimate lateral load for a single pile, Q_u

(a) Short free or fixed-headed pile in cohesive soil
Use D/b (and e_c/b for the free-headed case), enter Figure 5.32, select the corresponding value of $Q_u/c_u b^2$, and solve for Q_u (kN)

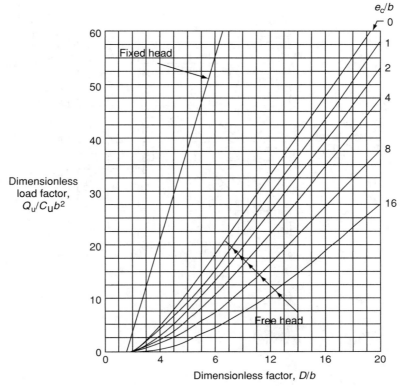

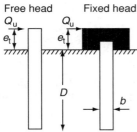

FIGURE 5.32
Ultimate lateral load capacity of short piles in cohesive soils. (From *LRFD Bridge Design Specifications, Customary U.S. Units*, 2nd ed., American Association of State Highway and Transportation Officials, Washington, DC, 1998 (with 1999 interim revisions). With permission.)

(b) Long free or fixed-headed pile in cohesive soil
   Using $M_y/c_u b^3$ (and e_c/b for the free-headed case), enter Figure 5.33, select the corresponding value of $Q_u/c_u b^2$, and solve for Q_u (kN)

(c) Short free or fixed-headed pile in cohesionless soil
   Use D/b (and e_c/D for the free-headed case), enter Figure 5.34, select the corresponding value of $Q_u/K_p b^3 \gamma$, and solve for Q_u (kN)

(d) Long free or fixed-headed pile in cohesionless soil
   Using $M_y/b^4 \gamma K_p$ (and e_c/b for the free-headed case), enter Figure 5.35, select the corresponding value of $Q_u/K_p b^3 \gamma$, and solve for Q_u (kN)

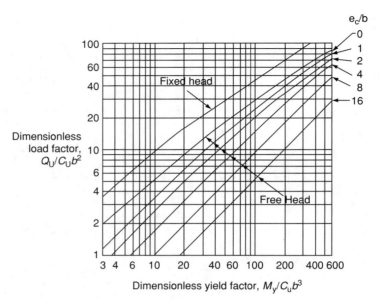

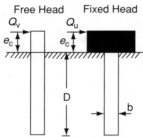

FIGURE 5.33
Ultimate lateral load capacity of long piles in cohesive soils. (From *LRFD Bridge Design Specifications, Customary U.S. Units*, 2nd ed., American Association of State Highway and Transportation Officials, Washington, DC, 1998 (with 1999 interim revisions). With permission.)

(e) Intermediate free or fixed-headed pile in cohesionless soil
    Calculate Q_u for both short pile (Step 9c) and long pile (Step 9d) and use the smaller value.

Step 10: Calculate the maximum allowable working load for a single pile Q_m. Calculate Q_m, from the ultimate load Q_u determined in step 9 as shown in Figure 5.36.

$$Q_m = \frac{Q_u}{2.5} \text{ (kN)}$$

Step 11. Calculate the working load for a single pile, Q_a to (kN)
Calculate Q_a corresponding to a given design deflection at the ground surface y (m) or the deflection corresponding to a given design load (Figure 5.36). If Q_a and y are not given, substitute the value of Q_m (kN) from Step 10 for Q_a in the following cases and solve for Y_m (m):

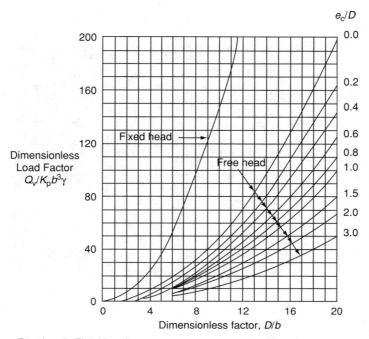

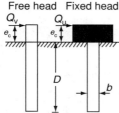

FIGURE 5.34
Ultimate lateral load capacity of short piles in cohesionless soils. (From *LRFD Bridge Design Specifications, Customary U.S. Units*, 2nd ed., American Association of State Highway and Transportation Officials, Washington, DC, 1998 (with 1999 interim revisions). With permission.)

(a) Free or fixed-headed pile in cohesive soil
Using $\beta_h D$ (and e/D for the free-headed case), enter Figure 5.37, select the corresponding value of $yK_h bD/Q_a$, and solve for Q_a (kN) or y (m)

(b) Free or fixed-headed pile in cohesionless soil
Using nD (and e/D for the free-headed case), enter Figure 5.38, select the corresponding value of $y(EI)^{3/5}K_h^{2/5}/Q_a D$, and solve for Q_a (kN) or y (m)

Step 12. Compare Q_a to Q_m

If $Q_a > Q_{m1}$ use Q_m and calculate y_m (Step 11)

If $Q_a < Q_m$ use Q_a and y

If Q_a and y are not given, use Q_m and y_m

Step 13. Reduce the allowable load from Step 12 for pile group effects and the method of pile installation

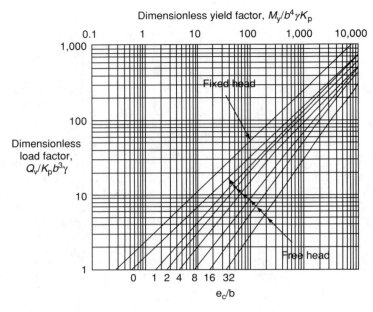

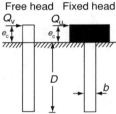

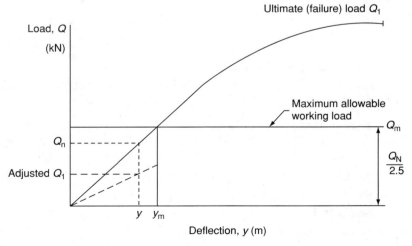

FIGURE 5.35
Ultimate lateral load capacity of long piles in cohesionless soils. (From *LRFD Bridge Design Specifications, Customary U.S. Units,* 2nd ed., American Association of State Highway and Transportation Officials, Washington, DC, 1998 (with 1999 interim revisions). With permission.)

FIGURE 5.36
Load deflection relationship used in determination of Brom's maximum working load. (From *LRFD Bridge Design Specifications, Customary U.S. Units,* 2nd ed., American Association of State Highway and Transportation Officials, Washington, DC, 1998 (with 1999 interim revisions). With permission.)

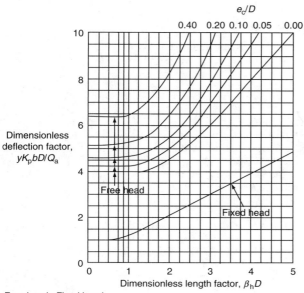

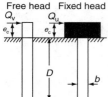

FIGURE 5.37

Lateral deflection at ground surface of piles in cohesive soils. (From *LRFD Bridge Design Specifications, Customary U.S. Units,* 2nd ed., American Association of State Highway and Transportation Officials, Washington, DC, 1998 (with 1999 interim revisions). With permission.)

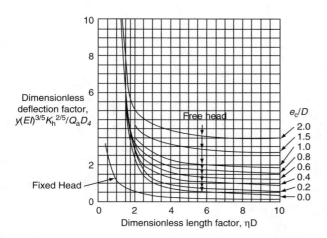

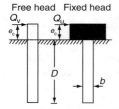

FIGURE 5.38

Lateral deflection at ground surface of piles in cohesionless soils. (From *LRFD Bridge Design Specifications, Customary U.S. Units,* 2nd ed., American Association of State Highway and Transportation Officials, Washington, DC, 1998 (with 1999 interim revisions). With permission.)

TABLE 5.12

Group Reduction Factors

Z	Reduction Factor
8b	1.0
6b	0.8
4b	0.65
3b	0.5

Source: From *LRFD Bridge Design Specifications, Customary U.S. Units,* 2nd ed., American Association of State Highway and Transportation Officials, Washington, DC, 1998 (with 1999 interim revisions). With permission.

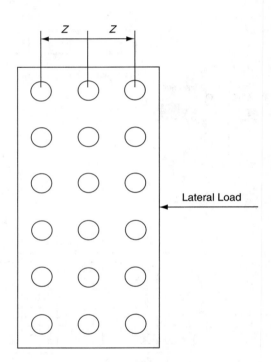

FIGURE 5.39
Guide for Table 5.12.

(a) Group reduction factor determined by the center-to-center pile spacing, z, in the direction of load (Table 5.12 and Figure 5.39)

(b) Method of installation reduction factor
   1. For driven piles use no reduction
   2. For jetted piles use 0.75 of the value from Step 13a

Step 14. Determine pile group lateral capacity
The total lateral load capacity of the pile group equals the adjusted allowable load per pile from Step 13b times the number of piles. The deflection of the pile group is the value selected in Step 12. It should be noted that no provision has been made to include the lateral resistance offered by the soil surrounding an embedded pile cap.

Example 5.8
Drill shaft design

You have been engaged as a foundation engineering consultant to design a drilled shaft for a building. Geotechnical engineers have recommended a drilled shaft or a group of piles. The value engineering analysis has indicated the drill shaft will be the most cost-effective solution. The structural engineer analyzing the building has given the following loading data that need to be transferred to the ground:

Working DL = 520 kips
         LL = 314 kips
Working DL moment = 2,550 kip-ft
         LL moment = 1,120 kip-ft
Working horizontal load = 195 kip

The attached borehole data (Figure 5.40) were given by the geotechnical engineer. You are required to design a single reinforced concrete drill shaft with concrete $f'_c = 4$ ksi and $f_y = 60$ ksi.

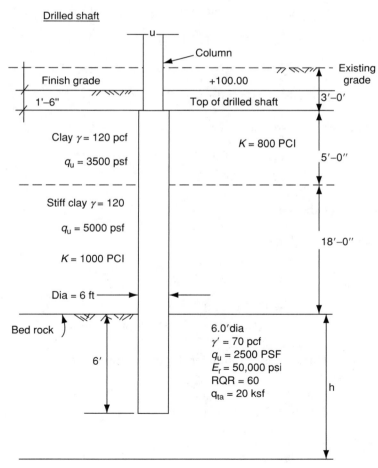

FIGURE 5.40
Soil profile for Example 5.8.

Drill shaft
Given:

DL = 520 kips
DLM = 2,550 kip-ft   DLM = 2550.00
LL = 314 kips   LLM = 1120.00
LLM = 1,120 kip-ft   Total M = 3670.00 kip-ft
V = 834 kips
M = 3,670 kips
H = 195 kips
Vertical load = 834 kips
Let us assume a 6-ft diameter drilled shaft:
$Q_u = Q_s + Q_T - W$ using a safety factor of

$$Q_a = \frac{1}{2.5}[Q_s + Q_T] - \frac{W}{2.5}$$
$$= q_{sa}A_s + q_{ta}A_T - \frac{W}{2.5}$$

Neglect resistance from the top layer of 5 ft. We will consider the second layer and assume the rock layer is the cohesive layer to determine the length of drilled shaft:

$$A_s = \pi D = 6\pi = 18.85 \text{ ft}^2/\text{ft}$$
$$A_{T2} = \frac{\pi D^2}{4} = \frac{5^2 \cdot 6^2 \pi}{4} = 9\pi = 28.27 \text{ ft}^2$$
$$A_{s2} = \pi_2 = 5.5\pi = 17.28 \text{ ft}^2/\text{ft}$$

Skin friction from stiff clay (second layer)

$$q_{sa} = 2.0 \text{ ksf}(\Leftarrow 5/2.5)$$
$$Q_s = 18.85 \times 2 \times 18 = 678 \text{ kips}$$

End bearing from bed shocket
  Dia = 6.0 dia
$Q_T = 20 \times 28.27 = 565$ kips

$$Q = 678 + 565 - 23 \times \left(\frac{\pi 6^2}{4}\right)\left(\frac{0.15}{2.5}\right)$$
$$= 1243 - 39.0$$
$$= 1204 \text{ kips} > V = 834 \text{ kips}$$

Therefore, take drill shaft at least 1 diameter depth into the rock, say 6 ft.
Now check for lateral loads

$$H = 195 \text{ ksf}, \quad M = 3,670 \text{ kips-ft} \quad e = \frac{M}{H} + 5 = \frac{3670}{195} + 5 = 23.82 \text{ ft}$$

From the top of stiff clay

Solution

Following the step-by-step procedure:

   Step 1. Soil type within $(4 \times D =)$ 24 ft depth
            = cohesive stiff clay

   Step 2. Computation of coefficient of horizontal subgrade reaction, K_h, with the critical
     depth

$$q_u = \text{average value} \left(\frac{5 \times 3.500 + 18 \times 5000}{23} \right)$$

$$= 4674 \text{ psf}$$

$$= 224 \text{ kPa}$$

Concrete drilled shaft from Table 5.10

$$n_1 = 0.4; \; n_2 = 1.15$$

$$K_h = \frac{n.n_2 80 q_u}{b}$$

where $q_u = 224$ kPa

$$b = \frac{6}{3.281} = 1.8287 \text{ m}$$

$$K_h = \frac{0.4 \times 1.15 \times 80 \times 224^3}{1.8287}$$

$$= 4507.7 \text{ kN/m}$$

$$= 28.85 \text{ kips/ft}^3$$

   Step 3. Adjust K_h for loading and soil conditions for stiff clays

$$K_h = \tfrac{1}{2} K_h (\text{from step 2})$$

$$= \tfrac{1}{2} \times 4507.7$$

$$= 2253.8 \text{ kN/m}^3$$

$$= 14.5 \text{ kips/ft}^3$$

   Step 4. Determine shaft parameters (Figure 5.41)
     (a) Modulus of elasticity

$$E_c = 57,000 \sqrt{f_c^1} = 57,000 \sqrt{4000}$$

$$= 3604996.5 \text{ psi}$$

$$E_c = 24,856.5 \text{ MPa}$$

$$E_s = 199,955 \text{ MPa} \qquad n = \frac{E_s}{E_c} = 8.0$$

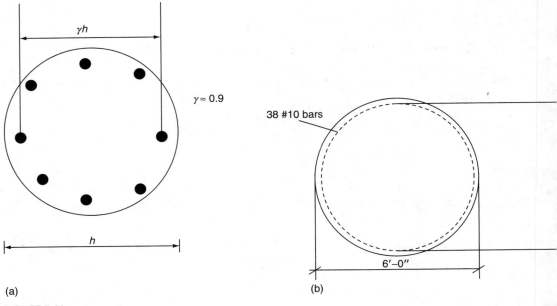

FIGURE 5.41
Shaft section with reinforcement layout: (a) schematic; (b) actual design.

(b) Moment of inertia

$$I_g = \frac{\pi D4}{64} = \frac{\pi \times 1.8287^4}{64} = 0.549 \text{ m}^4$$
$$A_s = 48.2b \text{ in}^2$$
$$= 0.03113 \text{ m}^2$$
$$I_{se} = 0.125 A_{sc}(\gamma D)^2$$

$$= 0.125 \times 0.03113 \times 1.6^2$$
$$= 0.010 \text{ m}^4$$
$$I_e = I_g + (n-1)I_{se}$$
$$= 0.549 + (g-1)(0.010) \text{ m}^4$$
$$I_e = 0.619 \text{ m}^4$$

(c) Section modulus $= \dfrac{I}{D/2} = \dfrac{0.619}{0.9144}$

$$S = 0.677 \text{ m}^3$$
$$= 677 \times 10^{-3} \text{ m}^3$$

(d) Yield stress of drilled shaft

$$\text{Concrete} = 4{,}000 \text{ psi}$$
$$= 27.58 \text{ MPa}$$

(e) Embedded shaft length
   $D = 23$ ft $= 7.02$ m

(f) Diameter $= b = 1.8287$ m
(g) Eccentricity of applied load e_c
   $e_c = 23.82$ ft $= 7.265$ m

(h) Resisting moment of pile M_y

$$\begin{aligned} M_y &= f'_c \cdot s \\ &= 27.58 \times 677 \times 10^{-3} \\ &= 18.671b \text{ kN m} \\ &= 13{,}584 \text{ kip} - \text{ft} \gg M_u \end{aligned}$$

Step 5. Determine βh for cohesive soils

$$\begin{aligned} \beta h &= \sqrt[4]{khb/4EI} \\ &= \left[\frac{2253.8 \times 1.8287}{4 \times (24.9 \times 10^6)(0.6122 \text{ m}^4)} \right]^4 \\ &= \left[\frac{4121.5}{60.975 \times 10^6} \right]^{1/4} \\ &= 0.091 \text{ m}^{-1} \end{aligned}$$

Step 6. Determine the dimensionless length factor

$$\begin{aligned} \beta hD &= 0.091 \times 7.265 \\ &= 0.66 \end{aligned}$$

Step 7. Determine if the shaft is long or short (cohesive soil)
$\beta hD > 2.25$ (long)
$\beta hD < 2.25$ (short)
Since $\beta hD = 0.66$, it is a short drilled shaft

Step 8. Determine the soil parameters
Rankine passive pressure coefficient cohesionless soil
Since the soil that is of concrete, this design is a cohesive one

$$C_u = \tfrac{1}{2}q_u = \frac{224}{2} = 112 \text{ kPa}$$

Or $C_u = 2.35$ ksf

Step 9. Determine the dimensionless factor D/b

$$D/b = \frac{5 + 18}{6} = 3.833 \simeq 4.0$$

This is a fixed head (the building column is fixed at base). From Figure 5.32 (for cohesive soil),
$Q_u/c_u b^2 = 20.0$ (dimensionless b factor)

$$Q_u = b^2 c_u (20)$$
$$= 6^2 \times 2.35 \times 20$$
$$= 1692 \text{ kips} \quad \text{(horizontal force)}$$

Step 10. Maximum allowable working load $= 1692/2.5 = 676.8$ kips
which is > 195 kips
Step 11. Calculate the deflection y
Dimensionless factor $\beta_h D = 0.66$
From Figure 5.37 (lateral deflection at ground surface for cohesive soil),
Dimensionless factor $Y \times K_h (bd/Q_a) = 1.05$ (fixed head)
Replace $Q_a \cdot Q_m$ by the applied load

$$Y = \frac{1.05 Q_m}{K_h b D} = \frac{1.05 \times 195.0}{14.5 \times 6 \times 29}$$
$$= 0.081 \text{ ft}$$
$$\approx 0.97 \text{ in.} \quad \text{OK}$$

Shaft Design (Figure 5.42 and Figure 5.43; Table 5.13)

$$f_c' = 4.0 \text{ ksi}$$
$$f_y = 60$$
$$A_s = 38\#10 \text{ bars}$$
$$= 48.26 \text{ m}^2$$
$$A_g = \frac{\pi B^2}{4} = \frac{\pi \times 72^2}{4} = 4071.5$$
$$\rho_g = 48.26/4071.5 = 0.012 = 1.2\%$$
$$\gamma = \frac{(72 - 2(3 + 0.625) - 1.25)}{72} = 0.88$$

$$\text{Min } P = 0.1 f_c' A_c$$
$$= 0.1 \times 4 \times 4071.5$$
$$= 1628.6 \text{ kips} \gg 1261.8 \text{ kips}$$
$$P_u = 1.4 \text{ DL} + 1.7 \text{ LL}$$
$$= 1.4 \times 520 + 1.7 \times 314$$
$$= 1261.8 \text{ kips}$$
$$M_u = 1.4 \times 2550 + 1.7 \times 1120$$
$$= 5474 \text{ kip-ft}$$

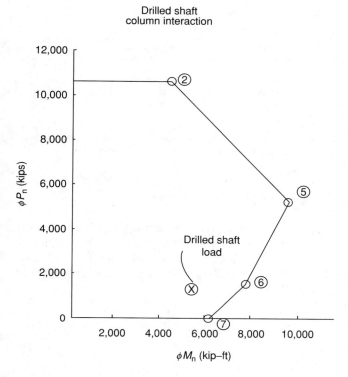

Drilled shaft
column interaction

FIGURE 5.42
Drilled shaft column interaction.

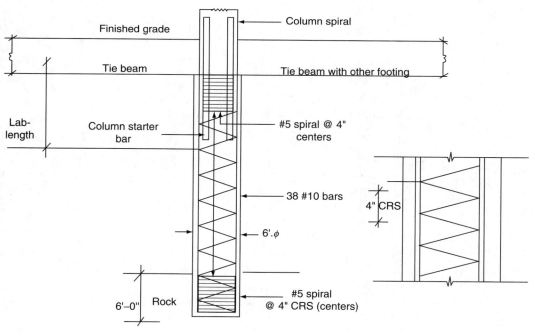

FIGURE 5.43
Rebar details of drilled shaft.

TABLE 5.13

Strength of Reinforced Column Sections from ACI Column Chart

ρ_g	p_f	$QP_n/4g$ (ksi)	ϕP_n (kips)	QM_n/Agh (ksi)	QM_n (k.ft)
0.012	2	2.60	10586	0.181	4422
0.012	5	1.318	5366	0.392	9576
0.012	6	0.400	1628.4	0.317	7743
0.012	7	0	0	0.254	6205

Source: LRFD Bridge Design Specifications, Customary U.S. Units, 2nd edn, American Association of State Highway and Transportation Officials, Washington, DC, 1998 (with 1999 interim revisions). With permission.

Example 5.10: Additional footing design example (rigid footing)

Machine Foundation Problem

As a foundation engineer you have been asked to design a machine foundation footing for a bakery mixer. The mixer loads are given below. The rear legs are subjected to additional shock load of 16 kips/ft^2. All four legs are identical with 100 in.^{2} area. Given that f_c = 4 ksi and f_y = 60 ksi, design the footing. Maximum allowable bearing pressure = 2.5 ksf (Figure 5.44–Figure 5.47).

Data from the manufacturer or mixer:

Net weight/leg DL = 10 kips/ft^2
Floor load, FL = 10 kips/ft^2
Shock load, SL = 16 kips/ft^2

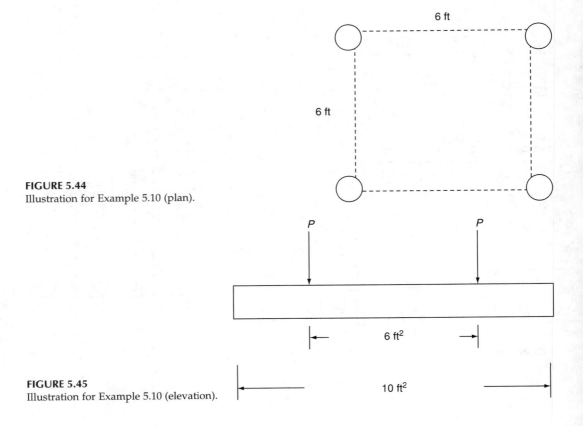

FIGURE 5.44
Illustration for Example 5.10 (plan).

FIGURE 5.45
Illustration for Example 5.10 (elevation).

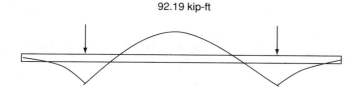

92.19 kip-ft

FIGURE 5.46
Bending moment diagram for Example 5.10.

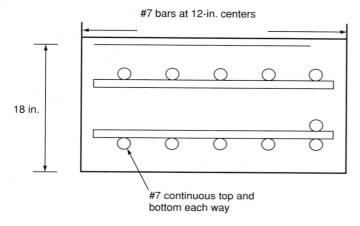

#7 bars at 12-in. centers

18 in.

#7 continuous top and
bottom each way

FIGURE 5.47
Reinforcement details (Example 5.10).

Load on rear legs $= 36$ kips/ft^2
Area of a rear leg $= 100$ in.^{2}
Area of a leg $= 0.69$ ft^2
Design load $= 1.6(\text{FL} + \text{SL} + \text{DL})$ for rear legs. Also note that a load factor of 1.6 is used

$\quad P = 40.00$ kips
By considering the case with rear leg loading, we will design the footing for this load case:

Total load/rear legs $= 2P = 80$ kips
Width of the footing $= 4.00$ ft (4 ft strip) for worse condition
Length of the footing $= 10.00$ ft
Depth $= 1.50$ ft
$d = 13.50$ in.
Area $= 40.00$ ft^2, assume 3-in. cover
Pressure under the footing $= 2.225$ ksf < 2.5 ksf.   OK

References

1. American Concrete Institute: ACI-318 Building Code Requirements for Reinforced Concrete and Commentary.
2. ACI 209 publication on shrinkage, Creep and Thermal Movements.
3. Notes on ACI 318-95 Building Code Requirements for Structural Concrete with Design Applica-

tion by Portland Cement Association.
4. AASHTO LRFD Bridge Design Specifications.
5. Florida Department of Transportation: Bridge Design Guidelines.
6. FHWA, 1998, *Design and Construction of Driven Pile Foundations*, Workshop Manual, vol. I, Publication Number FHWA HI-97-013, Revised, November.

6

Design of Driven Piles and Pile Groups

Manjriker Gunaratne

CONTENTS

6.1 Introduction

A pile foundation can be employed to transfer superstructure loads to stronger soil layers deep underground. Hence, it is a viable technique for foundation construction in the presence of undesirable soil conditions near the ground surface. However, owing to the high cost involved in piling, this foundation type is only utilized after other less costly alternatives, such as (1) combined footings and (2) ground modifications, have been considered and ruled out for the particular application. On the other hand, piles may be the only possible foundation construction technique in the presence of subgrades that are prone to erosion and in offshore construction involving drilling for petroleum.

6.2 Design of Pile Foundations

6.2.1 Selection of Pile Material for Different Construction Situations

Depending on applicability in a given construction situation, one of three different pile types, timber, concrete, or steel, is selected to construct a pile foundation.

6.2.1.1 Timber Piles

Timber is a relatively inexpensive material to be used in construction and its durability against rotting can be improved using preservatives and advanced techniques such as that are available in the market. However, the main drawback of timber piles is the limited structural capacity and length. Hence, timber piles are mostly suitable for construction of residential buildings in marshy areas and for stabilization of slopes (Figure 6.1a).

6.2.1.2 Concrete Piles

Concrete piles can be selected for foundation construction under the following circumstances:

1. The need to support heavy loads in maritime areas where steel piles easily corrode.
2. Existence of stronger soil types located at relatively shallow depths that are accessible to concrete piles.
3. Design of bridge piers and caissons that require large-diameter piles.
4. Design of large pile groups is needed to support heavy extensive structures so that the total expense can be minimized.
5. The need for minipiles to support residential buildings on weak and compressible soils.

(a)

(b)

(c)

FIGURE 6.1
(a) Groups of timber piles in construction. (From www.timberpilingcouncil.org. With permission.) (b) Production of precast concrete piles. (From www.composite-piles-marine-pilings.com. With permission.) (c) Steel sheet piles in a cofferdam application. (From www.dissen-juhn.com. With permission.)

The disadvantages of concrete piles are that they can be damaged by acidic environments or organic soils and they undergo abrasion due to wave action when used to construct offshore foundations.

Concrete piles are in wide use in construction due to their relatively high capacity and reasonable cost. The two most common types of concrete piles are (1) precast and (2) cast-*in-situ*. Of these, precast piles may be constructed to specifications at a separate casting yard or at the pile construction site itself if a large number of piles are needed for the particular construction. In any case, handling and transportation can cause intolerable tensile stresses in precast concrete piles. Hence, one should be cautious in handling and transportation so as to minimize the bending moments in the pile. Two other important issues that have to be addressed with precast piles that have to be driven are the ground displacement that they cause and the possible damage due to driving stresses. Therefore, driving of precast piles would not be suitable for construction situations where soil-displacement-sensitive structures are located in the proximity. Preaugering or jetting would be alternative installation techniques to suit such construction situations.

Cast-*in-situ* piles are of two types:

1. Cased type, which are piles that are cast inside a steel casing that is driven into the ground.
2. Uncased type, which are piles that are formed by pouring concrete into a drilled hole or into a driven casing before the casing is gradually withdrawn.

A detailed discussion of the use of casings in cast-*in-situ* pile construction is found in Bowles (2002).

Auger-cast concrete piles have the following properties:

- Higher capacity having larger diameter (tall building foundation)
- Low vibration during construction (business districts with high-rise buildings)
- Higher depth (load transfer into deeper strong soil).
- Replacement pile (no lateral soil movement). No compression of surrounding soil

6.2.1.3 Steel Piles

Steel piles offer excessive strength in both compression and tension. In addition, they are highly resistant to structural damage during driving. Furthermore, they can be spliced very conveniently to suit any desired length. On the other hand, the main disadvantages of steel piles are (1) high expense and (2) vulnerability to corrosion in marine environments. Therefore, steel piles are ideal for supporting excessively heavy structures such as multistorey buildings in soft ground underlain by dense sands, stiff clays, or bedrock in nonmarine environments.

6.2.2 Selection of the Method of Installation

Piles can be classified into three categories depending on the degree of soil displacement during installation: (i) large volume displacement piles; (ii) small volume displacement piles; and (iii) replacement piles. Driven precast solid concrete piles, close ended pipe piles, and driven and cast in-place concrete piles fall into the large volume displacement category in which a large volume of soil is displaced during installation. Steel piles with thin cross sections, for example, H and open-ended pipe piles, fall into the small volume displacement pile category where the amount of soil displaced during installation is small. All bored and cast in-place concrete piles and caissons fall into the replacement pile category, in which the soil is removed and replaced with concrete. Installation of large volume displacement piles obviously causes disturbance to the soil surrounding the pile.

6.2.3 Design Criteria

Failure of a structurally intact pile can be caused due to two reasons: (1) shear failure of the soil surrounding the pile and (2) excessive settlement of the foundation. Therefore, the task of the foundation designer is to find out an economical pile to carry the working load with a low probability of shear failure, while keeping the resulting settlement to within allowable limits. In designing a single pile against shear failure, it is customary to estimate the maximum load that can be applied to a pile without causing shear failure, generally referred to as the ultimate carrying capacity.

As in the case of shallow footings, two design approaches, (1) allowable stress design (ASD) method and (2) load resistance factor design (LRFD) method, are available for piles. The following sections will mostly elaborate the ASD method and basics of the LFRD method will be presented in Section 6.9. The ASD requires the following conditions:

6.2.3.1 *Allowable Loads*

$$R_n/FS = Q_{all} \qquad (6.1)$$

where R_n is the ultimate resistance of the pile, Q_{all} is the allowable design load, and FS is the factor of safety.

6.2.3.2 *Allowable Deflections*

$$\delta_{est} \leq \delta_{tol} \qquad (6.2)$$

where δ_{est} is the estimated deflection (settlement) of a pile foundation component and δ_{tol} is the deflection (settlement) that can be tolerated by that component.

6.3 Estimation of Static Pile Capacity of a Single Pile

Piles are usually placed in service as a group rather than on an individual basis to meet loading demands and ensure stability. In addition, if some probability of nonvertical loading also exists and the designer is uncertain of the lateral capacity of the piles, then it is common to include some battered piles as well in the group (Figure 6.2).

As one realizes from Figure 6.2, the structural load ($P_{structural}$) is transferred to each individual pile in the group ($P_{pile,i}$) through the pile cap. The relation between $P_{structural}$ and $P_{pile,i}$ is determined by considering the pile cap as a statically determinate or a statically indeterminate structure depending on the pile configuration. The primary objective of designing a pile is to ascertain that the foundation of a given pile, i, in the group, or the individual pile capacity, can meet the demand of the load imposed on it, i.e., $P_{pile,i}$.

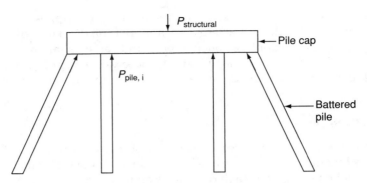

FIGURE 6.2
Piles in a typical service condition.

The pile designer must be knowledgeable of the capacity of a pile (1) under normal working conditions (static capacity) and (2) while it is driven (dynamic capacity). Since the dynamic pile capacity is addressed in detail in Chapter 8, discussions in this chapter would be limited to the static pile capacity only (Figure 6.3).

The ultimate working load that can be applied to a given pile depends on the resistance that the pile can produce in terms of side friction and point bearing (Figure 6.2). Hence, the expression for the allowable load P_a on a pile would take the following form:

$$P_a = \frac{P_{pu} + P_{su}}{FS} \tag{6.3}$$

where P_{pu} is the ultimate point capacity, P_{su} is the ultimate side friction, and FS is the safety factor.

A suitable factor of safety is applied to the ultimate carrying capacity to obtain the allowable load on a pile, subject to the allowable settlement. The magnitude of the safety

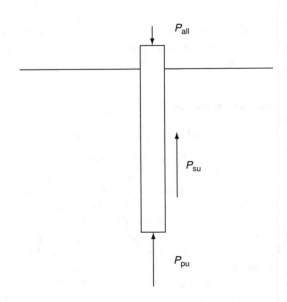

FIGURE 6.3
Illustration of pile capacity.

factor depends on the confidence of the designer on the design, and a factor of safety between 3 and 4 is very often used.

6.3.1 Estimation of Point Capacity

6.3.1.1 Meyerhoff's Method

The ultimate point capacity component in Equation (6.3) corresponds to the bearing capacity of a shallow footing expressed by Equation (3.1), and is a modified form of Equation (3.2):

$$P_{p,ult} = A_p[cN_c^* + q(N_q^* - 1)] \tag{6.4}$$

where A_p is the area of the pile cross section, q is the vertical effective stress at the pile tip, c is the cohesion of the bearing layer, and N_c^* and N_q^* are the bearing capacity factors modified for deep foundations (and a B/L ratio of 1.0).

It is noted that the surcharge component ($0.5BN_\gamma.\gamma$) of Equation (3.2) has been omitted due to the insignificance of the surcharge zone of the pile compared to the entire stress regime along the depth of the pile.

The bearing capacity factors for deep foundations can be found in Figure 6.4. However, use of the bearing capacity factors mentioned above is more complex than in the case of shallow footings since, in the case of deep foundations, the mobilization of shear strength also depends on the extent of the pile's penetration into the bearing layer. In granular soils, the depth ratio at which the maximum strength is mobilized is called the critical depth ratio $(L_b/D)_{cr}$ for the mobilization of N_c^* and N_q^* for different values of Φ (Figure 6.5).

According to Meyerhoff (1976), the maximum values of N_c^* and N_q^* are usually mobilized at depth ratios of 0.5 $(L_b/D)_{cr}$. Hence, one has to follow an interpolation process to evaluate the bearing capacity factors if the depth ratio is less than 0.5 $(L_b/D)_{cr}$. This is illustrated in Example 6.1.

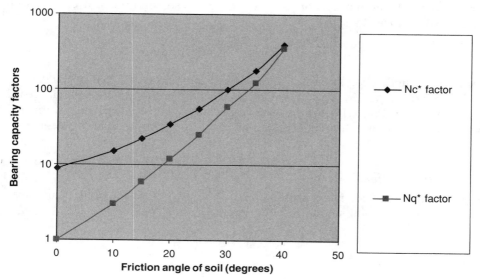

FIGURE 6.4
Bearing capacity factors for deep foundations.

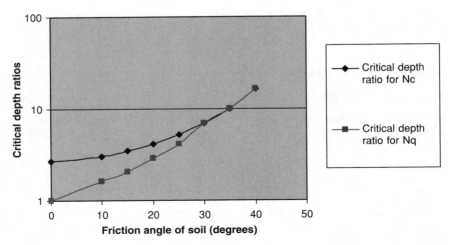

FIGURE 6.5
Variation of critical depth ratio with friction angle of soil.

Sandy Soils
In the case of sandy soils, where the cohesive resistance is negligible, Equation (6.4) can be reduced to

$$P_{p,ult} = A_p[q(N_q^* - 1)] \tag{6.5}$$

where the limiting point resistance is

$$P_{p,ult} = 50A_pN_q^*(\tan\phi)kN \tag{6.6}$$

Clayey Soils
The most critical design condition in clayey soils is the undrained condition where the apparent angle of internal friction is zero. Under these conditions, it can be seen that Equation (6.4) reduces to

$$P_{p,ult} \approx 9.0A_pc_u \tag{6.7}$$

where c_u is the undrained strength of the clay. It must be noted that in the case of steel piles (HP or pipe type) A_p is usually computed as the entire cross-sectional area due to plugging of the section with bearing soil, especially when it is driven to firm bearing. However, when piles are driven to bearing on rock, A_p is computed as the steel area of the cross section.

6.3.1.2 Vesic's Method

Based on the theory of cavity expansion, Vesic (1977) expressed the point bearing capacity of a pile by an expression similar to Equation (6.4):

$$P_{p,ult} = A_p[cN_c^* + \eta q N_q^*] \tag{6.8}$$

where

$$\eta = \frac{1 + 2K_0}{3}$$

c and q are defined as in Equation (6.4). K_0 is the coefficient of earth pressure at rest. N_c^* and N_q^* can be obtained from Table 6.1(a) and (b) based on the rigidity index I_r and the reduced rigidity index I_{rr} defined as follows:

$$I_r = \frac{G_s}{c + q \tan \phi} \tag{6.9}$$

and

$$I_{rr} = \frac{I_r}{1 + I_r \varepsilon_v} \tag{6.10}$$

where G_s is the shear modulus of the foundation soil and ε_v is the average plastic volumetric strain undergone by the foundation soil due to the imposed stresses. The following values are recommended for I_r (Bowles, 2002):

It is noted that in the case of sand that does not exhibit volumetric dilation or undrained clay

$$\varepsilon_v \to 0$$

Then, $I_r = I_{rr}$.

TABLE 6.1(a)

N_c^* Factors for Vesic's Bearing Capacity Evaluation Method

Φ (°)	I_{rr}				
	10	50	100	200	500
0	6.97	9.12	10.04	10.97	12.19
5	8.99	12.82	14.69	16.69	19.59
10	11.55	17.99	21.46	25.43	31.59
20	18.83	34.53	44.44	56.97	78.78
30	30.03	63.21	86.64	118.53	178.98
35	37.65	84.00	118.22	166.15	260.15
40	47.04	110.48	159.13	228.97	370.04
45	53.66	144.11	211.79	311.04	516.60

Source: From Bowles, J.E., 2002, *Foundation Analysis and Design*, McGraw-Hill, New York. With permission.

TABLE 6.1(b)

N_q^* Factors for Vesic's Bearing Capacity Evaluation Method

Φ (°)	I_{rr}				
	10	50	100	200	500
0	1.00	1.00	1.00	1.00	1.00
5	1.79	2.12	2.28	2.46	2.71
10	3.04	4.17	4.78	5.48	6.57
20	7.85	13.57	17.17	21.73	29.67
30	18.34	37.50	51.02	69.43	104.33
35	27.36	59.82	83.78	117.34	183.16
40	40.47	93.70	134.53	193.13	311.50
45	59.66	145.11	212.79	312.04	517.60

Source: From Bowles, J.E., 2002, *Foundation Analysis and Design*, McGraw-Hill, New York. With permission.

6.3.2 Skin-Friction Capacity of Piles

The skin-friction capacity of piles can be evaluated by means of the following expression:

$$P_{s,u} = \int_0^L pf \, dz \qquad (6.11)$$

where p is the perimeter of the pile section, z is the coordinate axis along the depth direction, f is the unit skin friction at any depth z, and L is the length of the pile.

6.3.2.1 Unit Skin Friction in Sandy Soils

Since the origin of skin friction in granular soils is due to the frictional interaction between piles and granular material, the unit skin friction (skin-frictional force per unit area) can be expressed as

$$f = K\sigma_v' \tan \delta \qquad (6.12)$$

where K is the earth pressure coefficient (K_0 for bored piles and 1.4 K_0 for driven piles), δ is the angle of friction between the soil and the pile material (usually assumed to be 2/3 if one looks for a generic value; if a more appropriate value for interaction between a particular pile material and a soil is needed, one can use the values suggested in Chapter 10), and σ_v' is the vertical effective stress at the point of interest (i.e., where f is computed).

It can be seen from the above expression that the unit skin friction can increase linearly with depth. However, practically, a depth of 15B (where B is the cross-sectional dimension) has been found to be the limiting depth for this increase. K_0, the coefficient of lateral earth pressure at rest, is typically expressed by

$$K_0 = 1 - \sin \phi \qquad (6.13)$$

6.3.2.2 Skin Friction in Clayey Soils

In clayey soils, on the other hand, skin friction results from adhesion between soil particles and the pile. Hence, the unit skin friction can be simply expressed by

$$f = \alpha c_u \qquad (6.14)$$

where the adhesion factor α can be obtained from Table 6.2 based on the undrained shear strength. Table 6.2 has been developed based on information from Peck (1974).

TABLE 6.2

Adhesion Factors

Undrained Strength (kPa)	α
0	1.0
50	0.95
100	0.8
150	0.65
200	0.6
250	0.55
300	0.5

Example 6.1

Estimate the maximum allowable static load on the 200-mm^2 driven pile shown in Figure 6.6. Assume the following soil properties:

	Loose Sand	Clay	Dense Sand
Unit weight (kN/m^3)	17.0	17.5	18.0
Undrained cohesion (kPa)		40	
Friction angle	28°		38°

Solution
Computation of skin friction in loose sand
Applying Equations (6.12) and (6.13), one would obtain

$$f = 1.4K_0(17.0)z \tan \delta$$

up to a depth of −3.0 m (i.e., 15.0 × 0.2) and constant thereafter.
   Assume that $\delta = 2/3(\phi) = 19°$ and $K_0 = (1 - \sin \phi)$ OCR
   Also assume that the surficial loose sand is normally consolidated. Hence the over consolidation ratio (OCR) = 1.0, $K_0 = 0.574$

$$f = 4.203z \text{ kPa} \quad \text{for } z < 3m$$
$$f = 12.6 \text{ kPa} \quad \text{for } z > 3m$$

Computation of skin friction in clay
Applying Equation (6.14), one obtains

$$f = \alpha(40)$$

where $\alpha = 1.0$ from Table 6.2

$$f = 40 \text{ kPa}$$

The dense sand layer can be treated as an end-bearing layer and hence its skin-frictional contribution cannot be included. Since the pile perimeter is constant throughout the depth, the total skin-frictional force (Equation (6.11)) can be computed by multiplying the area of the skin-friction distribution shown in Figure 6.6 by the pile perimeter of 0.8 m. Hence,

$$P_{sf} = (0.8)[0.5(3)(12.6) + 12.6(1) + 40(6)] = 217.2 \text{ kN}$$

Computation of the point resistance in dense sand
From Figure 6.5, $(L/D)_{cr} = 15$ for $\phi = 30°$. For the current problem, $L/D = 1/0.2 = 5$. Since in this case, $L/D < 0.5(L/D)_{cr}$, N_q^*

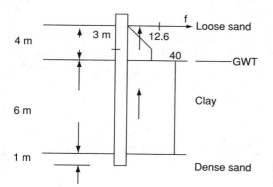

FIGURE 6.6
Illustration for the Example 6.1.

$$N_q = [(L/D)/0.5(L/D)_{cr}](N_q^*) = 5/7.5 \times 300 = 208$$

Note that an N_q^* value of 300 is obtained from Figure 6.4. Also

$$A_p = 0.2 \times 0.2 = 0.04 \text{ m}^2$$

$$q = \sigma_v' = 17.0(4) + (17.5 - 9.8)(6) + (18.0 - 9.8)(1) = 122.4 \text{ kPa}$$

Then, by substituting in Equation (6.5)

$$P_{pu} = 0.04(208 - 1)(122.4) \text{ kN} = 974.3 \text{ kN}$$

But $P_{pu \text{ max}} = 0.04(50)(208) \tan 38° = 271.8 \text{ kN}$. Therefore,

$$P_{pu} = 271.8 \text{ kN}$$

Finally, by applying Equation (6.3), one can determine the maximum allowable load as

$$P_{all} = (271.8 + 217.2)/4 = 122.3 \text{ kN}$$

In practice, the ultimate carrying capacity is estimated using the static bearing capacity methods and then often verified by pile load tests.

6.3.2.3 The α Method

In the author's opinion, the pile capacity evaluations outlined above are generalized in the method popularly known as the α method expressed as follows:

$$f = \alpha c + \sigma_v' K \tan \delta \tag{6.15a}$$

in which the mathematical symbols have been defined based on Equations (6.12) and (6.14).

Sladen (1992) derived the following analytical expression that explains the dependence of the α factor on the undrained shear strength of saturated fine grained soils

$$\alpha = C_1 \left(\frac{\sigma_v'}{s_u}\right)^{0.45} \tag{6.15b}$$

where s_u is the undrained shear strength described in the Section 1.4.2.2 and $C_1 = 0.4$ to 0.5 for bored piles and greater than 0.5 for driven piles.

6.3.2.4 The β Method

The β method suggested by Burland (1973) for the computation of skin-friction derives from the concepts used in the formulation of Equation (6.12) that is used for the determination of skin friction in granular soils. It can be expressed in the following general formulation:

$$f = \beta \sigma_v' \tag{6.16}$$

Comparison of Equations (6.12) and (6.16) shows that the factor β represents the term $K \tan \theta$, which is completely dependent on the angle of friction ϕ. Bowles (2002) shows that, for most granular soils, the factor β is in the range of 0.27 to 0.3, providing a convenient practical way of evaluating the skin friction of piles in granular soils.

6.3.2.5 The λ Method

A semiempirical approach for prediction of skin-friction capacity of piles in clayey soils was presented by Vijayvergia and Frocht (1972) based on load tests conducted on long piles that support offshore oil production structures. The corresponding expression for skin-friction capacity is given in Equation (6.17)

$$f = \lambda[\sigma'_v + 2s_u] \tag{6.17}$$

Based on back calculation of observed capacities of static pile load tests, the nondimensional coefficient λ has been presented as a function of the depth as shown in Figure 6.7.

6.3.3 Pile Capacity Estimation from *In Situ* Tests

6.3.3.1 Pile Capacity Estimation from Standard Penetration Test Results

Meyerhoff (1976) proposed a relationship (Equation (6.18)) to determine the point capacity of a pile in coarse sand and gravel, in kPa, using standard penetration test (SPT) data:

$$q_{pu} = (40N)\frac{L_b}{D} \tag{6.18}$$

where N is the weighted SPT average in an influence zone between $8B$ below and $3B$ above the pile tip, q_{pu} is in kPa (Bowles, 2002 suggests the use of N_{55} for N in this relationship), L_b is the pile penetration in the bearing layer, and D is the pile diameter (or the equivalent diameter).

As pointed out in Section 6.3.1, the point resistance reaches a limiting value at a critical L_b/D. For the above outlined q_{pu} vs. N relationship (Equation (6.18)), the suggested critical L_b/D is about 90. Meyerhoff (1976) also proposed the following alternative relationship for nonplastic silt:

$$q_{pu} = 300N \tag{6.19}$$

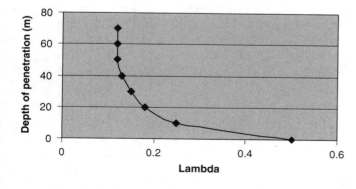

FIGURE 6.7
Dependence of the λ factor on pile penetration.

On the other hand, for the ultimate unit skin friction in sands, the following relationships were proposed by Meyerhoff (1976):

$$f_{su} = 2\bar{N} \quad (6.20a)$$

for moderate to large displacement piles and

$$f_{su} = \bar{N} \quad (6.20b)$$

for small displacement piles such as steel H piles, $\bar{N}$ where is the weighted average SPT of soil layers within the embedded length.

6.3.3.2 Pile Capacity Estimation from Cone Penetration Test Results

AASHTO (1996) recommends the following technique proposed by Nottingham and Schmertmann (1975) to determine the point bearing capacity in clay based on cone penetration data:

$$q_p = R_1 R_2 \frac{q_{c1} + q_{c2}}{2} \quad (6.21)$$

where q_{c1} and q_{c2} are minimum averages (excluding sudden peaks and troughs) of q_c values in the influence zones below the pile tip and above the pile tip, respectively. These influence zones are shown in Figure 6.8. R_1 is a reduction factor evaluated from Table 6.3. R_2 is 1.0 for the electrical cone and 0.5 for the mechanical cone.

A similar expression is available for the evaluation of point bearing resistance of sands (DeRuiter and Beringen, 1979):

$$q_p = \frac{q_{c1} + q_{c2}}{2} k_b' \quad (6.22)$$

where q_{c1} and q_{c2} are minimum averages (excluding sudden peaks and troughs) of q_c values in the influence zones below the pile tip and above the pile tip, respectively. $k_b' = 1.0$ for normally consolidated sand and 0.67 for overconsolidated sand.

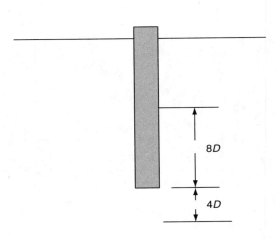

FIGURE 6.8
Tip influence zone.

TABLE 6.3

C_u vs. R_1

C_u (kPa)	R_1
<50	1
75	0.64
100	0.53
125	0.42
150	0.36
175	0.33
200	0.30

Nottingham and Schmertmann (1975) also developed a correlation between skin friction and the sleeve resistance obtained from cone penetration test (CPT) as expressed in Equation (6.23):

$$f_{su} = \alpha' f_s \tag{6.23}$$

In the case of electrical cone penetrometers, α', the frictional resistance modification factor can be evaluated from Table 6.4 based on the depth of embedment, Z/B.

Tomlinson (1994) advocates the use of the cone resistance in evaluating the skin friction developed in piles since the former is found to be more sensitive to variations in soil density than the latter. Tomlinson (1994) provides the empirical data in Table 6.5 for this evaluation.

Example 6.2

The SPT profile of a site is shown in Figure 6.9. Estimate the depth to which a HP 360 × 108 pile must be driven at this site if it is to carry a load of 1500 kN. Assume that the SPT test was performed in silty clay in the absence of water and the unit weights of peat, silty clay (dry), saturated silty clay, and saturated medium-dense sand are 10.5, 16.0, 17.5, and 17.2 kN/m^3, respectively. Use Meyerhoff's method for estimating point bearing and the α method for estimating skin-friction capacity.

TABLE 6.4

Frictional Resistance Modification
Factors Applied to CPT Results (α')

Z/B	Timber	Concrete	Steel
5	2.5	1.4	2.0
10	1.7	1.1	1.25
15	1.25	0.85	0.9
20	1.0	0.8	0.82
25	0.85	0.7	0.8
30	0.8	0.7	0.75
35	0.8	0.7	0.75
40	0.8	0.7	0.75

TABLE 6.5

Relationships between Pile Shaft Friction and Cone Resistance

Pile Type	Ultimate Unit Shaft Friction
Timber	$0.012q_c$
Precast concrete	$0.012q_c$
Precast concrete with enlarged base	$0.018q_c$
Steel displacement	$0.012q_c$
Open-ended steel tube	$0.0008q_c$
Open-ended steel tube driven into fine to medium sand	$0.0033q_c$

Source: From Tomlinson, M.J., 1994, *Pile Design and Construction Practices*, 4th ed., E & FN Spon, London. With permission.

Dimensions of HP 360 × 108 pile

Depth = 346 mm, width = 371 mm, flange thickness = 12.8 mm, web thickness = 12.8 mm

Plugged area = 0.371 m × 0.346 m = 0.128 m^2 = A_p

Pile perimeter (assuming no plugging for skin friction) = 371(2) + 12.8(4) + 346(2) + (371 − 12.8)(2) = 2.2 m = p

Minimum pile dimension = 0.346 m

Limiting skin-friction depth in sand = $L_{s,lm}$ = 15(0.346) = 5.19 m (assumed to apply from the clay–sand interface)

Critical end-bearing penetration = $L_{p,cr}$ (Figure 6.5) = 3(0.346) for clays = 1 m = 10(0.346) for sand = 3.46 m

The following soil strength properties can be obtained based on the SPT values (Table 6.6):

N_q' values are obtained from Figure 6.4

$\delta = 2/3 N$

$K = 1.4K_0 \tan \delta = 1.4(1 - \sin N) \tan \delta$

From Equations (6.7) and (6.14), the maximum total ultimate resistance produced by the clayey layers = point bearing + skin friction = 9(0.128)(25) + (2.2)[1.03(19)(1) + 0.92(50)(4) + (1.0)(25)(2)] = 28.8 + 556.6 = 585.4 kN.

Hence, the pile has to be driven into sand (say up to a depth of L m).

TABLE 6.6

Soil Parameters Related to the Pile Design in Example 6.2

Depth (m)	ϕ	C_u (kPa)	N_c'	N_q'	α	δ (°)	K	$L_{s,lm} = 5.2$ Applies?	$L_{p,cr}$
0–1		19	9		1.03				1
1–5		50	9		0.92				1
5–7		25	9		1.0				1
7–10.1	34			100		23	0.26	No	3.46
10.1–12.3	36			150		24	0.26	No	3.46
12.3–	38			200		25	0.25	Yes	3.46

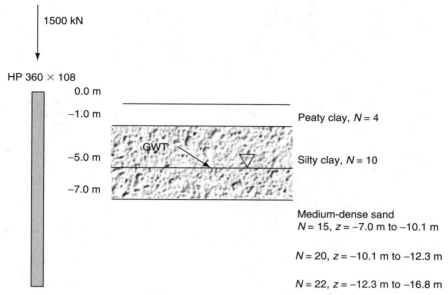

1500 kN

HP 360 × 108

0.0 m

−1.0 m　　　Peaty clay, $N = 4$

GWT

−5.0 m　　　Silty clay, $N = 10$

−7.0 m

Medium-dense sand
$N = 15$, $z = -7.0$ m to -10.1 m

$N = 20$, $z = -10.1$ m to -12.3 m

$N = 22$, $z = -12.3$ m to -16.8 m

FIGURE 6.9
Illustration for Example 6.2.

Since the critical embedment is 3.46 m, one can assume that the pile needs to be driven passing a 12.3 m depth for complete mobilization of point capacity and skin friction.

Assume that for depths greater than 16.8 m the soil properties are similar to those from 12.3 to 16.8 m.

Effective clay overburden $= (10.5)(1) + 16(4) + (17.5 - 9.8)(2) = 89.9\,\text{kPa}$

Effective sand overburden $= (L - 7)(17.2 - 9.8) = 7.4L - 51.8$

From Equation (6.5) for net ultimate point resistance

$P_{pu} = (200 - 1)(0.128)[89.9 + 7.4L - 51.8] = (188.5L + 970.5)\,\text{kN}$

$P_{su} = $ (in sand) $= 2.2\{(0.26)(3.1)(1/2)[89.9 + 89.9 + (3.1)(17.2 - 9.8)] +$
　　　　　　$(0.26)(2.2)(1/2)[89.9 + (3.1)(17.2 - 9.8) + 89.9 + (3.1 + 2.2)$
　　　　　　$(17.2 - 9.8)] + 0.25(L - 12.3)[89.9 + (5.3)(17.2 - 9.8)]\}$

$= 2.2\{81.7 + 69 + 32.3(L - 12.3)\} = 2.2(32.3L - 246.6)\,\text{kN}$

Total ultimate resistance $= 556.6 + 188.5L + 970.5 + 2.2(32.3L - 246.6)$
　　　　　　　　　　　　$= 259.6L + 984.6$

Applying Equation (6.3),

$1500 = (259.6L + 984.6)/2.5$

$L = 10.7\,\text{m}$

Hence, it must be driven to only 13.5 m below the ground.

Example 6.3
A cased concrete pile is required to carry a safe working load of 900 kN in compression at a site where the CPT results are given in Figure 6.10. Recommend a suitable pile size and a depth of penetration.

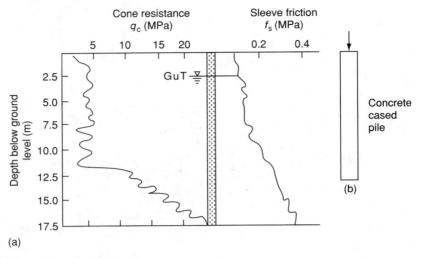

FIGURE 6.10
CPT results for Example 6.3.

From Figure 6.10(a), it is seen clearly that the immediate subsurface consists of a loose sand layer up to a depth of 11.0 m underlain by a denser sand layer.

Based on the cone resistance (Figure 6.10), and Equation (6.22), the maximum end-bearing resistance that can be obtained from the loose fine sand layer is

$$q_P = \frac{6+3}{2}(1.0) = 4.5 \text{ MPa}$$

Assume that a 400 mm diameter pile is employed in order not to overstress the concrete as shown later in the solution. The tip area of this pile = 0.126 m^2 and the pile perimeter = 1.26 m.

Then, the maximum working load that can be carried at the tip is computed as 4.5(1000)(0.126)/2.5 = 226 kN.

Hence, it is advisable to set this pile in the dense sand with an embedment of 13.0 m as shown in Figure 6.10(b).

By applying Equation (6.22) again,

$$q_P = \frac{13+7}{2}(1.0) = 10 \text{ MPa}$$

$$P_{p,u} = 1260 \text{ kN}$$

TABLE 6.7

Computational Aid for Example 6.4

Depth Interval (m)	f_s (MPa)	αf_s (kPa)	P_{su} (kN), Equation (6.11)
0–2.0	0.1	0.07	176
2.0–4.0	0.11	0.077	194
4.0–6.0	0.12	0.084	211
6.0–8.0	0.14	0.098	246
8.0–10.0	0.16	0.112	282
10.0–12.0	0.2	0.14	352
12.0–13.0	0.22	0.154	194

For a depth of embedment (z/B) ratio of $13.5/0.4 = 33.75$, from Table 6.4, $\alpha = 0.7$. As shown in Table 6.7, Equation (6.23) can be applied on incremental basis,

Total ultimate skin friction $= 1654\,\text{kN}$
Static pile capacity $= (1260 + 1654)/2.5 = 1165.6\,\text{kN}$

Hence, the load can be carried safely at a pile embedment of 13.0 m. The same design is repeated under LRFD guidelines in Section 6.9.

6.4 Pile Load Transfer

When a structural load is applied on a pile, it will be supported by certain amounts of skin friction and point bearing resistance that are mobilized as required. The degree of mobilization of both skin friction and point bearing resistance depends on the relative displacement undergone by the pile at the particular location of reference with respect to the surrounding soil. This condition is graphically illustrated in Figure 6.11, that shows the mobilization of skin friction through shear stress along points on the embedded pile surface governed by the shear strain undergone by the pile with respect to the surrounding soil (slip) at those locations. Although the magnitude of slip needed to mobilize the ultimate shear resistance depends on the soil type, typically it would be within a few millimeters (e.g., <10 mm). Similarly, mobilization of point (tip) resistance depends on the axial strain or penetration of the pile tip in the bearing layer and for complete mobilization, a penetration of 10 to 25% of the pile diameter would be required.

The discussion in the previous section enables one to evaluate the ultimate or the maximum resistance that can be mobilized at the tip or the shaft. It is quite typical of many soil types to reach a critical state at a much higher strain than is required for the mobilization of ultimate or peak strength, especially in shear. The shear strength at the critical state is known as the residual shear strength (Figure 6.11).

Based on the above discussion, one realizes that when a certain structural load is applied on a pile that has been already installed by driving or *in situ* casting, the following conditions must be satisfied:

1. $P_{\text{working}} = P_s + P_p$
2. P_s and P_p cause an immediate settlement of the pile with respect to the surrounding layers and the bearing layer producing slip at the frictional interface and penetration of the bearing layer, respectively. The above slip induces shear strains, γ, and the penetration induces an axial strain, ε, on the pile tip.
3. The magnitudes of γ and ε, respectively, determine the levels of interfacial shear stress, τ, and normal stress at the tip, σ, based on the deformation characteristics

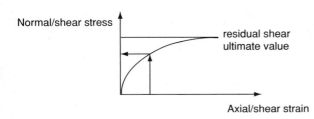

FIGURE 6.11
Mobilization of pile resistance.

shown in Figure 6.11. On the other hand, it is the mobilization of τ and σ that finally determine the magnitudes of P_s and P_p.

It is realized how an interplay between forces P_s and P_p occurs under conditions 1 to 3 until an equilibrium is finally reached. This process is known as the pile load transfer process. A typical load transfer curve at the equilibrium is illustrated in Figure 6.12. A load transfer curve such as in Figure 6.12 depicts the axial load carried by the pile at any given depth. Hence, the difference between the applied load and the axial load at that depth indicates the cumulative frictional resistance mobilized up to that depth. The axial load in the pile effective at any depth z can be experimentally determined by installing strain gages at that depth. When the longitudinal strain, ε_z at a depth z, is electronically monitored, the axial force at that point, $P(z)$, can be estimated as follows:

$$P(z) = EA_p\varepsilon_z \tag{6.24}$$

where E is the elastic modulus of the pile material.

The plot of $P(z)$ vs. z (the load transfer curve) in Figure 6.12 corresponds to the applied working load of P_{w1} where the mobilized point resistance is shown as P_{p1}. The above technique also provides one with the means of observing the variation of the load transfer curve as P_w is increased, for instance, to P_{w2} (Figure 6.12).

6.5 Time Variation of Pile Capacity (Pile Setup)

Due to the initial disturbance caused by pile installation and the consequent stabilization of the surrounding soil and pore water; in most soils, the axial or lateral pile capacity changes with time. A number of researchers have proposed analytical methods to estimate the change in pile capacity with time. Thilakasiri et al. (2003) have performed a case study comparing many previously established methods of predicting variation of pile capacity with time.

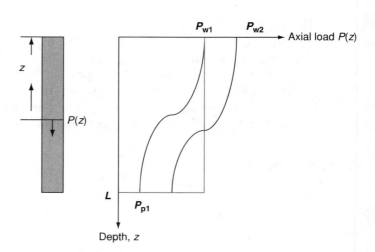

FIGURE 6.12
Pile load transfer curves.

Flaate (1972) identified three different zones surrounding a driven pile in clay: (i) remolded zone of 100 to 150 mm thickness from the pile surface; (ii) transition zone; and (iii) unaffected zone outside the transition zone. Pile driving can set up high pore pressure in the remolded zone and the soil is remolded under constant water content. The pore pressure generated during the installation process would be dissipated with time depending on the permeability of the surrounding soils. Further, the structure of the soil disturbed due to driving may also be recovered with time. The process of recovery of the soil structure with time and the consolidation of surrounding soil with time due to dissipation of the excess pore pressure is termed "thixotropic recovery." It is believed that the time taken for the recovery to be complete depends on the amount of disturbance caused by the pile installation process and the properties of the surrounding soil.

Due to the thixotropic recovery, the ultimate carrying capacity of the pile will vary with time. If the ultimate carrying capacity of the pile is increased due to thixotropic recovery, it is termed "set up", whereas if it is decreased, it is termed "relaxation." The phenomenon of time-dependent strength gain in piles driven into cohesive soil deposits is well established (Fellenius et al., 1989; Skove et al., 1989; Svinkin and Skov, 2002).

Due to the rapid dissipation of excess pore pressure, the increase in bearing capacity of piles driven into sandy deposits is expected to be complete within a few hours or at most within a few days after installation. However, substantial increases in capacity of driven piles in sand over a long period of time have been reported (Tavenas and Audy, 1972; York et al., 1994; Tomlinson, 1994; Chun et al., 1999). Since a substantial increase in the ultimate carrying capacity of a pile driven into sand over a long period of time cannot be attributed to dissipation of excess pore pressure, Chun et al. (1999) suggested the possibility of other reasons for such an increase in the ultimate carrying capacity. Some of these are: (i) bonding of sand particles to the pile surface; (ii) increase in strength due to soil aging; and (iii) long-term changes in the stress state surrounding the pile due to breakdown of arching around the pile resulting from the creep behavior of sand particles.

Svinkin and Skov (2002) modified an earlier relationship suggested by Skov and Denver (1989) for cohesive soils to include the pile capacity at the end of initial driving (EOID)

$$\frac{R_u(t)}{R_{EOID}} - 1 = B[\log_{10}(t) + 1] \tag{6.25a}$$

where $R_u(t)$ is the bearing capacity of the pile at time t, t is the time since end of initial driving (EOID), R_{EOID} is the bearing capacity of the pile at the end of initial driving, B is a factor, depending on soil type, pile type, and size evaluated by fitting field data.

Svinkin and Skov (2002) suggested that the pile capacity gain relationship given by Equation (6.25a) should be used only as a guide for assessment of pile capacity with time. The pile capacity vs. time relationships proposed by Skov and Denver (1989) and Svinkin and Skov (2002) indicate that when plotted on a log time scale the capacity gain continues indefinitely. Chun et al. (1999) showed that the capacity gain was not infinite but eventually converged to a constant value (long-term capacity). Chun et al. (1999) observed that there is a linear relationship between the ratio of bearing capacity gain ($R_u(t)/R_{EOID}$) and the rate of the ratio of capacity gain $d/dt[R_u(t)/R_{EOID}]$, where R_{EOID} and $R_u(t)$ are EOID capacity and the capacity at time t after installation, respectively, and proposed the following general relationship to estimate the capacity gain regardless of the soil type:

$$\frac{R_u(t)}{R_{EOID}} = C - B\frac{d}{dt}(R_u(t)/R_{EOID}) \tag{6.26a}$$

where $C = (R_u(\infty)/R_{EOID})$, which is the long-term ratio of capacity gain, B is a constant $(= G/K)$, K is the dissipation factor, and G is the aging factor.

Factor K depends on the permeability of the soil while factor G depends on the aging potential associated with soil properties. A higher B value could be expected for sandy soils, where the aging effect is predominant, and a lower value of B could be associated with clayey soils as the pore pressure dissipation is predominant. The ratio of capacity gain can be obtained by solving the above differential equation:

$$\frac{R_u(t)}{R_{EOID}} = C\left[1 - \left(\frac{C-1}{C}\right)e^{t/B}\right] \tag{6.26b}$$

6.5.1 Reported Results from Field Studies

Tomlinson (1994) reports the results of pile load tests carried out on 200×215 mm piles into soft clay at different times after installation. Figure 6.13(a) shows the measured and estimated (Skov and Denver, 1989) capacity gain ratio for one pile; Figure 6.13(b) shows the measured and estimated (Svinkin and Skov, 2002; Chun et al., 1999) capacity gain ratio for the same pile.

It is evident from Figures 6.13(a) that for large lapsed times after EOID the relationship proposed by Chun et al. (1999) predicts the capacity gain ratio over the entire time duration better than the method proposed by Svinkin et al. (2002) for the case studies considered above.

Two material parameters, B and C, are needed for the Chun method of capacity prediction where B and C are the long-term capacity gain ratio and a material constant, respectively. The relevant value of B is obtained by considering the time capacity variation of the measured capacity gain ratio whereas parameter C is obtained by matching the measured with the predicted values of the capacity gain ratio. The values of parameters B and C estimated by Thilakasiri et al. (2003) and available values obtained from the literature are shown in Table 6.8.

Also indicated in Table 6.8 are the times taken to develop 90% of the long-term capacity and the percentage of the long-term capacity developed 1 week after the EOID. Table 6.8 shows that the 1-week wait period from the EOID is sufficient for piles in sand whereas the 1-week period is not enough for piles driven into clay deposits for which a minimum wait period of 2 to 3 weeks may be required.

More recently, Bullock et al. (2005) published their test findings on a Florida test pile program. The Florida DOT commonly uses 457 mm (18 in.), square, prestressed, concrete piles to support low-level bridges. Bullock et al. (2005) provided the instrumentation and installed dedicated test piles of this type at four bridge construction sites in northern Florida. Each pile included an O-cell cast into the tip, strain gauges at soil layer boundaries, and total stress cells and pore pressure cells centered in one pile face between adjacent strain gauge elevations. They calculated the shear force and average shear stress acting on the face of the pile from the difference in load between adjacent strain gauge levels. The strain gauges defined a total of 28 side shear segments, of which 18 also included pore pressure and total horizontal stress instrumentation. Subsequent "staged" (repeated) tests over time provided data to investigate the Side Shear Setup (SSS) for each segment. Bullock et al. (2005) expressed Equation (6.25a) as

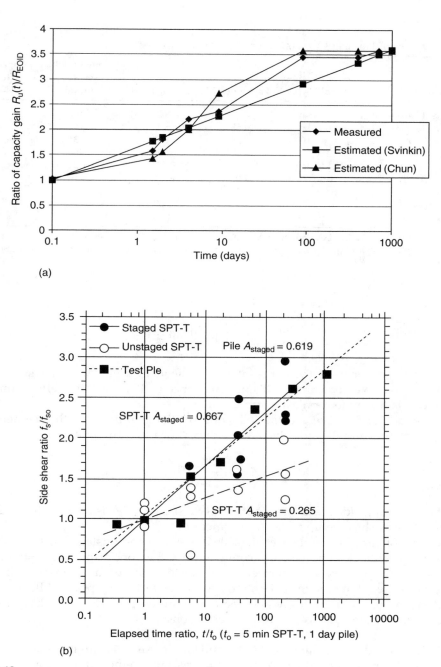

(a)

(b)

FIGURE 6.13

(a) Measured and estimated capacity from Svinkin et al. (2002) and Chun et al. (1999) for the pile I of Tomlinson (1994). (From Svinkin, M.R. and Skov, R., 2002, Setup effect of cohesive soils in pile capacity, 2002, http://www.vulcanhammer.net/svinkin/set.htm. With permission.) (b) Side shear setup results from the Florida pile testing program. (From Bullock, P.J., Schmertmann, J.H., Mcvay, M.C., and Townsend, F., 2005, *Journal of Geotechnical and Geoenvironmental Engineering*, 131(3): 292–300. With permission.)

TABLE 6.8

Available Values of Parameters *B* and *C* for Chun (1999) Method

Source	Soil Type	Parameter C	Parameter B	Time for 90% of R_α (days)	Capacity Gain After 7 Days (% of R_α)
Tomlinson (1994) — pile I	Soft clay	3.01	8.2	16	72
Tomlinson (1994) — pile II	Soft clay	3.58	8.2	16	69
Chun et al. (1999)	Clay	5.34	1.64	3	99
Chun et al. (1999)	Clay	6.13	9.29	20	60
Chun et al. (1999)	Stiff clay	2.8	1.7	3	99
Chun et al. (1999)	Sand	2.08	0.57	1	100
Chun et al. (1999)	Sand	1.41	4.33	5	94
Chun et al. (1999)	Sand	2.8	1.7	3	100
Chun et al. (1999)	Sand	1.6	0.21	1	100

Source: From Thilakasiri, H.S., Abeyasinghe, R.M., and Tennakoon, B.L., 2003, A study of strength gain of driven piles, *Proceedings of the 9th Annual Symposium*, Engineering Research Unit, University of Moratuwa, Sri Lanka. With permission.

$$\frac{Q_u(t)}{Q_0} = A \, \log_{10}(t/t_0) + 1 \tag{6.25b}$$

where Q is the capacity of the entire pile in subsequent segmental analysis, Q_0 is the capacity at initial reference time t_0, t is the time since EOID, t_0 is the reference time since EOID, and A is the dimensionless setup factor.

Bullock et al. (2005) presented the following relationship between the segmental side shear setup factors and the side shear setup factor for the entire pile:

$$A = \frac{\sum A_i f_{s0i} L_i}{\sum f_{s0i} L_i} \tag{6.25c}$$

where f_{s0i} is the unit side shear stress at time t_0 for segment i, L_i is the length of segment i, and A_i is the side shear setup factor for segment i.

Values of A obtained by Bullock et al. (2005) for staged and unstaged tests are shown in Figure 6.13(b). Bullock et al. (2005) recommend a reduction factor of $C_{st} = (A_{unstaged}/A_{staged}) = 0.4$ for all soil types to correct setup A factors measured using staged field tests, including repeated dynamic re-strikes, repeated static tests, or repeated SPT-Ts.

Based on the relevant literature and the above study, Bullock et al. (2005) reach the following general conclusions and recommendations:

1. Using staged tests of unloaded piles, and an accurate measurement of side shear obtained by the O-cell test method, this research demonstrated SSS similar to that observed by others in prior research.

2. All pile segments showed setup, with similar average magnitudes in all soils and at all depths, continuing long after the dissipation of pore pressures, and with postdissipation setup due to aging effects at approximately constant horizontal effective stress. The pile tests (all soil types) and the SPT-T predictor tests (cohesive soils only) confirm the approximately semi-log-linear time setup behavior previously observed by others.

3. For soils similar to those tested in this research or known to exhibit SSS, a default $A = 0.1$ is recommended without performing predictor tests, and higher values when supported by dynamic or static testing of whole piles, or staged SPT-Ts in clay and mixed soils. Reduce A-values measured during staged tests (pile or SPT-T) by the factor $C_{st} = 0.4$. Reduce pile segment A_i and SPT A by the factor $C_{pile} = 0.5$ for movement compatibility with whole-pile side shear capacity (if unknown). If the SPT-T $A_{staged} \leq 0.5$, use the default $A_i = 0.2$ and $A = 1$.

4. A conservative method is proposed for including SSS in pile capacity design. The appendix in Bullock et al. (2005) provides some idealized, but realistic, examples to show the methods recommended for including SSS in design. Depending on the percentage of capacity due to side shear, the final design time, and the applicable setup factors, SSS may significantly increase design pile capacity.

5. Dynamic tests during initial driving and subsequent re-strikes provide a method, after applying the 0.4 reduction factor for stage testing, by which to check the design A value. Repeated re-strikes also allow SSS behavior to occur at the increased rate of staged testing, and may permit the acceptance of a pile that initially does not demonstrate adequate capacity.

The research program by Bullock et al. (2005b) confirms the approximate semi-log-linear time relationship of SSS and extends it to instrumented pile segments as well as the entire pile. Short-term dynamic tests and long-term static tests produced similar SSS behavior (Bullock et al., 2005a), apparently with no significant change before and after the dissipation of excess pore pressure. The measured side shear and horizontal effective stresses seemed reasonable, with negligible adhesion at the pile–soil interface and an increase in the interface friction coefficient (tan δ) of 40% during SSS. All depths had about the same range of pile segment A_1 values, with a minimum $A_1 = 0.2$ and with no apparent depth dependency. These findings apply to all of the soil types tested, ranging from plastic (plasticity index $\leq 60\%$) clays to shelly sands.

6.6 Computation of Pile Settlement

In contrast to shallow footings, a pile foundation settles not only because of the compression the tip load causes on the underlying soil layers, but also because of the compression caused by the skin friction on the surrounding layers. The elastic shortening of the pile itself is another source of settlement. In addition, if an underlying saturated soft clay layer is stressed by the pile, the issue of consolidation settlement will also have to be addressed. In this section, only the immediate settlement components will be treated analytically, as the consolidation settlement computation of a pile group is provided in Example 6.4.

6.6.1 Elastic Solution

According to Poulos and Davis (1990), the immediate settlement of a single pile (with a load of p) can be estimated from the following expressions:

Floating piles:

$$s = \frac{PI}{E_s d} \tag{6.27a}$$

$$I = I_0 R_k R_h R_v \tag{6.27b}$$

End-bearing piles:

$$I = I_0 R_k R_b R_\nu \tag{6.27c}$$

where I_0 is the influence factor for an incompressible pile in a semi-infinite medium with $v_s = 0.5$ (Figure 6.14), R_k is the correction factor for pile compressibility K ($= E_p / E_s$) (Figure 6.15), R_h is the correction factor for a finite medium of thickness h (Figure 6.16), R_v is the correction factor for the Poisson ratio (v_s) of soil (Figure 6.17), R_b is the correction factor for stiffness of bearing medium (Figure 6.18), E_s is the elastic modulus of soil, E_p is the elastic modulus of the pile material, d is the minimum pile dimension (pile diameter), d_b is the diameter of the pile base, and h is the total depth of the soil layer.

Equations (6.27a)–(6.27c) account for all of the following components of settlement:

1. Immediate settlement occurring at the tip
2. Immediate settlement due to the stressing of surrounding soil
3. Elastic pile shortening.

Example 6.4
Figure 6.19 shows the configuration of a single concrete pile (300 mm × 300 mm) embedded 1 m in a clay layer and loaded by 500 kN. Estimate the immediate settlement of the pile top.
Assume that $E_{conc} = 27{,}600$ MPa
Based on the SPT values (assumed as corrected for the overburden)

For granular soils
$E = 250(N + 15)$ kPa (Table 2.11)
$E_{sand} = 5.25$ MPa, $<_s = 0.3$

For clayey soils
$E = 300(N + 6)$ kPa
$E_{clay} = 12.3$ MPa, $<_s = 0.3$
$K = E_p / E_s = 27{,}600/5.25 = 5257$
$L/d = 20/0.3 = 66.67$
$d_b/b = 1/1 = 1.0$ (No base enlargement)
$h/L = 20/20 = 1.0.$

Based on the above, the following parameters can be extracted from Figure 6.14–Figure 6.17:

$I_0 = 0.038$ (Figure 6.14)
$R_k = 1.1$ (Figure 6.15)
$R_h = 0.7$ (Figure 6.16)
$R_\nu = 0.93$ (Figure 6.17)

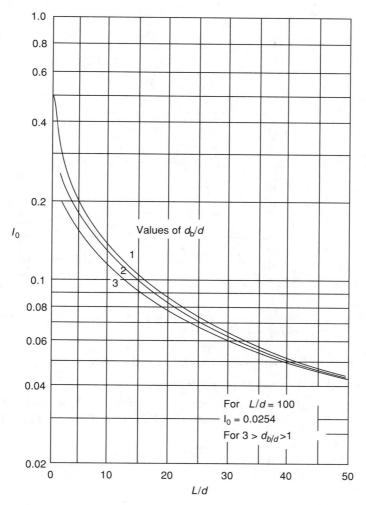

FIGURE 6.14.
Settlement-influence factor, I_0. (From Poulos, H.G. and Davis, E.H., 1990, *Pile Foundation Analysis and Design*, Krieger, Melbourne, FL. With permission.)

Therefore, if the pile is considered as a floating (friction) pile, the settlement of pile top (Equations (6.27a) and (6.27b))

$$= (500)(0.038)(1.1)(0.7)(0.93)/(5250)/0.3 \text{ m}$$
$$= 8.63 \text{ mm}$$

On the other hand, if the pile was driven to bearing in the stiff clay,

$$E_{clay}/E_{sand} = 12.3/5.25 = 2.34$$
$$R_b = 0.95 \text{ (Figure 6.18a)}$$

Thus, the new settlement of pile top would be (Equations (6.27a) and (6.27c))
$$= (500)(0.038)(1.1)(0.95)(0.93)/(5250)/0.3 \text{ m} = 11.72 \text{ mm}$$

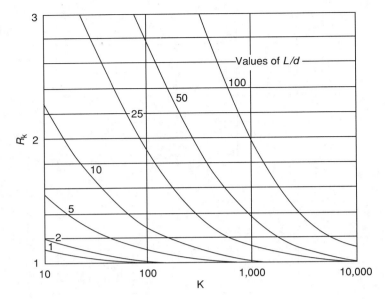

FIGURE 6.15
Compressibility correction factor for settlement, R_k. (From Poulos, H.G. and Davis, E.H., 1990, *Pile Foundation Analysis and Design*, Krieger, Melbourne, FL. With permission.)

6.6.2 Computation of Pile Settlement Using Approximate Methods

6.6.2.1 Elastic Method for End-Bearing Piles

One can use the Timoshenko and Goodier (1951) method to estimate the immediate settlement undergone by the tip of an individual point bearing pile (Equation (4.8)).

$$s_i = qB\frac{1 - \nu_s^2}{E_s}\left[I_1 + \frac{1 - 2\nu_s}{1 - \nu_s}I_2\right]I_F \tag{6.28}$$

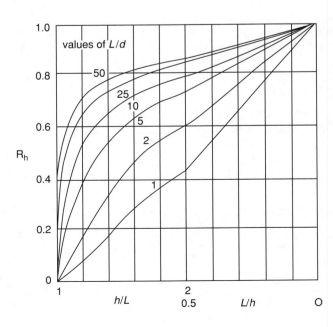

FIGURE 6.16
Depth corrector for settlement, R_h. (From Poulos, H.G. and Davis, E.H., 1990, *Pile Foundation Analysis and Design*, Krieger, Melbourne, FL. With permission.)

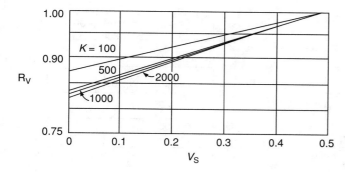

FIGURE 6.17
Poisson's ratio correction factor for settlement, R_v. (From Poulos, H.G. and Davis, E.H., 1990, *Pile Foundation Analysis and Design*, Krieger, Melbourne, FL. With permission.)

where B is the pile diameter (or equivalent diameter)

$I_1 = 1.0$ and $I_2 = 0$ (Table 4.1, for $N \to \infty$ and $M = 0$)
$I_F = 0.50$ for $L/B \to \infty$
$q = P_p/A_p$

If the point load, P_p, is not known, one can make an assumption on the ratio of P_p/P_w to obtain P_p from P_w.

6.6.2.2 SPT-Based Method for End-Bearing Piles

Meyerhoff (1976) suggested a simple expression to determine the tip settlement of a pile based SPT data. Bowles (2002) modified this expression to obtain the following form:

$$s_i \, (mm) = \frac{q_P}{2N_{55}B} \tag{6.29}$$

where N_{55} is the weighted SPT average in the influence zone that extends $2B$ below the tip and B above the tip, q_P is the tip stress in kPa.

6.6.2.3 Elastic Shortening of Piles

An axially loaded pile will undergo elastic shortening due to the compressive axial stress it carries throughout its length. Using Hooke's law, the magnitude of elastic shortening can be expressed in the following expressions:

For a pile with a uniform cross section

$$s_e = E_p A_p \int_0^L P(z) \, dz \tag{6.30a}$$

For a tapered pile

$$s_e = E_p \int_0^L \frac{P(z)}{A_P} \, dz \tag{6.30b}$$

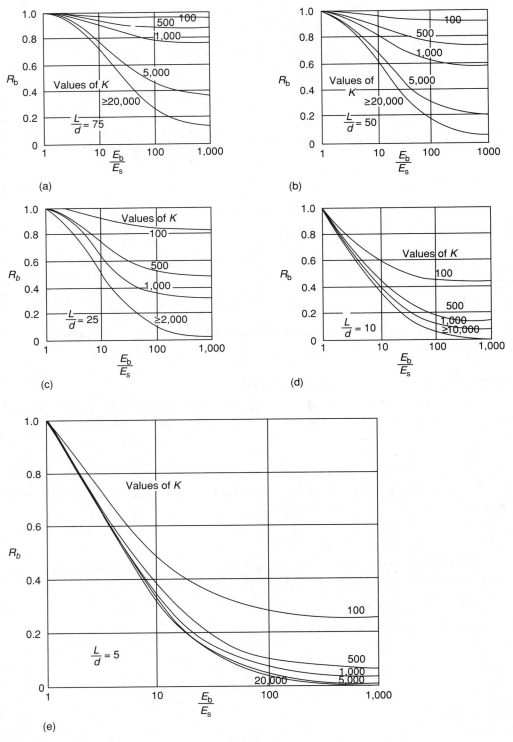

FIGURE 6.18
Base modulus correction factor for settlement, R_b. (From Poulos, H.G. and Davis, E.H., 1990, *Pile Foundation Analysis and Design*, Krieger, Melbourne, FL. With permission.)

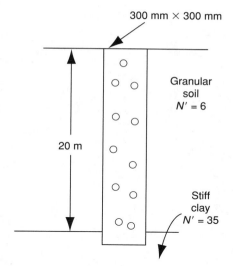

300 mm × 300 mm

Granular
soil
$N' = 6$

20 m

Stiff
clay
$N' = 35$

FIGURE 6.19
Illustration for Example 6.4.

where E_P is the elastic modulus of the pile material, A_P is the cross section area of the pile, $P(z)$ is the axial load at a depth of z from the pile top, and L is the length of the pile.

Thus, one can estimate the elastic settlement from the area under the stabilized load transfer curve. If the actual load transfer curve is not known, one has to make an assumption of the load distribution along the pile length to estimate the elastic pile shortening.

6.7 Pile Groups

For purposes of stability, pile foundations are usually constructed of pile groups that transmit the structural load through a pile cap, as shown in Figure 6.20. If the individual piles in a group are not ideally placed, there will essentially be an overlap of the individual influence zones, as shown in Figure 6.20. This will be manifested in the following group effects, which must be considered when designing a pile group:

1. The bearing capacity of the pile group will be different (generally lower) than the sum of the individual capacities owing to the above interaction.
2. The group settlement will also be different from individual pile settlement owing to additional stresses induced on piles by neighboring piles.

6.7.1 Bearing Capacity of Pile Groups

The efficiency of a pile group is defined as

$$\eta = \frac{P_g(u)}{\sum P_{u,i}}$$

(6.31)

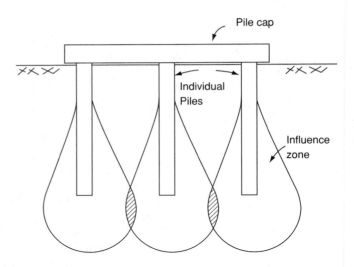

FIGURE 6.20
Illustration of pile group effect. (From *Concrete Construction Engineering Handbook*, CRC Press. With permission.)

or

$$\eta = \frac{\text{Ultimate capacity of the pile group}}{\text{Sum of the individual ultimate capacities}}$$

where η is the group efficiency.

Owing to the complexity of individual pile interaction, the literature does not indicate any definitive methodology for determining the group efficiency in a given situation other than the following common the converse Labarre equation that is appropriate for clayey soils:

$$\eta = 1 - \frac{\xi}{90}\left[\frac{(n-1)m + (m-1)n}{mn}\right] \tag{6.32}$$

where $\xi = \tan^{-1}$ (diameter–spacing ratio), n is the number of rows in the group, and m is the number of columns in the group.

Although the above expression indicates that the maximum achievable group efficiency is about 90%, reached at a spacing–diameter ratio of 5, the results from experimental studies (Das, 1995; Bowles, 2002; Poulos and Davis, 1990) have shown group efficiency values of well over 100% being reached under certain conditions, especially in dense sand. This may be explained by possible densification usually accompanied by pile driving in medium-dense sands. Computation of group capacity will be addressed in Example 6.5.

According to FHWA (1998) guidelines, the following group effects can be included:

Cohesive soils:

Stiff cohesive soils — no loss in resistance due to group effects
Soft cohesive soils — when the pile cap is not touching the ground
Group efficiency $= \eta = 0.7$ for center-to-center spacing of 3.0D
Group efficiency $= \eta = 1.0$ for center-to-center spacing of 6.0D

where D is the diameter of a single pile.

Generally, in order to determine the group capacity, one would first determine the pile capacity based on the individual pile capacities and the group efficiency. Then a second estimate is obtained by considering the group as a single pier. The ultimate group capacity is considered as the more critical (or lesser) estimate.

Cohesionless soils

Recommended group efficiency is 100% (or $\eta = 1.0$) irrespective of the pile spacing and the interaction between the pile cap and the ground.

Both topics are discussed in the subsequent sections and illustrative examples are provided.

6.7.2 Settlement of Pile Groups

One simple method of determining the immediate settlement of a pile group is by evaluating the interaction factor, α_F, defined as follows:

$$\alpha_F = \frac{s_{j,i}}{s_i} \tag{6.33}$$

or

$$\alpha_F = \frac{\text{Additional settlement caused by adjacent pile } j \text{ or pile } i \ (s_{j,i})}{\text{Settlement of pile } i \text{ under its own load } (s_i)}$$

where s_i is the settlement of a pile (i) under its own load and $s_{j,i}$ is the additional settlement in pile i caused by the adjacent pile j.

The settlement of individual piles can be determined on the basis of the method described in Section 6.6. Then, once α_F is estimated from Figure 6.21a, b, or c, based on the length–diameter ratio, the relative stiffness k, the spacing–diameter ratio, and soil elasticity properties, one can easily compute the settlement of each pile in a group configuration. At this point, the issue of flexibility of the pile cap has to be considered. This is because if the pile cap is rigid (thick and relatively small in area), it will ensure equal settlement throughout the group by redistributing the load to accommodate equal settlements. On the other hand, if the cap is flexible (thin and relatively extensive in area), all of the piles will be equally loaded, which results in piles undergoing different settlements.

Under conditions where the consolidation settlement under a pile group is significant, one can assume that the pile group acts as a large rigid single footing and use the consolidation settlement principles discussed in Section 1.5, compressibility, and settlement. However, in this case, the difference in load attenuation between a shallow footing and a rigid pile group with substantial skin friction is accounted for by assuming that the load attenuation originates from the lower middle-third point of the pile length, as shown in Figure 6.22.

Example 6.5

This problem is solved in British and US units. Hence, the reader is referred to Table 4.1 for conversion of these units to SI units. The pile group (6, 1 ft × 1 ft piles) shown in Figure 6.23 is subjected to a load of 80 kips. A compression test performed on a representative clay sample at the site yielded an unconfined compression strength of 3 psi and an elastic

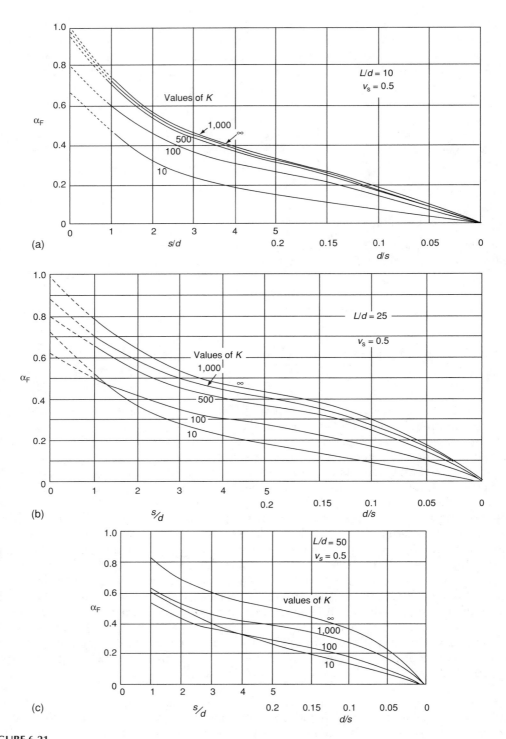

FIGURE 6.21
Determination of α_F factor for: (a) $L/d = 10$; (b) $L/d = 25$; and (c) $L/d = 50$. L is the pile length and d is the pile diameter; K and V_s are as defined in Equation (6.33). (From Poulos, H.G. and Davis, E.H., 1990, *Pile Foundation Analysis and Design*, Krieger, Melbourne, FL. With permission.)

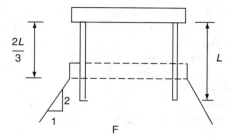

FIGURE 6.22
Illustration of pile group load attenuation.

modulus of 8000 psi, while a consolidation test indicated no significant overconsolidation, with a compression index of 0.3 and a water content of 15%. Estimate the safety of the pile foundation and its total settlement. The saturated unit weight is 115 psf.

Computation of skin friction of a single pile. Using Equation (6.14), $f = 1.0(0.5)(3)(144)(216) = 216$ psf. From Equation (6.11), the resultant skin-frictional force $= \frac{1}{2}(3.0)(144)(4)(50) = 43.2$ kips.

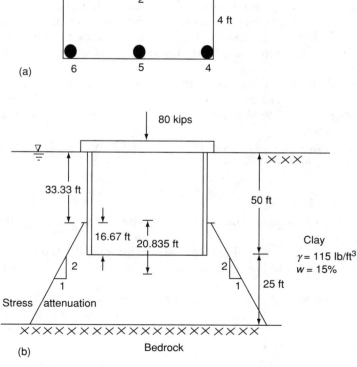

FIGURE 6.23
Illustration for Example 6.5: (a) plan; (b) elevation. (From *Concrete Construction Engineering Handbook*, CRC Press. With permission.)

Computation of end bearing of a single pile. Using Equation (6.7), $P_{pu} = (1)(9)(216) = 1.944$ kips (in fact, one could expect the insignificance of this contribution owing to the "frictional" nature of the pile).

Thus, the ultimate capacity of the pile is 45.14 kips.

Estimation of group efficiency. Using Equation (6.32),

$\eta = 1 - [(3 - 1)(2) + (2 - 1)(3)](\xi)/[90(3)(2)] = 0.8$.

Then, the group capacity can be obtained as $0.8(45.14)(6) = 216$ kips (Equation (6.31)) and the safety factor can be computed as $216/80$ kips $= 2.7$.

Estimation of single pile immediate settlement. The relative stiffness factor of the pile, K, is $E_{concrete.}/E_s = 4,000,000/8,000 = 500L/D = 50/1 = 50$. Then

$\quad I_0 = 0.045$ (Figure 6.14)

$\quad R_k = 1.85$ (Figure 6.15)

$\quad R_h = 0.8$ (Figure 6.16)

$\quad R_\nu = 1.00$ (undrained $\nu = 0.5$) (Figure 6.17)

Substituting the above parameters in Equation (6.27a)

$$s = P(0.045)(1.85)(0.8)(1.00)/(8000 \times 12) = 0.032 \times 10^{-5} P \text{ in.}$$

where P is the load on a single pile in kips.

Analysis of group settlement. If the cap is assumed to be rigid, then the total settlement of all six piles must be identical. The total settlement consists of both immediate settlement and consolidation settlement. However, only an average consolidation settlement can be computed for the entire pile group based on the stress attenuation method (Figure 6.22) assuming equal consolidation settlement. Thus, one has to assume equal immediate settlements as well.

Owing to their positions with respect to the applied load, it can be seen that piles 1, 3, 4, and 6 can be considered as one type of pile (type 1) carrying identical loads, while piles 2 and 5 can be categorized as type 2. Thus, it will be sufficient to analyze the behavior of pile types 1 and 2 only.

Assume that the loads carried by piles of types 1 and 2 are P_1 and P_2, respectively. Then, for vertical equilibrium

$$4P_1 + 2P_2 = 80 \text{ kips} \tag{6.34}$$

Using Figure 6.21, the interaction factors for pile types 1 and 2 due to other piles can be obtained as follows:

Then, using Equation (6.33), the total settlement of pile type 1 is estimated as $(1+1.7)(0.032 \times 10^{-5} P_1)$, and the total settlement of pile type 2 would be $(1+1.9)(0.032 \times 10^{-5} P_2)$.

By equating the settlement of pile types 1 and 2 (for equal immediate), one obtains

$$2.7P_1 = 2.9P_2$$
$$P_1 = 1.074P_2$$

PILE TYPE 1

Pile i	s/d for pile i	α_F from pile i
1	0	—
2	4	0.4
3	8	0.3
4	4	0.4
5	5.67	0.35
6	8.94	0.25

$E\alpha F = 1.7$

PILE TYPE 2

Pile i	s/d for pile i	α_F from pile i
1	4	0.4
2	0	—
3	4	0.4
4	5.67	0.35
5	4	0.4
6	8.94	0.35

$E\alpha F = 1.9$

By substituting in Equation (6.34),

$$2(1.074P_2) + P_2 = 40$$
$$P_2 = 12.706 \text{ kips}$$
$$P_1 = 13.646 \text{ kips}$$

Hence, the immediate settlement of the pile group is equal to $2.9(0.032)(10^{-5})(12.706)(10^3)$ $= 0.012$ in., $S_{ult} = (0.3)(41.67)(1/1.4) \log[1 + 111.71/2{,}849] = 0.149$ ft $= 1.789$ in.

Computation of consolidation settlement. On the basis of the stress attenuation shown in Figure 6.22, the stress increase on the midplane of the wet clay layer induced by the pile group can be found as

$$\Delta_\sigma = 80{,}000/[(8 + 20.835)(4 + 20.835)] = 111.71 \text{ psf}$$

The initial effective stress at the above point is equal to $(115 - 62.4) \times (33.33 + 20.835) = 2849$ psf. For a saturated soil sample from Section 1.7, $e = wG_s = 0.15 \times (2.65) = 0.4$ (assuming the solid specific gravity, G_s, of 2.65). Then, by applying Equation (1.18), one obtains the consolidation settlement as

$$S_{ult} = (0.3)(41.67)(1/1 + 0.4)\log[1 + 111.71/2849] = 0.149 \text{ ft} = 1.789 \text{ in.}$$

It is seen that in this case, the consolidation settlement is predominant.

6.7.3 Approximate Methods for Computation of Immediate Settlement of Pile Groups

6.7.3.1 Vesic's Pile Group Interaction Factor

Vesic (1977) suggested the following group factor to convert the single pile settlement to that of a pile group:

$$s_i)_g = s_i)_P \sqrt{\frac{B_g}{B}} \tag{6.35}$$

where $S_i)_g$ = immediate settlement of the pile group, $S_i)_g$ = immediate settlement of a single pile, B_g is the least lateral group dimension and B is the pile dimension. Equation (6.35) is mostly recommended for cohesionless soils.

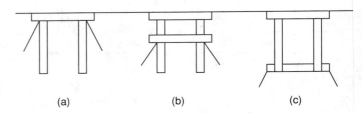

FIGURE 6.24
Approximate stress distributions due
to pile groups.   (a)   (b)   (c)

6.7.3.2 CPT-Based Method for Pile Groups

Bowles (2002) provides an expression (Equation (6.36)) to evaluate the settlement of a pile group

$$s_i(mm) = \frac{k\Delta_q B}{2q_c} \tag{6.36}$$

where q_c is the weighted cone resistance in the influence zone that extends $2B$ below the tip and B above the tip and Δ_q is the vertical pressure at the pile tip, B is the width of the group; k is defined as follows:

$$k = 1 - \frac{L}{8B} \geq 0.5$$

where L is the pile length.

6.7.3.3 Load Distribution Method for Pile Groups

Bowles (2002) outlines a simplified method to compute the settlement of pile groups. In order to obtain the settlement due to skin friction a 2:1 distribution of the load is used from the pile cap in the case of piles completely embedded in friction layers (Figure 6.24a) or a fictitious pile cap starting at the friction layer (Figure 6.24b). The tip settlement is similarly estimated based on a 2:1 distribution of the load from a fictitious pile cap at the tip (Figure 6.24c).

The approximate stress distributions shown in Figure 6.24 can be used to predict the immediate settlement as well as the consolidation settlement (Example 6.5). In the case of computation of immediate settlement, one can use the method outlined in Section 4.5.1 (Equation (4.8)) considering the footing to be the pile cap or the fictitious pile cap in Figure 6.24. On the other hand, in the case of computation of consolidation settlement, the above distribution can be used to evaluate the stress increase Δp in Equations (1.13)–(1.15).

6.8 Downdrag (Negative Skin Friction)

According to FHWA (1997), the potential for downdrag loading must be considered when the indicators in Table 6.9 are present. In terms of performance limits, downdrag presents a foundation settlement concern for friction piles and for end-bearing piles founded on a very stiff layer such as very dense sand or rock.

6.9 Load and Resistance Factor Design Criteria

According to FHWA (1998), the general LRFD pile design criteria can be expressed as given below.

TABLE 6.9

Conditions Where Downdrag Is Significant in Design

1	Total settlement of the ground surface $>10\,\text{mm}$
2	Settlement of ground surface after pile driving $>1\,\text{mm}$
3	Height of embankment filling on ground surface $>2\,\text{m}$
4	Thickness of soft compressible layer $>10\,\text{m}$
5	Water table drawn down $>4\,\text{m}$
6	Piles length $>25\,\text{m}$

Source: From Federal Highway Administration, 1997, *Load and Resistance Factor Design (LRFD) for Highway Bridge Substructures*, Washington, DC. With permission.

6.9.1 Strength Limit States

$$\phi R_n \geq \eta \sum \gamma_i Q_i \tag{6.37}$$

where R_n is the ultimate nominal resistance of pile, ϕ is the resistance factor, Q_i is the load effect, γ_i is the load factor, and η is the load modifier, which is given as

$$\eta = \eta_D \eta_R \eta_I \geq 0.95$$

where δ_{est} = estimated deflection (settlement) and δ_{tol} = tolerable deflection.

where η_D is the effect of ductility, η_R is the effect of redundancy, and η_I is the operational importance.

For a driven pile foundation design,
$\eta_D = \eta_R = 1.00$

$\eta_I = 1.05$ for structures deemed operationally important

$\quad = 1.00$ for typical structures

$\quad = 0.95$ for relatively less important structures

The following design considerations must be evaluated for piles at the strength limit state:

1. Bearing resistance of single pile groups
2. Pile group punching
3. Tensile resistance of uplift loaded piles
4. Structural capacity of axially or laterally loaded piles

6.9.2 Service I Limit State

$$\delta_{est} = f(\gamma_i Q_i) \leq \delta_{tol} \tag{6.38}$$

where δ_{est} = estimated deflection (settlement) and δ_{tol} = tolerable deflection.

Similarly, the following must be evaluated at the service I limit states:

TABLE 6.10

Load Factors

Limit State	Dead Load of Structural Components and Nonstructural Attachments	Dead Load of Wearing Surfaces and Utilities	Vehicular Live Load
Strength I	1.25	1.5	1.75
Service I	1.00	1.00	1.00

Source: From Federal Highway Administration, 1998, *Load and Resistance Factor Design (LRFD) for Highway Bridge Substructures*, Washington, DC. With permission.

1. Settlement of piles
2. Structural capacity of axially or laterally loaded piles

FHWA (1998) recommends the load factors provided in Table 6.10 to be used in LRFD.

6.9.3 Design Criteria for Axially Loaded Piles

According to AASHTO LRFD Specification, the ultimate geotechnical resistance of piles subjected to axial loading can be expressed by

$$\phi R_n = A_p \phi_{qp} q_p + A_s \phi_{qs} q_s \tag{6.39}$$

where ϕ_{qp} and ϕ_{qs} are resistance factors (Table 6.11), q_p is the ultimate unit point resistance, q_s is the ultimate unit skin friction, and A_p and A_s are cross-sectional area and embedded surface area of pile, respectively.

On the other hand, according to AASHTO LRFD Specification, the ultimate structural resistance of piles subjected to axial loading can be expressed by

$$P_r = \sum \phi P_n \tag{6.40}$$

where P_n is the ultimate structural resistance of the pile, P_r is the factored structural resistance of the pile, and is the resistance factors (Table 6.14).

Example 6.6

Figure 6.25 shows a bridge pier supported by a steel pile (HP 360 × 108) group that has to be designed to carry a dead load of 5000 kN and a live load of 4000 kN. The CPT results for the site are also illustrated in Figure 6.25 in an idealized form. Use the LRFD method to estimate the number of piles needed in the group assuming that the driving conditions are severe.

Step 1: Geotechnical resistance. For end bearing, Equation (6.21)

For $Z = 10$ m, $R_1 = 1.0$, $R_2 = 1.0$
$\quad q_p = (1.0)(1.0)(16{,}000 + 16{,}000)/2 = 16000$ kPa
$\quad Q_p = 16{,}000(0.346)(0.371) = 2{,}053$ kN

For skin-friction
$\quad F = \alpha C_u$ and $f = \alpha' f_s$

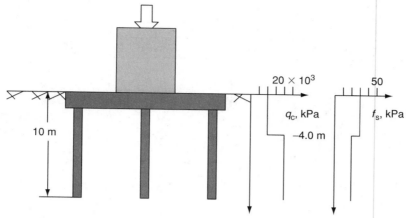

FIGURE 6.25
Illustration for Example 6.6.

So $C_u = (\alpha'/\alpha)f_s$

C_u can be determined from Table 6.2 and Table 6.4 using trial and error. It is recorded in the column 4 of Table 6.12.

Referring to Table 6.12 for determination of segmental frictional contributions.

Total $Q_{side} = 583.6$ kN

TABLE 6.11

Resistance Factors for Geotechnical Resistance of Piles — ASD-Based Calibration (Safety Factor of 2.75)

Method/Soil/Condition		Resistance Factor
Ultimate bearing resistance of single piles	Skin friction: clay	
	α-Method	0.70
	β-Method	0.50
	λ-Method	0.55
	End bearing: clay and rock	
	Clay	0.70
	Rock	0.50
	Skin friction and end bearing: sand	
	SPT method	0.45
	CPT method	0.55
	Skin friction and end bearing: all soils	
	Load test	0.80[a]
	Pile driving analyzer	0.70
Block failure	Clay	0.65
Uplift resistance of single piles	α-Method	0.60
	β-Method	0.40
	λ-Method	0.45
	SPT method	0.35
	CPT method	0.45
	Load test	0.80
Group uplift resistance	Sand	0.55
	Clay	0.55

[a] ASD safety factor of 2.0.

Source: From Federal Highway Administration, 1997, *Load and Resistance Factor Design (LRFD) for Highway Bridge Substructures*, Washington, DC. With permission.

TABLE 6.12

Illustration of Computations for Example 6.6

Depth (m)	α' Steel (Table 6.4)	f_s (kPa)	C_u (kPa)	f_{pile} (kPa)	$F_{segmental}$ (kPN)
0					0
1.73	2	30	75	60	114
3.46	1.25	30	39.87	37.5	185
4	1.14	30	35.73	34.2	42.6
5.19	0.9	20	17.47	18	68.3
6.92	0.82	20	15.81	16.4	65.5
8.65	0.8	20	15.40	16	61.7
10	0.75	20	14.38	15	46
					583.6 kN

LRFD
$$Q_{total} = 0.55(2{,}053+583) = 1{,}449 \text{ kN}$$

ASD
$$Q_{total} = 2{,}636 \text{ kN}$$

Step 2: Structural resistance
$$A_g = 13.9(10^{-3}) \text{ m}^2$$

LRFD
$$\phi F_g A_g = 0.35(210)(10)^3(13.9)(10)^{-3}$$
$$= 1{,}021.7 \text{ kN}$$

ASD
$$F_g A_g = (210)(10)^3(13.9)(10)^{-3}$$
$$= 2{,}919 \text{ kN}$$

Step 3: Compute applied loads
$$P_u = 1.25(5000) + 1.75(4000)$$
$$= 13{,}250 \text{ kN}$$

$$P = 5{,}000+4{,}000$$
$$= 9{,}000 \text{kN}$$

Step 4: Determine the number of piles required.
  Geotechnical criterion
  LRFD
$$= 13{,}250/1{,}449$$
$$= 9.1 = 10 \text{ piles}$$

  ASD
$$= 9{,}000(2.75)/2{,}636$$
$$= 9.4 = 10 \text{ piles}$$

Structural criterion
$$= 13{,}250/1{,}021.7$$
$$= 13 \text{ piles}$$
Then the number of piles required is 13.

$$= 9000(2.75)/2919$$
$$= 8.47 = 9 \text{ piles}$$

Example 6.7
For Example 6.3, the factored axial resistance can be obtained from Equation (6.39) as follows:

$$\phi R_n = \phi_{qp}(1260) + \phi_{qs}(1654)$$

From Table 6.13, for CPT results
  $\phi_{qp} = 0.55$ and $\phi_{qs} = 0.55$
  Hence, $\phi R_n = 1602.7$ kN
Assuming that the load of 900 kN is desynthesized as follows:

  Weight of structural components $= 700 \text{ kN}$

  Weight of vehicular traffic $= 200 \text{ kN}$

The LHS of Equation (6.37) can be used to compute the factored load as

TABLE 6.13

Resistance Factors for Geotechnical Resistance of Piles (Reliability-Based Calibration)

		ϕ Values by Method of Axial Pile Capacity Estimation							
		A				Λ			
Pile Length (m)	β_T	Type I	Type II	β	Type I	Type II	CPT	SPT
10	2.0	0.78	0.92	0.79	0.53	0.65	0.59	0.48
30	2.0	0.84	0.96	0.79	0.55	0.71	0.62	0.51
10	2.5	0.65	0.69	0.68	0.41	0.56	0.48	0.36
30	2.5	0.71	0.73	0.68	0.44	0.62	0.51	0.38
Average ϕ			0.78	0.74	0.56		0.55	0.43
Selected ϕ			0.70	0.50	0.55		0.55	0.45

Source: From Federal Highway Administration, 1997, *Load and Resistance Factor Design (LRFD) for Highway Bridge Substructures*, Washington, DC. With permission.

TABLE 6.14

Resistance Factors for Structural Design of Axially Loaded Piles

Pile Type	Resistance Factor
Steel	
Severe driving conditions	0.35
Good driving conditions	0.45
Prestressed concrete	0.45
Concrete-filled pipe	
Steel pipe	0.35
Concrete	0.55
Timber	0.55

Source: From Federal Highway Administration, 1998, *Load and Resistance Factor Design (LRFD) for Highway Bridge Substructures*, Washington, DC. With permission.

$\Sigma \gamma_i Q_i = 1.25(700) + 1.75(200) = 1225 \, \text{kN}$

It is seen that Equation (6.37) is satisfied for the geotechnical strength.

As for the structural strength, Table 6.14 provides the structural resistance factor for a concrete pile with a steel casing as 0.35.

Assuming that the compressive strength of concrete is 20 MPa

Factored resistance $= 20,000(0.35) = 7000 \, \text{kPa}$

Factored load $= (\frac{1}{4})B(0.4)^2(7000) = 879 \, \text{kN} < 1225 \, \text{kN}$

Since Equation (6.37) is not satisfied from a structural perspective, the diameter of the pile has to be increased to about 0.5 m, which would improve the geotechnical strength further.

6.10   Static Capacity of Piles on Rock

In the case of piles driven into rock, the static capacity can be estimated in a manner similar to that followed for soils. According to Kulhawy and Goodman (1980), the ultimate point bearing capacity of a pile driven in rock would be given by

tion factor. In pullout calculations, caution must also be exerted in checking the tensile strength of the pile material (Table 6.15).

6.12 Screw Piles

Two types of screw piles are available in construction: (1) steel screw piles and (2) concrete screw piles. Screw piles made of steel are circular hollow sections of shaft with one or more tapered steel plates (helices) welded to the outside of the tube at the base. The steel pile is screwed into the ground as giant self-tapping screws with a suitable torque rating by using planetary drive or rotary hydraulics attached to earth moving equipment such as mini excavators, bobcats, proline crane borers, or large excavators. The purpose of providing additional helices is to reduce the slenderness ratio of the pile.

In the case of concrete screw piles, the hollow tube with an auger head is screwed into the ground until it reaches the base depth. Then, the hollow cavity is filled with reinforced concrete while the tube and the auger are screwed back. During and at completion of the screw pile installation, the installer monitors the installing torques to ensure that a sufficient load capacity is achieved. Constant torque monitoring provides an accurate indication of ground profile and founding soil capacity that can be compressive or tensile. Some effect of friction on the pile shaft may be considered as well. Screw piles generally work in both sand and clay conditions. From the design point of view, soil types and profiles play a crucial role in the design and performance of a screw pile.

The displacement screw piles are of a diameter of 0.40 to 0.70 m, a length of 10 to 22 m, and a range of bearing capacity of 1000 to 2000 kN. Since the bearing capacity depends on the type of pile, the method of installation, and the nature of the soil, it is prudent to

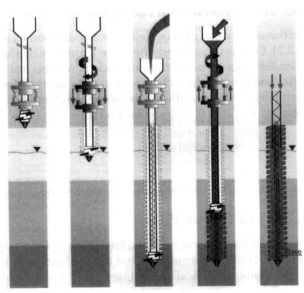

FIGURE 6.27
Installation procedure Atlas screw pile. (From Van Impe, W.F., 2004, *Two Decades of Full Scale Research on Screw Piles*. With permission.)

FIGURE 6.28
Installation of the reinforcement cage. (From Van Impe, W.F., 2004, *Two Decades of Full Scale Research on Screw Piles*. With permission.)

FIGURE 6.29
Filling of funnel and tube with concrete. (From Van Impe, W.F., 2004, *Two Decades of Full Scale Research on Screw Piles*. With permission.)

obtain the relevant experimental coefficients from load tests carried out until large displacements are obtained or up to the region of ultimate pile capacity. Compared to the driven, vibrated, and bored piles, the screw piles have some advantages such as no "reflux" of cut soil, no vibrations, and no noise during the installation (Van Impe, 2004). The following sections illustrate the installation are experienced procedure.

6.12.1 Atlas Screw Piles

A schematic overview of the installation procedure is shown in Figure 6.27–Figure 6.29.

6.12.2 Omega Screw Pile

A schematic overview of the installation procedure of the Omega screw pile is shown in Figure 6.30–Figure 6.32.

6.12.3 Application of Screw Piles

Due to their quick and vibration-free installation facility screw piles can be used in renovation work including operations within existing buildings, close to adjacent buildings, old quay walls, and piling to strengthen or renew foundations (underpinning). They are also useful in the construction of new homes, garages, room additions, mobile home anchoring systems, and commercial construction. Furthermore, they find wide application in reconstruction work in industry, road works, or structures in close proximity to water-retaining, vibration-sensitive installations and buildings containing vibration-sensitive equipments.

Screw piles can also be used for underpinning of damaged structures since it is one of the easiest methods to achieve a deep and thus secure foundation depth.

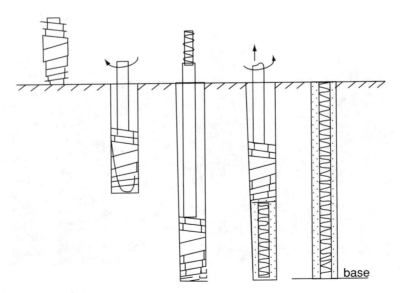

FIGURE 6.30
Installation procedure Omega screw pile. (From Van Impe, W.F., 2004, *Two Decades of Full Scale Research on Screw Piles*. With permission.)

FIGURE 6.31
Detail of the Omega displacement auger. (From Van Impe, W.F., 2004, *Two Decades of Full Scale Research on Screw Piles*. With permission.)

6.13   Pile Hammers

The process and the method of installation are just as important as the design of pile foundations. When the hammers are used to install a pile, the following important factors must be considered:

1. Weight and size of the pile
2. Driving resistance and net transferred energy
3. Available space at site
4. Crane facility
5. Noise control or restrictions

Hammer-operating principle: The driving criteria required to achieve a certain pile capacity can be evaluated based on concepts of work or energy (Chapter 9). The hammer energy is equated with the work done when the hammer forces the pile into the ground. The energy rating of hammers operated by gravity is assigned based on the potential energy at full stroke. More recently, accurate wave analysis has been implemented to derive dynamic formulae (Chapter 9). Pile hammers can be categorized into two main types, such as impact hammers and vibratory hammers.

FIGURE 6.34
Single acting air or steam hammer.
(From www.vulcanhammer.net. With
permission.)

6.14 Additional Examples

Example 6.8
Figure 6.38(a) shows the subsurface profile of a site where a new tall building will be
constructed in downtown Boston, MA. It is decided to use a steel pile group foundation
(HP 250 × 62), of which a single pile is shown in Figure 6.38(d). Each pile in the group is
required to carry a vertical load of at least 1000 kN (Table 6.18):

Saturated unit weight of silty sand = 17.5 kN/m³

SPT variation for sand = $5 + z$

Saturated unit weight of Boston Blue clay = 17.5 kN/m³

Dry unit weight of clay = 16.5 kN/m³

SPT variation for clay = $1 + 2z$ (for $z < 1.0$ m)

$\qquad = 2 + z$ (for $z > 1.0$ m)

Moisture content of clay = 15%

Compression index of clay = 0.3

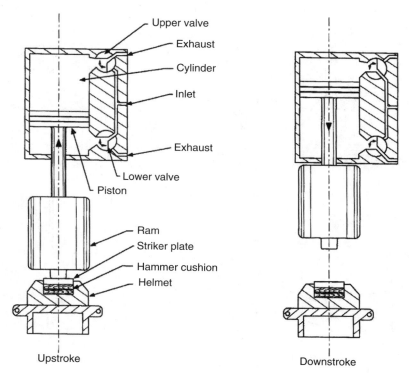

Upstroke

Downstroke

FIGURE 6.35

Schematic of double acting air or steam hammer. (From www.vulcanhammer.net. With permission.)

(a) Plot the SPT profile for the site.

(b) Determine the depth to which the pile has to be driven.

(c) Plot the load transfer curve assuming that it is linear and that 25% of the load is transmitted to the tip.

(d) The total settlement of the pile top.

(a) SPT variation with the depth is plotted in Figure 3.8(b)

(b) Steel pile foundation: HP 250 × 62

Depth $= 246\,\text{mm}$

Width $= 256\,\text{mm}$

Flange thickness $= 10.7\,\text{mm}$

Web thickness $= 10.5\,\text{mm}$

Plugged area, $A_p = (0.246 \times 0.256) = 0.063\,\text{m}^2$

Pile perimeter, $p = 0.256 \times 2 + 0.246 \times 2 + (0.246 - 0.0105) \times 2$
$= 1.475\,\text{m}$

Min. pile dimension $= 0.246\,\text{mm}$

Limiting skin friction depth in sand, $L_{s,\text{lm}} = 15D = 15(0.246) = 3.69\,\text{m}$

Critical end-bearing penetration $= L_{p,\text{cr}}//D = 8$, $L_{p,\text{cr}} = 8(0.246) = 1.97\,\text{m}$

(from Figure 6.5 for $N \approx 10\text{–}25$, $\phi \approx 35$)

For clay, $\phi = 0$; from Figure 6.5, $L_{p,\text{cr}}//D = 3$, $L_{p,\text{cr}}^{(\text{clay})} = 3 \times 0.246 = 0.738\,\text{m}$

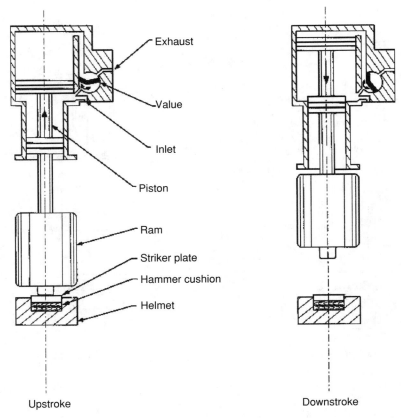

FIGURE 6.36
Schematic of differential air or steam hammer. (From www.vulcanhammer.net. With permission.)

The following soil strength properties can be obtained based on the SPT values:

$\delta = \frac{2}{3}\phi$

$K = 1.4K_0 \tan \delta = 1.4(1 - \sin \phi) \tan \delta$

For $\phi = 32°$, $K = 1.4(1 - \sin 32) \tan 21.3 = 0.257$

FIGURE 6.37
Vibratory pile driving. (From www.apevibro.com. With permission.)

From Equation (6.7), $P_{p,ult} \approx 9.0A_pC_u$, and Equation (6.14),

$$f = \alpha C_u \rightarrow P_{s,u} = \int_0^L Pf \, dZ$$

The maximum total ultimate resistance by clay layer

$= $ point bearing $+$ skin friction

$= 9.0(0.063) \, 35 + 1.475 \, [1 \times 1.0 \times 18 + 4 \times 0.98 \times 35]$

$= 248.7 \, kN \ll 1000 \, kN$

So, the pile has to driven into sand layer. Say, up to a depth of "L" m

Effective clay overburden $= 16.5 \times 1 + (17.5 - 9.8)4 = 47.9 \, kPa$
Effective sand overburden $= (L - 5)(17.5 - 9.8) = (7.7L - 38.5) \, kPa$
Pile point capacity in sand $= P_{p,ult} = A_p[q(N_q - 1)]$ \hfill (6.5)

$= 0.063(210 - 1)(7.7L - 38.5 + 47.9)$
$= (101.39L + 123.77) \, kPa$

Pile ultimate skin resistance, $P_{s,ult} = P_{s,clay} + P_{s,sand}$

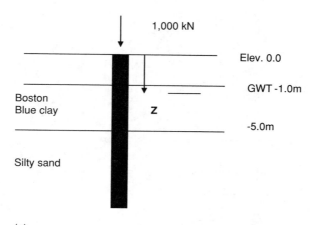

(a)

FIGURE 6.38
Illustration for Example 6.8.

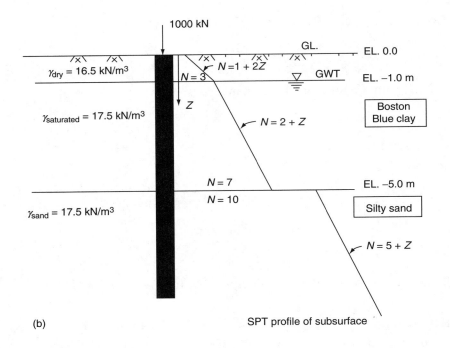

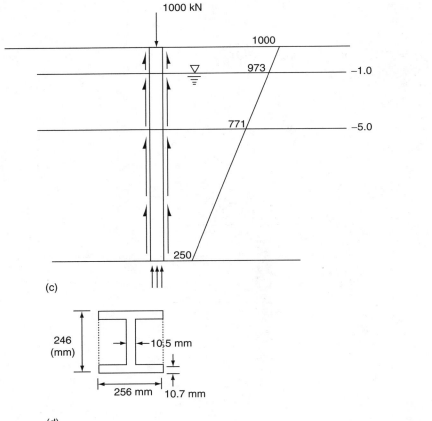

FIGURE 6.38
Continued

TABLE 6.18

Worksheet for Example 6.8

	Depth (m)	Φ (°)	C_u (kPa)	From Figure 6.4 N'_c	From Figure 6.4 N'_q	α	$\delta = \frac{2}{3}\phi$ (°)	K	$L_{s,lm} = 3.69$ m Applies?	$L_{p,cr}$ (m)
Boston clay	0–1	—	18	9	—	1.0	—	—		0.738
	1–5	—	35	9	—	0.98	—	—		0.738
Silty sand	5–10	32	—	—	70	—	21.3	0.257	Yes	1.97
	10–15	34	—	—	110	—	22.6	0.257	Yes	1.97
	15–20	36	—	—	140	—	24	0.257	Yes	1.97
	20+	38	—	—	210	—	25.3	0.254	Yes	1.97

$$P_{s,clay} = 1.475[1 \times 1.0 \times 18 + 4 \times 0.98 \times 35] = 228.92 \text{ kN}$$

$$\begin{aligned}
P_{s,sand} = {}& 1475[(0.257)(5)(1/2)\{47.9 + 47.9 + 5(17.5 - 9.8)\} \\
& + (0.257)(5)(1/2)\{47.9 + 5(17.5 - 9.8) + 47.9 + 10(17.5 - 9.8)\} \\
& + (0.257)(5)(1/2)\{47.9 + 10(17.5 - 9.8) + 47.9 + 15(17.5 - 9.8)\} \\
& + (0.254)(L - 20)\{47.9 + 15(17.5 - 9.8)\}] \\
= {}& 1.475[86.29 + 135.76 + 185.23 + 41.5L - 830]
\end{aligned}$$

Total ultimate resistance $= P_{p,ult} + P_{s,ult}$

$$\begin{aligned}
&= 101.39L + 123.77 + 228.92 + 61.2L - 623.5 \\
&= 162.59L - 270.81
\end{aligned}$$

Therefore, $162.59L - 270.81 = 1000 \times 2.5$ where FS $= 2.5$

Therefore, $L = 17.04$ m < 20 m

Revise, $P_{s,sand} = 1.475[86.29 + 135.76 + (0.257) (L - 15)\{47.9 + 10(17.5 - 9.8)\}]$

$$= 47.35L - 382.67$$

Therefore, $101.39L + 123.77 + 228.92 + 47.35L - 382.67 = 1000 \times 2.5$

Therefore, $L = 17.00$ m within the 15–20 m layer

So, it is adequate.

(c) $P_{s,clay} = 228.92$ kN(26.55 kN(1st 1 m) $+ 137.2$ kN(next 4 m))

$$P_{s,sand} = 47.35 \times 17 - 382.67 = 422.3 \text{ kN}$$
$$P_{p,ult} = 101.39 \times 17 + 123.77 = 1847.4 \text{ kN}$$

(d) Total settlement of pile top

$$\text{Tip settlement: } S_i = qB \frac{1 - \nu_s^2}{E_s}\left[I_1 + \frac{1 - 2\nu_s}{1 - \nu_s} I_2\right] I_F \qquad (6.28)$$

$$q = \frac{P_p}{A_p} = \frac{(1000 - 228.92 - 422.8)}{0.063} = 5528.25 \text{ kPa}$$

$$B = \text{pile equivalent diameter} = \sqrt{\frac{0.063}{\pi/4}} = 0.283 \text{ m}$$

$$\nu_s = 0.3, \ I_1 = 1.0, \ I_2 = 0, \ I_F = 0.50$$

$$S_i = (5528.25)(0.283)\frac{1 - 0.3^2}{30 \times 10^3}\left[1.0 + \frac{1 - 0.3 \times 2}{1 - 0.3}(0)\right]0.50$$

$$= 0.0237 \text{ m}$$

Elastic shortening of pile,   (6.30)

$$S_e = \frac{1}{E_p A_p}\int_0^L P(Z)\, dZ$$

$$E_{p,\text{steel}} = 29 \times 10^6 \text{ kPa}$$

$$A_p = \text{pile section area}$$

$$= 8.0 \times 10^{-3}\, \text{m}^2$$

$$S_e = \frac{1}{29 \times 10^6 \times 8.0 \times 10^{-3}}$$
$$[\tfrac{1}{2}(1000 + 973) \times 1 + \tfrac{1}{2}(973 + 771) \times 4 + \tfrac{1}{2}(771 + 348.28) \times 12]$$

$$= 0.0482 \text{ m}$$

Total settlement = tip settlement + elastic shortening

$$= 0.0237 \text{ m} + 0.0482 \text{ m}$$
$$= 0.0719 \text{ m} \approx 72 \text{ mm}$$

Example 6.9
Figure 6.39 shows the subsurface profile of a site where a pile group is to be designed as the foundation of a high rise. The following are the soil properties:

Dry unit weight of silty sand = 16.5 kN/m^3

Saturated unit weight of silty sand = 17.5 kN/m^3

Average cone resistance of silty sand = 1.5 MPa

Friction ratio of silty sand = 0.01

Saturated unit weight of clay = 18.5 kN/m^3

Moisture content of clay = 15%

Compression index of clay = 0.3

The envisioned pile group contains 16 piles arranged in a square configuration with a spacing of four diameters. The piles to be used are concrete ones with a diameter of 12 in.

(a) Find the maximum load that can be allowed on the group.

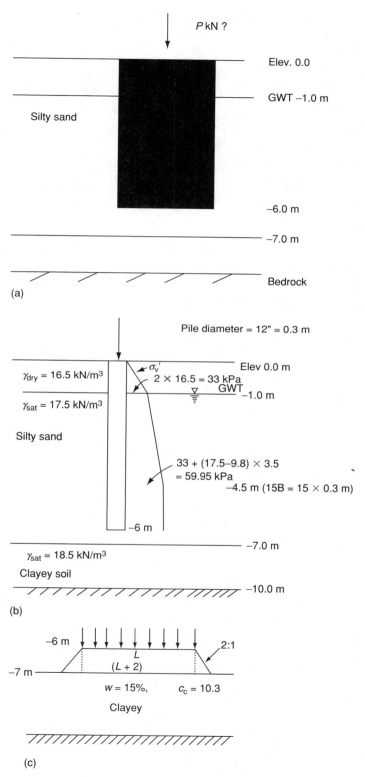

FIGURE 6.39
Illustration for Example 6.9.

(b) Assuming an appropriate load distribution, compute the settlement of the building. (You may neglect the elastic shortening.)

Single pile point capacity, $P_p = A_p[q(N_q - 1)]$ (6.5)

A_p = pile cross-sectional area = $(\Pi/4)(0.3)^2 = 0.0707\,m^2$

Cone tip resistance = $(1 - 0.01)1.5\,MPa = 1.48\,MPa$

Effective overburden pressure at pile bottom

$Q = 16.5 \times 1 + (17.5 - 9.8) \times 5 = 55$ kPa

For $q = 55$ kPa, cone tip = $1.48\,Mpa \rightarrow \Phi = 35°$

For $\Phi = 35°$, $N_q = 120$ (Figure 6.4)

Therefore, $P_p = 0.0707[55(120 - 1)] = 463$ kN

For concrete-silty sand, $\delta = \frac{2}{3}\phi = \frac{2}{3}(35) = 23.3°$

$$K = 1.4K° \tan \delta = 1.4\,(1 - \sin\,\phi)\tan\delta = 1.4\,(1 - \sin 35)\,\tan 23.3$$
$$= 0.257$$

Ultimate side (skin) friction in sand:

$P_{su} = \int_0^L pf\,dZ$

p = perimeter of pile = $\Pi D = \Pi(0.3)$

F = unit skin friction at any depth Z

L = length of pile in silty sand layer

$f = K\sigma_v' = 0.257\sigma_v'$

Therefore, $P_{su} = \dfrac{1}{2}\,(0.257)(33)\pi(0.3)(2) + \dfrac{1}{2}\,(0.257)(33$
$$+ 59.95)\pi(0.3)(3.5) + (0.257)(59.95)\pi(0.3)(1.5)$$
$$= 67.2\text{ kN}$$

Total single pile capacity = $P_s + P_{su}/FS = 463 + 67.2/4 = 132.5$ kN

(a) Maximum load allowed on the group = $16 \times$ single pile capacity

$= 16 \times 132.5\,kN = 2120\,kN$

L = width of pile group

$= 3 \times 4 \times 0.3 + 0.3 = 3.9\,m$

Load = 2120 kN

Stress area = $(L + 2)(L + 2) = 5.9\,m \times 5.9\,m$ (Figure 6.39c)

Applied stress on clay,

$$\Delta\sigma = \frac{2120}{5.9 \times 5.9} = 60.9\text{ kPa}$$

Consolidation settlement for clayey layer

$S_c = \frac{H}{1 + e_0}\,C_c\,\log\left[\frac{\sigma_{v0}' + \Delta\sigma}{\sigma_{v0}'}\right]$

H = clay thickness = $3\,m$

$e_0 = wG_s = 0.15 \times 2.6 = 0.39$

C_c = compression index = 0.3

σ'_{v0} = Effective stress at clay center

$\quad = 2 \times 16.5 + 6(17.5 - 9.8) + 1.5(18.5 - 9.8)$

$\quad = 92.25 \, \text{kPa}$

$\therefore$ (b) consolidation settlement = 0.475 m

Example 6.10

Figure 6.40(a) shows the subsurface profile of a site where a new tall building that is to be constructed in downtown Boston. It is decided to use a pile foundation of which a single pile is also shown:

Saturated unit weight of sand = 17.5 kN/m^3

Dry unit weight of sand = 16.5 kN/m^3

Angle of internal friction of sand = 22°

Saturated unit weight of Boston Blue clay = 17.5 kN/m^3

Undrained strength of Boston Blue clay = 20 kPa

Elevation of Boston Blue clay = −7 m

Depth of pile = −10 m

(a) Select a pile type for this foundation with appropriate justification.

(b) If it is decided to use piles of 25 cm in diameter, determine the maximum load that can be carried by each pile.

Allowable pile load capacity,

$$P_a = \frac{P_{pu} + P_{su}}{\text{FS}}$$

Ultimate point capacity in clay (by Meherhoff's method):

$$
\begin{aligned}
&= P_{pu} \approx 9.0 \, A_p c_u \\
&= 9.0 \times \frac{\pi}{4} (0.25)^2 \times 20 \\
&= 8.84 \, \text{kN}
\end{aligned}
\tag{6.7}
$$

where A_p is the base cross-sectional area, C_u is undrained cohesion, and S_u is undrained shear strength $\approx 20 \, \text{kPa}$.

Ultimate side (skin) friction in sand:

$$P_{su} = \int_0^L pf \, dZ \tag{6.11}$$

where p is the perimeter of the pile section, f is the unit skin friction at any depth Z, and L is the length of the pile in the sand layer

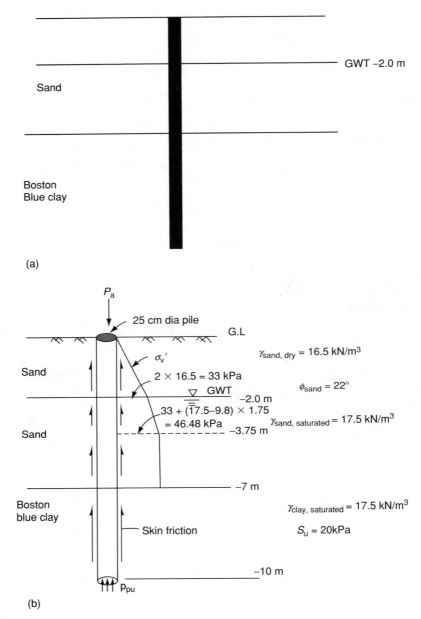

FIGURE 6.40
Illustration for Example 6.10.

$$f = 1.4K_0\sigma'_v \tan\delta \quad \text{(assume a driven pile)}$$
$$= 1.4 \times 0.625 \ \tan 14.67 \ (\sigma'_v) \quad\quad (6.12)$$
$$= 0.23\sigma'_v$$

$K_0 = 1 - \sin \phi = 1 - \sin 22° = 0.625$

$\delta = \dfrac{2}{3} \phi = \dfrac{2}{3} (22°) = 14.67°$

σ'_v = vertical effective stress (increases

linearly up to a depth of $15B = 15 \times 0.25 = 3.75$ m) $\hfill (6.13)$

$$P_{\text{su,sand}} = \frac{1}{2} (0.23)(33) \pi (0.25)(2) + \frac{1}{2}(0.23)(33 + 46.48) \pi (0.25)(1.75)$$
$$+ (0.23)(46.48) \pi (0.25)(3.25)$$
$$= 45.81 \text{ kN}$$

Ultimate Side (Skin) Friction in Clay:

$f = \alpha C_u$ $\hfill (6.14)$

α = adhesion factor ≈ 0.98 (interpolation, Table 6.2)

C_u = undrained cohesion $\approx S_u \approx 20 \text{ kPa}$

$P_{\text{su,clay}} = pfL_{\text{clay}} = \pi (0.25)(0.98)(20)(3) = 46.18 \text{ kN}$ $\hfill (6.11)$

Total skin friction $= P_{\text{su,sand}} + P_{\text{su,clay}} = 45.81 + 46.18 = 91.99 \text{ kN}$

Therefore, $P_a = \dfrac{P_{\text{pu}} + P_{\text{su}}}{\text{FS}} = \dfrac{8.84 + 91.99}{4} = 25.2 \text{ kN}$ $\hfill (6.3)$

Auger-cast concrete piles are preferred due to their higher capacity (larger diameters) and low vibration during construction.

References

AASHTO, 1996, *Standard Specifications for Highway Bridges*, 16th ed., American Association of State Highway and Transportation Officials, Washington, DC.

Bowles, J.E., 2002, *Foundation Analysis and Design*, McGraw-Hill, New York.

Bullock, P.J., Schmertmann, J.H., Mcvay, M.C., and Townsend, F., 2005a, Side shear setup I: Test piles driven in Florida, *Journal of Geotechnical and Geoenvironmental Engineering*, 131(3): 292–300.

Bullock, P.J., Schmertmann, J.H., Mcvay, M.C., and Townsend, F., 2005b, Side shear setup II: Results from Florida test piles, *Journal of Geotechnical and Geoenvironmental Engineering*, 131(3): 301–310.

Burland, J.B., 1973, Shaft friction piles in clay — a simple fundamental approach, *Ground Engineering*, 6(3): 30–42.

Chun, B.S., Cho, C.W., and Lee, M.W., 1999, Prediction of increase in pile bearing capacity with time after driving, *Eleventh Asian Regional Conference on Soil Mechanics and Geotechnical Engineering*, Balkema, Rotterdam.

Das, B.M., 1995, *Principles of Foundation Engineering*, PWS Publishing, Boston, MA.

DeRuiter, J. and Beringen, F.L., 1979, Pile foundations for large North Sea structures, *Marine Geotechnology*, 3(3): 267–314.

Federal Highway Administration, 1998, *Load and Resistance Factor Design (LRFD) for Highway Bridge Substructures*, Washington, DC.

Fellenius, B.H., Riker, R.E, O'Brien, A.J., and Tracy, G.R., 1989, Dynamic and static testing in soil exhibiting set-up, *Journal of Geotechnical Engineering*, 115(7): 984–1001.

Goble, G.G., Raushe, F., Likens and Associates, Inc. (GRL), 1996, *Design and Construction of Driven Pile Foundations*, Volume I, NHI Course 13221 and 132222, US Department of Transportation, Federal Highway Administration, Washington, DC.

Kulhawy, F.H. and Goodman, R.E., 1980, Design of foundations in discontinuous rock, *Proceedings of the International Conference of Structural Foundations on Rock*, Sydney, Balkema, Rotterdam, Vol. 1, 209–220.

Meyerhoff, G.G., 1976, Bearing capacity and settlement of pile foundations, *Journal of Geotechnical Engineering, ASCE*, 102(GT3): 197–227.

Nottingham, L. and Schmertmann, J., 1975, An Investigation of Pile Design Procedures, Final Report D629 to Florida Department of Transportation, Department of Civil Engineering, University of Florida.

Peck, R.B., 1974, *Foundation Engineering*, 2nd ed., John Wiley, New York.

Poulos, H.G. and Davis, E.H., 1990, *Pile Foundation Analysis and Design*, Krieger, Melbourne, FL.

Seed, H.B. and Reese, L.C., 1955, The action of soft clay along friction piles, *Proceedings ASCE*, 81, Paper 842.

Skov, R. and Denver, H., 1989, Time-dependence of bearing capacity of piles, *Proceeding of the 3rd International Conference on the application of Stress Wave Theory to Piles*, Bitech Publishers, Ottawa, Canada, pp. 879–888.

Sladen, J.A., 1992, The adhesion factor, applications and limitations, *Canadian Geotechnical Journal*, 29(2): 322–326.

Svinkin, M.R., 1995, Pile-soil dynamic system with variable damping, *Proceedings of the 13th International Modal Analysis Conference*, IMAC-XIII, Beyond the Modal Analysis, SEM, Bethel, Connecticut, Vol. 1, pp. 240–247.

Svinkin, M.R. and Skov, R., 2002, Setup effect of cohesive soils in pile capacity, 2002, http://www.vulcanhammer.net/svinkin/set.htm.

Tavenas, F. and Audy, R., 1972, Limitations of the driving formulas for predicting the bearing capacities of piles in sand, *Canadian Geotechnical Journal*, 9(1): 47–62.

Thilakasiri, H.S., Abeyasinghe, R.M., and Tennakoon, B.L., 2003, A study of strength gain of driven piles, *Proceedings of the 9th Annual Symposium*, Engineering Research Unit, University of Moratuwa, Sri Lanka.

Tomlinson, M.J., 1994, *Pile Design and Construction Practices*, 4th edn, E & FN Spon, London.

Van Impe, W.F., 2004, *Two Decades of full scale research on screw piles*, Ghent University Laboratory of Soil Mechanics Technologiepark 9, B-9052 Ghent, Zwijnaarde, Belgium.

Vesic, A.S., 1977, *Design of Pile Foundations*, National Cooperative Highway Research Program, Synthesis of Practice, No. 42, Transportation Research Board, Washington, DC.

Vijayvergiya, V.N. and Frocht, J.A., 1972, A new way to predict capacity of piles in clay, *4th Offshore Technology Conference*, Paper 1718, Houston, Texas.

York, D.L., Brusey, W.G., Clemente, F.M., and Law, S.K., 1994, Setup and relaxation in glacial sand, *Journal of Geotechnical Engineering, ASCE*, 120(9): 1498–1513.

Websites

http://www.screw-piling.com/
http://www.uslgroup.com.au/stlscw.html
http://www.piling.com.au
http://www.vulcanhammer.net
http://sbe.napier.ac.uk/projects/piledesign/guide/chapter8.htm
http://www.apevibro.com

7

Design of Drilled Shafts

Gray Mullins

CONTENTS

casings can induce necking due to low pressure developed at the base of the extracted casing.

With the exception of full-length temporary casing methods, the practical upper limit of shaft length is on the order of 30D (i.e., 90 ft for 3 ft diameter shafts) but can be as much as 50D in extraordinary circumstances using special excavation methods.

7.2.3 Concreting and Mix Design

Drilled shaft concrete is relatively fluid concrete that should be tremie placed (or pumped to the base of the excavation) when using any form of wet construction to eliminate the possibility of segregation of fine and coarse aggregate or mixing with the *in situ* slurry. A tremie is a long pipe typically 8 to 12 in. in diameter used to take the concrete to the bottom of the excavation without being altered by the slurry (i.e., mixing or aggregate segregation). Prior to concreting, some form of isolation plug should be placed in-line or at the tip of the tremie to prevent contamination of the concrete flow as it passes through the initially empty tremie. During concrete placement, the tremie tip elevation should be maintained below the surface of the rising concrete (typically 5 to 10 ft). However, until a concrete head develops at the base of the excavation, the potential for initial mixing (and segregation) will always exist. In dry construction, free-fall concrete placement can be used although it is restricted by some State agencies in the United States. The velocity produced by the falling concrete can induce higher lateral pressure on the excavation walls, increase concrete density, and decrease porosity or permeability. However, velocity-induced impacts on reinforcing steel may misalign tied steel stirrups and the air content (if specified) of the concrete can be reduced.

The concrete mix design for drilled shafts should produce a sufficient slump (typically between 6 and 9 in.) to ensure that lateral fluid concrete pressure will develop against the excavation walls. Further, the concrete should maintain a slump no less than 4 in. (slump loss limit) for several hours. This typically allows enough time to remove the tremie and any temporary casing while the concrete is still fluid enough to replace the volume of the tremie or casing and minimize suction forces (net negative lateral pressure) during extraction. However, recent studies suggest that a final slump in the range of 3.5 to 4 in. (or less) at the time of temporary casing extraction can drastically reduce the side shear capacity of the shaft (Garbin, 2003). As drilled shaft concrete is not vibrated during placement, the maximum aggregate size should be small enough to permit unrestricted flow through the steel-reinforcing cage. The ratio of minimum rebar spacing to maximum aggregate diameter should be no less than 3 to 5 (FHWA, 1998).

7.3 Design Capacity of Drilled Shafts

The capacity of drilled shafts is developed from a combination of side shear and end bearing. The side shear is related to the shear strength of the soil and in sands can be thought of as the lesser of the friction ($F_r = \mu N$) that develops between the shaft concrete and the surrounding soil or the internal friction within the surrounding soil itself. Although a coefficient of friction (μ) can be reasonably approximated, the determination of the normal force (N) is more difficult due to lateral stress relaxation during excavation. In clayey soils or rock side shear is most closely related to the unconfined compressive strength, q_u. The end bearing is analogous to shallow foundation bearing capacity with a very large depth of footing. However, it too is affected by construction-induced disturbances and like the side shear, it has been empirically incorporated into the design methods discussed in the ensuing sections. Most of the design charts and tables in this chapter are developed in British units. Hence, the reader is referred to Table 7.9 for the appropriate conversions to relevant SI units.

The design approach for drilled shafts can be either allowable stress design (ASD) or load and resistance factor design (LRFD) as dictated by the client, local municipality, or State agency in the United States. In either case, the concept of usable capacity as a function of ultimate capacity must be addressed. This requires the designer to have some understanding of the capacity versus displacement characteristics of the shaft. Likewise, a permissible displacement limit must be established to determine the usable capacity rather than the ultimate capacity that may be unattainable within a reasonable displacement. The permissible displacement (or differential displacement) is typically set by a structural engineer on the basis of the proposed structure's sensitivity to such movement. To this end, design of drilled shafts (as well as other foundation types) must superimpose displacement criteria onto load-carrying capability even when using an LRFD approach. This is divergent from other nongeotechnical LRFD approaches that incorporate design limit states independently (discussed later).

The designer must be aware of the difference in the required displacements to develop significant capacity from side shear and end bearing. For instance, in sand the side shear component can develop 50% of ultimate capacity at a displacement of approximately 0.2% of the shaft diameter (D) (AASHTO, 1998), and develops fully in the range of 0.5 to 1.0%D (Bruce, 1986). In contrast, the end bearing component requires a displacement of 2.0%D to develop 50% of its capacity (AASHTO, 1998), and fully develops in the range of 10 to 15%D (Bruce, 1986). Therefore, a 4 ft diameter shaft in sand can require up to 0.5 in. of displacement to develop ultimate side shear and 7.2 in. to develop ultimate end bearing. Other sources designate the displacement for ultimate end bearing to be 5%D but recognize the increase in capacity at larger displacements (Reese and Wright, 1977; Reese and O'Neill, 1988).

In most instances, the side shear can be assumed to be 100% usable within most permissible displacement criteria but the end bearing may not. This gives rise to the concept of mobilized capacity. The mobilized end bearing is the capacity that can be developed at a given displacement. Upon determining the permissible displacement, a proportional capacity can then be established based on a capacity versus displacement relationship as determined by either load testing or past experience. A general relationship will be discussed in the section discussing end bearing determination methods.

7.3.1 ASD versus LRFD

In geotechnical designs, both ASD and LRFD methods must determine an ultimate capacity from which a usable capacity is then extracted based on displacement criteria. As such the ultimate capacity is never used, but rather a displacement-restricted usable capacity is established as the effective ultimate capacity. For drilled shafts, this capacity typically incorporates 100% of ultimate side shear and the fraction of end bearing mobilized at that displacement. Once this value has been determined, the following generalized equations represent the equality that must be satisfied when using either an ASD approach or an LRFD approach, respectively:

$$Service\ Load \leq \frac{Effective\ Ultimate\ Strength}{F}N \quad (ASD) \tag{7.1a}$$

or

$$P_u = \sum \gamma_i P_i \le \phi P_n N \quad (LRFD) \tag{7.1b}$$

where F is the safety factor, P_u represents the sum of factored or inflated service loads based on the type of loads, P_n represents the effective ultimate shaft capacity, ϕ is the number of shafts, and N (the resistance factor) reduces the effective ultimate capacity based on the reliability of the capacity determination method. The use of LRFD in geotechnical designs is relatively new and as such present methods have not yet completely separated the various limit states.

Typically there are four LRFD limit states: strength, service, fatigue, and extreme event. These limit states treat each area as mutually exclusive issues. Strength limit states determine if there is sufficient capacity for a wide range of loading conditions. Service limit states address displacement and concrete crack control. Fatigue addresses the usable life span of steel in cyclic or stress reversal regions. Extreme event limit states introduce less probable but more catastrophic occurrences such as earthquakes or large vessel impacts. Any of the four limit states can control the final design. The ASD method lumps all load types into a single service load and assumes the same probability for all occurrences.

Although LRFD strength limit states should be evaluated without regard to the amount of displacement required to develop full ultimate capacity (P_n), present LRFD methods establish geotechnical ultimate capacity based on some displacement criteria. As a result, LRFD geotechnical service limit states are relatively unused. To this end, this chapter will emphasize the design methods used to determine ultimate capacity and will denote (where applicable) the displacement required to develop that capacity. The following design methods are either the most up-to-date or the most widely accepted for the respective soil type or soil exploration data.

7.3.2 Standard Penetration Test Data in Sand

Standard penetration test (SPT) (Section 2.4.1) results are most commonly used for estimating a drilled shaft capacity in sandy soils. For some design methods direct capacity correlations to the SPT blow count (N) have been developed; in other cases correlations to soil properties such as unit weight or internal angle of friction are necessary. Where the unit weight or the internal friction angle (sands) of a soil is required the relationships shown in Figure 7.2 can be used.

7.3.3 Estimation of Side Shear

The side shear developed between a shaft and surrounding sandy soils can be estimated using the methods given in Table 7.1. The ultimate load-carrying capacity from side shear (Q_s) can be expressed as the summation of side shear developed in layers of soil to a given depth containing n layers:

$$Q_s = \pi \sum_{i=1}^{n} f_{si} L_i D_i \tag{7.2}$$

where f_{si} is the estimated unit side shear for the ith soil layer, L_i is the thickness of (or length of shaft in) the ith soil layer, and D_i is the diameter of the shaft in the ith soil layer.

Using the above methods, the variation in estimated side shear capacity is illustrated for a 3 ft diameter shaft and the given SPT boring log in sandy soil in Figure 7.3. Although

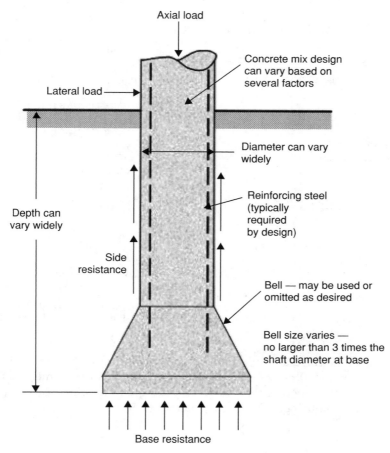

Axial load

Lateral load

Concrete mix design
can vary based on
several factors

Diameter can vary
widely

Depth can
vary widely

Reinforcing steel
(typically
required
by design)

Side
resistance

Bell — may be used or
omitted as desired

Bell size varies —
no larger than 3 times the
shaft diameter at base

Base resistance

FIGURE 7.1
Details of a drilled shaft. (From www.dot.state.fl.us/construction. With permission.)

any of these methods may correlate closely to a given site or local experience, the author recommends the O'Neill and Hassan (1994) approach in spite of its less conservative appearance.

7.3.4 Estimation of End Bearing

Recalling the importance of the mobilized end bearing capacity concept, a parameter termed the tip capacity multiplier (TCM) will be used to quantify the relationship between ultimate and usable end bearing capacity. Four design methods using two different approaches to mobilized capacity are discussed. The first and second assume ultimate end bearing occurs at 1.0 in. displacement (Touma and Reese, 1974; Meyerhof, 1976). The others assume ultimate end bearing occurs at a 5% displacement as shown in Figure 7.4 (*Reese and Wright, 1977*; Reese and O'Neill, 1988). Figure 7.4 shows the latter relationship in terms of the permissible displacement expressed as a percentage of the shaft diameter. Therein, the TCM for convention shafts tipped in sand is linearly proportional to the displacement where the TCM = 1 at 5% displacement. This concept can be extended to the first two design methods as well where TCM = 1 at 1.0 in. displacement. Table 7.2 lists the four methods used to estimate the ultimate end bearing to which a TCM should be applied.

TABLE 7.1

Drilled Shaft Side Shear Design Methods for Sand

Source	Side Shear Resistance, f_s (in tsf)
Touma and Reese (1974)	$f_s = K\sigma_v \tan\phi < 2.5$ tsf where $K = 0.7$ for $D_b < 25$ ft $K = 0.6$ for 25 ft $< D_b \leq 40$ ft $K = 0.5$ for $D_b > 40$ ft
Meyerhof (1976)	$f_s = N/100$
Quiros and Reese (1977)	$f_s = 0.026N < 2.0$ tsf
Reese and Wright (1977)	$f_s = N/34$, for N ≤ 53 $f_s = (N - 53)/450 + 1.6$, for $53 < N \leq 100$ $f_s \leq 1.7$
Reese and O'Neill (1988) *Beta method*	$f_s = \beta\sigma_v' < 2.0$ tsf for $0.25 \leq \beta < 1.2$ where $\beta = 1.5 - 0.135\,z^{0.5}$, z in ft
O'Neill and Hassan (1994) *Modified beta method*	$f_s = \beta\sigma_v' < 2.0$ tsf for $0.25 \leq \beta \leq 1.2$ where $\beta = 1.5 - 0.135\,z^{0.5}$ for $N > 15$ $\beta = N/15\,(1.5 - 0.135\,z^{0.5})$ for $N \leq 15$

Source: AASHTO, 1998, *LRFD Bridge Design Specifications, Customary U.S. Units,* 2nd edn, American Association of State Highway and Transportation Officials, Washington, DC, with 1999 interim revisions. With permission.

Figure 7.5 shows the calculated ultimate end bearing using each of the four methods in Table 7.2. The Reese and Wright (1977) or Reese and O'Neill (1988) methods are recommended by the author for end bearing analysis. Using the combined capacity from 100% side shear and TCM*q_p using O'Neill and Hassan (1994) and Reese and O'Neill (1988)

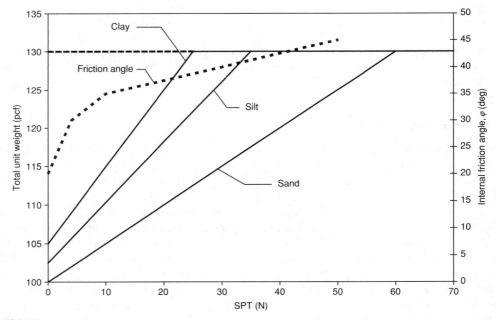

FIGURE 7.2
Estimated soil properties from SPT blow count.

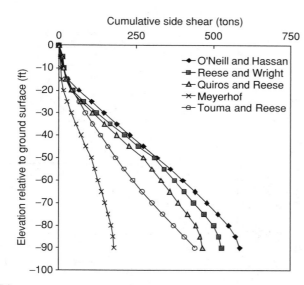

FIGURE 7.3

Comparison of estimated side shear capacities in sandy soil (3 ft diameter).

methods, respectively, the effective ultimate capacity of a 3 ft diameter drilled shaft can be estimated as a function of depth, Figure 7.6. This type of curve is convenient for design as it is a general capacity curve independent of a specific design load. However, when using an LRFD approach, the factored load (P_u) should be divided by the appropriate resistance factor before going to this curve.

7.4   Use of Triaxial or SPT Data in Clay

Unconsolidated, undrained (UU) triaxial test results are preferred when estimating the side shear or end bearing capacity of drilled shafts in clayey soil. The mean undrained shear strength (S_u) is derived from a number of tests conducted on Shelby tube specimens where $S_u = \frac{1}{2}\sigma_{1,\,max}$. In many instances, both UU and SPT data can be obtained from which local SPT (N) correlations with S_u can be established. In the absence of any UU test results, a general correlation from Kulhawy and Mayne (1990) can be used

$$S_u = 0.0625N, \text{ in units of tsf} \tag{7.3a}$$

7.4.1   Side Shear (Alpha Method)

The alpha method of side shear estimation is based on correlations between measured side shear from full-scale load tests and the clay shear strength as determined by UU test results. Therein, the unit side shear f_s is directly proportional to the product of the adhesion factor (Table 7.3) and S_u

$$f_s = \alpha S_u \tag{7.3b}$$

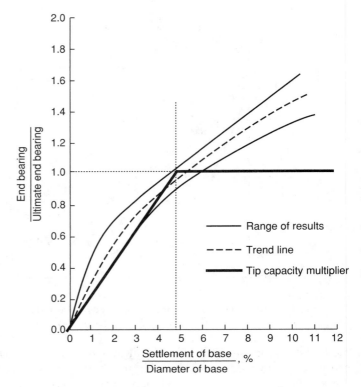

FIGURE 7.4
End bearing response of sands as a function of displacement. (Based on Reese, L.C. and O'Neill, M.W., 1988, *Drilled Shafts: Construction and Design*, FHWA, Publication No. HI-88-042. *With permission.*)

The side shear developed around drilled shafts in clayey soil has several limitations that were not applied previously applied to shafts cast in sand. Specifically, the top 5 ft of the shaft sides are considered noncontributing due to cyclic lateral movements that separate

TABLE 7.2

Drilled Shaft End Bearing Design Methods for Sands

Source	End Bearing Resistance, q_p (in tsf)[a]
Touma and Reese (1974)	Loose sand, $q_p = 0.0$ Medium dense sand, $q_p = 16/k$ Very dense sand, $q_p = 40/k$ where $k = 1$ for $D_p = 1.67$ ft $k = 0.6D_p$ for $D_p \leq 1.67$ ft Only for shaft depths $> 10D$
Meyerhof (1976)	$q_p = (2N_{corr}D_b)/(15D_p)$ $q_p < 4/3N_{corr}$ for sand $q_p < N_{corr}$ for nonplastic silts
Reese and Wright (1977)	$q_p = 2/3N$ for $N \leq 60$ $q_p = 40$ for $N > 60$
Reese and O'Neill (1988)	$q_p = 0.6N$ for $N \leq 75$ $q_p = 45$ for $N > 75$

[a]For $D > 4.17$ ft, the end bearing resistance should be reduced to $q_{pr} = 4.17q_p/D$.
Source: From AASHTO, 1998, *LRFD Bridge Design Specifications, Customary U.S. Units*, 2nd edn, American Association of State Highway and Transportation Officials, Washington, DC, with 1999 interim revisions. With permission.

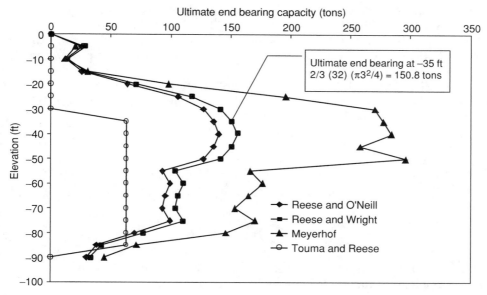

FIGURE 7.5
Comparison of end bearing methods in sand (3 ft diameter, boring B-1).

the shaft from the soil as well as potential desiccation separation of the surficial soil. Additionally, the bottom 1*D* of the shaft side shear is disregarded to account for lateral stresses that develop radially as the end bearing mobilizes.

Although rarely used today, belled ends (Figure 7.1) also affect the side shear near the shaft base. In such cases, the side shear surface area of the bell as well as that area 1*D* above the bell should not be expected to contribute capacity.

7.4.2 End Bearing

The end bearing capacity of shafts tipped in clay is also dependent on the mean undrained shear strength of the clay within two diameters below the tip, S_u. As discussed with shafts tipped in sands, a TCM should be applied to estimated end bearing capacities

TABLE 7.3

Adhesion Factor for Drilled Shafts in Clayey Soils

Adhesion Factor (Dimensionless)	Undrained Shear Strength, S_u (tsf)
0.55	<2.0
0.49	2.0–3.0
0.42	3.0–4.0
0.38	4.0–5.0
0.35	5.0–6.0
0.33	6.0–7.0
0.32	7.0–8.0
0.31	8.0–9.0
Treat as rock	>9.0

Source: From AASHTO, 1998, *LRFD Bridge Design Specifications, Customary U.S. Units*, 2nd edn, American Association of State Highway and Transportation Officials, Washington, DC, with 1999 interim revisions. With permission.

using the relationship shown in Figure 7.7. At displacements of 2.5% of the shaft diameter, shafts in clay mobilize 75 to 95% of ultimate capacity. Unlike sands, however, there is little reserve bearing capacity beyond this displacement. Therefore, a maximum TCM of 0.9 is recommended for conventional shafts at displacements of 2.5%D and proportionally less for smaller permissible displacements.

Similar to shallow foundation analyses, the following expressions may be used to estimate the ultimate end bearing for shafts with diameters less than 75 in. (AASHTO, 1998):

$$q_p = N_c S_u \leq 40 \text{ tsf} \tag{7.4a}$$

where

$$N_c = 6[1 + 0.2(Z/D)] \leq 9 \text{ for } S_u > 0.25 \text{ tsf} \tag{7.4b}$$

$$N_c = 4[1 + 0.2(Z/D)] \leq 9 \text{ for } S_u < 0.25 \text{ tsf} \tag{7.4c}$$

and Z/D is the ratio of the shaft diameter to depth of penetration. For shafts greater than 75 in. in diameter a reduction factor should be used as follows:

$$q_{pr} = q_p F_r \tag{7.5}$$

where

$$F_r \quad \frac{2.5}{12aD_p + 2.5b} \leq 1.0 \tag{7.6}$$

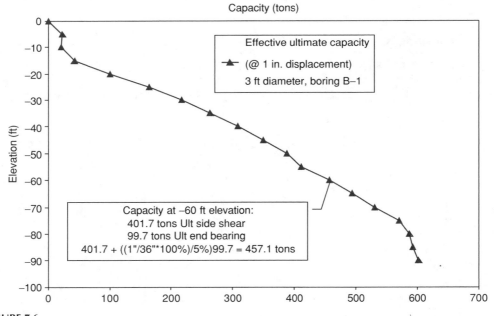

FIGURE 7.6
Example design curve using boring B-1 from Figure 7.3.

and

$$a = 0.0071 + 0.0021 \, Z/P \leq 0.015 \tag{7.7}$$

$$b = 0.45 \, (2 \, S_u)^{0.5} \tag{7.8}$$

where S_u is in tsf
  for

$$0.5 \leq b \leq 1.5$$

7.5 Designing Drilled Shafts from CPT Data

Cone penetration test data are considered to be more reproducible than SPT data and can be used for shaft designs in cohesionless and cohesive soils using correlations developed by Alsamman (1995). Although that study provided design values for both mechanical and electric cone data, a single approach is presented below that can conservatively be used for either based on that work.

7.5.1 Estimation of Side Shear

This method for determining side shear resistance in cohesionless soils is divided into two soil categories: gravelly sand or gravel and sand or silty sand. In each case (as given in Table 7.4), the side shear is correlated to the cone tip resistance, q_c, instead of the sleeve friction due to the absence of that data from some case studies at the time of the study. In cohesive soils, a single expression is given, which is also dependent on the total vertical stress, σ_{v0}. The same regions of the shaft should be discounted (top 5 ft and bottom 1D) when in cohesive soils as discussed earlier.

The upper limits for side shear recommended by Alsamman (1995) are somewhat less than those cited from AASHTO (1998) (e.g., 2.0 tsf for sands using the "beta method"). However, CPT data can also be used to estimate the internal friction and soil density necessary for the Touma and Reese (1974) or beta methods (Table 7.1).

TABLE 7.4

Side Shear Resistance from CPT Data

Soil Type	Ultimate Side Shear Resistance, q_s (tsf)
Gravelly sand/gravel	$f_s = 0.02q_c$ for $q_c > 50$ tsf
	$f_s = 0.0019q_c + 0.9 \leq 1.4$ for $q_c > 50$ tsf
Sand/silty sand	$f_s = 0.015 \, q_c$ for $q_c \leq 50$ tsf
	$f_s = 0.012q_c + 0.7 \leq 1.0$ for $q_c > 50$ tsf
Clay	$f_s = 0.023 \, (q_c - \sigma_{vo}) \leq 0.9$

Source: From AASHTO, 1998, *LRFD Bridge Design Specifications, Customary U.S. Units,* 2nd edn, American Association of State Highway and Transportation Officials, Washington, DC, with 1999 interim revisions. With permission.

TABLE 7.5

End Bearing Resistance from CPT Data

Soil Type	Ultimate End Bearing Resistance, q_p (tsf)
Cohesionless soils	$q_p = 0.15\, q_c$ for $q_c \leq 100$ tsf
	$q_p = 0.05\, q_c + 10 \leq 30$ for $q_c > 100$ tsf
Cohesive soils	$q_p = 0.25\,(q_c - \sigma_{vo}) \leq 25$

Source: From AASHTO, 1998, *LRFD Bridge Design Specifications, Customary U.S. Units*, 2nd edn, American Association of State Highway and Transportation Officials, Washington, DC, with 1999 interim revisions. With permission.

7.5.2 Estimation of End Bearing

Expressions for estimating the end bearing using CPT data were also recommended in the same study (Alsamman, 1995). Therein, the end bearing categories were limited to cohesionless and cohesive soils. Table 7.5 provides correlations based on those findings.

The capacities estimated from Table 7.5 expressions are ultimate values that should be assigned a proportionally less usable capacity using the general relationships shown in Figure 7.6 and Figure 7.7 for sands and clays, respectively.

7.6 Designing from Rock Core Data

A common application for drilled shaft is to be socketed in a rock formation some distance, H_s. In these cases, the side shear of softer overlying materials is disregarded due to the mismatch in the displacement required to mobilize both material types. Rock sockets require relatively small movements to develop full capacity when compared to sand or clay strata. Further, although the end bearing strength of a rock socket can be quite considerable, it too is often discounted for the same reason. Alternately, a rock socket may be designed for all end bearing instead of side shear knowing that some side shear capacity will always be available in reserve.

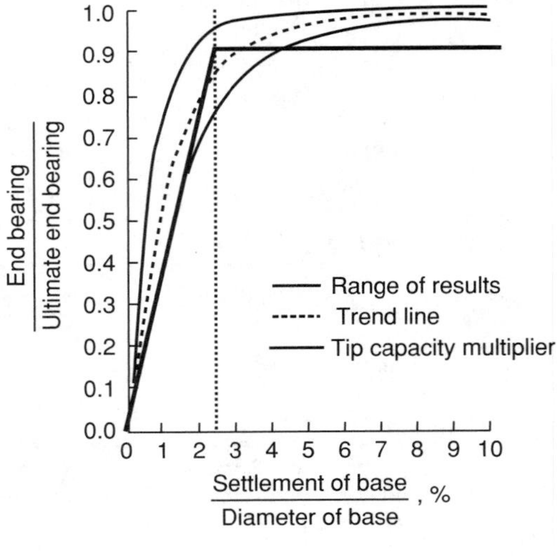

FIGURE 7.7
End bearing response of shafts tipped in clays. (Based on Reese, L.C. and O'Neill, M.W., 1988, *Drilled Shafts: Construction and Design*, FHWA, Publication No. HI-88-042. With permission.)

TABLE 7.6

Drilled Shaft Side Shear Design Methods for Rock Sockets

Source	Side Shear Resistance, f_s (tsf)	
Carter and Kulhawy (1988)	$f_s = 0.15\, q_u$	for $q_u \leq 20$ tsf
Horvath and Kenney (1979)	$f_s = 0.67\, q_u^{0.5}$	for $q_u > 20$ tsf
McVay and Townsend (1990)	$f_s = 0.5\, q_u^{0.5}\, q_s^{0.5}$	

Source: From AASHTO, 1998, *LRFD Bridge Design Specifications, Customary U.S. Units*, 2nd edn, American Association of State Highway and Transportation Officials, Washington, DC, with 1999 interim revisions. *With permission.*

7.6.1 Estimation of Side Shear

The side shear strength of rock-socketed drilled shafts is similar to that of clayey soils in that it is dependent on the *in situ* shear strength of the bearing strata. In this case, rock cores are taken from the field and tested using various methods. Specifically, mean failure stresses from two tests are commonly used: the unconfined compression test, q_u, and the splitting tensile test, q_s. The test results from these tests can be used to estimate the side shear of a rock socket using the expressions in Table 7.6. The estimated side shear capacity can be reduced by multiplying q_s by either the rock quality index, RQD, or the percent sample recovered from the rock core. Local experience and results from load tests can provide the best insight into the most appropriate approach.

7.6.2 Estimation of End Bearing

When determining the end bearing resistance (as well as side shear) of drilled shafts in rock, the quality of rock and type of rock can greatly affect the capacity. In competent rock the structural capacity of the concrete will control the design. In fractured, weathered rock or limestone, the quality of the formation as denoted by the RQD or percentage recovery should be incorporated into the capacity estimate. However, these parameters are influenced by drilling equipment, driller experience, and the type of core barrel used to retrieve the samples. The designer should make some attempt to correlate the rock quality to load test data where possible. The Federal Highway Administration recommends the following expression for estimating the end bearing resistance in rock (FHWA, 1988):

$$q_b = 2.5\, q_u \%\text{Rec} \leq 40 \text{ tsf} \tag{7.9}$$

The value of 40 tsf is undoubtedly conservative with respect to ultimate capacity, but when used in conjunction with a rock socket side shear it may be reasonable. Under any circumstances, load testing can verify much higher capacities even though they are near impossible to fail in competent rock.

7.7 Designing from Load Test Data

The use of an instrumented load test data for design is thought to be the most reliable approach and is given the highest resistance factor (LRFD) or lowest safety factor (ASD) as a result. This method involves estimating the shaft capacity using one of the previously discussed methods (or similar) and verifying the estimated capacity using a full-scale prototype shaft loaded to ultimate capacity. These tests can be conducted prior to construction or during construction (denoted as design phase or construction phase load testing, respectively). In either event, the shaft should be loaded well in excess of the design load while monitoring the response (i.e., axial displacement, lateral displacement, or internal strains).

An instrumented load test is one that incorporates strain gages along the length of the foundation to delineate load-carrying contributions from various soil strata. The test can merely distinguish side shear from end bearing or additional information from discrete shaft segments or soil strata can be obtained. Any test method capable of applying the ultimate load can provide useful feedback to the designer. Tests conducted on lesser loads are still useful, but provide only a "proof test" to the magnitude of the maximum load and can only provide a lower bound of the actual capacity. As such, the designer should realize that a test shaft that fails geotechnically, thus providing the ultimate capacity, is desirable in such a program so that the upper limit of capacity can be realized. The challenge then is to design a shaft that fails at a load reasonably close to the desired ultimate without being too conservative. However, the loading apparatus should have sufficient reserve to account for a slightly conservative capacity estimate.

7.7.1   Estimation of Side Shear

The ultimate side shear can be determined from load testing by evaluating the response from embedded strain gages at various elevations in the shaft. It is desirable to delineate bearing strata by placing these gages at the interface between significantly different soil strata (e.g., clay–sand interface). At a minimum, one level of gages should be placed at the tip of the shaft to separate the load-carrying contributions from the side shear and end bearing. By monitoring the strain at a given level, the corresponding load and difference in load between levels can be determined. It is further desirable to use four gages per level to help indicate eccentricities in the loading as well as provide redundancy.

The load at a particular level can be evaluated using strain gage data using the following expression:

$$P_i = \epsilon_i \, E_i \, A_i \tag{7.10}$$

where P_i is the load at the ith level, ϵ_i is the strain measured at the ith level, E_i is the composite modulus of the ith level, and A_i is the cross-sectional area of the ith level.

The side shear from a given shaft segment can then be calculated from the difference in measured load from the two levels bounding that segment

$$f_s = (P_i - P_{i+1})/(L\pi D) \tag{7.11}$$

where L and D are the length and diameter of the shaft segment, respectively. If only using a single gage level at the toe of the shaft, P_i is the applied load to the top of the shaft and P_{i+1} is the load calculated from strain at the toe.

7.7.2   Estimation of End Bearing

The end bearing can be similarly determined from strain data. However, the ultimate end bearing is not necessarily established. Rather, the effective ultimate capacity (usable capacity) is determined on the basis of permissible displacement. Although several

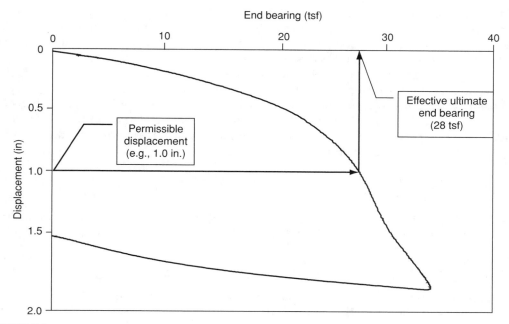

FIGURE 7.8
End bearing load test results.

approaches do exist that attempt to extract a single capacity value from test data, the entire load versus displacement response should be noted. Figure 7.8 shows the end bearing response as measured from a load test. A comparison between the measured and predicted values should be prepared so that the original design approach can be calibrated. The end bearing strength (ϵ_b) is determined from strain gage data (ϵ_{toe}) using the following expression:

$$q_b = P_{toe}/A = \epsilon_{toe}E_{toe} \tag{7.12}$$

7.8   Design of Postgrouted Shafts

The end bearing component of drilled shafts is only fractionally utilized in virtually all design methods (TCM < 1.0) due to the large displacement required to mobilize ultimate capacity. Consequently, a large portion of the ultimate capacity necessarily goes unused. In an effort to regain some of this unusable capacity, mechanistic procedures to integrate its contribution have been developed using pressure grouting beneath the shaft tip (also called postgrouting or base grouting). Pressure grouting the tips of drilled shafts has been successfully used worldwide to precompress soft debris or loose soil relaxed by excavation (Bolognesi and Moretto, 1973; Stocker, 1983; Bruce, 1986; Fleming, 1993; Mullins et al., 2000 a, b; Dapp and Mullins, 2002). The postgrouting process entails: (1) installation of a grout distribution system during conventional cage preparation that provides grout tube-access to the bottom of the shaft reinforcement cage and (2) after the shaft concrete has cured, injection of high-pressure grout beneath the tip of the shaft, which both densifies the *in situ* soil and compresses any debris left by the drilling process. By essentially preloading the soil beneath the tip, higher end bearing capacities can be realized within the service displacement limits.

Although postgrouting along the sides of the shaft has been reported to be effective, this section will only address the design of postgrouted shaft tips. The overall capacity of the shaft is still derived from both side shear and end bearing where the available side shear is calculated using one or a combination of the methods discussed earlier. Further, the calculation of the available side shear is an important step in determining the pressure to which the grout can be pumped.

7.8.1   Postgrouting in Sand

The design approach for postgrouted drilled shaft tips makes use of common parameters used for a conventional (ungrouted) drilled shaft design. This methodology includes the following seven steps:

(1) Determine the ungrouted end bearing capacity in units of stress.

(2) Determine the permissible displacement as a percentage of shaft diameter (e.g., for a 4-ft diameter shaft, $1''/48'' \times 100\% \simeq 2\%$).

(3) Evaluate the ultimate side shear resistance for the desired shaft length and diameter (in units of force).

(4) Establish a maximum grout pressure that can be resisted by the side shear (ultimate side shear divided by the tip cross-sectional area).

(5) Calculate the grout pressure index, GPI, defined as the ratio of grout pressure to the ungrouted end bearing capacity (Step 4/Step 1).

(6) Using design curves from Figure 7.9, determine the tip capacity multiplier, TCM, using the GPI calculated in Step 5.

(7) Calculate the grouted end bearing capacity (effective ultimate) by multiplying the TCM by the ungrouted end bearing (TCM × Step 1).

The ungrouted capacity (GPI = 0) is represented by these curves at the *y*-intercept where TCM = 1 for a 5% displacement (no improvement). The 1% and 2% intercepts reduce the end bearing according to the normal behavior of partially mobilized end bearing. Interestingly, the grouted end bearing capacity is strongly dependent on available side shear capacity (grout pressure) as well as the permissible displacement. However, it is relatively independent of the ungrouted end bearing capacity when in sandy soils. As such, the end bearing in loose sand deposits can be greatly improved in both stiffness and ultimate capacity given sufficient side shear against which to develop grout pressure. In dense sands and clays significant improvement in stiffness can be realized with more modest effects on ultimate capacity. Figure 7.10 shows the effective ultimate capacity that can be expected from a grouted shaft similar to that from example in Figure 7.3.

7.8.2   Postgrouting in Other Formations

Postgrouting shaft tips in other formations such as clays, silts, and rock can be advantageous for the same reasons as in sand. However, the degree of improvement may be more modest. In clays and plastic silts, the TCM can be assumed to be 1.0 although studies have shown it to be as high as 1.5 if sufficient side shear can be developed (Mullins and O'Neill,

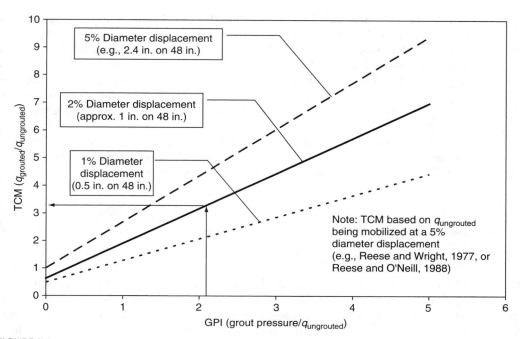

FIGURE 7.9
Correlations used in step 6 (Section 7.8.1) to establish TCM. (From Mullins, G., Dapp, S., Frederick, E., and Wagner, R., 2001, Pressure Grouting Drilled Shaft Tips, Final Report submitted to Florida Department of Transportation, April, 257 pp. With permission.)

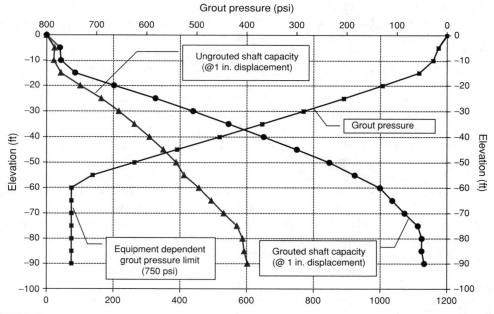

FIGURE 7.10
Postgrouted shaft capacity extended from example in Figure 7.3.

2003). In nonplastic silts, the TCM can be assumed to be 1.0 for initial designs but a verification load test program is recommended as much higher values may be reasonable. In rock, postgrouted shafts have the potential to engage both the side shear and end bearing simultaneously. In all soil types, the achieved grout pressure can be used as a lower bound for usable end bearing and the attainable grout pressure is always dependent on the available side shear against which to react. In contrast, sufficient side shear capacity does not assure that grout pressure can be developed without excessive volumes of grout.

Postgrouting shaft tips provide capacity verification for every shaft grouted. To optimize its use and design, a full load test program should be scheduled at the onset to confirm the TCM most appropriate for a given site and soil type.

7.9 Economy of Load Testing

Although the cost of foundations is most closely linked to the presence of an adequate bearing strata and the applied load, it is also directly affected by the design approach and the diameter of the shaft selected. As such, a designer may employ a range of safety factors (or resistance factors) given the level of confidence that can be assigned to a particular scenario. The most common method of establishing a particular level of certainty is via some form of testing. This testing can range from applying the full anticipated load (static or statnamic tests) to a minimum of a subsurface investigation to estimate *in situ* soil properties. Load tests result in the highest increase in designer confidence and can be incorporated into the design in the form of adjusted or calibrated unit strengths, reduced safety factors, or increased resistance factors. The effects of design uncertainty can be illustrated by the AASHTO (1998) specifications for driven piles where the designer must select from nine different resistance factors ranging from 0.35 to 0.80 based on the design methodology. Four of these conditions are selected based on the level or quality of testing that is anticipated. Therein, the highest resistance factor (0.8) and confidence is associated with a load test. The next highest resistance factor (0.65) is assigned to test methods related to installation monitoring. In contrast, the lowest confidence and resistance factor (0.35 to 0.45) is assigned when a design is based solely on capacity correlations with SPT data. Although some resistance factors for drilled shafts are not given by AASHTO (1998), the resistance factors most commonly range from 0.5 to 0.8 for no testing to load testing, respectively.

The following two examples will use estimated costs to illustrate the impact of shaft size (diameter) and design approach (ϕ factor) on cost effectiveness. The cost of shaft construction and testing can vary significantly based on the number of shafts and type of material excavated as well as the physical conditions and location of the site. Even though a typical unit price of a drilled shaft includes each of these parameters, this approach can be used for comparisons using updated site-specific values.

Given:

3 ft diameter shaft	$100/lineal foot	Excavation and concreting
4 ft diameter shaft	$200/lineal foot	Excavation and concreting
6 ft diameter shaft	$400/lineal foot	Excavation and concreting
Static load test	$125/ton of test	1% of shafts tested (1 min)
Statnamic load test	$35/ton of test	1% of shafts tested (1 min)

Use: Boring log and effective ultimate capacity calculations from Example in Figure 7.3, as well as the following resistance values (slightly updated from most recent AASHTO):

Static load test	$\phi = 0.75$
Statnamic load test	$\phi = 0.73$
No testing (SPT only)	$\phi = 0.55$

Assume a maximum excavation depth of $30D$ where D is the shaft diameter.

7.9.1 Selecting the Most Economical Shaft Diameter

Many options are available to the designer when selecting the diameter of shaft to be used for a specific foundation. For instance, a long, small diameter shaft can provide equivalent axial capacity to a shorter, larger diameter shaft. Figure 7.11 shows the result of reevaluating example in Figure 7.3 for 3, 4, and 6 ft diameter shafts while incorporating the cost per ton of capacity using $100, $200, and $400 per ft of shaft, respectively. These curves are based on axial capacity and the cost may further vary given significant lateral loading and the associated bending moment requirements. In this case, the 3 ft diameter shaft is the most cost effective at all depths.

7.9.2 Selecting the Most Economical Design Method

The next comparison that can be made is that which evaluates the cost effectiveness of various design or testing methods. As additional testing (beyond soil exploration) incurs extra expense, a break-even analysis should be performed to justify its use. In this case, a 3 ft diameter shaft will be used due to the results shown in Figure 7.12 where it was consistently less costly. The maximum capacity that can be reasonable provided by a 3 ft

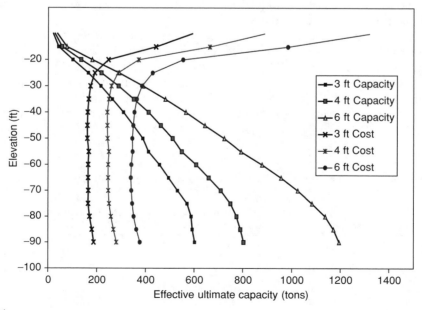

FIGURE 7.11
Effect of shaft size selection on cost.

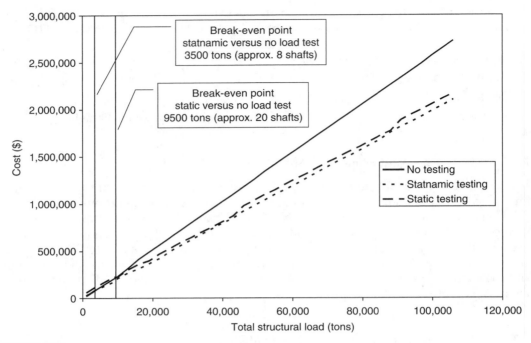

FIGURE 7.12
Break-even analysis of various design or testing methods.

diameter shaft will be calculated to be 602 tons at a depth of 90 ft (30*D*). (As deeper excavations are possible, the ultimate structural capacity based on concrete strength should not be exceeded.) The effective ultimate capacity is then reduced based on the presumption of testing (or no testing) and the appropriate resistance factor. Using these values a 3500 ton factored pier load (P_u) would require more or fewer shafts given various resistance factors as shown in Table 7.7.

The above shaft costs will also have to incorporate the cost of testing as well. As such, larger projects can justify more extensive testing, whereas very small projects may not warrant the expense. Figure 7.12 incorporates the cost of testing while extending the above example to a wide range of project sizes (expressed in terms of total structure load and not the number of shafts). The individual curves representing the various design approaches exhibit different slopes based on the permissible load-carrying capability per unit length of shaft.

TABLE 7.7

The Effect of Various Design Approaches on Required Number of Shafts

Design Method or Test Scheme	Resistance Factor	Eff. Ult. Capacity @ 90 in., P_n (tons)	Usable Capacity ϕP_n (tons)	Number of Shafts Required ($P_u = 3500$ tons)	Total Shaft Costs
Static	0.75	602	451.5	7.75 (8)	$69,750
Statnamic	0.73	602	439.5	7.96 (8)	$71,640
No testing	0.55	602	331.1	10.57 (11)	$116,640

Source: From AASHTO, 1998, *LRFD Bridge Design Specifications, Customary U.S. Units*, 2nd edn, American Association of State Highway and Transportation Officials, Washington, DC, with 1999 interim revisions. With permission.

Comparing the costs for each design and test approach, it can be seen that for smaller projects up to eight shafts (less than 3500 tons total), no testing (over and above SPT) is most cost effective. Above 3500 tons, the cost savings produced by statnamic or static load testing become significant with statnamic costs being slightly less in all cases. The selection of load test method and the associated cost are often based on the availability of test equipment capable of producing the ultimate geotechnical capacity.

Further, the disparity between testing and no testing can be even more drastic when design-phase testing can be implemented. Therein, the estimated ultimate capacity based on empirical design methods is often conservative and can be raised using the results of a test program that further widens the range of shaft numbers (testing versus no testing) required for a given pier.

In general, load test results typically show that predictions of ultimate capacity are conservative. This form of verification can be helpful in all instances: when under-predictions are severe, the design capacity of the foundations can be adjusted to provide cost savings; when over-predictions are encountered, more moderate design values can be incorporated to circumvent possible failures.

7.10 Pressure Injected Footings

Pressure-injected footings (PIF) are cast-*in situ* concrete footings containing an expanded base of concrete formed by ramming concrete into place. Installation of PIF needs only a minimum of site preparation. They can contain a cased or uncased concrete shaft with or without reinforcement to transmit the load from the superstructure to the expanded base. Due to high-energy driving during installation, the concrete can penetrate stiff soils and reach large depths laterally as well as vertically. The soil surrounding the pile base is improved due to the expulsion caused by the dry concrete plug, and thus the soil bearing capacity can be increased significantly and immediately. It is generally easier to form an expanded base in granular soil strata (Figure 7.13).

However, PIFs also have some drawbacks such as high cost and the induction of potentially disruptive vibrations at adjacent structures generated during installation. Even though the application of PIF is less popular due to cost and environmental considerations, this construction method is still competitive and is widely used when site conditions are suitable. Preexcavation and preaugering can be performed at proposed PIF locations to remove obstructions and reduce driving vibrations that could endanger existing buildings and adjacent structures.

Pressure injected footing may be vertical or battered and its bearing capacity obviously depends on the diameter of the pile base and the driving tube used. The driving tube can be chosen based on the specified loads. In cases where negative friction is encountered, a permanent steel casing or pipe may be placed without any difficulty to reduce the friction. Furthermore, a permanent steel casing of sufficient thickness, strength, and rigidity can be provided to prevent deformation, collapse, or distortion caused by driving adjacent PIF or by soil or hydrostatic pressure. Casings are made watertight in general.

When soil conditions indicate that it may be tedious or even impractical to fill the annular space between the shaft and the soil surrounding a single casing, the shaft can be supported laterally or the PIF can be reinforced. In situations where a single PIF is used as the foundation and the shaft is cased, the shaft can be supported at the top in at least two directions perpendicular to each other. On the other hand, in cases where two PIFs are

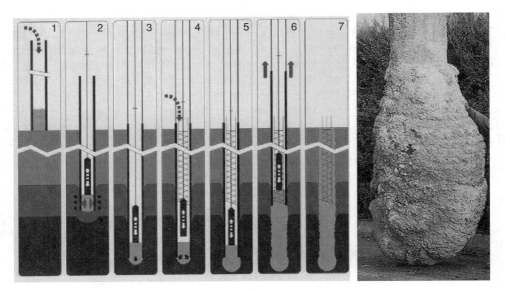

FIGURE 7.13
Illustration of the formation of pressure injected footings. (From http://www.geoforum.com. With permission.)

used in a group and their shafts are cased, the groups may be supported laterally at the top in a direction perpendicular to the line joining the centers of the footings. Shafts are reinforced only when they are required to withstand conditions other than compression, such as tension, moment, or shear. However, shaft reinforcement may also be required for compression or lateral loads for battered shafts.

Design of PIF depends upon the assumed subsurface elevations to which the PIF is expected to penetrate at various locations and the total energy required to drive them. Based upon results of PIF test loadings, as in the case of pile driving, one can generally specify the actual elevation to which PIF should penetrate and the total energy needed to drive the concrete into the base.

The following empirical formula is used to determine the allowable bearing capacity of PIF by ramming zero-slump concrete, in batches of 0.14 mm 5 cubic feet, into granular soil stratum by a drop hammer:

$$L = \frac{2}{3} \times \frac{B \times W \times H \times V}{K} \qquad (7.13)$$

where L is the safe bearing capacity of PIF in metric tons or tons, B is the average number of blows of hammer required to inject one cubic meter or one cubic foot of concrete in expanded base, during injection of the last batch, W is the weight of the drop hammer in metric tons, H is the height of the fall of drop hammer in meters feet, V is the total volume of concrete in expanded base measured in cubic meters or cubic feet, and K is a constant determined from the load test.

In the absence of a load test, K-values shown in the Table 7.8 (Norlund, 1982) can be used.

7.10.1 Construction of Pressure Injected Footings

During the construction procedure, a thick-walled steel casing is placed vertically on the ground. Using a special concrete bucket, a certain amount of almost dry concrete is poured into the bottom of the driving tube. Then the concrete is rammed into the ground using a 2

TABLE 7.8

Recommended *K* Values for PIF

Recommended *K* with Compacted Soil Description	Recommended *K* Concrete Shaft	With Cased Shafts
Gravel	9	12
Medium to coarse sand	11	14
Fine to medium sand	14	18

Note: N = number of blows from standard penetration test.
Source: From Norlund, R.L., 1982, Dynamic formula for PIF, *Proceedings of the American Society of Civil Engineers*, Vol. 108, March. With permission.

TABLE 7.9

Unit Conversion Table

From English	To SI	Multiply by	Quantity	From SI	To English	Multiply by
Ft	m	0.3048	Lengths	m	ft	3.28
In	m	0.0254		m	in	39.37
Tons	kN	8.9	Loads	kN	tons	0.114
Tsf	kPa	95.76	Stress/strength	kPa	tsf	0.0104
(pcf) lbs/ft^3	N/m^3	157.1	Force/Unit volume	N/m^3	lbs/ft^3	0.0064
kips/ft^3	kN/m^3	157.1		kN/m^3	Kips/ft^3	0.0064
Lb-inch	N-mm	112.98	Moment; or energy	N-mm	lb-inch	0.0089
kip-inch	kN-mm	112.98		kN-mm	kip-inch	0.0089
lb-ft	N-m	1.356		N-m	lb-ft	0.7375
kip-ft	kN-m	1.356		kN-m	Kip-ft	0.7375
ft-lb	Joule	1.356		Joule	ft-lb	0.7375
ft-kip	kJoule	1.356		kJoule	ft-kip	0.7375
s/ft	s/m	3.2808	Damping	s/m	s/ft	0.3048
Blows/ft	Blows/m	3.2808	Blow count	blows/m	blows/ft	0.3048

Source: Courtesy of New York Department of Transportation.

to 8 ton hammer dropping from a height of several meters while the tube is held in position by steel cables. Under the impact of the hammer, the concrete creates a plug at the bottom of the casing that penetrates slightly into the soil. Alternatively, the steel casing can be installed by "top-driving" using ordinary drop hammers. In such cases, the driving casing must be provided with a bottom plate that would be left in the ground. A large-stem auger can also be used to install the casing. When the tube has been driven to the required depth, the casing is very slightly lifted and held in position using steel cables. The plug is then removed by imparting heavy blows using the hammer and ascertaining that a certain amount of rammed concrete remains in the casing in order to prevent any future seepage of water or erosion of soil into the pile shaft. This operation is verified with marks made on the driving cable of the hammer and on the lifting cables. The expanded base of the pile is then formed by adding enough dry concrete to achieve a predetermined "driving set." Hence, an enlarged concrete bulb that would serve as a pile base would be formed in the foundation soil that is also heavily compressed and densified.

7.10.2 Concreting of the Shaft

The shaft of the pile is formed by ramming successive layers of "dry concrete" by raising the casing 0.2 to 0.5 m at a time. The hammer moves the concrete laterally into the soil that

FIGURE 7.14
Concreting of the shaft during construction of PIF. (From   http://www.tggonline.com/geotechnical/ projects. With permission.)

is already compressed by the driving action. Because of the ramming process, the concrete comes in close contact with the soil, thus forming a cylindrical shaft. As mentioned before, the shaft rests on an enlarged base formed to induce a highly densified "refusal" state in the bearing layer. Furthermore, it has been experienced that wet concrete accelerates the construction process (Figure 7.14).

Nomenclature

%R	percent recovery of rock coring (%)
α	adhesion factor applied to S_u (DIM)
β	coefficient relating the vertical stress and the unit skin friction of a drilled shaft (DIM)
β_m	SPT N corrected coefficient relating the vertical stress and the unit skin friction of a drilled shaft (DIM)
D	diameter of drilled shaft (FT)
D_b	depth of embedment of drilled shaft into a bearing stratum (FT)
D_p	diameter of the tip of a drilled shaft (FT)
ϕ, ϕ_f	angle of internal friction of soil (DEG)
f_s	nominal unit side shear resistance (TSF)
γ	unit weight (pcf)
K	empirical bearing capacity coefficient (DIM)
K	load transfer factor
N	average (uncorrected) SPT Standard Penetration Test blow count, SPT N (blows/FT)
N_c	bearing capacity factor (DIM)

N_{corr}	corrected SPT blow count
q_b	end bearing resistance (units of stress)
q_c	cone penetration tip resistance (units of stress)
q_s	average splitting tensile strength of the rock core (TSF)
Q_s	side shear capacity (units of force)
q_u	average unconfined compressive strength of the rock core (TSF)
σ'_v	vertical effective stress (TSF)
S_u	undrained shear strength (TSF)
ϵ	measured strain from embedded strain gage

References

AASHTO, 1994, *LRFD Bridge Design Specifications, SI*, 1st edn, American Association of State Highway and Transportation Officials, Washington, DC, with 1996 and 1997 interim revisions.

AASHTO, 1998, *LRFD Bridge Design Specifications, Customary U.S. Units*, 2nd edn, American Association of State Highway and Transportation Officials, Washington, DC, with 1999 interim revisions.

ASTM D 0000-99, submission 2000, *Standard Test Method for Piles Under Rapid Axial Compressive Load*, under review at American Society for Testing and Materials, Philadelphia, PA.

ASTM D 1194-94, *Standard Test Method for Bearing Capacity of Soil for Static Load and Spread Footings*, American Society for Testing and Materials, Philadelphia, PA.

ASTM D 1586-84, *Standard Method for Penetration Test and Split-Barrel Sampling of Soils*, American Society for Testing and Materials, Philadelphia, PA.

ASTM D-1143-98, 1998, Standard Test Method for Piles Under Static Axial Compressive Load, in *Annual Book of ASTM Standards*, Part 20, Philadelphia, PA.

Alsamman, O.M., 1995, The use of CPT for calculating axial capacity of drilled shafts, Ph.D. Dissertation, University of Illinois, Urbana-Champaign.

Azizi, F., 2000, *Applied Analyses in Geotechnics*, E & FN Spoon, New York.

Baker, A.C. and Broadrick, R.L., 1997, *Ground Improvement, Reinforcement, and Treatement: A Twenty Year Update and a Vision for the 21st Century*, Earth Tech, Clear Water, FL.

Bolognesi, A.J.L. and Moretto, O., 1973, Stage grouting preloading of large piles on sand, *Proceedings of 8th ICSMFE*, Moscow.

Brown, D., 2002, Effect of construction on axial capacity of drilled foundations in piedmont soils, *Journal of Geotechnical Engineering*, 128(12). 000–000.

Bruce, D.A., 1986, Enhancing the performance of large diameter piles by grouting, Parts 1 and 2, *Ground Engineering*, May and July, respectively.

Bruce, D.A., Nufer, P.J., and Triplett, R.E., 1995, *Enhancement of Caisson Capacity by Micro-Fine Cement Grouting — A Recent Case History*, ASCE Special Publication 57, Verification of Geotechnical Grouting.

Carter, J.P. and Kulhawy, F.H., 1987, Analysis and Design of Foundations Socketed into Rock, Research Report 1493-4, Geotechnical Engineering Group, Cornell University, Ithaca, NY.

Dapp, S. and Mullins, G., 2002, Pressure-grouting drilled shaft tips: full-scale research investigation for silty and shelly sands, *Deep Foundations 2002: An International Perspective on Theory, Design, Construction, and Performance*, ASCE Geo Institute, GSP No. 116, Vol. I, pp. 335–350.

FHWA, 1998, *Load and Resistance Factor Design (LRFD) for Highway Bridge Substructures*, U.S. Department of Transportation, Publication No. FHWA HI-98-032.

Flemming, W.G.K., 1993, The improvement of pile performance by base grouting, *Proceedings of the Institution of Civil Engineers*, London.

Florida Department of Transportation, 2002, Section 455: Structures and Foundations, Tallahassee, FL.

Garbin, E.J., 1999, Data Interpretation for Axial Statnamic Testing and the Development of the Statnamic Analysis Workbook, Master's Thesis, University of South Florida, Tampa, FL.

Horvath, R.G. and Kenney, T.C., 1979, Shaft resistance of rock socketed drilled piers, In *Proceedings of the Symposium on Deep Foundations*, ASCE, Atlanta, Georgia, 1979, pp. 182–214.

Kulhawy, F.H. and Mayne, P.W., 1990, *Manual on Estimating Soil Properties for Foundation Design*, Electric Power Research Institute, Palo Alto, CA.

Littlejohn, G.S., Ingle, J., and Dadasbilge, K., 1983, Improvement in base resistance of large diameter piles founded in silty sand, *Proceedings of the Eighth European Conference on Soil Mechanics and Foundation Engineering*, Helsinki, May.

McVay, M.C. and Townsend, F.C., and Williams, R.C., 1992, Design of socketed drilled shafts in limestone, *Journal of Geotechnical Engineering*, 118(10), 1626–1637 pp.

Meyerhof, G.G., 1976, Bearing capacity and settlement of piled foundations, *Journal of Geotechnical Engineering, ASCE*, 102(GT3): 197–227.

Mullins, A.G., Dapp, S., Fredrerick, E., and Wagner, R., 2000, Pressure Grouting Drilled Shaft Tips, Final Report submitted to Florida Department of Transportation, April, 357 pp.

Mullins, G., Dapp, S., and Lai, P., 2000, Pressure grouting drilled shaft tips in sand, *New Technological and Design Developments in Deep Foundations*, Dennis, N.D. et al. (eds.), ASCE, Geo Institute, Vol. 100, pp. 1–17.

Mullins, G., Dapp, S., and Lai, P., 2000, *New Technological and Design Developments in Deep Foundations, Pressure-Grouting Drilled Shaft Tips in Sand*, American Society of Civil Engineers, Denver, CO.

Mullins, G., Dapp, S., Frederick, E., and Wagner, R., 2001, Pressure Grouting Drilled Shaft Tips, Final Report submitted to Florida Department of Transportation, April, 257 pp.

Mullins, G. and O'Neill, M., 2003, Pressure Grouting Drilled Shaft Tips, Research Report submitted to A.H. Beck Foundation, Inc., May, 198 pp.

Norlund, R.L., 1982, Dynamic formula for PIF, *Proceedings of the American Society of Civil Engineers*, Vol. 108, March.

O'Neill, M.W. and Reese, L.C., 1970, Behavior of Axially Loaded Drilled Shafts in Beaumont Clay, Research Report No. 89-8, Center for Highway Research, University of Texas at Austin, December.

O'Neill, M.W. and Hassan, K.M., 1994, Drilled shafts: effects of construction on performance and design criteria, *Proceedings of the International Conference on Design and Construction of Deep Foundations*, December, Vol. 1, pp. 137–187.

O'Neill, M.W., 1998. Project 89 revisited, *Proceedings of the ADSC Drilled Shaft Foundation Symposium Held to Honor Dr. Lymon C. Reese*, ADSC, Dallas, TX, January, pp. 7–47.

O'Neill, M.W., 2002, Discussion of "side resistance in piles and drilled shafts," *Journal of Geotechnical and Geoenvironmental Engineering*, 127(1): 3–16.

Quiros, G.W. and Reese, L.C., 1977, Design procedures for axially loaded drilled shafts, research report 176-5F, project 3-5-72-176, center for Highway Research, University of Texas, Austin, 156 pp.

Reese, L.C. and Wright, S.J., 1977, *Construction Procedures and Design for Axial Loading*, Vol. 1, Drilled Shaft Manual, HDV-22, Implementation Package 77-21, Implementation Division, U.S. Department of Transportation, McLean, Virginia, 140 pp.

Reese, L.C. and O'Neill, M.W., 1988, *Drilled Shafts: Construction and Design*, FHWA, Publication No. HI-88-042.

Stocker, M.F., 1983, The influence of post grouting on t he load bearing capacity of bored piles, *Proceedings of the Eighth European Conference on Soil Mechanics and Foundation Engineering*, Helsinki, May.

Touma, F.T. and Reese, L.C., 1974, Behavior of bored piles in sand, *Journal of the Geotechnical Engineering Division, ASCE*, 100(GT7): 749–761.

Touma, F.T., 1972, The Behavior of Axially Loaded Drilled Shafts in Sand, Doctoral Dissertation, Department of Civil Engineering, The University of Texas at Austin, December.

TxDOT, 1993, *Standard Specifications for Construction of Highways, Streets and Bridges*, Texas Department of Transportation, Austin, March.

8

Design of Laterally Loaded Piles

Manjriker Gunaratne

CONTENTS

8.1   Introduction

Single piles such as sign-posts and lamp-posts and pile groups that support bridge piers and offshore construction operations are constantly subjected to significant natural lateral loads (such as wind loads and wave actions) (Figure 8.1). Lateral loads can be also introduced on piles due to artificial causes like ship impacts. Therefore, the lateral load capacity is certainly a significant attribute in the design of piles under certain construction situations.

Unlike in the case of axial load capacity, the lateral load capacity must be determined by considering two different failure mechanisms: (1) structural failure of the pile due to yielding of pile material or shear failure of the confining soil due to yielding of soil, and (2) pile becoming dysfunctional due to excessive lateral deflections. Although passive failure of the confining soil is a potential failure mode, such failure occurs only at relatively large deflections which generally exceed the tolerable movements.

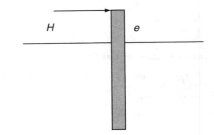

FIGURE 8.1
Laterally loaded pile.

One realizes that "short" piles embedded in relatively stiffer ground would possibly fail due to yielding of the soil while "long" piles embedded in relatively softer ground would produce excessive deflections. In view of the above conditions, this chapter is organized to analyze separately, the two distinct issues presented above. Hence the discussion will deal with two main issues: (1) lateral pile capacity from strength considerations, and (2) lateral pile capacity based on deflection limitations.

On the other hand, piles subjected to both axial and lateral loading must be designed for structural resistance of the piles as beam-columns.

8.2   Lateral Load Capacity Based on Strength

8.2.1   Ultimate Lateral Resistance of Piles

Broms (1964a,b) produced simplified solutions for the ultimate lateral load capacity of piles by considering both the ultimate strength of the bearing ground and the yield stress of the pile material. For simplicity, the Broms (1964a,b) solutions are presented separately for different soil types, namely, cohesive soils and cohesionless soils.

8.2.1.1   *Piles in Homogeneous Cohesive Soils*

When a pile is founded in a predominantly fine-grained soil, the most critical design case is the case where soil is in an undrained situation. The maximum load that can be applied on the pile depends on the the following factors:

1. Fixity conditions at the top (i.e., free piles or fixed piles). Most single piles can be considered as free piles under lateral loading whereas piles clustered in a group by a pile cap must be analyzed as fixed piles.

2. Relative stiffness of the pile compared to the surrounding soil. If the deformation conditions are such that the soil yields before the pile material then the pile is classified as a "short" pile. Similarly, if the pile material yields first, then the pile is considered a "long" pile.

8.2.1.1.1   *Unrestrained or Free-Head Piles*

Figure 8.2 and Figure 8.3 illustrate the respective failure mechanisms that Broms (1964a,b) assumed for "short" and "long" piles, respectively.

The ultimate lateral resistance P_u can be directly determined from Figure 8.4(a) and (b) based on the geometrical properties and the undrained soil strength. For short piles, M_{max}, g, P_u, and f can be determined from Equations (8.1) to (8.4).

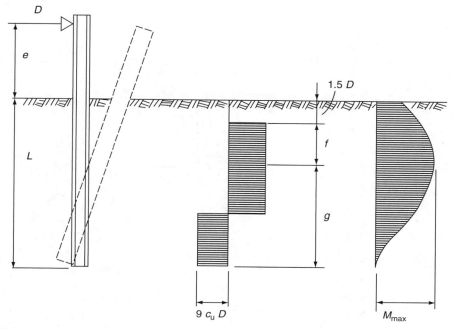

FIGURE 8.2
Deflection, soil reaction, and bending moment distributions for laterally loaded short piles in cohesive soil. (From Broms, B., 1964a, *J. Soil Mech. Found. Div., ASCE*, 90(SM3):27–56. With permission.)

Since the shear force is zero at the location of maximum moment, from the area of the soil reaction plot (Figure 8.2) one obtains

$$f = \frac{P_u}{9c_u D} \tag{8.1}$$

Similarly, by taking the first moments of Figure 8.2 about the yield point

$$M_{max} = 2.25 D g^2 c_u \tag{8.2}$$

$$M_{max} = H_u(e + 1.5D + 0.5f) \tag{8.3}$$

For the total length of the pile,

$$L = g + 1.5D + f \tag{8.4}$$

8.2.1.1.2 *Restrained or Fixed-Head Piles*

According to the Broms (1964a) formulations, restrained piles can reach their ultimate capacity through three separate mechanisms giving rise to (1) short piles, (2) long piles, and (3) intermediate piles. These failure mechanisms assumed by Broms (1964a) for restrained piles are illustrated in Figure 8.5(a)–(c). The assumption that leads to the analytical solutions is that the moment generated on the pile top can be provided by the pile cap to restrain the pile with the boundary condition at the top (i.e., no rotation).

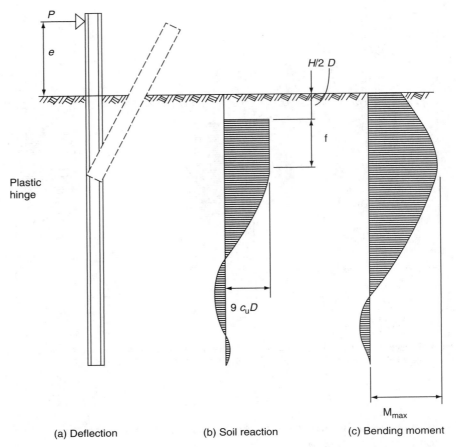

(a) Deflection   (b) Soil reaction   (c) Bending moment

FIGURE 8.3
Failure mechanism for laterally loaded long piles in cohesive soil. (From Broms, B., 1964a, *J. Soil Mech. Found. Div., ASCE*, 90(SM3):27–56. With permission.)

The ultimate lateral load, P_u, of short piles can be directly obtained from Figure 8.4(a). The reader would notice that this condition is presented through a single curve in Figure 8.4(a) due to the insignificance of the e parameter. M_{max} and KP_u can also be determined using the following equations:

$$P_u = 9c_uD(L - 1.5D) \tag{8.5}$$

$$M_{max} = P_u(0.5L + 0.75D) \tag{8.6}$$

For long piles, the ultimate lateral load, P_u, can be found from Figure 8.4(b). Then, the following equations can be used to determine f and hence the location of pile yielding:

$$P_u = \frac{2M_y}{(1.5d + 0.5f)} \tag{8.7}$$

On the other hand, for "intermediate" piles where yielding occurs at the top (Figure 8.5b), the basic shear moment and total length consideration in Equations (8.1), (8.4), and (8.8) can be used to obtain P_u:

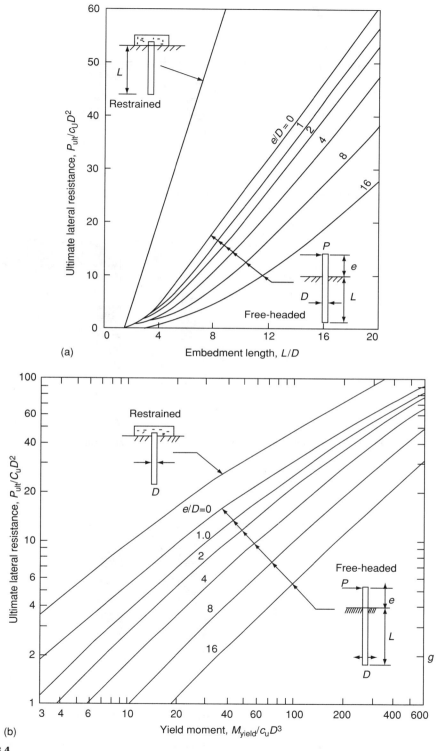

FIGURE 8.4
Ultimate lateral resistance of piles in cohesive soils: (a) short piles and (b) long piles. (From Broms, B., 1964a, *J. Soil Mech. Found. Div., ASCE*, 90(SM3):27–56. With permission.)

From steel section tables and Figure 8.6(b)

$S_{xx} = 0.711 \times 10^{-3}$ m^3, $d = 256$ mm
$M_y = S_{xx} \sigma_y = (0.711)(10^{-3})(300)$ MN m $= 213.3$ kN m.

From the q_c profile in Figure 8.6(a), q_c can be expressed as

$$q_c = 4.7 + 0.04z \text{ MPa}$$

From Robertson and Campanella (1983)

$$S_u = \frac{q_T - p_0}{N_{KT}}$$

$$q_T = q_c + u_c(1 - a)$$

From Bowles (1996)

$$N_{KT} = 13 + \frac{5.5}{\text{PI}}$$

where PI is the plasticity index of the soil.
One obtains the following s_u profile for PI $= 35$:

$$S_u = (1/13.16)[(4.7 + 0.04z) + 0.001\{(9.8z)(1 - 0.5) - (17.5 - 9.8)z\}]$$
$$= 0.357 + 0.0028z \text{ MPa}$$

s_u ranges along the length of the pile from 357 to 385 kPa showing the linear trend with depth that is typical for clays. Due to its relatively narrow range, it can be reasonably averaged along the pile depth to be about 371 kPa

$$c_u = 371 \text{ kPa}$$

Assume that the ground conditions and the pile stiffness are such that it behaves as a short pile.
Then from Figure 8.4(a) or Equation (8.5), for an embedment length of 10 m/0.256 m $= 39$, $P_u/c_u D^2$ can be extrapolated as $P_u/c_u D^2 = 337$
But $c_u D^2 = 24.314$ kN, and hence $P_u = 8.22$ MN.
Thus, if the pile does not yield, it can take 8.22 MN before the soil fails.
In order to check the maximum moment in the pile, Equation (8.6) can be applied.

$$M_{\max} = P_u(0.5L + 0.75D) = 8.22(0.5 \times 10 + 0.75 \times 0.256) \text{ MN m} = 42.68 \text{ MN m}$$

But $M_y = 213.3$ kN m. Hence the pile would yield long before the clay, and the pile has to be reanalyzed as a long pile.

$$\frac{M_y}{c_u D^3} = (213.3)/(371)/(0.256)^3 = 34.27$$

From Figure 8.4(b),

$P_u/c_u D^2 \approx 25$
But $c_u D^2 = 24.314$ kN
$P_u = 608$ kN.

Hence, the ultimate lateral load that can be applied on the given pile is about 600 kN.

8.2.1.2 Piles in Cohesionless Soils

Based on a number of assumptions, Broms (1964b) formulated analytical methodologies to determine the ultimate lateral load capacity of a pile in cohesionless soils as well. The most significant assumptions were: (1) negligible active earth pressure on the back of the pile due to forward movement of the pile bottom, and (2) tripling of passive earth pressure along the top front of the pile. Hence

$$p_{\mathrm{p}} = 3\sigma_{\mathrm{v}}' K_{\mathrm{P}} \tag{8.9}$$

where σ_{v}' is the effective vertical overburden pressure and $K_{\mathrm{P}} = (1 + \sin \phi')/(1 - \sin \phi')$. ϕ' is the angle of internal friction (effective stress).

8.2.1.2.1 Free-Head Piles

By following terminology similar to that in the case of cohesive soils, the failure mechanisms of short and long piles are illustrated in Figure 8.7 and Figure 8.8, respectively.

The ultimate lateral load for short piles can be estimated from Figure 8.9(a) or the following equation.

$$P_{\mathrm{u}} = \frac{0.5\gamma D L^3 K_{\mathrm{P}}}{e + L} \tag{8.10}$$

Then, the location of the maximum moment (f in Figure 8.7) can be determined by the following equation.

$$f = 0.82\sqrt{\frac{P_{\mathrm{u}}}{D K_{\mathrm{P}} \gamma}} \tag{8.11}$$

Finally, the maximum moment can be estimated by Equation (8.12)

$$M_{\max} = P_{\mathrm{u}}\left(e + \frac{2}{3}f\right) \tag{8.12}$$

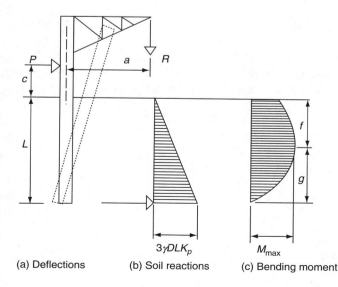

(a) Deflections (b) Soil reactions (c) Bending moment

FIGURE 8.7
Failure mechanism for laterally loaded short pile in cohesionless soil. (From Broms, B., 1964b, *J. Soil Mech. Found. Div.*, ASCE, 90(SM3):123–156. With permission.)

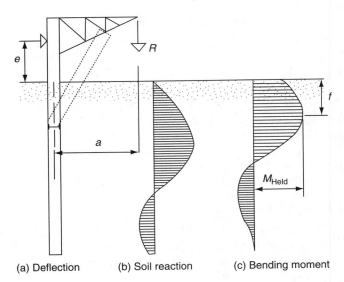

FIGURE 8.8
Failure mechanism for laterally loaded long piles in cohesionless soil (From Broms, B., 1964b, *J. Soil Mech. Found. Div., ASCE*, 90(SM3):123–156. With permission.)

(a) Deflection   (b) Soil reaction   (c) Bending moment

If the M_{max} value computed from Equation (8.12) is larger than M_{yield} for the pile material, then obviously the pile behaves as a long pile and the actual ultimate lateral load P_u can be computed from Equations (8.11) and (8.12) by setting $M_{max} = M_{yield}$.

On the other hand, Figure 8.9(b) enables one to determine the ultimate lateral load for long piles directly.

8.2.1.2.2   Restrained or Fixed-Head Piles

For restrained short piles, consideration of horizontal equilibrium in Figure 8.10(a) yields

$$P_u = 1.5\gamma L^2 DK_P \tag{8.13}$$

Hence P_u can be found either from Equation (8.13) or Figure 8.9(a). Also, from Figure 8.10(a) it follows that

$$M_{max} = \frac{2}{3}P_u L \tag{8.14}$$

If M_{max} computed from Equation (8.14) is larger that M_{yield} for the pile material, then the failure mechanism in Figure 8.10(b) applies. For this case, the following expression can be written for the moment about the pile bottom from which the ultimate lateral load can be computed:

$$M_y = (0.5\gamma DL^3 K_P) - P_u L \tag{8.15}$$

The above solution only applies if the moment M_{max} at a depth of f computed by

$$f = 0.82\sqrt{\frac{P_u}{DK_P\gamma}} \tag{8.11}$$

is less than M_{yield} for the pile material.

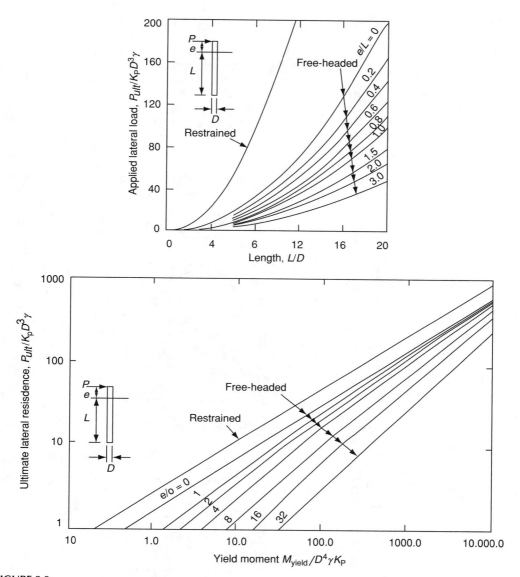

FIGURE 8.9
Ultimate lateral resistance of piles in cohesionless soils: (a) short piles, (b) long piles. (From Broms, B., 1964b, *J. Soil Mech. Found. Div.*, *ASCE*, 90(SM3):123–156. With permission.)

Finally, if the above M_{max} is larger than M_{yield}, then the failure mechanism in Figure 8.10(c) applies. Thus, the ultimate lateral load can be computed from the following equation or its nondimensional form in Figure 8.9(b).

$$P_u\left(e + \frac{2}{3}f\right) = 2M_y \tag{8.16}$$

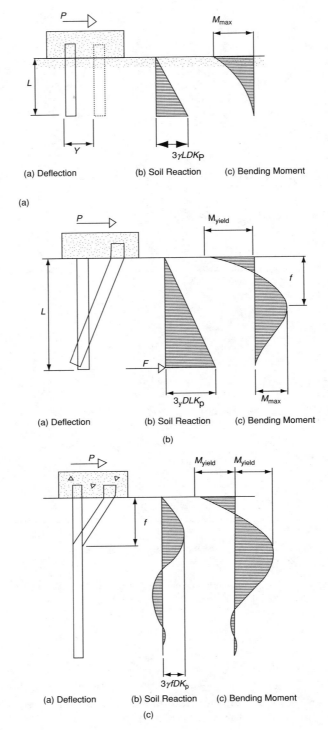

FIGURE 8.10
Failure mechanisms for restrained piles in cohesionless soils: (a) short piles, (b) intermediate piles, and (c) long piles. (From Broms, B., 1964b, *J. Soil Mech. Found. Div.*, *ASCE*, 90(SM3):123–156. With permission.)

8.3   Lateral Load Capacity Based on Deflections

The maximum permissible ground line deflection must be compared with the lateral deflection of a laterally loaded pile to fulfill one important criterion of the design procedure. A number of commonly adopted methods to determine the lateral deflection are discussed in the ensuing sections.

8.3.1   Linear Elastic Method

A laterally loaded pile can be idealized as an infinitely long cylinder laterally deforming in an infinite elastic medium (Pyke and Beikae, 1984) with the horizontal deformation governed by the following equation:

$$p = k_h y \tag{8.17}$$

But, from distributed load vs. moment relations,

$$pB = \frac{d^2 M}{dz^2} = -E_P I \frac{d^4 y}{dz^4} \tag{8.18}$$

where B is the width of pile and $E_P I$ is the pile stiffness.

Then the equation governing the lateral deformation can be expressed by combining (8.17) and (8.18) as

$$E_P I \frac{d^4 y}{dz^4} + B k_h y = 0 \tag{8.19}$$

The characteristic coefficient of the solution to y is defined by

$$\beta = \left(\frac{k_h D}{4 E_p I} \right)^{1/4} \tag{8.20}$$

$1/\beta$ is also known as the nondimensional length, where k_h is the coefficient of horizontal subgrade reaction.

Broms (1964a,b) showed that a laterally loaded pile behaves as an infinitely stiff member when the coefficient β is less than 2. Further, when $\beta L \geq 4$, it was shown to behave as an infinitely long member in which failure occurs when the maximum bending moment exceeds the yield resistance of the pile section.

For the simple situation where k_h can be assumed constant along the pile depth, Hetenyi (1946) derived the following closed-form solutions:

8.3.1.1   *Free-Headed Piles*

8.3.1.1.1   Case (1): Lateral Deformation due to Load H

The following expressions can be used in conjunction with Figure 8.11, for a pile of width d. Horizontal displacement

$$\Delta = \frac{2H\beta}{k_h d} K_{\Delta H} \tag{8.21a}$$

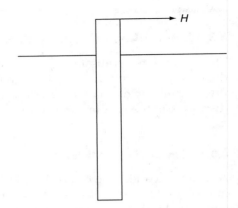

FIGURE 8.11
Aid for using Table 8.1 for lateral load.

Slope

$$\theta = \frac{2H\beta^2}{k_h d} K_{\theta H}$$

(8.21b)

Moment

$$M = -\frac{H}{\beta} K_{MH}$$

(8.21c)

Shear force

$$V = -H K_{VH}$$

(8.21d)

The influence factors $K_{\Delta H}$, $K_{\theta H}$, K_{MH}, and K_{VH} are given in Table 8.1.

8.3.1.1.2 Case (2): Lateral Deformation due to Moment **M**

The following expressions can be used with Figure 8.12.
Horizontal displacement

$$\Delta = \frac{2M_0\beta^2}{k_h d} K_{\Delta M}$$

(8.22a)

Slope

$$\theta = \frac{2M_0\beta^3}{k_h d} K_{\theta M}$$

(8.22b)

Moment

$$M = M_0 K_{MM}$$

(8.22c)

Shear force

$$V = -2M_o\beta K_{VM}$$

(8.22d)

The influence factors $K_{\Delta M}$, $K_{\theta M}$, K_{MM}, and K_{VM} are also given in Table 8.1.

TABLE 8.1

Influence Factors for the Linear Solution

βL	Z/L	$K(\Delta H)$	$K(\theta H)$	$K(MH)$	$K(VH)$	$K(\Delta M)$	$K(\theta M)$	$K(MM)$	$K(VM)$
2.0	0	1.1376	1.1341	0	1	−1.0762	1.0762	1	0
2.0	0.125	0.8586	1.0828	0.1848	0.5015	−0.6579	0.8314	0.9397	0.2214
2.0	0.25	0.6015	0.9673	0.262	0.1377	−0.2982	0.6133	0.7959	0.3387
2.0	0.375	0.3764	0.8333	0.2637	−0.1054	−0.0376	0.4366	0.6138	0.3788
2.0	0.5	0.1838	0.7115	0.218	−0.2442	0.1463	0.3068	0.4262	0.3639
2.0	0.625	0.0182	0.6192	0.1491	−0.2937	0.2767	0.222	0.2564	0.3101
2.0	0.75	−0.1288	0.5628	0.0776	−0.2654	0.3747	0.1757	0.1208	0.2282
2.0	0.875	−0.2659	0.5389	0.0222	−0.1665	0.4572	0.1578	0.0318	0.1241
2.0	1	−0.3999	0.5351	0	0	0.5351	0.1551	0	0
3.0	0.125	0.6459	0.8919	0.2508	0.3829	−0.3854	0.6433	0.8913	0.2514
3.0	0.25	0.3515	0.6698	0.3184	0.0141	−0.0184	0.3493	0.6684	0.3202
3.0	0.375	0.1444	0.4394	0.285	−0.1664	0.1607	0.1429	0.436	0.2887
3.0	0.5	0.0164	0.2528	0.2091	−0.2223	0.2162	0.0168	0.2458	0.215
3.0	0.625	−0.0529	0.1271	0.1272	−0.2057	0.2011	−0.0489	0.1148	0.1353
3.0	0.75	−0.0861	0.0584	0.0594	−0.1519	0.1524	−0.0763	0.0396	0.0684
3.0	0.875	−0.1021	0.0321	0.0154	−0.0807	0.0916	−0.0839	0.0069	0.0225
3.0	1	−0.113	0.0282	0	0	0.0282	−0.0847	0	0
4.0	0	1.0008	1.0015	0	−0.0000	0.0282	−0.0847	0.0000	0
4.0	0.1250	0.5323	0.8247	0.2907	0.2411	−0.2409	0.5344	0.8229	0.2910
4.0	0.2500	0.1979	0.5101	0.3093	−0.1108	0.1136	0.2010	0.5082	0.3090
4.0	0.3750	0.0140	0.2403	0.2226	−0.2055	0.2118	0.0178	0.2397	0.2200
4.0	0.5000	−0.0590	0.0682	0.1243	−0.1758	0.1858	−0.0558	0.0720	0.1176
4.0	0.6250	−0.0687	−0.0176	0.0529	−0.1084	0.1200	−0.0696	−0.0043	0.0406
4.0	0.7500	−0.0505	−0.0488	0.0147	−0.0475	0.0538	−0.0616	−0.0206	−0.0025
4.0	0.8750	−0.0239	−0.0552	0.0014	−0.0101	−0.0033	−0.0535	−0.0096	−0.0148
4.0	1.0000	0.0038	−0.0555	−0	0.0000	−0.0555	−0.0517	−0.0000	−0
5.0	0	1.0003	1.0003	0	1.0000	−1.0003	1.0002	1.0000	0
5.0	0.1250	0.4342	0.7476	0.3131	0.1206	−0.1210	0.4343	0.7472	0.3133
5.0	0.2500	0.0901	0.3628	0.2716	−0.1817	0.1818	0.0907	0.3620	0.2720
5.0	0.3750	−0.0466	0.1013	0.1461	−0.1919	0.1930	−0.0455	0.1002	0.1461
5.0	0.5000	−0.0671	−0.0157	0.0494	−0.1133	0.1163	−0.0654	−0.0161	0.0482
5.0	0.6250	−0.0456	−0.0435	0.0026	−0.0412	0.0461	−0.0444	−0.0409	−0.0012
5.0	0.7500	−0.0197	−0.0369	−0.0088	−0.0008	0.0055	−0.0221	−0.0276	−0.0159
5.0	0.8750	0.0002	−0.0279	−0.0044	0.0108	−0.0139	−0.0110	−0.0086	−0.0125
5.0	1.0000	0.0167	−0.0259	−0	0.0000	−0.0259	−0.0091	−0.0000	−0

8.3.1.2 Fixed-Headed Piles

Due to the elastic nature of the solution, lateral deformation of the fixed-headed piles can be handled by superimposing the deformations caused by: (1) the known deforming lateral force and the unknown restraining pile head moment, or (2) the known deforming moment and the unknown restraining pile head moment. Then, by setting the pile head rotation to zero (for fixed end conditions), the unknown restraining moment and hence the resultant solution can be determined.

Example 8.2

The 300 mm wide steel pile shown in Figure 8.13 is one member of a group held together by a pile cap that exerts a lateral load of 8 kN on the given pile and a certain magnitude of a moment required to restrain the rotation at the top. It is given that the coefficient of

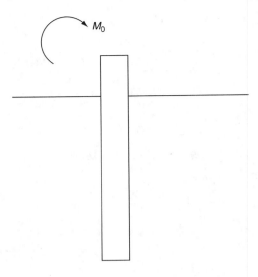

FIGURE 8.12
Aid for using Table 8.1 for moment.

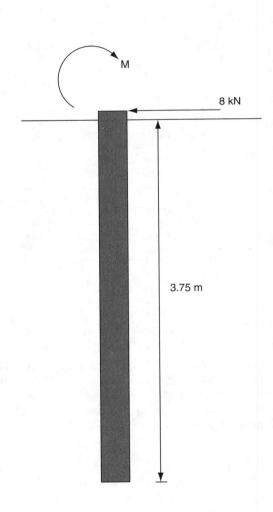

FIGURE 8.13
Illustration for Example 8.2.

horizontal subgrade modulus is $1000 \, \text{kN/m}^3$ and invariant with the depth. Determine the lateral deflection and the restraining moment at the top. Assume that the second moment of area (I) of the steel section is $2.2 \times 10^{-6} \, \text{m}^4$ and the elastic modulus of steel to be $2.0 \times 10^6 \, \text{kPa}$.

Solution
First determine $\beta = \left(\dfrac{k_h B}{4 E_p I} \right)^{1/4} = \left(\dfrac{(1000)(0.3)}{4(2,000,000)(0.0000022)} \right)^{1/4} = 2.03 \, \text{m}^{-1}$ (Equation 8.20)

But $L = 3.75 \, \text{m}$, therefore, $\beta L = 7.61$.

Then, determine the lateral displacement and the slope due to a force 8 kN (Equation 8.21)

$$\Delta_H = \frac{2H\beta}{k_h d} K_{\Delta H} = \frac{2(8)(2.03)}{(1,000)(0.3)}(1.0) = 0.108 \, \text{m} = 108 \, \text{mm}$$

$$\theta_H = \frac{2H\beta^2}{k_h d} K_{\theta H} = \frac{2(8)(2.03)^2}{(1,000)(0.3)}(1.0) = 0.219 \, \text{rad}$$

If the restraining moment needed at the top is M, then the lateral displacement and the slope due to M are evaluated as follows (Equation 8.22):

$$\Delta_M = \frac{2M_0\beta^2}{k_h d} K_{\Delta M} = \frac{2M(2.03)^2}{(1000)(0.3)}(-1.0) = -0.0275M \, \text{m}$$

$$\theta_M = \frac{2M_0\beta^3}{k_h d} K_{\theta M} = \frac{2M(2.03)^3}{(1000)(0.3)}(1.0) = 0.056M \, \text{rad}$$

For restrained rotation at the top,
$0.056M + 0.219 = 0; \quad M = -3.93 \, \text{kN m}$
Then $\Delta_M = 0.108 \, \text{m}$
Hence, the total lateral displacement is $\Delta_M + \Delta_H = 0.216 \, \text{m}$.

8.3.2 Nonlinear Methods

Several nonlinear numerical methods have become popular nowadays due to the availability of superior computational capabilities. Of them the most widely used ones are the stiffness matrix method of analysis and the lateral force–deflection (p–y) approach.

8.3.2.1 Stiffness Matrix Analysis Method

This method is also known as the finite element method due to the similarity in the basic formulation of the conventional finite element method and the stiffness matrix analysis method. First, the pile is discretized into a number of one-dimensional (beam) elements. Figure 8.14 shows a typical discretization of a pile in preparation for load–deflection analysis. The following notation applies to Figure 8.14:

1, 2, . . . ,*N* (in bold) — node number

P_i (i even) — internal lateral forces on pile elements concentrated (lumped) at the nodes

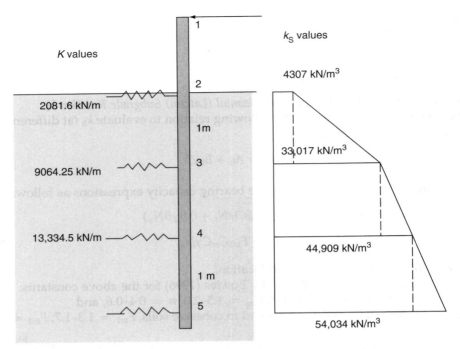

FIGURE 8.15
Illustration for Example 8.3.

Example 8.4
The 300 mm wide steel pile shown in Figure 8.16 is one member of a group held together by a pile cap that exerts a lateral load of 8 kN on the given pile and a moment of certain magnitude required to restrain the rotation at the top. It is given that the coefficient of horizontal subgrade modulus is $1000 \, \text{kN/m}^3$ and invariant with the depth. Determine the relevant force and deflection vectors assuming that the total number of nodes is 6. Also illustrate the solution procedure to obtain the lateral deflection of the pile and the moment required at the cap. Assume that the second moment of area (I) of the steel section is $2.2 \times 10^{-6} \, \text{m}^4$ and the elastic modulus of steel is $2.0 \times 10^{6} \, \text{kPa}$.

Solution
The equivalent spring stiffness has been computed as in Example 8.3 and indicated in Figure 8.16. As shown in Figure 8.16, the only external forces applied on the pile are the ones applied by the pile cap and the soil reactions at the bottom that assure fixity. It is also noted that the spring associated with the bottom-most node has been added to the unknown force P_{12}.

Hence, the external force vector is given by the following equation:

$$[P] = [M_1 \quad 8 \quad 0 \quad 0 \quad 0 \quad 0 \quad 0 \quad 0 \quad 0 \quad M_{11} \quad P_{12}]^\text{T} \tag{8.30}$$

On the other hand, the deflection vector is given by the following equation

$$[X] = [0 \quad \Delta \quad \theta_2 \quad \Delta_2 \quad \theta_3 \quad \Delta_3 \quad \theta_4 \quad \Delta_4 \quad \theta_5 \quad \Delta_5 \quad 0 \quad 0] \tag{8.31}$$

in which it is assumed that the rotation at the top is restrained due to the pile cap (i.e., $X_1 = \theta_1 = 0$) and the translation as well as the rotation at the bottom are retrained by the ground fixity (i.e., $X_{12} = \Delta_6 = 0$ and $X_{11} = \theta_6 = 0$). The required lateral deflection is Δ.

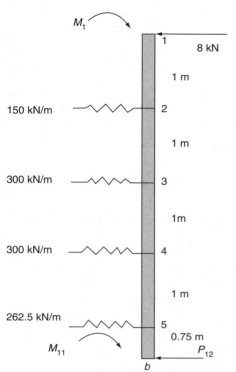

FIGURE 8.16

Illustration for Example 8.4.

The stiffness matrices for the first four elements and the fifth element are expressed by the following matrices:

$$
\begin{bmatrix}
17.6 & 26.4 & 8.8 & -26.4 \\
26.4 & 52.8 & 26.4 & -52.8 \\
8.8 & 26.4 & 17.6 & -26.4 \\
-26.4 & -52.8 & -26.4 & 52.8
\end{bmatrix}
\begin{bmatrix}
23.5 & 46.9 & 11.7 & -46.9 \\
46.9 & 125.2 & 46.9 & -125.2 \\
11.7 & 46.9 & 23.5 & -46.9 \\
-46.9 & -125.2 & -46.9 & 125.2
\end{bmatrix}
$$

Hence, the assembled and modified (for springs) global stiffness matrix would be

$$
[K] =
\begin{bmatrix}
17.6 & 26.4 & 8.8 & -26.4 \\
26.4 & 52.8 & 26.4 & -52.8 \\
8.8 & 26.4 & 35.2 & 0 & 8.8 & -26.4 \\
-26.4 & -52.8 & 0 & 255.6 & 26.4 & -52.8 \\
& & 8.8 & 26.4 & 35.2 & 0 & 8.8 & -26.4 \\
& & -26.4 & -52.8 & 0 & 405.6 & 26.4 & -52.8 \\
& & & & 8.8 & 26.4 & 35.2 & 0 & 8.8 & -26.4 \\
& & & & -26.4 & -52.8 & 0 & 405.6 & 26.4 & -52.8 \\
& & & & & & 8.8 & 26.4 & 41.1 & 20.5 & 11.7 & -46.9 \\
& & & & & & -26.4 & -52.8 & 20.5 & 440.5 & 46.9 & -125.2 \\
& & & & & & & & 11.7 & 46.9 & 23.5 & -46.9 \\
& & & & & & & & -46.9 & -125.2 & -46.9 & 125.2
\end{bmatrix}
$$

$$(8.32)$$

If $[K]$ in Equation (8.32) is rearranged and partitioned so that

$$
[P]_1 = [K]_{11}[X]_1 + [K]_{12}[X]_2 \tag{8.33a}
$$

and

$$[P]_2 = [K]_{21}[X]_1 + [K]_{22}[X]_2 \tag{8.33b}$$

where

$$[P]_1 = [M_1 \quad M_{11} \quad P_{12}]^T \tag{8.34a}$$

$$[P]_2 = [8 \quad 0 \quad 0 \quad 0 \quad 0 \quad 0 \quad 0 \quad 0 \quad 0]^T \tag{8.34b}$$

$$[X]_1 = [0 \quad 0 \quad 0] = \bar{0} \tag{8.34c}$$

$$[X]_2 = [\Delta \quad \theta_2 \quad \Delta_2 \quad \theta_3 \quad \Delta_3 \quad \theta_4 \quad \Delta_4 \quad \theta_5 \quad \Delta_5] \tag{8.34d}$$

and

$[K]_{11}, [K]_{12}, [K]_{21}$, and $[K]_{22}$ are the corresponding $3 \times 3, 3 \times 9, 9 \times 3$, and 9×9 partitions of $[K]$ as illustrated below:

$$\begin{bmatrix}
26.4 & 52.8 & 26.4 & | & -52.8 & 0 & 0 & 0 & 0 & 0 & 0 & 0 & 0 \\
0 & 0 & 0 & | & 0 & 0 & 0 & 0 & 0 & 11.7 & 46.9 & 23.5 & -46.9 \\
0 & 0 & 0 & | & 0 & 0 & 0 & 0 & 0 & -46.9 & -125.2 & -46.9 & 125.2 \\
- & - & - & + & - & - & - & - & - & - & - & - & - \\
17.6 & 26.4 & 8.8 & | & -26.4 & 0 & 0 & 0 & 0 & 0 & 0 & 0 & 0 \\
8.8 & 26.4 & 35.2 & | & 0 & 8.8 & -26.4 & 0 & 0 & 0 & 0 & 0 & 0 \\
-26.4 & -52.8 & 0 & | & 255.6 & 26.4 & -52.8 & 0 & 0 & 0 & 0 & 0 & 0 \\
0 & 0 & 8.8 & | & 26.4 & 35.2 & 0 & 8.8 & -26.4 & 0 & 0 & 0 & 0 \\
0 & 0 & -26.4 & | & -52.8 & 0 & 405.6 & 26.4 & -52.8 & 0 & 0 & 0 & 0 \\
0 & 0 & 0 & | & 0 & 8.8 & 26.4 & 35.2 & 0 & 8.8 & -26.4 & 0 & 0 \\
0 & 0 & 0 & | & 0 & -26.4 & -52.8 & 0 & 405.6 & 26.4 & -52.8 & 0 & 0 \\
0 & 0 & 0 & | & 0 & 0 & 0 & 8.8 & 26.4 & 41.1 & 20.5 & 11.7 & -46.9 \\
0 & 0 & 0 & | & 0 & 0 & 0 & -26.4 & -52.8 & 20.5 & 440.5 & 46.9 & -125.2
\end{bmatrix} \tag{8.35}$$

From Equation (8.33b) $[X]_2$ can be expressed as

$$[X]_2 = -[K]_{21}[X]_1 + [K]_{22}^{-1}[P]_2 = [K]_{22}^{-1}[P]_2 \tag{8.36}$$

Substituting the above result and Equation (8.34c) in Equation (8.33a),

$$[P]_1 = [K]_{12}[K]_{22}^{-1}[P]_2 \tag{8.37}$$

Hence, the unknown external forces can be determined from Equation (8.37). Accordingly, from Equation (8.36), $\Delta = 0.304$ m and $M_1 = 16$ kN m. Then, by substitution in Equation (8.36) the unknown deflections $[X]_2$ can be determined.

Finally, the moments and the shear forces along the pile length can be determined by substituting the nodal deflections in the individual **element equations** such as Equation (8.23).

8.3.2.2 Lateral Pressure–Deflection (p–y) Method of Analysis

The following form of Equation (8.18) is employed in the p–y curve approach (Figure 8.17) developed by Reese (1977):

$$\frac{d^2M}{dz^2} + P(z)\frac{d^2y}{dz^2} - p' = 0 \tag{8.38}$$

where $p' =$ soil reaction per unit pile length.

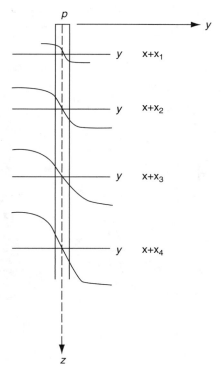

FIGURE 8.17
Set of *p–y* curves. (Reese, L.C., 1977, *J. Geotech. Eng., ASCE,*
103(GT4):283–305. With permission.)

It is observed that the difference between Equations (8.38) and (8.19) is that Equation
(8.38) accounts for the shear and moment effects induced by the axial force $P(z)$ due to the
finite curvature of the pile produced by lateral loading (Figure 8.18). Hence, the shear
force and the distributed soil reaction on the pile at any depth can be expressed as

$$S = EI\frac{d^3y}{dz^3} + P(z)\frac{dy}{dz} \tag{8.39}$$

$$p' = EI\frac{d^4y}{dz^4} + P(z)\frac{d^2y}{dz^2} \tag{8.40}$$

The finite difference (FD) form of the above equation is given as (Reese, 1977):

$$y_{m-2}R_{m-1} + y_{m-1}(-2R_{m-1} - 2R_m + P_zh^2) + y_m(R_{m-1} + 4R_m + R_{m+1} - 2P_zh^2 + k_mh^4)$$
$$+ y_{m+1}(-2R_m - 2R_{m+1} + P_zh^2) + y_{m+2}R_{m+1} = 0 \tag{8.41}$$

where

$$R_m = E_mI_m \tag{8.42}$$

is the stiffness of the mth mode, y_m is the lateral deflection at the mth node, h is the finite
difference step size (nodal distance along the pile), P_z is the axial force at the mth node
(depth z). The parameter k_m defined in Equation (8.43) can be evaluated for each node m
by predicting the $p–y$ curve corresponding to the depth, of that node, z.

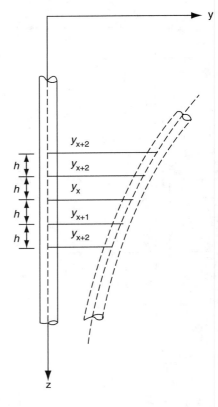

FIGURE 8.18
Representation of the deflected pile. (From Reese, L.C., 1977,
J. Geotech. Eng., ASCE, 103(GT4):283–305. With permission.)

$$p' = k_m y \qquad (8.43)$$

Finally, using the following boundary conditions:

1. shear and moment are zero at the bottom of the pile,
2. lateral load and the moment (or the slope or the rotational restraint) at the pile top are known,

the FD algorithm in Equation (8.41) can be solved and the lateral deflection, pile rotation, and moment and shear along the pile can be numerically determined at any location. According to Reese (1977), the *p–y* methodology implies that the behavior of the soil at any depth is independent of its behavior at other locations, which is strictly not true. However, experiments seem to indicate that the above implication is justified under practical circumstances.

8.3.2.3 Synthesis of p–y Curves Based on Pile Instrumentation

Strain gauge readings obtained along the length of a laterally tested pile can be employed to develop the lateral load transfer curves (*p–y* curves) at a finite number of points along the pile (Hameed, 1998). The values of *p* (horizontally distributed load intensity) and *y* (lateral deflection) at any pile location at a given lateral loading stage can be determined using the following analytical procedure. From the simple beam theory,

$$\frac{\mathrm{d}^2 y}{\mathrm{d}z^2} = \frac{\varepsilon}{h} \qquad (8.44)$$

where y is the lateral deflection, z is the vertical coordinate along the pile, h is the distance from the neutral axis of the pile cross-section to the strain gauge location, and ε is the strain gauge reading at z.

Hence, the lateral deflection (y) can be expressed as

$$y = \frac{1}{h} \iint \varepsilon \, dz \tag{8.45}$$

Similarly, by using Equations (8.19) and (8.44), the distributed soil load (p) can be expressed as

$$p = -\frac{E_p I}{hB} \frac{d^2\varepsilon}{dz^2} \tag{8.46}$$

Therefore, it can be seen that both p and y values can be found from a mathematically approximated (fitted) ε curve to measured flexural strains. This is usually achieved either by fitting a cubic spline function between successive strain data points (Li and Byrne, 1992) or by fitting a higher-order polynomial to all of the strain data points (Ting, 1987). The fitting procedure can be illustrated as follows.

Example 8.5
For the model pile shown in Figure 8.19, assuming that the measured longitudinal strains are given by Figure 8.20, illustrate the fitting procedure that can be used to generate the p–y curves at specific depths.

Solution
The distance z is measured from the pile tip which is located 1.0 m below the ground surface. In order to closely trace all of the strain data, the following polynomial with five coefficients (a_i) can be considered:

$$y = a_1 z^6 + a_2 z^7 + a_3 z^8 + a_4 z^9 + a_5 z^{10} \tag{8.47}$$

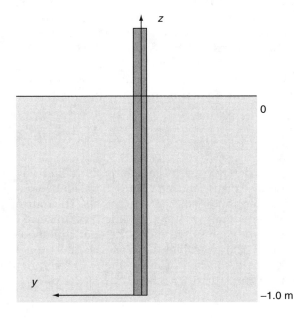

FIGURE 8.19
Illustration for Example 8.5.

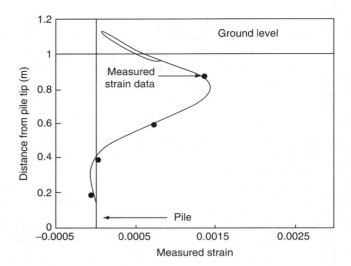

FIGURE 8.20
The measured strains in Example 8.5 and the polynomial fit. (From Hameed, R.A., 1998, lateral Load Behavior of Jetted and Preformed Piles, Ph.D. dissertation, University of South Florida, Tampa, FL. With permission.)

It can be seen that the terms up to z^5 have been discarded from Equation (8.47) since the pile deflection and all of its derivatives up to the fifth derivative are generally considered zero at the pile tip ($z = 0$) (Ting, 1987). This is because the deflection, slope, moment, shear, and the lateral pressure due to the applied lateral load are negligible at the pile tip. Thus, by combining Equations (8.44) and (8.47), the strain at any location within the embedded part of the pile can be expressed by the following function with five unknown coefficients a_i, $i = 1$–5:

$$\varepsilon = h\frac{d^2y}{dz^2} = h(30a_1z^4 + 42a_2z^5 + \varepsilon56a_3z^6 + 72a_4z^7 + 90a_5z^8) \tag{8.48}$$

Then four pairs of strain gauge readings and the known soil pressure ($p = 0$) at the soil surface ($z = z_0 = 1.0\,\text{m}$) can be used to determine the unknown a_i ($i = 1$–5). Furthermore, a third-degree polynomial was employed for approximating the deflection (y) of the free portion of the pile (above the ground level). This ensures that the $p = 0$ condition is satisfied all over the free portion since the fourth derivative of this polynomial (p in Equation 8.18) automatically drops out. Consequently, the deflection above the soil surface can be given by the following function with four unknown coefficients b_i, $i = 0$–3:

$$y = b_0 + b_1(z - z_0) + b_2(z - z_0)^2 + b_3(z - z_0)^3 \tag{8.49}$$

Three of the above constants (b_i, $i = 0$–3) were determined by matching the deflection, slope and moment of the free pile portion with the corresponding values of the embedded portion as determined by Equations (8.47) and (8.48), at the soil surface ($z = z_0$). The fourth b_i constant can be determined by setting the moment at the lateral loading level to zero.

The distributions of deflections and soil pressure computed using the above methodology are illustrated in Figure 8.21 and Figure 8.22, respectively. In this case, the model pile is assumed to be embedded in an unsaturated soil bed of unit weight $16.2\,\text{kN/m}^3$. Figure 8.23 shows the analytical predictions of the lateral load behavior of the piles at specified depth.

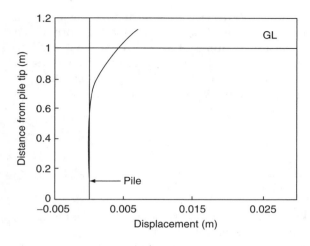

FIGURE 8.21
Computed lateral displacement vs. depth. (From Hameed, R.A., 1998, Lateral Load Behavior of Jetted and Preformed Piles, Ph.D. dissertation, University of South Florida, Tampa, FL. With permission.)

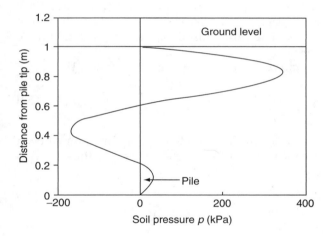

FIGURE 8.22
Computed soil pressure vs. depth. (From Hameed, R.A., 1998, Lateral Load Behavior of Jetted and Preformed Piles, Ph.D. dissertation, University of South Florida, Tampa, FL. With permission.)

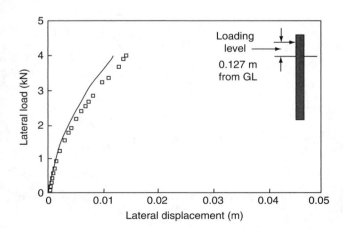

FIGURE 8.23
Analytically predicted load–displacement behavior. (From Hameed, R.A., 1998, Lateral Load Behavior of Jetted and Preformed Piles, Ph.D. dissertation, University of South Florida, Tampa, FL. With permission.)

In order to eliminate the depth dependency, p–y curves can be normalized using a soil parameter which depicts the mean normal stress level. p/E_sB is a suitable normalized parameter for this purpose since E_{max} (elastic modulus at very low strains) used to compute p/E_sB shows a strong mean normal stress dependence(Li and Byrne, 1992). Hameed et al. (2000) determined E_{max} from the measured coefficient of horizontal sub-grade reaction, K_{max}, using the following expressions (Glick, 1948; Bowles, 1996):

$$K'_s = BK_{max} \tag{8.50}$$

$$K'_s = \frac{22.4E_s(1 - \nu_s)}{(1 + \nu_s)(3 - 4\nu_s)[2 \ln(2L_pB) - 0.433]} \tag{8.51}$$

where K'_s and E_s have the same units (kPa) and K'_s is the horizontal subgrade modulus, L_p is the pile length, B is the pile width, and ν_s, is Poisson's ratio.

K_{max} at each depth can be obtained from the initial stiffness of the experimentally determined p–y curves (Figure 8.23) (Hameed, 1998). Similarly, the ultimate soil pressures (p_u) can be obtained from p–y curves at each depth by fitting the experimentally developed p–y curve with a hyperbolic function of the form $p = y/(a+by)$ (Kondner, 1963; Georgiadis et al., 1991). The p_u value for each fitted curve is expressed by the curve parameter, $1/b$, since $p_u = 1/b$ when $y \to \infty$. The variations of K_{max} and p_u with depth are shown in Figure 8.24 and Figure 8.25, respectively.

Based on the foregoing discussion p–y curve can be expressed as (Hameed, 1998):

$$p = \frac{y}{(1/K_{max}) + (y/p_u)} \tag{8.52}$$

On the other hand, two other popular mathematical formats for p–y curves have been provided by Reese et al. (1974) and Murchison and O'Neill (1984). These are illustrated in Figure 8.26 and Figure 8.27, respectively.

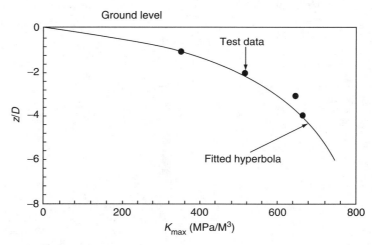

FIGURE 8.24
Variation of K_{max} with depth. (From Hameed, R.A., Gunaratne, M., Putcha, S., Kuo, C., and Johnson, S., 2000, *ASTM Geotech. Testing J.*, 23(3). With permission.)

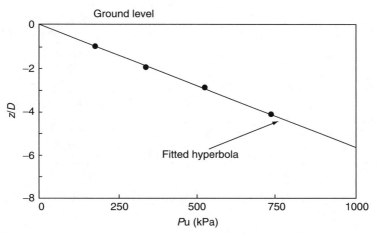

FIGURE 8.25
Variation of p_u with the depth. (From Hameed, R.A., Gunaratne, M., Putcha, S., Kuo, C., and Johnson, S., 2000, *ASTM Geotech. Testing J.*, 23(3). With permission.)

8.4 Lateral Load Capacity of Pile Groups

In most actual foundation applications, since piles installed as a cluster invoke group action, it is important for the foundation designer to be knowledgeable of the response of a group of piles to lateral loads. Ruesta and Townsend (1997) performed a field test involving an isolated single pile and a large-scale test group of 16 prestressed concrete piles spaced three (3) diameters apart to study how the lateral load characteristics of pile groups relate to those of individual piles in the group. Of the many *in situ* testing methods used to predict the p–y curves, SPT and pressuremeter test predictions were corroborated by the strain gage and inclinometer readings. Ruesta and Townsend (1997) concluded that an overall average p multiplier of 0.55 was needed for the individual p–y curves to predict the overall lateral response of the pile group.

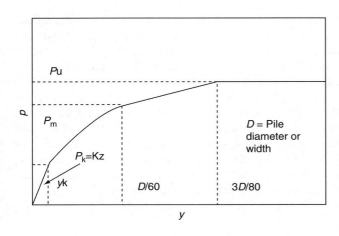

FIGURE 8.26
Analytically predicted load–displacement behavior. (From Reese, L.C., Cox, W.R., and Koop, F.D., 1974, Analysis of laterally loaded piles in sand, Proceedings of the 6th Offshore Technology Conference, Houston, TX, paper OTC 2080, pp. 473–483. With permission.)

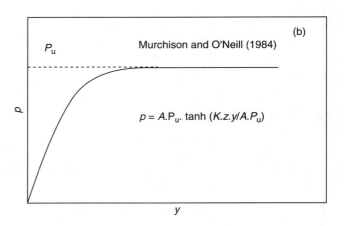

FIGURE 8.27
Analytically predicted load–displacement behavior. (From Murchison, J.M. and O'Neill, M.W., 1984, Evaluation of p–y relationship in cohesionless soils, in *Analysis and Design of Pile Foundation*, ASCE, New York, pp. 174–191. With permission.)

8.5 Load and Resistance Factor Design for Laterally Loaded Piles

Based on FHWA (1998) recommendations, design of laterally loaded piles involves determining the maximum lateral ground line deflections at the "service limit state" and the maximum moment at the "strength limit state" for an individual pile considering the installation method for the selected pile section and comparing it with the tolerable deformation and the maximum factored axial resistance of the pile, respectively, in order not to exceed both limits.

FHWA (1998) recommends that the allowable stress design (ASD) methods used to estimate the lateral resistance of a single pile or pile group can also be used for load and resistance factor design (LFRD) with the pile or the pile group subjected to the factored lateral loads, axial loads and moments, and the resulting factored axial and bending stresses are compared with the factored axial and bending capacities of the pile.

8.6 Effect of Pile Jetting on the Lateral Load Capacity

Water jetting can be utilized as an effective aid to impact pile driving when hard strata are encountered above the designated pile tip elevation. During jetting, the immediate neighborhood of the pile is first liquefied due to high pore pressure induced by the water jet and subsequently densified with its dissipation. In addition, the percolating water also creates a filtration zone further away from the pile. Hence, jetting invariably causes substantial disturbance to the surrounding soil, which results in a notable change in the lateral load behavior. Tsinker (1988) and Hameed et al. (2000) investigated the lateral load performance of driven and jetted–driven model piles installed under the same *in situ* soil conditions, by comparing the normalized p–y curves of driven piles to those of jetted–driven piles (Figure 8.28). They also explored the effect of jet water pressure, soil unit weight, and groundwater conditions on the p–y characteristics. Based on the above study, Hameed et al. (2000) developed approximate guidelines for predicting the lateral load behavior of *jetted* piles based on that of piles impact *driven* under similar soil conditions.

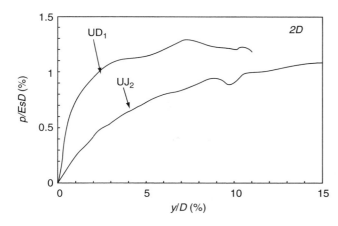

FIGURE 8.28
Comparison of p–y curves of driven (UD_1) and jetted (UJ_2) piles. (From Hameed, R.A., 1998, Lateral Load Behavior of Jetted and Preformed Piles, Ph.D. dissertation, University of South Florida, Tampa, FL. With permission.)

In the Hameed et al. (2000) study, K_{max} ratios (K_{jet}/K_{driven}) and p_u ratios ($p_{u,jet}/p_{u,driven}$) obtained from the model testing program were plotted against the nondimensional jetting pressure ($\pi_3 = P_0/k^2\gamma$) (k = permeability coefficient of the foundation soil) and are shown in Figure 8.29 and Figure 8.30, respectively. Each data point represents the mean of five ratio values. The K-ratio and p_u-ratio can be related to nondimensional jetting pressure by Equations (8.53a) and (8.53b). The foundation soil was a sand contaminated by bentonite clay.

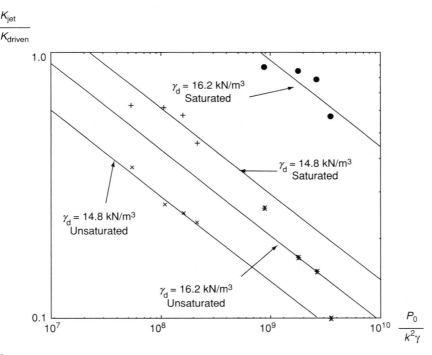

FIGURE 8.29
Effect of pile jetting on K_{max}. (From Hameed, R.A., Gunaratne, M., Putcha, S., Kuo, C., and Johnson, S., 2000, *ASTM Geotech. Testing J.*, 23(3). With permission.)

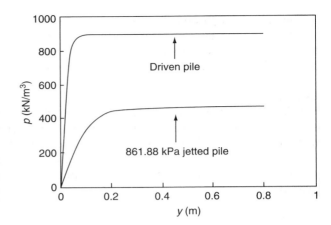

FIGURE 8.31
Synthesized p–y curve for the jetted pile. (From Hameed, R.A., Gunaratne, M., Putcha, S., Kuo, C., and Johnson, S., 2000, *ASTM Geotech. Testing J.*, 23(3). With permission.)

8.7 Effect of Preaugering on the Lateral Load Capacity

Hameed et al. (1998) developed similar relationships between the p–y curve parameters of preaugered piles and driven piles as shown in the following equations

$$\frac{K_{\text{pre}}}{K_{\text{driven}}} = \alpha_3 \left(\frac{d}{D}\right)^{\beta_3} \tag{8.54a}$$

$$\frac{p_{u,\text{pre}}}{p_{u,\text{driven}}} = \alpha_4 \left(\frac{P_0}{k^2 \gamma}\right)^{\beta_4} \tag{8.54b}$$

where α_3, α_4, β_3, and β_4 are soil type dependent constants, which can be evaluated from Table 8.4.

TABLE. 8.4

Parameters for Equations (8.54a and 8.54b)

Constant	$\gamma = 16.2\,\text{kN/m}^3$		$\gamma = 14.8\,\text{kN/m}^3$	
	Unsaturated	Saturated	Unsaturated	Saturated
(a) Equation (8.54a)				
α_3	0.14	0.69	0.38	0.69
β_3	−1.17	−1.17	−1.17	−1.17
(b) Equation (8.54b)				
α_4	0.64	0.39		
β_4	−0.68	−0.68		

References

Bowles, J.E., 1996, *Foundation Analysis and Design*, 5th edn, McGraw-Hill, New York.

Broms, B., 1964a, Lateral resistance of pile in cohesive soil, *Journal of Soil Mechanics Foundation Division, ASCE*, 90(SM3):27–56.

Broms, B., 1964b, Lateral resistance of pile in cohesionless soil, *Journal of Soil Mechanics Foundation Division, ASCE*, 90(SM3):123–156.

Federal Highway Administration, 1998, *Load and Resistance Factor Design (LRFD) for Highway Bridge Substructures*, Federal Highway Administration, Washington, DC.

Georgiadis, S.M., Anagnostopoulos, C., and Saflekou, S., 1991, Centrifugal testing of laterally loaded piles in sand, *Canadian Geotechnical Journal*, 27:208–216.

Glick, F.H., 1948, Influence of soft ground in the design of long piles, 2nd ICSMFE, Vol. 4, pp. 84–88.

Hameed, R.A., 1998, Lateral Load Behavior of Jetted and Preformed Piles, Ph.D. Dissertation, University of South Florida, USA.

Hameed, R.A., Gunaratne, M., Putcha, S., Kuo, C., and Johnson, S., 2000, Laterally loaded behavior of jetted piles, *ASTM Geotechnical Testing Journal*, 23(3), 358–368.

Kondner, R.L., 1963, Hyperbolic stress–strain response: cohesive soils, *Journal of Soil Mechanics Foundation Division, ASCE*, 89(1):115–143.

Li, Y. and Byrne, P.M., 1992, Lateral pile response to monotonic pile head loading, *Canadian Geotechnical Journal*, 29:955–970.

Murchison, J.M. and O'Neill, M.W., 1984, Evaluation of p–y relationship in cohesionless soils, In *Analysis and Design of Pile Foundation*, ASCE, New York, pp. 174–191.

Pyke, R. and Beikae, M., 1984, A new solution for the resistance of single piles to lateral loading, *Laterally Loaded Deep Foundation: Analysis and Performance*, ASTM STP 835, pp. 3–20.

Reese, L.C., 1977, Laterally loaded piles: program documentation, *Journal of Geotechnical Engineering, ASCE*, 103(GT4):283–305.

Reese, L.C, Cox, W.R., and Koop, F.D., 1974, Analysis of laterally loaded piles in sand, Proceedings of the 6th Offshore Technology Conference, Houston, TX, paper OTC 2080, pp. 473–483.

Robertson, P.K. and Campanella, R., 1983, Interpretation of cone penetration tests, Part I. Sand, *Canadian Geotechnical Journal*, 20(4):718–733.

Ruesta, P.F. and Townsend, F.C., 1997, Evaluation of laterally loaded pile group at Roosevelt bridge, *Journal of Geotechnical and Geoenvironmental Engineering, ASCE*, 123(12):1153–1161.

Ting, J.M., 1987, Full-scale dynamic lateral pile response, *Journal of Geotechnical Engineering, ASCE*, 113(1):30–45.

Tsinker, G.P., 1988, Pile jetting, *Journal of Geotechnical Engineering, ASCE*, 114(3):326–334.

Hetenyi, M., 1946, *Beams on Elastic Foundations*, University of Michigan Press, Ann Arbor, Michigan.

9

Construction Monitoring and Testing Methods of Driven Piles

Manjriker Gunaratne

CONTENTS

9.1   Introduction

Depending on the stiffness of subsurface soil and groundwater conditions, pile foundations can be constructed using a variety of construction techniques. The most common
techniques are (1) driving (Figure 9.1), (2) *in situ* casting and preaugering (Figure 9.2), and
(3) jetting (Figure 9.3). Due to the extensive nature of the subsurface mass that it influ-

FIGURE 9.1
Driven piles. (From www.vulcanhammer.com.
With permission.)

FIGURE 9.2
Cast-*in situ* piling. (From www.gdonaldson. com. With permission.)

ences, the degree of uncertainty regarding the actual working capacity of a pile foundation is generally much higher than that of a shallow footing. Hence, geotechnical engineers constantly seek more and more effective techniques of monitoring pile construction to estimate as accurately as possible the ultimate field capacity of piles.

In addition, pile construction engineers and contractors are also interested in innovative monitoring methods that would reveal information leading to (1) on-site determination of pile capacity as driving proceeds, (2) distribution of pile load between the shaft and the tip, (3) detection of possible pile or driving equipment damage, and (4) selection of effective driving techniques and equipment.

9.2 Construction Techniques Used in Pile Installation

9.2.1 Driving

The most common technique for installation of piles is driving them into strong bearing layers with an appropriate hammer (such as Vulcan, Raymond) system. In order for this technique to be effective, the hammer and the pile must be able to withstand the driving stresses. Although driving can be monitored using the specified penetration criteria (Section 9.3.1) to assure safe conditions, nowadays the technique of pile driving is commonly accompanied by the pile-driving analysis method of monitoring (Section 9.4). Specific details of hammers and hammer rating is found in Bowles (1995).

(a)

FIGURE 9.3
(a) Jetted piles. (From www.state.dot.nc.us.
With permission.) (b) Preaugered concrete pile.
(From www.iceusa.com. With permission.)   (b)

9.2.2 *In Situ* Casting

When the subsurface soil layers are relatively strong, it is common to install significantly large-diameter piles and using boring techniques. For caissons, this is the only viable installation method (Chapter 7). Depending on the collapsibility of the soils and availability of casings, *in situ* casting can be performed with or without casings. In cases where casing is desired, drilling mud (such as bentonite) is an economic alternative. More construction details of cast-*in situ* piles are found in Bowles (1995).

9.2.3 Jetting and Preaugering

Although driven piles are installed in the ground mostly by impact driving, jetting or preaugering can be used as aids when hard soil strata are encountered above the estimated tip elevation required to obtain adequate bearing. However, the final set is usually achieved by impact driving the last few meters, an exercise that somewhat restores the possible loss of axial load bearing capacity due to jetting or preaugering. Nonetheless, it has been reported (Tsinker, 1988) that impact-driven piles have better load bearing characteristics than jetted-driven piles under comparable soil conditions. This is possible due to the soil in the immediate neighborhood first liquefying as a result of the excessive jet water velocity and subsequently remolding with the dissipation of excess pore pressure. The original *in situ* soil structure and the skin-friction characteristics are significantly altered. During the jetting process, some water also infiltrates onto the neighborhood maintaining a high pore pressure there. Thus, the creation of liquefaction and filtration zones, known as the zone of combined influence of jetting, is expected to result in a reduction of the lateral load capacity. Consequently, although pile jetting may be effective as a penetration aid to impact driving in saving time and energy, the accompanying reduction in the lateral load capacity will be a significant limitation of the technique. Similar inferences can be made regarding preaugering as well.

9.3 Verification of Pile Capacity

There are several methods available to determine the static capacity of piles. The commonly used methods are (1) use of pile-driving formulae, (2) analysis using the wave equation, and (3) full-scale load tests. A brief description of the first two methods will be provided in the next two subsections.

9.3.1 Use of Pile-Driving Equations

In the case of driven piles, one of the very early methods available to determine the load capacity was the use of pile-driving equations. Hiley, Dutch, Danish, Janbu, Gates, and modified Gates are some of pile-driving formulae available for use. For more information on these, the reader is referred to Bowles (1995) and Das (2002). Of these equations, one of the formulae most popular ones is the engineering news record (ENR) equation, that expresses the pile capacity as follows:

$$P_u = \left[\frac{1.25 e_h E_h}{s + C}\right]\left[\frac{W_r + n^2 W_p}{W_r + W_p}\right] \tag{9.1}$$

where n is the coefficient of restitution between the hammer and the pile (<0.5 and >0.25), W_h is the weight of the hammer, W_p is the weight of the pile, s is the pile set per blow (in inches), C is a constant (0.1 in.), $E_h = W_h(h)$, h is the hammer fall, and e_h is the hammer efficiency (usually estimated by monitoring the free fall).

It is seen how one can use Equation (9.1) to compute the instant capacity developed at any given stage of driving by knowing the pile set (s), which is usually computed by the reciprocal of the number of blows per inch of driving. It must be noted that when driving has reached a stage where more than ten blows are needed for penetration of 1 in. ($s = 0.1$ or at "refusal"), further driving is not recommended to avoid damage to the pile and the equipment.

Example 9.1

(This example is solved in British units. Hence, please refer to Table 7.9 for appropriate conversion to SI units.) Develop a pile capacity versus set criterion for driving a 30 ft concrete pile of 10 in. diameter using a hammer with a stroke of 1 ft and a ram weighing 30 kips (kilopounds).

The weight of the concrete pile $= \frac{1}{4} \pi (10/12)^2 (30)(150)(0.001)$ kips $= 2.45$ kips

Assume the following parameters:

$n = 0.3$

Hammer efficiency $= 50\%$

Substituting in Equation (9.1),

$$P_u = \left[\frac{1.25(0.5)(30)(12)}{s + 0.1}\right] \left[\frac{30 + (0.3)^2(2.45)}{30 + 2.45}\right]$$

$$P_u = \left[\frac{210}{s + 0.1}\right] \text{ kips}$$

9.3.2 Use of the Wave Equation

With the advent of modern computers, the use of the wave-equation method for pile analysis, introduced by Smith (1960), became popular. Smith's idealization of a driven pile is elaborated in Figure 9.4.

The governing equation for wave propagation can be written as follows:

$$\rho A_p \frac{\partial^2 u}{\partial t^2} dz = A_p E \frac{\partial^2 u}{\partial z^2} dz - R(z) \tag{9.2}$$

where ρ is the mass density of the pile, E is the elastic modulus A_p is the area of cross section of the pile, u is the particle displacement, t is the time, z is the coordinate axis along the pile and $R(z)$ is the resistance offered by any pile slice, dz.

The above equation can be transformed into the finite-difference form to express the displacement (D), the force (F), and the velocity (v), respectively, of a pile element i at time t as follows:

$$D(i,t) = D(i,t - \Delta t) + V(i,t - \Delta t) \tag{9.3}$$

$$F(i,t) = [D(i,t) - D(i+1,t)]K \tag{9.4}$$

$$V(i,t) = V(i,t - \Delta t) + [\Delta tg/w(i)][F(i-1,t) - F(i,t) - R(i,t)] \tag{9.5}$$

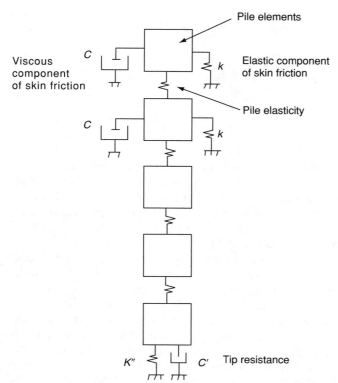

Viscous component of skin friction

C

Elastic component of skin friction

k

Pile elements

Pile elasticity

C

k

K'' C' Tip resistance

FIGURE 9.4
Application of the wave equation.

where

$K = EA_p/\Delta z$

$W = \rho \Delta z A_p$

Δt = selected time interval at which computations are made as the solution progresses with time.

Δz = selected pile segment size at which computation is performed along the pile length.

Idealization of soil resistance. In Smith's (1960) model, the point resistance and the skin friction of the pile are assumed to be viscoelastic and perfectly plastic in nature. Therefore, the separate resistance components can be expressed by the following equations:

$$P_p = P_{pD}(1 + JV_p) \tag{9.6}$$

and

$$P_s = P_{sD}(1 + J'V_p) \tag{9.7}$$

where P_{pD} and P_{sD} are static resistances at a displacement of D, V_p is the velocity of the pile, and J and J' are damping factors corresponding to the pile tip and the shaft.

The assumed elastic, perfectly plastic characteristics of P_{pD} and P_{sD} are illustrated in Figure 9.5.

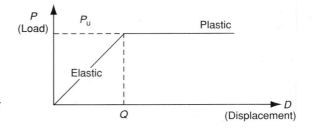

FIGURE 9.5
Assumed viscoelastic perfectly plastic behavior
of soil resistance.

In implementing this method, the user must assume a magnitude for the total resistance (P_u), a suitable distribution (or ratio) of the resistance between the skin friction and point resistance $(P_{pD}$ and $P_{sD})$, the quake (Q in Figure 9.5), and damping factors J and J'. Then, by using Equations (9.3)–(9.5), the pile set (s) can be determined. By repeating this procedure for other trial values of P_u, a useful curve between P_u and s (such as in Example 9.1), which can be eventually used to determine the resistance at any given set s, can be obtained.

The above system of equations ((9.3)–(9.5)) can be easily solved using a simple worksheet program, and the total static resistance to the pile movement during driving can be obtained. There are many commercially available wave-equation programs, such as GRLWEAP (Goble and Raushe, 1986), TTI and TNOWAVE, that are available for this purpose. However, the reliability of the above method depends on the estimation of soil damping constant along the pile shaft (J'), soil damping constant at the pile toe (J), soil quake along the pile shaft (Q_s), soil quake at the pile toe (Q_p), and the proportion of the force taken by pile toe (ξ). Smith (1960) suggested that 2.5 mm (or 0.1 in.) is a reasonable assumption for the skin quake (Q_s) and later it was suggested to take that the end quake at the pile bottom (Q_p) as $B/120$ where B is the pile diameter. Table 9.1 shows the range of the skin damping constants used for different soil types.

Example 9.2
(This example is solved in the British system of units. Hence, please refer to Table 7.9 for appropriate conversion to SI units.) For simplicity, assume that a model pile is driven into the ground using a 1000 lb hammer dropping 1 ft, as shown in Figure 9.6. Assuming the following data, predict the velocity and the displacement of the pile tip after three time steps:

$J = 0.0 \sec/\text{ft}, J' = 0.0 \sec/\text{ft}$
$Q = 0.1 \text{ in.}$
$\Delta t = 1/4000 \sec$
$R_{pu} = R_{su} = 50 \text{ kips } (\xi = 0.5)$
$K = 2 \times 10^6 \text{ lb/in.}$

TABLE 9.1

Some Typical Damping Constants

Soil Type	Damping Factor
Gravel	0.3–0.4
Sand	0.4–0.5
Silt	0.5–0.7
Clay	0.7–1.0

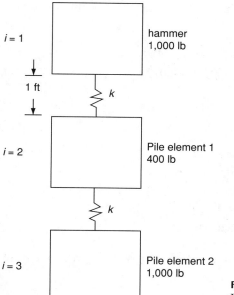

FIGURE 9.6
Illustration for Example 9.2.

As shown in Figure 9.6, assume the pile consists of two segments ($i = 2$ and 3) and the time step to be $1/4000$ sec. Then the following initial and boundary conditions can be written:

$$D(1, 0) = D(2, 0) = D(3, 0) = 0$$
$$F(1, 0) = F(2, 0) = F(3, 0) = 0$$
$$V(1, 0) = \sqrt{gh} = 96.6 \text{ in./sec}$$

After the first time step. From Equation (9.3),

$$D(1, 1) = D(1, 0) + V(1, 0)\Delta t = 1 + 96.6(1/4000) = 0.024 \text{ in.}$$
$$D(2, 1) = D(3, 1) = 0$$

From Equation (9.4),

$$F(1, 1) = [D(1, 1) - D(2, 1)]k = (0.024 - 0)(2)(10^6) = 48 \times 10^3 \text{ lb/in.}$$
$$F(2, 1) = F(3, 1) = 0$$

From Equation (9.5),

$$V(1, 1) = V(1, 0) + (1/4000)(388.8)(0 - 48,000)/1000 = 91.93 \text{ in./sec}$$
$$V(2, 1) = 0 + (1/4000)(388.8)[48.000 - 0 R(2, 1)]/400 = 11.664 \text{ in./sec}$$
$$V(3, 1) = 0.0$$

After the second time step. By repeating the above procedure, one obtains the following results:

$$D(1, 2) = D(1, 1) + V(1, 1)\Delta t = 0.024 + 91.93(1/4000) = 0.047\,\text{in.}$$
$$D(2, 2) = D(2, 1) + V(2, 1)\Delta t = 0 + 11.664(1/400) = 0.0029\,\text{in.}$$
$$D(3, 2) = 0$$
$$F(1, 2) = [D(1, 2) - D(2, 2)](2)(10^6) = 88{,}200\,\text{lb/in.}$$
$$F(2, 2) = [D(2, 2) - D(3, 2)](2)(10^6) = 5900\,\text{lb/in.}$$
$$F(3, 2) = [D(3, 2) - D(4, 2)](2)(10^6) = 0$$
$$V(1, 2) = V(1, 1) + 9.72(10^{-5})[F(0, 2) - F(1, 2)]$$
$$\qquad\quad = 91.93 + 9.72(10^{-5})(0 - 88{,}200) = 83.35\,\text{in./sec}$$
$$V(2, 2) = V(2, 1) + 24.3(10^{-5})(88{,}200 - 5900 - 1.450) = 31.3\,\text{in./sec}$$
$$V(3, 2) = 0 + 9.72(10^{-5})(5900 - 0 - 0) = 0.56\,\text{in./sec}$$

After the third time step. Again by repeating above steps, one obtains the following results:

$$D(1, 3) = D(1, 2) + V(1, 2)\Delta t = 0.047 + 83.35(1/4000) = 0.0678\,\text{in.}$$
$$D(2, 3) = D(2, 2) + V(2, 2)\Delta t = 0.0029 + 31.3(1/400) = 0.0078\,\text{in.}$$
$$D(3, 3) = 0 + 0.56(1/4000) - 0.00014\,\text{in.}$$
$$F(1, 3) = [D(1, 3) - D(2, 3)](2)(10^6) = 120{,}000\,\text{lb/in.}$$
$$F(2, 3) = [D(2, 3) - D(3, 3)](2)(10^6) = 15{,}320\,\text{lb/in.}$$
$$F(3, 3) = [D(3, 3) - D(4, 3)](2)(10^6) = 280\,\text{lb/in.}$$
$$V(1, 3) = V(1, 2) + 9.72(10^{-5})[F(0, 3) - F(1, 3) - R(1, 3)] = 71.69\,\text{in./sec}$$
$$V(2, 3) = V(2, 2) + 24.3(10^{-5})(120{,}000 - 15{,}320 - 3.900) = 55.79\,\text{in./sec}$$
$$V(3, 3) = 0.56 + 9.72(10^{-5})(15{,}320 - 70 - 70) = 2.04\,\text{in./sec}$$

The above computational procedure must be repeated on the computer until all of the pile segments cease to move during a given time step and their velocities approach zero.

Physically, this condition is identified as the stage where the effect of the stress pulse has expired due to damping.

9.4   Pile-Driving Analyzer

During pile driving, the stresses and accelerations imparted to the pile can be monitored and recorded to assess the quality of the installation. Although this information is also used to ascertain the load-carrying capacity of the pile, the quality assurance associated with type of equipment is perhaps its greatest contribution. Therein, the tensile and compressive stresses in piles can be monitored via strain gage instrumentation to prevent unnecessary damage while adjusting pile-driving hammer energy to maximize production rates. The movement is also monitored using integrated accelerometer data. Figure 9.7(a) shows the instrumentation and its position during pile driving.

In fact, wave-equation analysis of pile capacity can be supplemented by fabricating a pile driven by an impact or vibratory hammer as shown in Figure 9.7(a) to obtain records

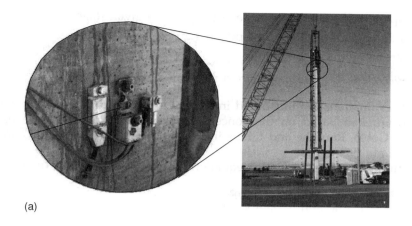

(a)

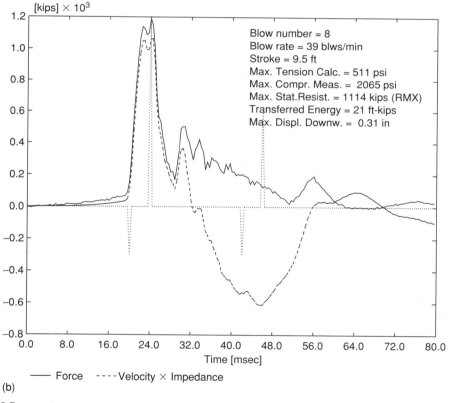

Blow number = 8
Blow rate = 39 blws/min
Stroke = 9.5 ft
Max. Tension Calc. = 511 psi
Max. Compr. Meas. = 2065 psi
Max. Stat.Resist. = 1114 kips (RMX)
Transferred Energy = 21 ft-kips
Max. Displ. Downw. = 0.31 in

—— Force - - - -Velocity × Impedance

(b)

FIGURE 9.7
(a) Strain gages and accelerometers attached to pile during pile driving. (Courtesy of Applied Foundation Testing, Inc.) (b) Field data showing pile-driving performance (1 kip = 4.45 kN, 1 psi = 6.9 kPa, 1 in. = 25.4 mm, 1 ft-kip = 1.36 kJ, 1 ft = 0.305 m). (Courtesy of Applied Foundation Testing, Inc.)

of the particle velocity and the longitudinal force at the pile top (Figure 9.7b). This technique known as pile-driving analysis has now gained worldwide popularity and application. When the above instrument records are used in conjunction with wave-equation analysis, one would be able to evaluate:

monitored pile is instrumented with an accelerometer and a longitudinal strain gage during pile-driving analyzer monitoring. The accelerometer record is converted to obtain the particle velocity of the pile top, in the longitudinal direction, as

$$V = \int f \, dt \tag{9.14}$$

On the other hand, the axial force on the pile top at a given instant in time can be obtained by the strain gage reading (ε) as

$$F = EA\varepsilon \tag{9.15}$$

Since both the force and the velocity records are typically plotted on the same scale in PDA, the particle velocity must be converted to an equivalent force (F^*) by the following conversion:

$$F^* = \frac{EA}{c} V \tag{9.16}$$

The EA/c term is denoted as the pile impedance or Z. Hence, it is necessary to know the elastic modulus of the pile material, the compression wave velocity in the pile material, and the cross-sectional area of the pile in order to plot the equivalent force record. Either these parameters can be included in the input data or the velocity record can be calibrated *a priori* against the force record to obtain the pile impedance.

If the pile is unrestrained or completely free of shaft friction and end bearing, using basic mechanics it can be shown that

$$F = F^* = \frac{EA}{c} V \tag{9.17}$$

Then it is understood that both the force (F) and the equivalent force (F^*) records due to a hammer blow would coincide. It is the above fact (Equation (9.17)) that is useful in calibrating the V record due to a hammer blow to coincide with the corresponding F record (and indicate F^*), before the pile is driven in.

When the pile is constrained particularly at the tip, the impact wave (downward) and the reflected wave (upward) together produce a resultant wave at a given location on the pile. Hence, what are recorded by the instrumentation are in fact the resultant force and the velocity at the top of the pile. The resultant longitudinal force on any pile section can be desynthesized as follows to reveal the respective force components due to the downward (H) and upward (G) waves:

$$F_H = \frac{1}{2}(F + ZV) \tag{9.18a}$$

and

$$F_G = \frac{1}{2}(F - ZV) \tag{9.18b}$$

Similarly, the particle velocities induced by the downward and the upward waves can be extracted from the PDA records as follows:

$$V_H = \frac{1}{2}\left(V + \frac{F}{Z}\right) \tag{9.19a}$$

and

$$V_G = \frac{1}{2}\left(V - \frac{F}{Z}\right)$$
(9.19b)

The following typical PDA records (Goble et al., 1996) are presented to illustrate the basic interpretations:

Case 1. Pile entering a hard stratum — This would be equivalent to Case 1 in Section 9.4.1 where the pile tip enters a relatively stiffer stratum. Hence, one would expect an almost negligible tip velocity and a relatively high compressive force on the tip in response to a given hammer blow. However, if the pile length is L, since it takes a time of L/c for the stress pulse induced by the hammer to reach the tip and an additional L/c time interval for the tip response to return to the top and get recorded by the instruments, the above response will be reflected on the PDA monitoring after a time period of $2L/c$ from the instant of hammer impact. This is illustrated in Figure 9.9.

Case 2. Pile entering a soft stratum — This would be equivalent to Case 2 in Section 9.4.1 where the pile tip is in a relatively softer stratum. Hence, one would expect an almost negligible tip stress (force) and a relatively high tip velocity in response to a given hammer blow. As explained above, these conditions will be reflected in the PDA monitoring equipment only after a time period of $2L/c$ from the instant of hammer impact. This is illustrated in Figure 9.10.

Case 3. Condition of high shaft resistance — Figure 9.9 and Figure 9.10 also clearly illustrate that if the pile shaft is relatively free, i.e., with a minimum shaft resistance, $R(z)$ (in Equation 9.2), then both the force and equivalent force (velocity) records gradually attenuate showing the expected decay of the hammer pulse at the pile top until the reflection of the tip condition reaches the top at a time of $2L/c$. In fact, this can be seen numerically in Example 9.2 as well.

On the other hand, if the shaft resistance is significantly high, one would expect the force pulse to be constantly replenished by the reflected force pulses from the shaft resistance $R(z)$. Under these conditions, using basis mechanics, Equation (9.17) can be modified to:

$$F = \frac{EA}{c}V + R(z)$$

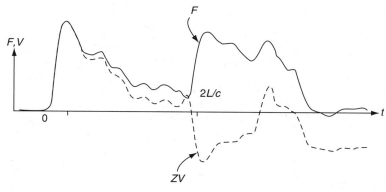

FIGURE 9.9
Illustration of large tip resistance condition.

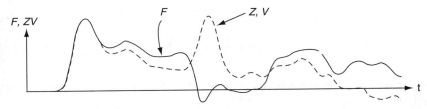

FIGURE 9.10
Illustration of minimal tip resistance condition.

This is illustrated in Figure 9.11 where the difference between F and F^* records until a time of $2L/c$ indicate the cumulative shaft resistance $R(z)$. A typical PDA record indicating significantly high shaft resistance is shown in Figure 9.12.

9.4.3 Analytical Determination of the Pile Capacity

Goble et al. (1988, 1996) presented a simple and approximate method of determining the pile capacity based on PDA records. This method is based on evaluating the parameters $RTL, RS1, RTL'$, and $RS1'$, which are defined as follows:

Total resistance (both static and dynamic components). The total resistance (static and dynamic) can be obtained from the following expression:

$$RTL = \frac{1}{2}[F1 + F2 + ZV1 - ZV2] \qquad (9.20a)$$

Static resistance. The static resistance can be obtained by subtracting the dynamic resistance component from the total resistance as

$$RS1 = (1 + J)RTL - J[F1 + ZV1] \qquad (9.20b)$$

where $(F1, ZV1)$ and $(F2, ZV2)$ are PDA records at $t = 0$ and $t = 2L/c$, respectively (Figure 9.13), and J is an empirical coefficient designated as the Case damping constant that accounts for damping action of soil both at the tip and the shaft.

The total resistance and its static component can be also evaluated by extending the $2L/c$ time window considered in Equation (9.20) to other times in the PDA record as well

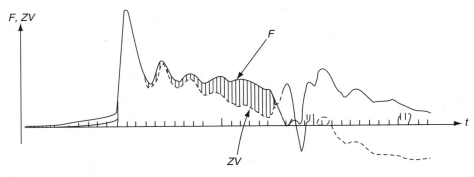

FIGURE 9.11
Wave effects of shaft friction and toe resistance.

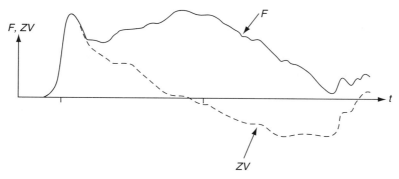

FIGURE 9.12
Illustration of significant shaft resistance condition.

$$RTL' = \frac{1}{2}[F1' + F2' + ZV1' - ZV2'] \tag{9.21a}$$

$$RS1' = (1 + J)RTL' - J[F1' + ZV1'] \tag{9.21b}$$

where $(F1', ZV1')$ and $(F2', ZV2')$ are PDA records at $t = t'$ and $t = t' + 2L/c$, respectively, and t' is a desired time selected on the record (Figure 9.13).

Typically, J is back-calculated based on correlation of PDA results with those of static load tests. Therefore, it must be noted that the Case damping constant cannot be considered as a soil property or a constant for a given soil. As seen in Table 9.2, it is seen to vary within a significant range of values even for the same type of soil depending on testing conditions.

Finally, the maximum static resistance based on the entire record can be obtained as

$$RMX = \text{Max}(RS1') \tag{9.22}$$

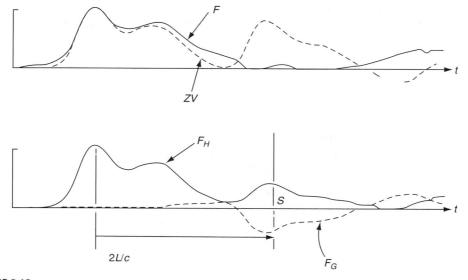

FIGURE 9.13
Illustration of the desynthesizing of PDA record.

TABLE 9.3

Assessment of Pile Damage

Cross-Section Reduction Factor (β)	Pile Condition
1.0	Uniform
0.8–1.0	Slightly damaged
0.6–0.8	Damaged
Below 0.6	Broken

Using Equations (9.21),

$$RTL' = \frac{1}{2}[3.91 + 0 + 5.54 + 3.75]$$
$$= 6.6 \text{ MN}$$

$$RS1' = (1 + 0.3)6.6 - 0.3[3.91 + 5.54]$$
$$= 5.745 \text{ MN} = R_{\text{MAX}}$$

Hence, the static pile capacity at the given instant can be predicted as 5.745 MN.

Example 9.4

Based on the PDA records indicated in Figure 9.16, compute the maximum tension force induced in the pile and its location. Assume that the pile length is 10 m and the compression wave velocity is 3300 m/sec.

Using Equation (9.18), the instant force records due to the upward and downward waves can be obtained as shown in Figure 9.17.

Based on Figure 9.17, it can be seen that the minimum compression pulse of 1.433 MN due to the downward compression wave occurred on the top at a time of $1.5L/c$. This compression pulse would move toward the pile tip at a velocity of c.

Similarly, it can also be seen that a maximum tension pulse of 2.225 MN reached the pile top at a time of $2L/c$ traveling upwards at a velocity of c. Hence, the two pulses (the minimum compression and the maximum tension) must have encountered each other at a time of T creating a net maximum tension of $2.225 - 1.433$ or 0.792 MN.

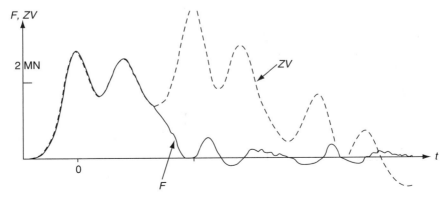

FIGURE 9.16
Illustration for Example 9.4.

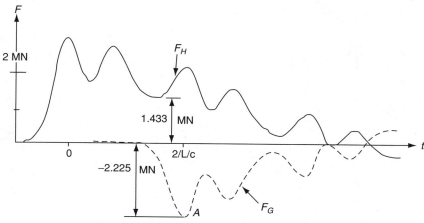

FIGURE 9.17
Desynthesized force components.

If the location where the two pulses encountered each other is at a depth of Z from the pile top, one can write the following expressions to compute Z:

$(T - 1.5L/c)$ = time taken for the minimum compression pulse to reach Z from the top

$(2L/c - T)$ = time taken for the maximum tension pulse to reach the top from Z

$c(T - 1.5L/c) = c(2L/c - T) = Z$

By solving, $T = 1.75L/c$ and $Z = 0.25L$.

Thus, it can be concluded that a maximum tension of 792 kN occurred at a distance of 2.5 m at a time of 5.3 msec after the input.

Example 9.5

Based on the PDA records indicated in Figure 9.18, assess the extent of concrete pile damage and the location of damage. Assume that the pile is of length 80 m.

The wave velocity can be estimated by using Equation (9.9) by knowing the elastic modulus of concrete as 27,600 MPa and the mass density as 2400 kg/m³. Therefore,

$c = (27,600,000,000/2400)^{0.5} = 3391 \, \text{m/sec}$

$L = 80 \, \text{m}$

$2L/c = 47.2 \, \text{msec}$

Therefore, the expected time of arrival of the return pulse $= 47.2 \, \text{msec}$.

The time of occurrence of the tension pulse (identified by the sudden increase of velocity) $= 15.7 \, \text{msec} \ll 47.2 \, \text{msec}$. Hence, one can assume that the pile is damaged.

If the effective length of the pile is L^* (up to the damaged location), then

$2L^*/c = 15.7 \, \text{msec} = 0.0157 \, \text{sec}$

Hence, $L^* = 26.6 \, \text{m}$

Using Equation (9.24) to determine the cross-section reduction factor β,

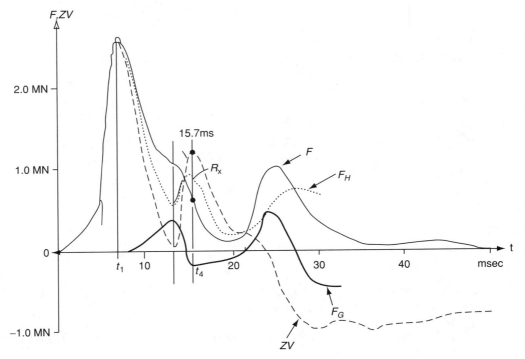

FIGURE 9.18
Illustration for Example 9.5.

$$\beta = \frac{[2600 - 1000 - 300]}{[2600 + 300]}$$

$$\beta = 0.45$$

Based on Table 9.3, it can be deduced that the tested pile is broken at a depth of 26.6 m. The allowable stresses for pile in common use are provided in Table 9.4.

A more precise evaluation of the pile capacity can be performed in conjunction with the wave-equation analysis. One of the popular methods currently used to perform this type of analysis is the Case Pile Wave Analysis program (CAPWAP) computational method (Goble and Raushe, 1986). Basically, in this technique one determines the set of soil resistance parameters (ultimate resistance, the quake and damping constants) that produces the best match between the instrument recorded and the wave equation based force and the velocity of the pile top. One of the two records (pile top velocity or force) is used as the top boundary condition and the complimentary quantity is computed using an analytical procedure similar to that presented in Section 9.3.2 and compared with the corresponding record. Further details of this technique can be found in Goble and Raushe (1986).

9.5 Comparison of Pile-Driving Formulae and Wave-Equation Analysis Using the PDA Method

Thilakasiri et al. (2002) report a case study in which the pile capacity predicted at the time of dynamic load testing of driven piles together with the measured sets were used to verify

TABLE 9.4

Allowable Stresses in Piles (FHWA, 1998)

Pile Type	Maximum Allowable Stress, σ all, (kPa)
Steel	
Driving damage likely	$0.25\,F_y$
Driving damage unlikely	$0.33\,F_y$
Concrete-filled steel pipe	$0.25\,F_y + 0.40\,f_c'$
Prestressed concrete	$0.33\,f_c' - 0.27\,f_p$
Round timber	
Douglas fir — coast	8.3
Douglas fir — interior	7.6
Lodgepole pine	5.5
Red oak	7.6
Southern pine	8.3
Western hemlock	6.9

Source: Federal Highway Administration, 1998, *Load and Resistance Factor Design (LRFD) for Highway Bridge Substructures*, Washington, DC. With permission.

the reliability of different dynamic prediction methods. For this purpose, a series of tests on driven cast-in-place concrete piles in residual formations of weathered rock were selected. Dynamic pile load testing was carried out according to ASTM D 4945. In dynamic pile load testing, using the pile-driving analyzer, a weight was dropped on to a pile instrumented with pairs of accelerometers and strain transducers. Variations of the acceleration and strain, during the application of the hammer blow were obtained in the field using the pile-driving analyzer. Subsequently, the acquired data is processed using the CAPWAP to obtain the static load–settlement curves from the measured force and velocity data. Moreover, the resulting penetration of the pile due to the hammer blow was independently measured as one parameter for checking the accuracy of pile-driving formulae.

The data collected in the field during the dynamic pile load testing program consisted of mobilized soil resistance, weight of the drop hammer, height of drop of the hammer, and the penetration of the pile per blow. In addition, the actual energy transferred to the pile, maximum compressive stress, and the maximum tensile stress developed during the hammer blow were also estimated from the PDA measurements. The skin frictional resistance and the end bearing resistance mobilized during the hammer blow were separated using the CAPWAP analysis.

The test results showed that for driven concrete piles in many residual formations, a significant part of load is carried by skin resistance. Test results also show that the efficiency of the hammer has varied between 15% and 60% for the piles tested. When crawler cranes were used with four-rope arrangement, where the hammer falls when the brakes are released, the efficiency factor was in the range of 60% to 40%. Similarly, when the hammer is raised and dropped using a mobile crane with a six-rope arrangement, where the hammer falls when the brakes are released, the efficiency dropped to the range of 30 to 15%. The measured efficiency factors are much smaller than the values quoted in literature. For example, Poulos and Davis (1980) recommend an efficiency of 75% for the drop hammer actuated by rope and friction winch.

In the estimation of the mobilized resistance using different driving formulae, the efficiency factors estimated from the PDA are used with the driving formulae containing such a factor. The mobilized resistance during the dynamic load testing was independently estimated using commonly used driving formulae and the measured set. For comparison purposes "Engineering news record" (ENR), Danish, Dutch, Hiley, Gates

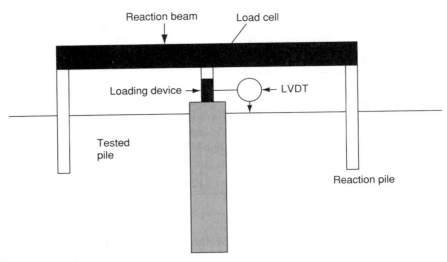

FIGURE 9.20
Schematic of pile load test setup.

FIGURE 9.21
Pile load testing. (From www.aecigeo.
com. With permission.)

TABLE 9.6

Comparison between Maintained Load and Quick Load Tests

Test Parameter	Maintained Load Test	Quick Load Test
Test load	200% of design load	300% of design load or up to failure
Load increment	25% of the design load	10–15% of the design load
Load duration	Up to a settlement rate of 0.001 ft/h or 2 h, whichever occurs first	2.5 min
Test duration	48 h	3–5 h

Although the above procedures are generally applicable for pile tension tests, additional loading procedures are found in ASTM Standard D-3689 (Standard Method of Testing Individual Piles Under Static Axial Tensile Load) for the pile tension test.

Two methods of pile load test interpretation are discussed in this handbook. They are

1. Davisson's (1972) offset limit method
2. De Beer's (1971) method

In Davisson's method, the failure load is identified as corresponding to the movement which exceeds the elastic compression of the pile, when considered as a free column, by a value of 0.15 in. plus a factor depending on the diameter of the pile. This critical movement can be expressed as follows: for piles of 600 mm or less in diameter or width,

$$S_f = S + (3.81 + 0.008D) \tag{9.25}$$

where S_f is the movement of the pile head (in mm), D is the pile diameter or width (in mm), and S is the elastic deformation of the total pile length (in mm).

For piles greater than 600 mm or less in diameter or width,

$$S_f = S + (0.033D) \tag{9.26}$$

In De Beer's method, the load and the movement are plotted on a double logarithmic scale, where the values can be shown to fall on two distinct straight lines. The intersection of the lines corresponds to the failure load.

As described in Section 6.6.2.3, the elastic displacement of a pile can be expressed as

$$\delta = \frac{1}{E_p A_p} \int_0^L P(z) \, dz \tag{9.27}$$

where E_p is the elastic modulus of the pile material and A_p is the cross-sectional area of the pile.

It is seen that the determination of elastic settlement can be cumbersome even if one has the knowledge of the elastic properties of the subsurface soil. This is because the actual axial load distribution mechanism or the load transfer mechanism is difficult to determine. Generally, one can instrument the pile with a number of strain gages to observe the variation of the axial load along the pile length and hence determine the load transfer.

If the reading on the ith strain gage is ε_i, then the axial load at the strain gage location can be expressed as

$$P_i = E_p A \varepsilon_i \tag{9.28}$$

This is illustrated in the load transfer curves in Figure 9.22.

Hence, the elastic displacement of the pile can be approximated by

$$s = \sum \varepsilon_i (\Delta L) \tag{9.29}$$

where Δ is the interval at which the strain gages are installed.

Example 9.6(a)

A static compression load test was performed on a 450 mm² prestressed concrete pile embedded 20 m below the ground surface in a sand deposit. The test pile was equipped

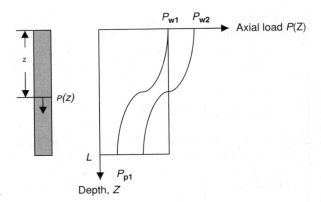

FIGURE 9.22
Load transfer curves based on strain gage data.

with two telltales (TT) extending to 0.3 m from the pile tip. The summary of data from the load test is shown in Table 9.7. Determine the failure load and the corresponding side friction and end bearing (Figure 9.23).

Solution
Based on Davisson's method (Figure 9.24), the failure load can be determined as 1.3 MN.
Based on De Beer's method (Figure 9.25), the failure load can be determined as 1.1 MN.

Example 9.6(b)
The static pile test was performed on a pile in Sarasota, FL. The pile was actually a test pile and was located in a parking lot near four other piles. The piles were auger cast piles, 13.85 m

TABLE 9.7

Load Test Data for Example 9.6(a)

Time (h:min)	Load (MN)	Top Δ (mm)	TT Δ (mm)	Tip Δ (mm)
12.55	0.0	0.0	0.0	0.0
13.00	0.05798	0.1016	0.0508	0.0508
13.05	0.082064	0.2032	0.1016	0.1016
13.10	0.146288	0.3048	0.2032	0.1016
13.15	0.22746	0.4572	0.3048	0.1524
13.19	0.299712	0.635	0.4064	0.2286
13.24	0.379992	0.8636	0.5334	0.3302
13.29	0.45938	1.0668	0.6604	0.4064
13.33	0.543228	1.3462	0.762	0.5842
13.38	0.619048	1.651	0.9144	0.7366
13.43	0.698436	2.0066	1.016	0.9906
13.48	0.781392	2.413	1.1684	1.2446
13.53	0.86524	2.8448	1.3208	1.524
13.57	0.940168	3.4036	1.4478	1.9558
14.02	1.015988	4.0132	1.5748	2.4384
14.07	1.09716	4.8768	1.7272	3.1496
14.11	1.184576	6.1214	1.8796	4.2418
14.16	1.263964	7.8232	1.9558	5.8674
14.21	1.340676	9.906	2.0828	7.8232
14.26	1.416496	12.7254	2.2352	10.4902
14.31	1.491424	16.6116	2.3622	14.2494
14.36	1.575272	23.7236	2.4892	21.2344

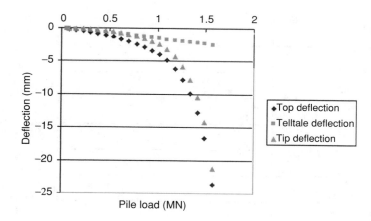

FIGURE 9.23
Pile load test results (Example 9.6a).

in length and were instrumented with gages at the top, middle, and bottom of the pile to determine the various loads at these points. The four adjacent piles were used as a reaction frame for the test pile. Two large steel beams were anchored into the support piles to provide the reaction frame. A large hydraulic jack was placed over the test pile and the test was started. The displacement was measured using a laser. The test was to last all day long.

Observations: It appears that there was a structural failure at an approximate load of 300 tons (2.66 MN). The reason that it appeared to be a structural failure opposed to a geotechnical failure was that the pile tip never realized any of the applied load. The actual design load for the piles was 90 tons (0.8 MN) so this pile is probably structurally sound for the design load. However, if this was a structural failure there is an indication of poor construction, which could mean that a different pile may not have the same capacity. The capacity was determined using Davisson's offset method, which is too conservative

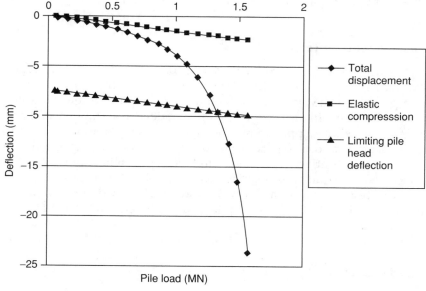

FIGURE 9.24
Determination of pile capacity — Davisson's method.

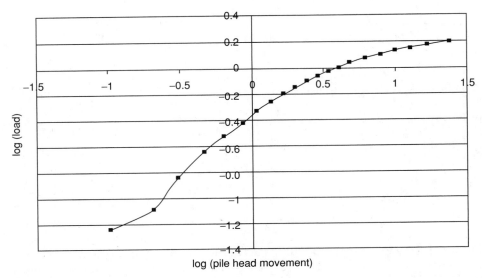

FIGURE 9.25
Determination of pile capacity — De Beer's method.

because the load–displacement curve that was formed peaked at 300 tons (2.66 MN) and dropped to about 225 tons (2 MN) for the offset value.

9.6.1 Advantages of Load Tests

This test provides very reliable data for pile capacity. The capacities are actual structural or geotechnical capacities, not calculated from idealized data. This can allow for a lower factor of safety in the design if the pile performs better than expected (and vice versa).

9.6.2 Limitations of Load Tests

The static load test can be very expensive to perform, especially when large loads are required because some sort of reaction frame must be constructed. They are often too expensive to perform if the structure to be built only requires a few piles. It would be more economical to use a higher factor of safety. Static load testing encompasses all test methods that systematically apply an increasing load to a foundation in multiple loading increments at such a rate so as to produce no dynamic movements as stated above. These tests include many applications (i.e., deep foundations or shallow foundations, tension or compression loads) with numerous loading configurations. With regard to full-scale *in situ* load tests, several test procedures are very prominent: plate load test (ASTM D 1195), pile load test in compression (ASTM D 1143), pile load test in tension (ASTM D 3689), and the Osterberg load test.

9.6.3 Kentledge Load Test

The load from structures to the foundations can be compression (downward), tension (upward), or lateral (sideways). The downward load-carrying resistance of a foundation encompasses most of the load conditions considered. In order to replicate these often enormous loads, several methodologies have been devised. The simplest form of a load test is the dead load or Kentledge method. This requires that the full test load be supplied in the form of dead weight stacked above the foundation on some framework. The framework must be capable of supporting the entire load at a single location where a hydraulic ram or jack can progressively transfer the load to the top of the foundation (Figure 9.26). This type of test accurately predicts the foundation response at full Kentledge load, but overestimates the stiffness of the foundation at lower loads due to the presence of the dead load overburden pressure applied to the ground surface. The practical upper limit of these tests is approximately 400 tons (3.56 MN) although physical site constraints may extend or restrict this limit drastically. Further, these tests are the most expensive and time consuming to perform from the standpoint of setup requirements. As with all static load tests, these tests are typically run in compliance with ASTM D-1143 or similar standard.

9.6.4 Anchored Load Tests

Static load tests with anchored reaction systems are the most common of the static load tests. These tests supply the full load to the foundation via a series of tension anchors (or adjacent deep foundations) in conjunction with a beam or truss (Figure 9.27). The beam must resist the load applied to the foundation by transferring it to the reaction anchors which are preferably no closer than five diameters of the foundation (center-to-center spacing). The reaction anchors must not displace significantly while developing the required load. Excessive upward movement from these anchors can alter the stress field surrounding the foundation being tested and decrease the resultant ultimate capacity. Due to the constraints in designing such a reaction system, rarely does an anchored static

FIGURE 9.26
Kentledge load test setup; 400 ton (35.6 MN). (Courtesy of Bermingham Construction, Ltd.)

FIGURE 9.27
Static anchored load test (using eight, H-type reaction piles; 1200 ton [10.68 MN]). (Courtesy of Bermingham Construction, Ltd.)

load test exceed 1500 tons (13.34 MN). However, anchored tests as large as 3500 tons (31.14 MN) are commonplace in some parts of the world. The analysis of static load testing requires no more than plotting the load–displacement response. As every foundation application can have a unique failure criterion, the design engineer must decide at what displacement the foundation capacity should be determined. In some instances this is based on a given fraction of ultimate load. In other cases, it may be based upon the some displacement offset method such as Davisson's method or the FHWA method. With load and resistance factor design type approaches, a fraction of the ultimate capacity is compared to the factored design load in a strength limit state, and displacement is considered separately in a service limit state. Figure 9.28 shows typical static load test results and three common approaches to determining capacity of (a) a maximum permissible displacement usually set by structural sensitivity, (b) displacement offset method, and (c) the load at which additional displacement is obtained without an increase in load.

9.7   Load Testing Using the Osterberg Cell

9.7.1   Bidirectional Static Load Test

The results of conventional static loading tests are limited to producing the load–deformation characteristics for the pile top (Section 9.6). However, designs concerned with settlements of pile foundations, or problems arising from the site conditions and construction procedures, require knowledge of the resistance distribution along the pile or at least the load–deformation characteristics of the pile toe. One way to obtain this information is to instrument the pile at a number of locations with strain gages (Section 6.4).

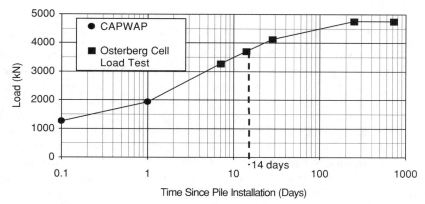

FIGURE 9.34
Illustration of the use of Osterberg cell in measuring pile setup. (From Titi et al., 1999. With permission.)

Example 9.7

(This example is solved in British units. Hence, please refer to Table 7.9 for appropriate conversion to SI units.) Assume that the results shown in Figure 9.33 were obtained during the O-cell test of a 3-ft diameter concrete shaft. If the O-cell is in close proximity to the shaft tip, which is at an elevation of 15 ft, estimate the shaft friction and tip bearing capacities assuming that the ground water table is at a depth of 5 ft.

Since the unit weight and elastic modulus of concrete are 150 lb/ft^3 and 4,000,000 psi, respectively, the total buoyant weight of the shaft = B$(3)^2$[(5)(150) + (10)(150 − 62.4)]/1000 kips = 45.97 kips.

From the upward load–displacement curve in Figure 9.33, the measured ultimate shaft friction = 7600 kips = buoyant weight of the pile + actual shaft friction mobilized in the downward direction (since the shaft is pushed upwards).

Hence, the actual ultimate shaft friction = 7550 kips.

Further, the ultimate tip resistance is greater than 8000 kips as the downward load–displacement curve in Figure 9.33 shows no definitive sign of "peaking out" until the unloading phase starts.

It is also seen that the displacement required for the mobilization of tip bearing is 0.2 in. for this diameter of a shaft. On the other hand, the measured total displacement for the mobilization of ultimate shaft friction is about 1.00 in., which also includes the elastic shortening of the pile. The elastic shortening of the pile can be computed using Equation (9.27)

$$\delta = \frac{1}{E_p A_p} \int_0^L P(z)\, dz \tag{9.27}$$

For the current example, if one assumes a linear distribution of the axial force along the pile = $\frac{1}{2}$(7550)(1000)(15)(1/4,000,000)(1/144)/(B$(3)^2$) × 12 in. = 0.0417 in.

It is seen that the elastic shortening does not contribute significantly to the total deflection.

In contrast to this specific case, generally, it is observed that the displacement needed for full mobilization of shaft friction is relatively small compared to that needed for the full mobilization of end bearing.

9.8 Rapid Load Test (Statnamic Pile Load Test)

The statnamic pile load test combines the advantages of both static and dynamic load tests. It is performed to test a pile's capacity and uses a rapid compressive loading method. The applied load, acceleration, and displacements are measured using load cells, accelerometers, and displacement transducers with a stationary laser reference.

The statnamic device consists of a large mass, combustion chamber, and a catch system of some sort. The force applied to the pile is produced by accelerating a mass upward. This is done by firing a rapid-burning propellant fuel within the combustion chamber, which applies equal force to the mass and to the pile. After the fuel is burned the gas port is opened, this allows the duration of the load pulse to be long enough to keep the pile in compression throughout the test (maintains rigid body). During the loading cycle, which is only a fraction of a second, over 2000 readings are taken of the load and displacement and the data are stored in a data-acquisition unit. The mass is caught as it falls by a gravel catch or mechanical tooth catch before it impacts the pile. The load–displacement curves generated are used to determine the equivalent static force from the measured statnamic force using the unloading point method.

9.8.1 Advantages of Statnamic Test

Statnamic load testing can apply much larger loads than possible with static load testing. The capacity of large-diameter foundations can be fully mobilized without risking damage. A controlled, predetermined load can be applied directly to the pile without introducing high-tension forces. Setting up and dismantling a statnamic test can be done very quickly. Considerable costs are saved since no reaction system is required. The load–displacement curve can be viewed immediately after test on a laptop, which indicates the performance of the test.

9.8.2 Limitations of Statnamic Test

This method is fairly new and the corresponding ASTM standard is still pending. The unloading point method (as well as other methods) used to evaluate the pile capacity is based on numerous idealized assumptions. These tests can be class field tests. A small-scale demonstration was performed by the Dr. Gray Mullins outside University of South Florida's Geotechnical laboratory. A pile in a pressurized cell was loaded to about 10 tons. Only a small amount of fuel (a few pellets) was required to achieve this loading. After the test, a mechanical catch caught the weights before they impacted the pile. This mini system was instrumented with an accelerometer and a load cell. The information was collected in a MagaeDec unit and the data could be viewed and interpreted using the SAW-R4 program. This test seems to be a quick and economical method for pile capacity evaluation. It seems to be advantageous over other methods of pile capacity testing. The SAW-R4 workbook is an excellent tool for regressing the data. Since its inception in 1988, the inertia loading technology called statnamic testing has gained popularity with many designers largely due to its time efficiency, cost effectiveness, data quality, and flexibility in testing existing foundations. Where large-capacity static tests may take up to a week to set up and conduct, the largest of statnamic tests (3500 tons or 31.14 MN) typically takes no more than a few days. Further, multiple smaller-capacity tests (up to 2000 tons or 17.8 MN) can easily be completed within a day. The direct benefit of this time efficiency is the cost savings to the client and the ability to conduct more tests within a given budget.

Additionally, this test method has boosted quality assurance by giving the contractor the ability to test foundations thought to have been compromised by construction difficulties without significantly affecting production and without requiring previous planning for its testing. Statnamic testing is designated as a rapid load test that uses the inertia of a relatively small reaction mass instead of a reaction structure to produce large forces. The duration of the statnamic test is typically 100 to 120 msec, but is dependent on the ratio of the applied force to the weight of the reaction mass. Longer-duration tests of up to 500 msec are possible but require more reaction mass. The statnamic force is produced by quickly formed high-pressure gases that in turn launch a reaction mass upward at up to 20 times the acceleration of gravity. The equal and opposite force exerted on the foundation is simply the product of the mass and acceleration of the reaction mass. It should be noted that the acceleration of the reaction mass is not significant in the analysis of the foundation; it is simply a by-product of the test. Secondly, the load produced is not an impact since the mass is in contact prior to the test. Further, the test is over long before the masses reach the top of their flight. The parameters of interest are only those associated with the movement of the foundation (i.e., force, displacement, and acceleration). Figure 9.35 shows the setup for both an axial compression and lateral statnamic test setup.

9.8.3 Procedure for Analysis of Statnamic Test Results

Typical analysis of statnamic data relies on measured values of force, displacement, and acceleration. A soil model is not required; hence, the results are not highly user dependent. The statnamic forcing event induces foundation motion in a relatively short period of time and hence acceleration and velocities will be present. The accelerations are typically small (1 to 2g), however the enormous mass of the foundation when accelerated resists movement due to inertia and as such the fundamental equation of motion applies,

$$F = ma + cv + kx \tag{9.3}$$

FIGURE 9.35
Axial statnamic test setup (left), lateral statnamic test in progress (right). (Courtesy of Bermingham Construction, Ltd.)

where F is the forcing event, m is the mass of the foundation, a is the acceleration of the displacing body, v is the velocity of the displacing body, c is the viscous damping coefficient, k is the spring constant of the displacing system, and x is the displacement of the body.

The equation of motion is generally described using four terms: forcing, inertial, viscous damping, and stiffness terms. The forcing term (F) denotes the load application that varies with time and is equated to the sum of remaining three terms. The inertial term (ma) is the force that is generated from the tendency of a body to resist motion, or to keep moving once it is set in motion (Young, 1992). The viscous damping term (cv) is best described as the velocity-dependent resistance to movement. The final term (kx) represents the classic system stiffness, which is the static soil resistance.

When this equation is applied to a pile or soil system the terms can be redefined to more accurately describe the system. This is done by including both measured and calculated terms.

The revised equation is displayed below:

$$F_{\text{Statnamic}} = (ma)_{\text{Foundation}} + (cv)_{\text{Foundation}} + F_{\text{Static}} \qquad (9.31)$$

where $F_{\text{Statnamic}}$ is the measured Statnamic force, m is the calculated mass of the foundation, a is the measured acceleration of the foundation, c is the viscous damping coefficient, v is the calculated velocity, and F_{static} is the derived pile or soil static response.

There are two unknowns in the revised equation, F_{static} and c; thus, the equation is underspecified. F_{static} is the desired value, so the variable c must be obtained to solve the equation. Middendorp (1992) presented a method to calculate the damping coefficient referred to as the unloading point method (UP). With the value of c known, the static force can be calculated. This force, termed "derived static," represents an equivalent soil response similar to that produced by a traditional static load test.

9.8.3.1 Unloading Point Method

The UP is a simple method by which the damping coefficient can be determined from the measured statnamic data. It uses a simple single degree of freedom model to represent the foundation–soil system as a rigid body supported by a nonlinear spring and a linear dashpot in parallel (see Figure 9.36). The spring represents the static soil response (F_{Static}), which includes the elastic response of the foundation as well as the foundation–soil interface and surrounding soil response. The dashpot is used to represent the dynamic resistance, which depends on the rate of pile penetration (Nishimura, 1995).

The UP makes two primary assumptions in its determination of "c." The first is the static capacity of the pile is constant when it plunges as a rigid body. The second is that the damping coefficient is constant throughout the test. By doing so a time window is defined in which to calculate the damping coefficient as shown in Figure 9.37. This figure shows a typical statnamic load–displacement curve which denotes points 1 and 2.

The first point of interest (1) is that of maximum statnamic force. At this point the static resistance is assumed to have become steady state for the purpose of calculating "c." Thus, any extra resistance is attributed to that of the dynamic forces (ma and cv). The next point of interest (2) is that of zero velocity, which has been termed the "unloading point." At this point the foundation is no longer moving and the resistance due to damping is zero. The static resistance, used to calculate "c" from (1) to (2), can then be calculated by the following equation:

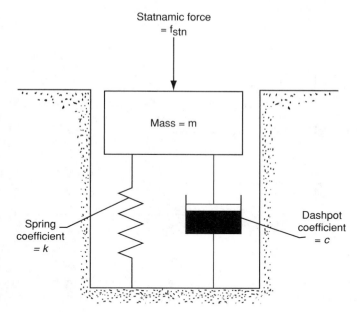

FIGURE 9.36
Single degree of freedom model.

$$F_{\text{Static UP}} = F_{\text{Statnamic}} - (ma)_{\text{Foundation}} \qquad (9.32)$$

where $F_{\text{Statnamic}}$, m, and a are all known parameters; $F_{\text{Static UP}}$ is the static force calculated at (2) and assumed constant from (1) to (2).

Next, the damping coefficient can be calculated throughout this range, from maximum force (1) to zero velocity (2). The following equation is used to calculate c:

$$c = \frac{F_{\text{Statnamic}} - F_{\text{Static UP}} - (ma)_{\text{Foundation}}}{v_{\text{Foundation}}} \qquad (9.33)$$

Damping values over this range should be fairly constant. Often the average value is taken as the damping constant, but if a constant value occurs over a long period of time it should be used (Figure 9.38).

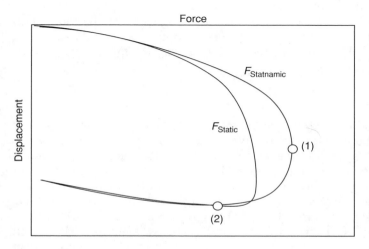

FIGURE 9.37
UP time window for determination of c.

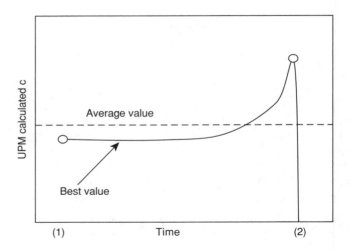

FIGURE 9.38
Variation in c between times (1) and (2).

Note that as v approaches zero at point (2), values of c can be different from that of the most representative value and therefore the entire trend should be reviewed. Finally, the derived static response can be calculated as follows:

$$F_{\text{Static}} = F_{\text{Statnamic}} - (ma)_{\text{Foundation}} - (cv)_{\text{Foundation}} \qquad (9.34)$$

Currently, software is available to the public that can be used in conjunction with statnamic test data to calculate the derived static pile capacity using the UP method (Garbin, 1999). This software was developed by the University of South Florida and the Federal Highway Administration and can be downloaded from www.eng.usf.edu/~gmullins under the Statnamic Analysis Workbook (SAW) heading.

The UP has proven to be a valuable tool in predicting damping values when the foundation acts as a rigid body. However, as the pile length increases an appreciable delay can be introduced between the movement of the pile top and toe, hence negating the rigid body assumption. This occurrence also becomes prevalent when an end bearing condition exists; in this case the lower portion of the foundation is prevented from moving jointly with the top of the foundation.

Middendorp (1995) defines the "wave number" (N_w) to quantify the applicability of the UP. The wave number is calculated by dividing the wave length (D) by the foundation depth (L). D is obtained by multiplying the wave speed c in length per second by the load duration (T) in seconds. Thus, the wave number is calculated by the following equation:

$$N_w = \frac{D}{L} = \frac{cT}{L} \qquad (9.35)$$

Through empirical studies Middendorp determined that the UP would predict accurately the static capacity from statnamic data, if the wave number was greater than 12. Nishimura (1995) established a similar threshold at a wave number of 10. Using wave speeds of 5000 and 4000 m/sec for steel and concrete, respectively, and a typical statnamic load duration, the UP is limited to piles shorter than 50 m (steel) and 40 m (concrete). Wave number analysis can be used to determine if stress waves will develop in the pile. However, this does not necessarily satisfy the rigid body requirement of the UP.

Statnamic tests cannot always produce wave numbers greater than 10, and as such there have been several methods suggested to accommodate stress wave phenomena in statnamically tested long piles (Middendorp, 1995). Due to space limitations these methods are not presented here.

9.8.3.2 Modified Unloading Point Method

Given the limitations of the UP, users of statnamic testing have developed a remedy for the problematic condition that arises most commonly. The scenario involves relatively short piles ($N_w > 10$) that do not exhibit rigid body motion, but rather elastically shorten within the same magnitude as the permanent set. This is typical of rock-socketed drilled shafts or piles driven to dense bearing strata that are not fully mobilized during testing. The consequence is that the top of pile response (i.e., acceleration, velocity, and displacement) is significantly different from that of the toe. The most drastic subset of these test results show zero movement at the toe while the top of pile elastically displaces in excess of the surficial yield limit (e.g., upwards of 25 mm). Whereas with plunging piles (rigid body motion) the difference in movement (top to toe) is minimal and the average acceleration is essentially the same as the top of pile acceleration; tip-restrained piles will exhibit an inertial term that is twice as large when using top of pile movement measurements to represent the entire pile.

The modified unloading point method (MUP), developed by Justason (1997), makes use of an additional toe accelerometer that measures the toe response. The entire pile is still assumed to be a single mass, m, but the acceleration of the mass is now defined by the average of the top and toe movements. A standard UP is then conducted using the applied top of pile statnamic force and the average accelerations and velocities. The derived static force is then plotted versus the top of pile displacement as before. This simple extension of the UP has successfully overcome most problematic data sets. Plunging piles instrumented with both top and toe accelerometers have shown little analytical difference between the UP and the MUP. However, MUP analyses are now recommended whenever both top and toe information is available.

Although the MUP provided a more refined approach to some of the problems associated with UP conditions, there still exists a scenario where it is difficult to interpret statnamic data with present methods. This is when the wave number is less than 10 (relatively long piles). In these cases the pile may still only experience compression (no tension waves) but the delay between top and toe movements causes a phase lag. Hence, an average of top and toe movements does not adequately represent the pile.

9.8.3.3 Segmental Unloading Point Method

The fundamental concept of the segmental unloading point (SUP) method is that the acceleration, velocity, displacement, and force from each segment of a pile can be determined using strain gage measurements along the length of the pile (Mullins et al., 2003). Individual pile segment displacements are determined using the relative displacement as calculated from strain gage measurements and an upper or lower measured displacement. The velocity and acceleration of each segment are then determined by numerically differentiating displacement and then velocity with respect to time. The segmental forces are determined by calculating the difference in force from two strain gage levels.

Typically, the maximum number of segments is dependent on the available number of strain gage layers. However, strain gage placement does not necessitate assignment of segmental boundaries; as long as the wave number of a given segment is greater than 10, the segment can include several strain gage levels within its boundaries. The number and the elevation of strain gage levels are usually determined based on soil stratification; as such, it can be useful to conduct an individual segmental analysis to produce the shear strength parameters for each soil strata. A reasonable upper limit on the number of segments should be adopted because of the large number of mathematical computations required to complete each analysis. Figure 9.39 is a sketch of the SUP pile discretization.

The notation used for the general SUP case defines the pile as having m levels of strain gages and $m + 1$ segments. Strain gage locations are labeled using positive integers starting from 1 and continuing through m. The first gage level below the top of the foundation is denoted as GL^1 where the superscript defines the gage level. Although there are no strain gages at the top of foundation, this elevation is denoted as GL^0. Segments are numbered using positive integers from 1 to $m + 1$, where segment 1 is bounded by the top of foundation (GL^0) and GL^1. Any general segment is denoted as segment n and lies between GL^{n-1} and GL^n. Finally, the bottom segment is denoted as segment $m + 1$ and lies between GL^m and the foundation toe.

9.8.3.4 Calculation of Segmental Motion Parameters

The SUP analysis defines average acceleration, velocity, and displacement traces that are specific to each segment. In doing so, strain measurements from the top and bottom of each segment and a boundary displacement are required. Boundary displacement may come from the statnamic laser reference system (top), top of pile acceleration data, or from embedded toe accelerometer data.

The displacement is calculated at each gage level using the change in recorded strain with respect to an initial time zero using Equation (9.36). Because a linearly varying strain distribution is assumed between gage levels, the average strain is used to calculate the elastic shortening in each segment.

Level displacements

$$x_n = x_{n-1} - \Delta\varepsilon_{\text{average seg n}} L_{\text{seg n}} \tag{9.36}$$

where x_n is the displacement at the nth gage level, $\Delta\varepsilon_{\text{average seg}}$ n is the average change in strain in segment n, and $L_{\text{seg } n}$ is the length of the nth segment.

To perform an unloading point analysis, only the top-of-segment motion needs to be defined. However, the MUP analysis, which is now recommended, requires both top and bottom parameters. The SUP lends itself naturally to providing this information. Therefore, the average segment movement is used rather than the top-of-segment; hence, the SUP actually performs multiple MUP analyses rather than standard UP. The segmental

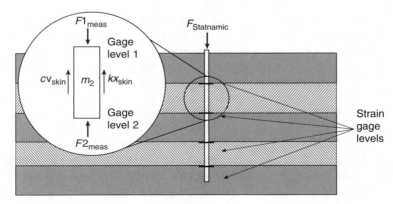

FIGURE 9.39
Segmental free body diagram.

displacement is then determined using the average of the gage level displacements from each end of the segment as shown in the following equation:

$$x_{\text{seg } n} = \frac{x_{n-1} + x_n}{2} \tag{9.37}$$

where $x_{\text{seg } n}$ is the average displacement consistent with that of the segment centroid.

The velocity and acceleration, as required for MUP, are then determined from the average displacement trace through numerical differentiation using Equations (9.38) and (9.39), respectively:

$$a_n = \frac{v_{n_t} - v_{n_{t+1}}}{\Delta t} \tag{9.38}$$

$$v_n = \frac{x_{n_t} - x_{n_{t+1}}}{\Delta t} \tag{9.39}$$

where v_n is the velocity of segment n, a_n is the acceleration of segment n, and Δt is the time step from time t to $t + 1$.

It should be noted that all measured values of laser displacement, strain, and force are time-dependent parameters that are field recorded using high-speed data-acquisition computers. Hence the time step, Δt, used to calculate velocity and acceleration is a uniform value that can be as small as 0.0002 sec. Therefore, some consideration should be given when selecting the time step to be used for numerical differentiation.

The average motion parameters (x, v, and a) for segment $m + 1$ cannot be ascertained from measured data, but the displacement at GL^m can be differentiated directly providing the velocity and acceleration. Therefore, the toe segment is evaluated using the standard UP. These segments typically are extremely short (1 to 2 m) producing little to no differential movement along its length.

9.8.3.5 Segmental Statnamic and Derived Static Forces

Each segment in the shaft is subjected to a forcing event that causes movement and reaction forces. This segmental force is calculated by subtracting the force at the top of the segment from the force at the bottom. The difference is due to side friction, inertia, and damping for all segments except the bottom segment. This segment has only one forcing function from GL^m and the side friction is coupled with the tip bearing component. The force on segment n is defined as

$$S_n = A_{(n-1)}E_{(n-1)}\varepsilon_{(n-1)} - A_n E_n \varepsilon_n \tag{9.40}$$

where S_n is the applied segment force from strain measurements, E_n is the composite elastic modulus at level n, A_n is the cross-sectional area at level n, ε_n is the measured strain at level n.

Once the motion and forces are defined along the length of the pile, an unloading point analysis on each segment is conducted. The segment force defined above is now used in place of the statnamic force in Equation (9.31). Equation (9.41) redefines the fundamental equation of motion for a segment analysis:

$$S_n = m_n a_n + c_n v_n + S_{n\,\text{Static}} \tag{9.41}$$

where $S_{n\,\text{Static}}$ is the derived static response of segment n, m_n is the calculated mass of segment n, and c_n is the damping constant of segment n.

The damping constant (in Equation (9.42)) and the derived static response (Equation (9.43)) of the segment are computed consistent with standard UP analyses:

$$c_n = \frac{S_n - S_{n\,Static}}{v_n} \tag{9.42}$$

$$S_{n\,Static} = S_n - m_n a_n - c_n v_n \tag{9.43}$$

Finally the top-of-foundation derived static response can be calculated by summing the derived static response of the individual segments as displayed in the following equation:

$$F_{Static} = \sum_{n=1}^{m+1} S_{n\,Static} \tag{9.44}$$

Software capable of performing SUP analyses (SUPERSAW) has been developed at the University of South Florida in cooperation with the Federal Highway Administration (Winters, 2002). It can be downloaded from www.eng.usf.edu/~gmullins under the Statnamic Analysis Software heading.

Example 9.8
Figure 9.40 contains data from a statnamic rapid load test on a precast concrete pile with mass of 9111 kg. Typical measured test values include acceleration, displacement, and applied load. The velocity shown can be calculated by numerically integrating the acceleration trace or by differentiating the displacement trace (procedure not shown herein). The unloading point method is applied to obtain the unknown damping coefficient, C, as discussed previously using the values marked from point (2) in Figure 9.40.

At the unloading point (Point [2] where $V = 0$), the equation of motion can be solved for F_{Static}.

$$F_{Static(2)} = F_{STN(2)} - ma_{(2)} = -2950 \text{ kN} - (9111 \text{ kg})(243 \text{ m/sec}^2) = -5164 \text{ kN}$$

Using that value of F_{Static}, the damping coefficient, C, can be determined for all times between points (1) and (2), maximum load and unloading point, respectively

$$C_i = (F_{STNi} - ma_i - F_{Static(2)})/V_i$$

This gives a range of values between points (1) and (2) as shown in Figure 9.41.

A median value is then selected from these values and used to determine the derived static capacity for the entire test duration (Figure 9.42).

This information is far more pertinent when expressed as a function of displacement to assure service limits are not being exceeded at a particular load (Figure 9.43).

9.9 Lateral Load Testing of Piles

The standard method of testing piles under lateral loads is found in ASTM Designation D 3966 (Standard method of Testing piles Under Lateral Loads). Typically, piles are tested up to 200% of the design lateral load with load increments of 12.5% of the test load for standard loading schedule (or 25% of the test load for cyclic loading schedule) for a loading duration of 30 min. Although ASTM standard emphasizes the determination of the lateral capacity of a pile, the routine practice is to evaluate the response of the pile to lateral loads in terms of lateral pressure versus lateral deflection (p–y) behavior (Figure 9.44).

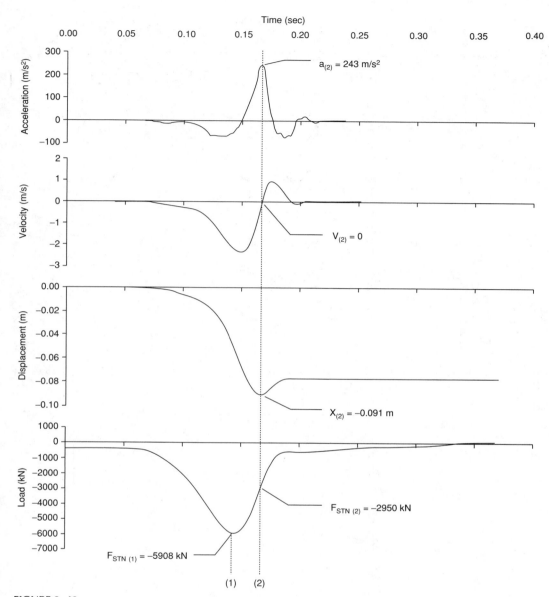

FIGURE 9. 40
Raw data from statnamic rapid load test.

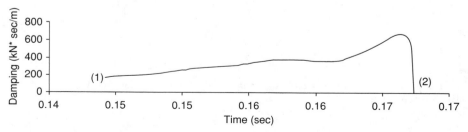

FIGURE 9.41
Damping coefficient calculated between points (1) and (2).

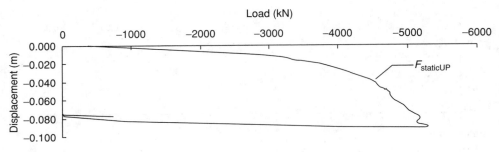

FIGURE 9.42
Derived static capacity expressed as a function of displacement.

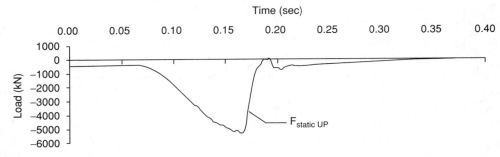

FIGURE 9.43
Derived static capacity as a function of time.

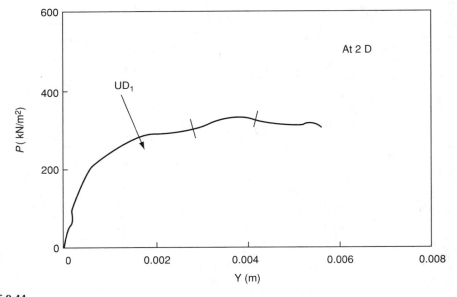

FIGURE 9.44
Typical *P–Y* curves for a laterally loaded pile at different depths. (From Hameed, 1998. University of South Florida. With permission.)

9.10 Finite Element Modeling of Pile Load Tests

As mentioned in Section 1.7, powerful numerical simulation tools such as the finite element method can be used to know more of the behavior of foundations under complex loading and geometric conditions, which would be extremely difficult to perceive under model or actual field experimental conditions. In this regard, engineers have been successful in modeling the behavior of pile foundations as well using the finite element method.

Titi and Wathugala (1999) presented a fully rational approach where the complete life history of the pile: (1) pile installation, (2) subsequent consolidation, and (3) axial loading is simulated using a two-dimensional finite element procedure based on the fully coupled formulation (extended Biot's) for porous media. The aim of this study was to predict the variation over time of the pile capacity at different degrees of consolidation after installation. The reader is referred to Section 1.7 for the technical details of the analytical concepts used in finite element modeling.

Titi et al. (1999) used the coupled theory of nonlinear porous media to determine the effective stresses and pore water pressures in the surrounding soil at the end of pile installation. Some of the basic concepts of flow in porous media are discussed briefly in Chapter 13. The variables obtained from this step simultaneously satisfy the equilibrium equations, strain compatibility, constitutive equations, and boundary conditions. The soil was assumed to remain under undrained conditions during the analysis involved in this step.

Pile load tests are simulated in Titi et al. (1999) by applying an incremental displacement at the pile–soil interface nodes for the pile segment models used in the verification. For the piles used in the numerical experiments, an incremental load or displacement is applied to the pile head until failure. The failure load for each pile load test represents the pile capacity corresponding to the degree of consolidation at which the load test is simulated. The stress–strain relationships used by Titi et al. (1999) were based on the nonassociative anisotropic HiSS-δ_2^* model (Wathugala et al., 1994), which characterizes soil behavior at the pile–soil interface as well as at the far field. For comparison, the reader is referred to Section 1.8.1.3, where an alternative but simpler stress–strain relationship of modified Cam-clay model is described which is based on the assumptions of isotropy and an associative flow rule.

Titi et al. (1999) used the coupled theory of nonlinear porous media through the general-purpose finite element program ABAQUS (HKS, Inc., 1995) to simulate the subsequent consolidation phase and the pile load tests. This formulation allows for (1) advanced constitutive models to characterize the deformation of the soil skeleton due to effective stresses, (2) Darcy's law to govern the movement of water through the porous medium, (3) linear elastic material model for the deformation of soil solids and water. In this respect, the reader would be able to visualize this analytical formulation by comparing it with Equation (1.38) of Section 1.7.3.

Titi et al. (1999) also compared the finite element model predictions with field measurements using pile segment models and pile load tests. The results of these comparisons, shown in Figure 9.45–Figure 9.47, seem to authenticate the innovative pile load test modeling techniques developed by Titi et al. (1999).

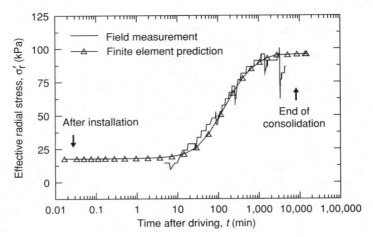

FIGURE 9.45
Comparison of measured and predicted radial effective stress with time. (From Titi et al., 1999. With permission.)

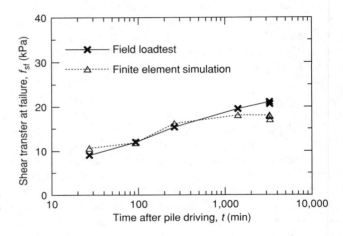

FIGURE 9.46
Comparison of measured and predicted pile setup. (From Titi et al., 1999. With permission.)

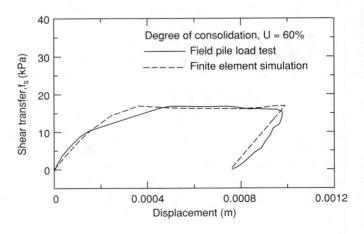

FIGURE 9.47
Comparison of measured and predicted response in pile load test #3. (From Titi et al., 1999. With permission.)

9.11 Quality Assurance Test Methods

The construction of a foundation is plagued with unknowns associated with the integrity of the as-built structure. This is particularly problematic with deep foundations that are installed without visual certainty of the actual conditions or configuration. This section will discuss several methods used to raise the confidence of the design with regard to concrete quality or capacity verification.

9.11.1 Pile Integrity Tester

The pile integrity tester (PIT) (Figure 9.48) is less sophisticated and informative than the PDA (Section 9.4) in that the required instrumentation only consists of a sensitive accelerometer and the amount of information obtained is also limited. In this nondestructive test, the accelerometer is attached to the top of the pile to be tested and a low strain hammer impact is imparted on the pile (Figure 9.49). The velocity records of the low strain compressive waves generated by the impact and their reflection from the pile toe or any other discontinuities are conditioned, processed, and finally graphically displayed.

However, the interpretation of PIT results is similar to that employed in pile integrity testing using the PDA (Section 9.4). When the pile is undamaged throughout its entire length, the compressive pulse induced by the hammer blow is reflected back by the toe resistance at a time t_{TR} equal to

$$t_{TR} = 2L/c \tag{9.45}$$

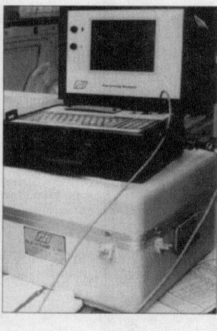

FIGURE 9.48
Dynamic testing of both the new and the existing timber piles conducted with PDA. (Courtesy of Pile Dynamic Inc.)

FIGURE 9.49
Pile integrity testing. (Courtesy of Pile Dynamic Inc.)

where L is the length of the pile and c is the velocity of compression waves in the pile material.

Similarly, reflected pulses also return to the pile top due to the soil resistance on the pile shaft, reduction in pile cross section due to damage, and change in the material characteristics (downgrading of the quality of concrete). Since it is known that the returning pulses due to shaft resistance and those due to cross-sectional reductions are of opposite signs (compression and tension, respectively), a tension pulse with an early return time, i.e., $t < t_{TR}$ indicates damage at a distance given by either one of the following expressions:

$$s = \frac{t}{t_{TR}} L \qquad (9.46)$$

$$s = \frac{tc}{2} \qquad (9.47)$$

9.11.1.1 Limitations of PIT

The following limitations affect the use of PIT in damage testing of piles:

1. Because of the attenuation of compression waves by skin friction, pile toe reflections can be generally identified only when the embedment length is less than 30 pile diameters.
2. In piles and caissons with highly varying cross sections, it is difficult to distinguish between pile defects and construction anomalies.
3. Mechanical splices would generally appear as gaps.

These limitations can be overcome in the case of a pile group where truly damaged piles can be distinguished based on their abnormal response to pile integrity testing with respect to the group.

9.11.2 Shaft Integrity Test

The shaft or pile integrity test (SIT) is an impact echo test that uses the reflections of anomalous cross-sectional shaft or pile dimensions to determine the quality of a drilled

FIGURE 9.50
Equipment used for sonic echo test (left), impact hammer struck on shaft head (right). (Courtesy of Applied Foundation Testing, Inc.)

shaft, auger-cast-*in situ*, or driven pile. The reflected sound waves from within the concrete are plotted as a function of arrival times which can then be correlated to the depth from which the reflection emanated. Figure 9.50 and Figure 9.51 show the equipment used to conduct the test as well as the output results.

FIGURE 9.51
Sonic echoes from three consecutive hammer impacts. (Courtesy of Applied Foundation Testing, Inc.)

This test is well suited for determining the depth of the foundation as well as the depth to anomalous features. However, it cannot determine the magnitude of anomalous features, as it requires access to the pile top to minimize confounding signals, and it is generally limited to depths on the order of 50 times the pile diameter.

9.11.3 Shaft Inspection Device

The inspection device (Figure 9.52) is a visual inspection system for evaluating bottom cleanliness of drilled shaft excavations. A special video camera contained in a weighted, trapped-air bell housing is lowered into the shaft excavation prior to concreting to record the condition of the bottom. This is particularly helpful in slurry excavations where quality assurance is difficult to maintain. The bell housing is outfitted with gages in clear sight of the video camera that are capable of registering the thickness of accumulated debris or sediment at the shaft excavation. The system is capable of testing shafts with depths in excess of 200 ft (61 m). Several generations of this device exist that range in size from less than a foot in diameter to over 3 ft (0.9 m) in diameter. The inspection is viewed in real time on a color video monitor and recorded on a standard VHS tape. Voice annotations are recorded simultaneously during the inspection process similar to standard camcorders.

FIGURE 9.52
Miniature shaft inspection device. (Courtesy of Applied Foundation Testing, Inc.)

9.11.4 Crosshole Sonic Logging

Crosshole sonic logging is a geophysical test method used to determine the compression wave velocity between two parallel, water-filled tubes or slurry filled boreholes. By using two geophones (one emitting and one receiving) the sound wave arrival times can be logged at various depths within the tubes. From this information the *in situ* properties of the materials between the tubes can be inferred, thus identifying various strata. More recently, this test has become a nondestructive method for evaluating the quality of newly placed drilled shaft concrete. Therein, the arrival times are measured between logging tubes attached peripherally to the reinforcing cage allowing concrete quality between the tubes to be assessed. As only the concrete in a direct line between the tubes can be tested, multiple access tubes can be installed. Typically, one tube for every foot of diameter is required to satisfactorily survey a representative portion of the shaft concrete. Data are viewed in the field on a special data-acquisition system (Figure 9.53).

9.11.5 Postgrout Test

The postgrout test is a by-product of an end bearing enhancement technique used during the construction of drilled shafts. This test is relatively simple in concept yet confirms the performance of every grouted shaft up to a lower limit of shaft capacity. During the process of tip grouting, the upward displacement, grout pressure, and grout volume are recorded. This information provides the design engineer the response of the shaft to loading. Therein, the side shear and the end bearing of the shaft are verified up to the level of the applied grout pressure. The product of the grout pressure and tip area produces the tip load; this preloading is afforded by an equivalent reaction from the side shear component. Therefore, the proven capacity of the shaft is established as twice the tip load. The upper limit of capacity can be shown to be on the order of two to three times the proven capacity when verified by downward load testing. The design of

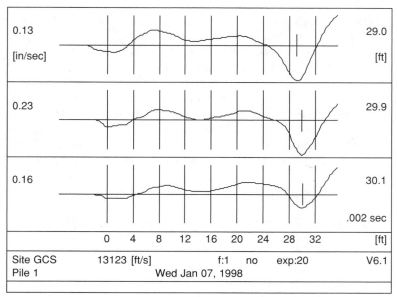

FIGURE 9.53
Crosshole sonic logging of 4 ft diameter shaft. (Courtesy of Applied Foundation Testing, Inc.)

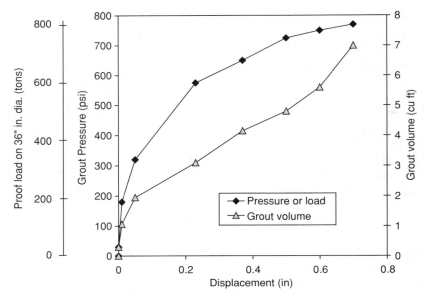

FIGURE 9.54
Field data used to confirm shaft performance.

postgrouted shafts is discussed in Chapter 7. Figure 9.54 shows the standard field data obtained from every grouted shaft. Figure 9.55 shows the performance for each of 76 shafts grouted on a bridge project in West Palm Beach, Florida. (Unit conversion: 1 ton = 8.9 kN, 1 in = 2.54 mm, 1 cu.ft = 0.0283 m³.)

9.11.6 Impulse Response Method

In this relatively novel technique, a low-frequency compression wave is generated at the top of a pile or a drilled shaft by a hammer impact and the reflected wave is recorded at the top (Gassman, 1997). Subsequent analysis of the frequency content of the reflected response can identify changes in the impedance of the deep foundation due to structural and material anomalies.

The velocity and force records of the reflected pulse are analyzed using a fast Fourier transform. The resulting velocity spectrum divided by the force spectrum is defined as the mobility. The average mobility N_c is defined as the geometric mean of the resonant peaks identified in the mobility curve. Therefore, if P and Q are the local maximum and minimum resonant peaks respectively, N_c can be expressed as (Figure 9.56):

$$N_c = \sqrt{PQ} \qquad (9.48)$$

On the other hand, the theoretical mobility is defined as (Stain, 1982)

$$N_T = \frac{1}{\rho V A_p} \qquad (9.49)$$

where ρ is the density of the pile material, V is the compression wave velocity in the pile material, and A_p is the cross-sectional area of the pile.

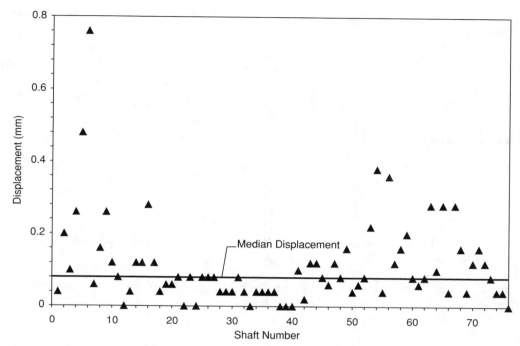

FIGURE 9.55

The displacement observed for every shaft on a project at design pressure.

If $N_c \geq N_T$, a defect likely exists due to an unexpectedly smaller cross section or subquality material within the pile (low ρ or low V).

In addition, if the frequency change between peaks (Δf) is measured from the mobility curve (Figure 9.56), the distance from the pile top (location of the monitoring device) to

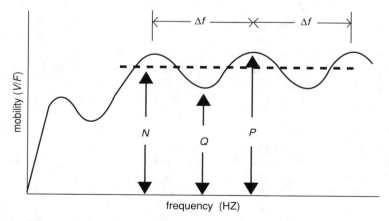

FIGURE 9.56

Typical mobility curve. (From Baxter, S.C., Islam, M.O., and Gassman, S.L., 2004, *Canadian Journal of Civil Engineering*. With permission.)

the source of reflection (pile–soil interface or a structural defect) can be determined from the following expression:

$$s = \frac{V}{2\Delta f} \tag{9.50}$$

9.12   Methods of Repairing Pile Foundations

Pile or shaft foundations could lose their functionality due to two main reasons:

1. The pile or the shaft can lose its structural integrity
2. The ground (or soil) support is inadequate causing excessive settlement problems

In the case of structural damage the pile can be repaired by a variety of methods of which one popular technique is illustrated below.

9.12.1   Pile Jacket Repairs

Reinforced concrete pilings located in or near sea water are prone to corrosion of the steel reinforcement. The most severe corrosion rate occurs in the splash zone, which is located immediately above the sea level and hence is subjected to alternating wet and dry cycles. Above the splash zone and toward the pile cap moderate corrosion rates can be expected. On the other hand, low corrosion rates are encountered in the submerged zone. Concrete pilings in hot tropical marine environments are especially disposed to deterioration as corrosion rates are greatly influenced by humidity, temperature, and resistivity. It is generally found that in a majority of pile damage cases due to corrosion the damage is located above the low tide level and extends upwards to include the splash zone. One popular and effective remedy for this deficiency is pile jacket repairs.

Pile jacketing is a repair technique that usually consists of a stay-in-place form (Figure 9.57), which is filled with a cementitious or polymer material. Preparations for the repair consist of the removal of deteriorated concrete, the cleaning of the steel and the bonding interface, and the installation of the form. A seal is provided at the bottom of the form. Water within the form is either pumped out or displaced by the placement of the grout deposited in the bottom of the form. A popular form type is a two-part fiberglass form that is placed around a damaged area and sealed along the connecting seams. Zippered nylon and 55 gallon drums have also been used as pile jacket forms.

Strength considerations for concrete pile repairs are generally secondary to serviceability issues. Perhaps the most pronounced concern on the strength side of a corroded pile involves lateral loading as in a vessel impact scenario. Repairs performed on scaled models are generally seen to restore a significant portion of the lateral capacity lost to the effects of corrosion. Under axial loading, repaired piles are seen to behave compositely until the bond is compromised. Attempts to improve this bond with powder actuated nails have proved to be futile, presumably due to damage induced on the parent concrete. The use of epoxied dowel bars viewed as less invasive have enhanced the

Goble, G.G. and Likins, G.E., 1996, On the Application of PDA Dynamic Pile Testing, STRESSWAVE '96 Conference: Orlando, FL, 263–273 pp.

Goble, G.G., Rausche, F., and Likins, G.E., 1988, The analysis of pile driving – A state-of-the-art, Third International Conference on the Application of Stress-Wave Theory to Piles: Ottawa, Canada; 131–161 pp.

Goble, G.G., Rausche, F., and Moses, F., 1970, Dynamic Studies on the Bearing Capacity of Piles – Phase III, Final report to the Ohio Department of Highways, Case Western Reserve University, Cleveland, Ohio, August.

Goble, G.G. and Raushe, F., 1986, *Wave Equation Analysis of Pile Driving — WEAP86 Program*, Vols. I–IV, U.S. Department of Transportation, Federal Highway Administration, Implementation Division, McLean, VA.

Goodman, R.E., 1989, *Rock Mechanics*, 2nd edn, John Wiley, New York.

Hameed, R.A., 1998, Lateral Load Behavior of Jetted and Preformed Piles, Ph.D. dissertation, University of South Florida, USA.

Hameed, R.A., Gunaratne, M., Putcha, S., Kuo, C., and Johnson, S., 2000, Lateral load behavior of jetted piles, *ASTM Geotechnical Testing Journal*.

Hartt, W. and Rapa, M., 1998, Condition Assessment of Jackets Upon Pilings for Florida Bridge Substructures, Final Report to FDOT, April.

HKS, Inc., 1995, *ABAQUS/Standard User's Manual*, Vols. I–II, Version 5.4, Hibbitt, Karlsson & Sorensen, Inc., Pawtucket, RI.

Janes, M.C., Justason, M.D., and Brown, D.A., 2000, Long period dynamic load testing ASTM standard draft, Proceedings of the Second International Statnamic Seminar, Tokyo, October, 1998, pp. 199–218.

Justason, M.D., 1997, Report of Load Testing at the Taipei Municipal Incinerator Expansion Project, Taipei City, Taipei.

Kusakabe O., Kuwabara F, and Matsumoto T, 2000, Statnamic load test, Draft of "Method for Rapid Load Test of Single Piles (JGS 1815–2000)," Proceedings of the Second International Statnamic Seminar, Tokyo, October 1998, pp. 237–242.

Middendorp, P. and Bielefeld, M.W., 1995, Statnamic load testing and the influence of stress wave phenomena, Proceedings of the First International Statnamic Seminar, Vancouver, Canada, pp. 207–220.

Middendorp, P., Bermingham, P., and Kuiper, B., 1992, Statnamic load testing of foundation pile, Proceedings of the 4th International Conference on Application of Stress-Wave Theory to Piles, The Hague, pp. 581–588.

Mullins, G., Lewis, C., and Justason, M., 2002, Advancements in statnamic data regression techniques, *Deep Foundations 2002: An International Perspective on Theory, Design, Construction, and Performance*, ASCE Geo Institute, GSP No. 116, Vol. II, pp. 915–930.

Mullins, G., Fischer, J., Sen, R., and Issa, M., 2003, Improving interface bond in pile repair, In: *System-Based Vision for Strategic and Creative Design*, Franco Bontempi, ed., Proceedings of the Second International Conference on Structural and Construction Engineering, September 23–26, Rome, Italy, Vol. 2, pp. 1467–1472, AA. Balkema Publishers, Lisse, The Netherlands.

Murchison, J.M. and O'Neill, M.W., 1984, Evaluation of *P–Y* relationship in cohesionless soils, In: *Analysis and Design of Pile Foundation*, ASCE, New York, pp. 174–191.

Nishimura, S. and Matsumoto, T., 1995, Wave propagation analysis during statnamic loading of a steel pipe pile, Proceedings of the First International Statnamic Seminar, Vancouver, Canada, September.

Osterberg, J.O., 1994, *Recent Advances in Load Testing Driven Piles and Drilled Shafts using the Osterberg Load Cell Method*. Geotechnical Lecture Series, Geotechnical Division of the Illinois Section, ASCE, Chicago, IL.

Osterberg, J.O, 1998, The Osterberg load test method for bored and driven piles, the first ten years, Proceedings of the Seventeenth International Conference on Piling and Deep Foundations, Vienna, Austria, June.

Poulos, H.G. and Davis, E.H., 1980, *Pile Foundation Analysis and Design*, John Wiley, New York.

Reese, L.C, Cox, W.R., and Koop, F.D., 1974, Analysis of laterally loaded piles in sand, Proceedings of the 6th Offshore Technology Conference, Houston, Texas, paper OTC 2080, pp. 473–483.

Rausche, F., Goble, G.G., and Likins, G., 1985, Dynamic determination of pile capacity, *Journal of the Geotechnical Division, ASCE*. 111(3): 367–383.

Rausche, F., Moses, F., and Goble, G.G., 1979, Soil resistance predictions from pile dynamics, *Journal of the Soil Mechanics and Foundation Division, ASCE*, 98(SM9): 917–937.

Smith, A.E.L., 1960, Pile driving analysis by wave equation, *Journal of SMFD, ASCE*, 86(4): 35–61.

Stain, R.T., 1982, *Integrity Testing, Civil Engineering*, London, April and May, pp. 55–59 and 71–73.

Tsinker, G.P., 1988, Pile jetting, *Journal of Geotechnical Engineering, ASCE*, 114(3): 326–334.

Thilakasiri, H.S., Abeyasinghe, R.K., and Tennekoon, B.L., 2002, A study of ultimate carrying capacity estimation of driven piles using pile driving equations and the wave equation method, Annual Transactions of the Institution of Engineers, Sri Lanka.

Thilakasiri, H.S., Abeysinghe, R.M., and Tennakoon, B.L., 2003, A study of static capacity estimation methods for bored piles on residual soils, Presented at the Annual Sessions of the Sri Lanka Association for Advancement of Science (SLAAS).

Titi, H.H. and Wathugala, G.W., 1999, Numerical Procedure for Predicting Pile Capacity — Setup/ Freeze, *Journal of the Transportation Research Board*, TRR 1663, Transportation Research Board, National Research Council, National Academy Press, Washington, DC, pp. 25–32.

Winters, D., 2002, SUPERSAW Statnamic Analysis Software, Master's Thesis, University of South Florida, Tampa, FL, May.

Wathugala, G.W., Desai, C.S., and Matlock, H., 1994, Field verification of pile load tests by numerical methods with a general constitutive model, Proceedings of the XIII International Conference on Soil Mechanics and Foundation Engineering, Vol. 1, New Delhi, India, January, pp. 433–436.

Young, H.D., 1992, *University Physics*, 8ht edn, Addison Wesley, Reading, MA.http://www.handygeotech.com/main.html

Websites

http://www.saberpiering.com/
http://www.e-foundationrepairs.com/underpinning.html
http://www.judycompany.com/underpinning.htm

10

Retaining Walls: Analysis and Design

Alaa Ashmawy

CONTENTS

10.1 Introduction

Retaining walls are soil-structure systems intended to support earth backfills. Construction of retaining walls is typically motivated by the need to eliminate slopes in road widening projects, to support bridges and similar overpass and underpass elements, or to provide a level ground for shallow foundations. Retaining walls belong to a broader class of civil engineering structures, *earth retaining structures*, which also encompass temporary support elements such as sheet-pile walls, concrete slurry walls, and soil nails. Evidence

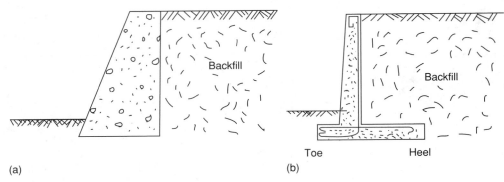

(a)   (b)

FIGURE 10.1
Conventional types of retaining walls: (a) gravity, (b) cantilever.

of stone blocks and rockfill retaining walls is found at archeological sites around the world. The western (wailing) wall in Jerusalem was built by King Herod as a retaining wall for the city, and the hanging gardens of Babylon are believed to have been stepped terraces supported by brick and stone walls.

Retaining walls have traditionally been constructed with plain or reinforced concrete, with the purpose of sustaining the soil pressure arising from the backfill. From an analysis and design standpoint, classical references categorize such walls into two types: gravity walls and cantilever walls (Figure 10.1). The basic difference lies in the mechanisms and forces contributing to the wall stability; gravity walls rely on their own weight to provide static equilibrium while cantilever walls derive a portion of their stabilizing forces and moments from the backfill soil above the heel. From a construction standpoint, gravity walls are typically made of plain (unreinforced) concrete or stone blocks, whereas cantilever walls require the use of steel reinforcement to resist the large moments and shear stresses.

With the advent of reinforced earth technologies in the 1960s and geosynthetic materials in the 1980s, gravity and cantilever walls are becoming largely obsolete. New technologies such as mechanically stabilized earth (MSE) walls and soil nailing are becoming increasingly popular due to their high efficiency, adaptability, and low cost. Figure 10.2 shows typical cross sections in such earth retaining structures. Where larger and deeper excavations are needed, sheet piles and tie-back anchored walls (Figure 10.3) are the structures of choice. Although such integrated soil-inclusion systems have only begun to be used in conventional civil engineering projects in recent years, the concept of soil

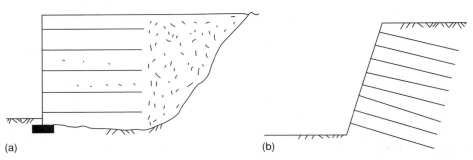

(a)   (b)

FIGURE 10.2
Cross section in (a) mechanically stabilized earth (MSE) wall and (b) soil-nailed wall.

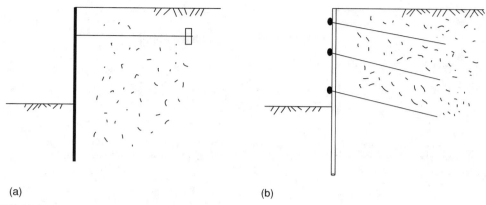

FIGURE 10.3
Alternative earth retaining systems: (a) sheet pile with a tie-back anchor, (b) tie-back anchored retaining wall.

reinforcement has surprisingly been around for thousands of years. The "ziggurats" built by the Babylonians, in what is modern day Iraq, were constructed from an outer wall of fired brick, with the inside filled with clay reinforced with cedar beams. Despite the low durability of the bricks and clay, archeological remains of many of the ziggurats are still in existence today, which reflects the high strength and durability of such systems.

The migration to MSE systems was initiated by the introduction of Reinforced Earth®, a proprietary technology that relies on reinforcing the backfill with galvanized steel strips. Since then, a broad range of similar technologies have emerged, relying on the same reinforcement mechanisms while utilizing other types of materials. The basic idea is to reinforce the soil with horizontal inclusions that extend back into the earth fill to form a monolithic mass that acts as a self-contained earth support system. Today, reinforcement elements include products ranging from natural fibers (e.g., coir and bamboo) to geosynthetics (e.g., geogrids and geotextiles). With progress made over the past decades in polymer science and engineering, new species of polymers have become available that exhibit relatively high strength and modulus, and excellent durability. As a result, MSE walls, specifically those reinforced with geosynthetics, have become increasingly popular in transportation and geotechnical earthworks such as slope stabilization, highway expansion, and, more recently, bridge abutments. Such bridge abutments can support higher surcharges, and loads concentrated near the facing of the wall.

10.2 Lateral Earth Pressure

In order to design earth retaining structures, it is necessary to have a thorough understanding of lateral earth pressure concepts and theory. Although a comprehensive review of lateral earth pressure theories is beyond the scope of this chapter, we will present an overview of the classical and commonly accepted theories. Because soils possess shear strength, the magnitude of stress acting at a point may be different depending on the direction. For instance, the horizontal pressure at a point within a soil mass is typically different from the vertical pressure. This is unlike fluids, where the pressure at a point is independent of direction (Figure 10.4). The ratio between horizontal effective stress, σ'_h, and the vertical effective stress, σ'_v is known as the coefficient of lateral earth pressure, K.

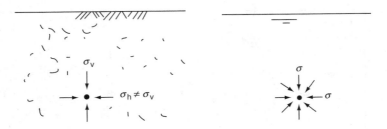

FIGURE 10.4
Illustration of the concept of lateral earth pressure. The diagram to the left shows a difference between vertical and horizontal earth pressures ($\sigma_v \neq \sigma_h$). The diagram to the right illustrates an equal fluid pressure in all directions.

$$K = \frac{\sigma'_h}{\sigma'_v} \tag{10.1}$$

Typically, vertical stresses in a soil mass can be reliably calculated by multiplying the unit weight of the soil by the depth. In contrast, the horizontal stresses cannot be accurately predicted. The magnitude of the coefficient of lateral earth pressure depends not only on the soil physical properties, but also on construction or deposition processes, stress history, and time among others. For a given vertical stress value, the ability of a soil to resist shear stresses results in a range of possible horizontal stresses (range of K values) where the soil remains stable. From a retaining earth structures design perspective, two limits or conditions exist where the soil fails: *active* and *passive*. The corresponding coefficients of lateral earth pressure are denoted K_a and K_p, respectively. Under "natural" *in situ* conditions, the actual value of the lateral earth pressure coefficient is known as the coefficient of lateral earth pressure at rest, K_0.

According to Rankine's (1857) theory, an active lateral earth pressure condition occurs when the horizontal stress, σ'_h, decreases to the minimum possible value required for soil stability. In contrast, a passive condition takes place when σ'_h increases to a point where the soil fails due to excessive lateral compression. Figure 10.5 shows practical situations where active and passive failures may occur. To further illustrate the relationship between the coefficient of lateral earth pressure and the soil's shear strength, we consider the retaining wall shown in Figure 10.6. Assuming the friction between the soil and the

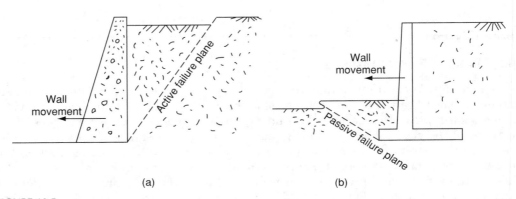

(a)   (b)

FIGURE 10.5
Idealized lateral earth pressure conditions leading to failure due to (a) active and (b) passive earth pressure.

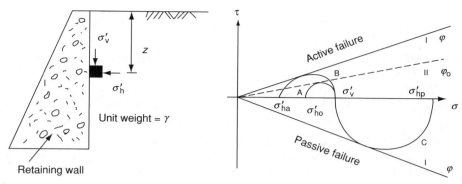

FIGURE 10.6
Schematic illustration of the relationship between lateral earth pressure and shear strength.

wall to be negligible, the vertical effective stress, σ'_v, at a depth z behind the wall is equal to γz. It follows that the horizontal effective stress is equal to $K\sigma'_v$ (Equation 10.1). Under at-rest conditions, the soil is far from failure, and the stress condition is represented in Mohr's stress space by circle A. Here, the coefficient of earth pressure at rest, K_0, is equal to the ratio between σ'_{h0} and σ'_v. Next, we assume that the wall "deforms" or moves away from the backfill, thereby gradually reducing the horizontal pressure. Throughout this process, the vertical pressure remains constant ($\sigma'_v = \gamma z$) since no changes are made in vertical loading conditions. The horizontal stress may be reduced up to the point where the stress conditions correspond to circle B in Mohr space. At this point, the soil will have failed under active conditions. The corresponding coefficient of lateral earth pressure, K_a, is related to the soil's angle of internal friction, φ, through the following equation:

$$K_a = \frac{1 - \sin\varphi}{1 + \sin\varphi} = \tan^2\left(45° - \frac{\varphi}{2}\right) \tag{10.2}$$

The angle of the shear plane with respect to horizontal is $(45° + \varphi/2)$, measured from the heel of the wall.

Now, let us consider the opposite scenario where, starting from at-rest conditions, the wall moves toward the backfill. While the vertical stress remains constant, the horizontal stress will gradually increase, until it reaches a value of σ'_{hp} at which the soil fails under passive conditions. The corresponding stresses are represented by Mohr circle C, and the coefficient of lateral earth pressure, K_p, is equal to the inverse of K_a:

$$K_p = \frac{1 + \sin\varphi}{1 - \sin\varphi} = \tan^2\left(45° + \frac{\varphi}{2}\right) \tag{10.3}$$

In this case, the angle of the shearing plane measured from the heel with respect to horizontal is $(45° - \varphi/2)$.

The illustrative example given above is a very powerful tool in understanding the concepts of lateral earth pressure. In conjunction, a number of important observations are noted:

1. The mobilized angle of internal friction at rest, φ_0, is related to the *in situ* horizontal and vertical stresses, and thus is a function of the coefficient of earth pressure at rest:

$$\varphi_0 = \sin^{-1}\left(\frac{1-K_0}{1+K_0}\right) = 2\left[45° - \tan^{-1}\sqrt{K_0}\right] \tag{10.4}$$

2. Although the soil remains within the failure limits between active and passive conditions, deformation does occur in conjunction with any changes in loading conditions.
3. Because active failure is reached through a "shorter" stress path compared to a passive condition, smaller deformations are associated with active failure.
4. When transitioning from active to passive and vice versa, a $K = 1$ condition must occur where the horizontal and vertical stresses are equal, and Mohr circle collapses into a point. At that instance, the soil is at its most stable condition.

It is very difficult to determine the *in situ* coefficient of lateral earth pressure at rest through measurement. Therefore, it is not uncommon to rely on typical values and empirical formulas for that purpose. A commonly used empirical formula for expressing K_0 in uncemented sands and normally consolidated clays as a function of φ was developed by Jáky (1948):

$$K_0 = 1 - \sin \varphi \tag{10.5}$$

Equation (10.5) was modified by Schmidt (1966) to include the effect of overconsolidation as follows:

$$K_0 = (1 - \sin \varphi)\text{OCR}^{\sin \varphi} \tag{10.6}$$

where OCR is the overconsolidation ratio. The coefficient of lateral earth pressure at rest, K_0, has also been correlated with the liquidity index of clays (Kulhawy and Mayne, 1990), the dilatometer horizontal stress index (Marchetti, 1980; Lacasse and Lunne, 1988), and the Standard Penetration Test N-value (Kulhawy et al., 1989). Table 10.1 lists typical values of K_0 for various soils.

For design purposes, two classical lateral earth pressure theories are commonly used to estimate active and passive earth pressures. Rankine's theory was described above, and relies on calculating the earth pressure coefficients based on the Mohr–Coulomb shear strength of the backfill soil. Although approximate solutions have been proposed in the literature for inclined backfill, they violate the frictionless wall–soil interface assumption and are therefore not presented here.

TABLE 10.1

Deformation (Δ_x) Corresponding to Active and Passive Earth Pressure, as a Function of Wall Height, H

Soil Type	Δ_x/H	
	Active	Passive
Dense sand	0.001	0.01
Medium-dense sand	0.002	0.02
Loose sand	0.004	0.04
Compacted silt	0.002	0.02
Compacted clay	0.01	0.05

Coulomb's theory, on the other hand, dates back to the 18th century (Coulomb, 1776) and considers the stability of a soil wedge behind a retaining wall (Figure 10.7). In the original theory, line AB is arbitrarily selected, and the weight of the wedge, W, is calculated knowing the unit weight of the soil. The directions of the soil resistance, R, and the wall reaction, P_A, are determined based on the soil's internal friction angle, φ, and the soil–wall interface angle, δ. The stability of the wedge ABC is satisfied by drawing the free-body diagram, and the magnitudes of R and P_A are determined accordingly. In order to determine the most critical condition, the direction of line AB is varied until a maximum value of P_A is obtained. The theory only gives the total magnitude of the resultant force on the wall, but the lateral earth pressure may be assumed to increase linearly from the top to the bottom of the wall. Therefore, it becomes possible to calculate an equivalent coefficient of lateral earth pressure for such conditions as follows:

$$P_A = \frac{1}{2} K_A \gamma H^2 \tag{10.7}$$

where γ is the unit weight of the backfill soil, H is the wall height, and K_A is Coulomb's active earth pressure coefficient. For simple geometries, such as the one shown in Figure 10.7, the inclination angle resulting in the maximum value of P_A under active conditions can be determined analytically, and the coefficient of active earth pressure may be calculated from the following expression:

$$K_A = \left(\frac{\sin(\alpha - \varphi)/\sin\alpha}{\sqrt{\sin(\alpha + \delta)} + \sqrt{\dfrac{\sin(\varphi + \delta)\sin(\varphi - \beta)}{\sin(\alpha - \beta)}}} \right)^2 \tag{10.8}$$

Similarly, Coulomb's coefficient of passive earth pressure, K_P, is expressed as:

$$K_P = \left(\frac{\sin(\alpha + \varphi)/\sin\alpha}{\sqrt{\sin(\alpha - \delta)} - \sqrt{\dfrac{\sin(\varphi + \delta)\sin(\varphi + \beta)}{\sin(\alpha - \beta)}}} \right)^2 \tag{10.9}$$

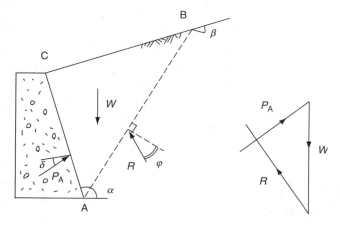

FIGURE 10.7
Coulomb's active earth pressure determination from the stability wedge.

For horizontal backfills ($\beta = 0$), vertical walls ($\alpha = 90°$), and smooth soil–wall interface ($\delta = 0$), Coulomb's earth pressure coefficients, as expressed by Equations (10.8) and (10.9), reduce to their corresponding Rankine equivalents (Equations 10.2 and 10.3).

In fine-grained soils, the lateral earth pressure is affected by the soil's cohesive strength component. Under active conditions, the lateral earth pressure decreases due to the ability of the soil to withstand shear stresses without confinement. The horizontal stress at depth z is, therefore, calculated from

$$\sigma_{h,a} = K_a \gamma H - 2c\sqrt{K_a} \qquad (10.10)$$

A critical depth, z_c, can be calculated from the ground surface, where the horizontal stress is equal to zero

$$z_c = \frac{2c}{\gamma\sqrt{K_a}} \qquad (10.11)$$

Above this depth, and because of the soil's inability to resist tension, no horizontal stresses develop. It has often been argued that tension cracks develop in the ground and may be even filled with water, which adds to the lateral pressure on the wall. However, it is now widely accepted that no such tension cracks develop due to the soil's ability to swell.

Under passive conditions, cohesive soils impose relatively high lateral earth pressures due to the soil's ability to resist shearing. The horizontal earth pressure is calculated from Equation 10.12

$$\sigma_{h,p} = K_p \gamma H + 2c\sqrt{K_p} \qquad (10.12)$$

Figure 10.8 illustrates the active and passive lateral earth pressure distributions in cohesive soils.

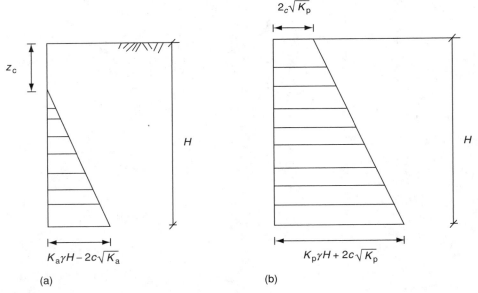

FIGURE 10.8
Lateral earth pressure distribution in cohesive soils: (a) active case; (b) passive case.

Example 10.1

Calculate the lateral earth pressure against the retaining wall shown in Figure 10.9 under both active and passive conditions. The backfill consists of coarse sand with a unit weight of 17.5 kN/m³ and an internal friction angle of 30°. The angle of interface friction, δ, between the wall and the soil is 15°.

Solution

Since the wall–soil interface is rough, Coulomb's theory must be used since Rankine's solution is limited to smooth interfaces. We first calculate Coulomb's coefficients of active and passive earth pressure from Equations (10.8) and (10.9), respectively:

$$K_A = \left(\frac{\sin(90° - 30°)/\sin 90°}{\sqrt{\sin(90° + 15°)} + \sqrt{\dfrac{\sin(30° + 15°)\sin(30° - 10°)}{\sin(90° - 10°)}}} \right)^2 = 0.343$$

$$K_P = \left(\frac{\sin(90° + 30°)/\sin 90°}{\sqrt{\sin(90° - 15°)} - \sqrt{\dfrac{\sin(30° + 15°)\sin(30° + 10°)}{\sin(90° - 10°)}}} \right)^2 = 8.14$$

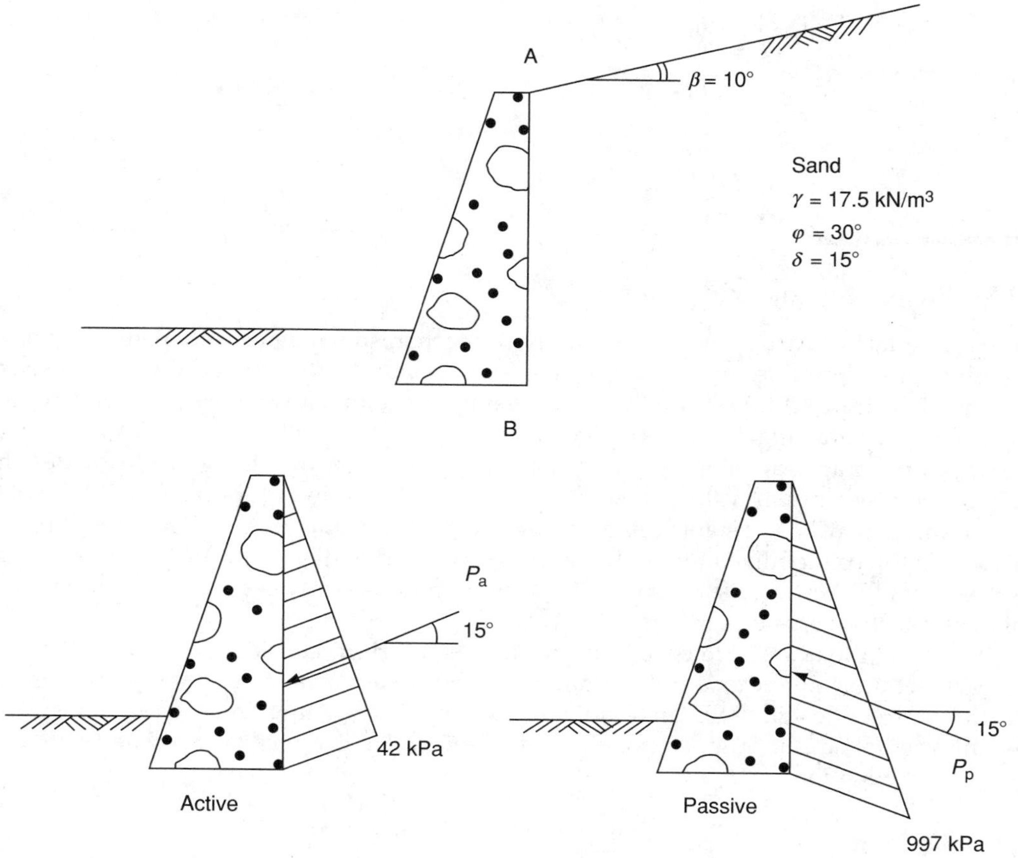

FIGURE 10.9
Illustration for Example 10.1.

Next, we calculate the vertical effective stress at points A and B. Here, since there is no water table behind the wall, the total and effective stresses are equal. At the top of the wall, since there is no surcharge, the vertical stress is equal to zero. At point B, the vertical stress is calculated by multiplying the unit weight of the soil by the height of the wall

$$(\sigma_v)_A = 0$$
$$(\sigma_v)_B = \gamma H = 17.5 \times 7 = 122.5 \text{ kPa}$$

Multiplying the vertical stress by the corresponding coefficient of lateral earth pressure, we obtain the active and passive lateral earth pressure

$$(\sigma_{h,active})_A = (\sigma_{h,passive})_A = 0$$
$$(\sigma_{h,active})_B = (\sigma_v)_A \times K_A = 122.5 \times 0.343 = 42 \text{ kPa}$$
$$(\sigma_{h,passive})_B = (\sigma_v)_A \times K_P = 122.5 \times 8.14 = 997 \text{ kPa}$$

Because of the friction that develops between the wall and the soil, the active and passive pressures act at downward and upward 15° angles, respectively, measured from horizontal. The pressure increases linearly with depth. Accordingly, the resultant force acts at a distance of $H/3$ from the bottom of the wall. The resultant force per unit width of the wall can be calculated by computing the area of the triangular pressure distribution

$$P_A = \frac{1}{2}(\sigma_{h,active})_B H = 0.5 \times 42 \times 7 = 147 \text{ kN/m}$$
$$P_P = \frac{1}{2}(\sigma_{h,passive})_B H = 0.5 \times 997 \times 7 = 3490 \text{ kN/m}$$

10.3   Basic Design Principles

In resisting lateral earth pressure, a variety of mechanisms may act independently or in combination to provide the stability of the earth retaining structure. Gravity and cantilever walls (Figure 10.10) rely on their own weight for stability, with the self-weight of the structure counteracting the external forces acting on the wall surface (Figure 10.10a). Tie-back anchorage, developing along the grouted portion of the anchor, provides the bulk of the resistance in walls and sheet piles, as illustrated in Figure 10.10(b). MSE walls are monolithic internally stable reinforced earth structures that derive their strength from the tensile forces mobilized along the reinforcement strips (Figure 10.10c). It is important to stress that for MSE walls, the role of the facing units is mainly aesthetic, with secondary functions such as erosion control.

   Most design methods are based on limiting equilibrium considerations, with little or no consideration given to the deformation of the system. Most commonly used is the allowable stress design (ASD) method, in which the forces acting on or within the system are analyzed at equilibrium. A global factor of safety, FS, is typically calculated based on the generic equation:

$$\text{FS} = \frac{\text{Stabilizing forces or moments}}{\text{Applied forces or moments}} \tag{10.13}$$

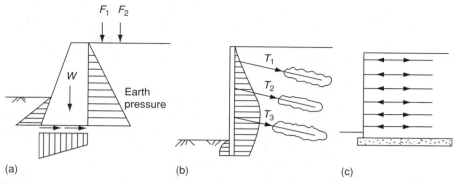

FIGURE 10.10

Stability analysis of retaining walls: (a) gravity walls, (b) tie-back anchored walls, and (c) MSE walls.

The global factor of safety essentially lumps all design uncertainties into a single quantity, with no consideration to the relative uncertainty of each of the parameters. More recently, load and resistance factor design (LRFD) has been introduced as an alternative to account for such differences (Witham et al., 1998). The main concept behind LRFD is that different levels of uncertainty are associated with different load and resistance components within a given system. For instance, consider the gravity retaining wall shown in Figure 10.10(a). Each of the load components, such as active earth pressure and surface loads, is multiplied by a specific load factor, which is greater than 1.0, in order to amplify the distress and account for uncertainties in loads. Similarly, the resisting forces are multiplied each by a reduction factor smaller than 1.0 to account for soil and geometric variability. The main difference between ASD and LRFD is that the latter design takes into consideration the different levels of uncertainty in each component, as opposed to lumping all the system uncertainties into a single parameter. For instance, the resistance factor associated with the self-weight of the wall, a highly reliable quantity, may be close to unity. In contrast, a larger reduction factor may be imposed on the passive earth pressure component if erosion of the toe soil is to be expected. The goal in LRFD is to achieve a combined factored resistance that is greater than the combined factored load:

$$\eta R_n \geq \sum \lambda_i Q_i \tag{10.14}$$

In Equation (10.14), η is a statistically based resistance factor associated with the nominal resistance of the system, R_n, and λ_i is the load factor associated with load Q_i. Because of the relatively recent introduction of LRFD, not much data exist with respect to the recommended or accepted values of η and λ_i. It is also important to note that, in the vast majority of references, the symbols φ and γ_i are used to denote the resistance and load factors, respectively. However, the terms η and λ_i have been adopted here to avoid confusion with other conventional geotechnical parameters.

While LRFD offers a sound and rational approach for designing geotechnical structures by taking into account the difference in reliability between different loading components, the resistance factors are lumped in a single quantity, namely η. In order to assess the redundancy in an existing design, or to analyze a system under new loading conditions, the available resistance is simply compared to the factored loads (right-hand term in Equation 10.14). The values are then compared, and the greater the difference the greater the design redundancy. LRFD procedures in geotechnical design are still in the development phase, and appropriate resistance and load factors are not yet available to the geotechnical engineer. ASD is still widely accepted among the geotechnical community, and is the specified method in most current design codes. Therefore, in this chapter, we will focus the attention on the ASD method in solving example problems.

10.3.1 Effect of Water Table

In many instances, the soil behind an earth retaining structure is submerged. Examples include seawalls, sheet-pile walls in dewatering projects, and offshore structures. Another reason for saturation of backfill material is poor drainage, which leads to an undesirable buildup of water pressure behind the retaining wall. Drainage failure often results in subsequent failure and collapse of the earth retaining structure.

In cases where the design considers the presence of a water table, the lateral earth pressure is calculated from the effective soil stress. Oddly enough, this leads to a *reduction* in effective horizontal earth pressure since the effective stresses are lower than their total counterpart. However, the total stresses on the wall increase due to the presence of the hydrostatic water pressure. In other words, while the effective horizontal stress decreases, the total horizontal stress increases. The next example illustrates this concept.

Example 10.2
Due to clogging of the drainage system, the water table has built up to a depth of 4 m below the ground surface behind the retaining wall shown in Figure 10.11. The soil above the water table is partially saturated and has a unit weight of 17 kN/m^3. Below the water table, the soil is saturated and has a unit weight of 19 kN/m^3. Calculate the total and effective vertical and horizontal stresses at point A under active conditions.

Solution
First, we calculate the total and effective vertical stress at point A

$$\text{Total vertical stress} = (\sigma_v)_A = 17 \times 4 + 19 \times 2 = 106 \text{ kPa}$$
$$\text{Pore water pressure} = u_A = \gamma_w z_w = 9.8 \times 2 = 19.6 \text{ kPa}$$
$$\text{Effective vertical stress} = (\sigma'_v)_A = 106 - 19.6 = 86.4 \text{ kPa}$$

Next, we calculate the coefficient of lateral earth pressure using Equation (10.2)

$$K_a = \frac{1 - \sin 33°}{1 + \sin 33°} = 0.295$$

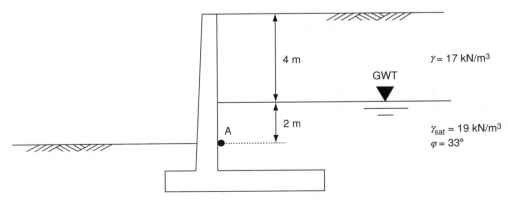

FIGURE 10.11
Illustration for Example 10.2.

We then calculate the effective horizontal stress using Equation (10.1)

$$\text{Effective horizontal stress} = (\sigma'_h)_A = K_a(\sigma'_v)_A = 0.295 \times 86.4 = 25.5 \text{ kPa}$$

The total horizontal stress is calculated by adding the pore water pressure to the effective horizontal stress

$$\text{Total horizontal stress} = (\sigma_h)_A = (\sigma'_h)_A + u_A = 25.5 + 19.6 = 45.1 \text{ kPa}$$

It is important to note that the total horizontal stress cannot be correctly calculated by multiplying the total vertical stress by the coefficient of lateral earth pressure

$$(\sigma_h)_A \neq K_a(\sigma_v)_A$$

Such calculation will result in significant underestimation of the horizontal stresses acting on a retaining structure under active conditions.

10.3.2 Effect of Compaction on Nonyielding Walls

In the case of rigid (nonyielding) walls, at-rest conditions are considered in the structural design. In addition, locked-in passive earth pressures can develop near the top of the wall if heavy compaction equipment is used. The passive condition, caused by a line load P from the roller, develops from the ground surface up to a depth of z_p and remains constant to a depth of z_r where:

$$z_p = \sqrt{\frac{2PK_aK_0}{\pi\gamma}}, \quad z_r = \sqrt{\frac{2P}{\pi\gamma K_aK_0}}$$

Below a depth of z_r, at-rest earth pressure conditions prevail. In the case of flexible walls, such conditions are not believed to occur due to wall displacement. Instead, active conditions are assumed. In addition, light compaction equipment is typically used to compact the backfill behind most retaining walls in order to reduce the lateral earth pressures. As a result, the additional earth pressure due to compaction may not need to be considered, depending on the construction method.

10.4 Gravity Walls

In the past, gravity and cantilever walls constituted the vast majority of earth retaining structures. However, in recent years, these structures have given way to MSE walls, which are more economical, easier to construct, and better performing. A small number of projects, however, still rely on gravity walls and their closely related support system of modular block walls. Traditionally, gravity walls are cast in place of plain or reinforced concrete structures that rely on their own weight for stability. They may be constructed in a wide range of geometries, some of which are illustrated in Figure 10.12.

Modular block walls, on the other hand, are constructed by stacking rows of interlocking blocks and compacting the soil in successive layers. Masonry or cinder blocks can also be used in conjunction with mortar binding to form limited height walls, typically no

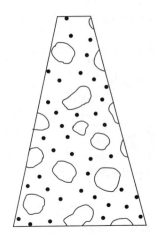

FIGURE 10.12
Typical geometries of gravity retaining walls.

taller than 2 m. Interlocking blocks are available commercially in a wide variety of shapes and materials, some of which are proprietary. They provide a greater level of stability than masonry walls and are sometimes manufactured so that the resulting facing is battered (see Figure 10.13).

In designing gravity walls, external stability, which is the equilibrium of all external forces, is more critical than the internal structural stability of the wall. This is mostly due to the massive nature of the structure, which usually results in conservative designs for internal stability. In analyzing or designing for external stability, all the forces acting on the structure are considered. These forces include lateral earth pressures, the self-weight of the structure, and the reaction from the foundation soil. The stability of the wall is then evaluated by considering the relevant forces for each potential failure mechanism.

The four potential failure mechanisms typically considered in design or analysis are shown in Figure 10.14 and are summarized next:

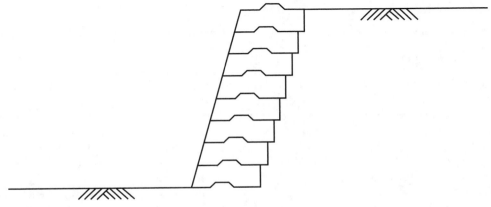

FIGURE 10.13
Modular block wall with battered facing.

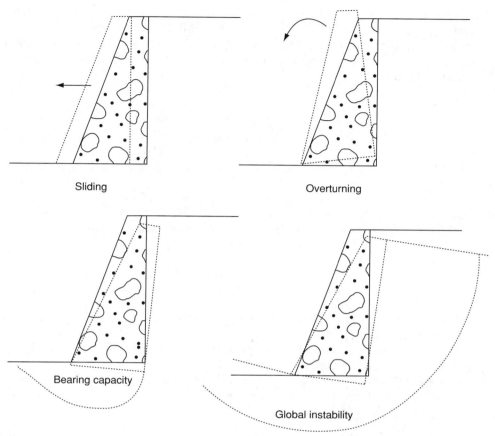

FIGURE 10.14
Potential failure modes due to external instability of gravity walls.

1. *Sliding resistance.* The net horizontal forces must be such that the wall is prevented from sliding along its foundation. The factor of safety against sliding is calculated from:

$$\text{FS}_{\text{sliding}} = \frac{\sum \text{Resisting forces}}{\sum \text{Sliding forces}} \qquad (10.15)$$

The minimum acceptable limit for $\text{FS}_{\text{sliding}}$ is 1.5. The most significant sliding force component usually comes from the lateral earth pressure acting on the active (backfill) side of the wall. Such force may be intensified by the presence of vertical or horizontal loads on the backfill surface. In the unlikely event where a water table is present within the backfill, the water pressure may reduce the lateral earth pressure due to the reduction in effective stresses, but greater lateral forces are generated on the wall from the hydrostatic pressure of the water itself. The main component resisting the sliding is the friction along the wall base. Due to the potential for erosion, the passive earth pressure in front of the toe of the wall is conservatively ignored in design. If such passive earth pressure is included, then the minimum acceptable limit for $\text{FS}_{\text{sliding}}$ increases to 2.0.

2. *Overturning resistance.* The righting moments must be greater than the overturning moments to prevent rotation of the wall around its toe. The righting moments result mainly from the self-weight of the structure, whereas the main source of overturning moments is the active earth pressure. The factor of safety against overturning is calculated from:

$$FS_{overturning} = \frac{\sum Righting\ moments}{\sum Overturning\ moments} \qquad (10.16)$$

The factor of safety against overturning must be equal to or greater than 1.5.

3. *Bearing capacity.* The bearing capacity of the foundation soil must be large enough to resist the stresses acting along the base of the structure. The factor of safety against bearing capacity failure, FS_{BC}, is calculated from:

$$FS_{BC} = \frac{q_{ult}}{q_{max}} \qquad (10.17)$$

where q_{ult} is the ultimate bearing capacity of the foundation soil, and q_{max} is the maximum contact pressure at the interface between the wall structure and the foundation soil. The minimum acceptable value for FS_{BC} is 3.0. In addition to "traditional" bearing capacity considerations, the movement of the wall due to excessive settlement of the underlying soil must also be limited. The components of the foundation settlement include immediate, consolidation, and creep settlement, depending on soil type.

4. *Global stability.* Overall stability of the wall system within the context of slope stability must also be assessed to ensure that no failure occurs either in the backfill or the native soil. As such, a separate analysis for slope stability must be performed on the zone in the vicinity of the wall using conventional limit equilibrium slope stability methods.

When considering the active and passive earth pressures on either side of the wall for sliding and overturning calculations, caution must be exercised. The wall movement needed to fully mobilize an active condition on one side of the wall is much smaller than that needed to mobilize the passive pressure on the other side. For sands, a horizontal deformation of approximately 0.0025 to 0.0075 of the wall height is required to reach the minimum earth pressure on the active side, with lower displacement corresponding to stiff (dense) sand. The horizontal displacement needed to develop the full passive resistance is approximately 10 times that amount, which raises the issue of displacement compatibility. Even though the soil on the passive side is typically looser due to the lack of overburden confinement, a fully passive condition rarely develops within typical acceptable displacement ranges in retaining walls. This lack of displacement compatibility may even be more significant if the rotation mechanism of the wall is considered. It is, therefore, advised to neglect the passive earth pressure in wall stability calculations.

If a design proves to be inadequate, remedial action must be taken to increase the corresponding factor of safety (see Figure 10.15). In the case of potential sliding failure, additional soil may be compacted in front of the wall toe, but provisions are needed to ensure that such soil does not erode with time. Another solution is the inclusion of a

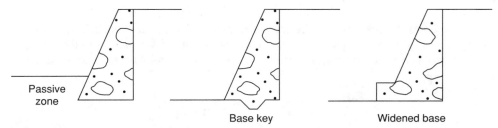

FIGURE 10.15
Typical provisions to increase the stability of gravity walls.

"key" across the base of the wall. In the case of overturning, the weight of the structure can be increased, the base widened, or the center of gravity moved further back from the wall face. Bearing capacity and global stability concerns may be addressed through conventional solution for such problems, such as geometric modification, soil improvement, or choice of a deep foundation alternative.

Example 10.3
Calculate the factor of safety against sliding, overturning, and bearing capacity failure for the retaining wall shown in Figure 10.16. The ultimate bearing capacity of the foundation soil is 500 kPa and the coefficient of base friction, $\mu = 0.3$. Assume that the wall is smooth, and include the passive earth pressure at the toe when applicable.

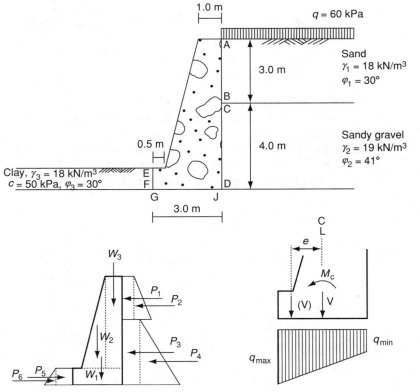

FIGURE 10.16
Illustration for Example 10.3.

Solution

Based on the conditions shown in Figure 10.16, active and passive earth pressures act on the right and left side of the wall, respectively. Accordingly, we calculate the coefficients of active earth pressure for the sand and the sandy gravel layers, and passive earth pressure for the clay layer:

$$K_{a1} = \frac{1 - \sin 30°}{1 + \sin 30°} = 0.333$$

$$K_{a2} = \frac{1 - \sin 41°}{1 + \sin 41°} = 0.208$$

$$K_{p3} = \frac{1 + \sin 10°}{1 - \sin 10°} = 1.42$$

We then calculate the vertical and horizontal stresses at points A through F. Within each soil layer, the horizontal stresses increase linearly since the soil is uniform and homogeneous. It is also important to note that the horizontal stress at point B is different from point C since there is an "abrupt" change in the coefficient of lateral earth pressure at that location. It is also noted that the passive pressure at points E and F is calculated from Equation (10.12) due to the presence of cohesion in the clay.

Point	σ'_v (kPa)	K	σ'_h (kPa)
A	60	0.333	$60 \times 0.333 = 20$
B	$60 + 18 \times 3 = 114$	0.333	$114 \times 0.333 = 38$
C	114	0.208	$114 \times 0.208 = 23.7$
D	$114 + 19 \times 4 = 190$	0.208	$190 \times 0.208 = 39.5$
E	0	1.42	$2 \times 50 \times (1.42)^{0.5} = 119.2$
F	$16 \times 1 = 16$	1.42	$119 + 16 \times 1.42 = 141.7$

The next step is to calculate the resultant vertical and horizontal forces by subdividing the lateral earth pressure diagram and the wall cross section into rectangles and triangles. The earth pressure forces are calculating from the area of the diagram, while the weights of the concrete wall are calculated by multiplying the area by the unit weight of concrete (23.5 kN/m^3). It is also prudent at this stage to compute the moment arm associated with each force (measured from the wall toe F) in anticipation of the overturning stability calculations.

Force	Magnitude (kN/m)	Moment Arm (m)
P_1	$20 \times 3 = 60$	$4 + 3/2 = 5.5$
P_2	$\frac{1}{2}(38 - 20) \times 3 = 27$	$4 + 3/3 = 5$
P_3	$23.7 \times 4 = 94.8$	$4/2 = 2$
P_4	$\frac{1}{2}(39.5 - 23.7) \times 4 = 31.6$	$4/3 = 1.33$
P_5	$119.2 \times 1 = 119.2$	$1/2 = 0.5$
P_6	$\frac{1}{2}(141.7 - 119.2) \times 1 = 11.25$	$1/3 = 0.33$
W_1	$3 \times 1 \times 23.5 = 70.5$	$3/2 = 1.5$
W_2	$\frac{1}{2} \times 1.5 \times 6 \times 23.5 = 105.8$	$0.5 + 1.5 \times 2/3 = 1.5$
W_3	$1 \times 7 \times 23.5 = 164.5$	2.5
F_b	$\Sigma W_i \mu = (70.5 + 105.8 + 164.5) \times 0.3 = 102.2$	0

The factor of safety against sliding is calculated from Equation (10.15):

$$\text{FS}_{\text{sliding}} = \frac{P_5 + P_6 + F_b}{P_1 + P_2 + P_3 + P_4} = \frac{232.7}{213.4} = 1.09$$

The factor of safety against overturning is calculated from Equation (10.16):

$$\text{FS}_{\text{overturning}} = \frac{119.2 \times 0.5 + 11.25 \times 0.33 + 70.5 \times 1.5 + 105.8 \times 1.5 + 164.5 \times 2.5}{60 \times 5.5 + 27 \times 5 + 94.8 \times 2 + 31.6 \times 1.33} = \frac{739}{697} = 1.06$$

In calculating the resisting moments, the passive earth pressure at the toe of the wall was included in this example. This should only be done in cases where it is guaranteed that such soil will not erode. Otherwise, the moments resulting from the passive earth pressure at the toe of the wall should be ignored.

In order to calculate the factor of safety against bearing capacity failure, it is necessary to determine the maximum and minimum base contact pressures. Due to the eccentricity generated by the moment, the maximum pressure will typically occur at the toe of the wall (point G) while the minimum will occur at the heel (point J). The total vertical force, $V = \Sigma W$, and the moment, M_c, about the center of the base are equivalent to a vertical force V acting at an eccentric distance e from the center of the base. An easy method for calculating e relies on the righting and overturning moments about the toe, which are available from the $\text{FS}_{\text{overturning}}$ calculations:

$$e = \frac{b}{2} - \frac{M_R - M_O}{V}$$

where b is the width of the base, M_R and M_O are the righting and overturning moments, respectively, and $V = \Sigma W$ is the summation of the vertical forces. Therefore,

$$e = \frac{3}{2} - \frac{739 - 697}{70.5 + 105.8 + 164.5} = 1.38 \text{ m}$$

The maximum and minimum pressures are then calculated from basic mechanics of materials concepts:

$$q_{\max} = \frac{V}{b}\left(1 + 6\frac{e}{b}\right), \quad q_{\min} = \frac{V}{b}\left(1 - 6\frac{e}{b}\right)$$

As such,

$$q_{\max} = \frac{340.8}{3}\left(1 + 6 \times \frac{1.38}{3}\right) = 427.1 \text{ kPa}$$

The factor of safety against bearing capacity failure is calculated from Equation (10.17):

$$\text{FS}_{\text{BC}} = \frac{q_{\text{ult}}}{q_{\max}} = \frac{500}{427.1} = 1.2$$

It is evident in this problem that the wall is marginally safe, since the factors of safety against sliding, overturning, and bearing capacity are slightly above 1.0. However, these values are much lower than the recommended values of 2.0, 1.5, and 3.0, respectively. Therefore, the design modifications should be introduced to increase the factors of safety to their minimum acceptable limits.

10.5   Cantilever Walls

Like gravity walls, cantilever retaining walls have also become largely obsolete, but are constructed in cases where MSE walls are not feasible. They also rely on their self-weight to resist sliding and overturning, but derive part of their stability from the weight of the backfill above the heel of the wall. Cantilever walls are made of reinforced concrete, and come in different geometries. They are often easier to erect than gravity wall, since they can be prefabricated in sections and transported directly to the site. Figure 10.17 shows isometric views of simple cantilever walls and counterfort walls.

In addition to the external stability, cantilever walls must also satisfy internal structural stability requirements. As shown in Figure 10.18, the wall section should be able to withstand the shear stresses and bending moments resulting from the lateral earth pressure as well as the difference in pressure between the top and bottom faces of the base. To this end, steel reinforcement is placed as shown in the figure. The size and density of the reinforcement are decided by the structural engineer, based on the structural design of the cross section. Counterforts are used to reduced shear forces and bending moments at the critical section where higher walls are needed.

When considering the external stability of the wall, the backfill section above the cantilever wall heel is assumed to be part of the wall, with Rankine or Coulomb conditions acting along the vertical line originating at the heel (line AB in Figure 10.19). This assumption is largely accurate, provided that the width of the heel is larger than $[H \tan(45° - \varphi/2)]$, which is typically true except for tall walls. The weight of the soil block above the heel is then added to the weight of the reinforced concrete wall in all stability calculations. All procedures for stability checks are identical to those described for gravity walls.

Example 10.4

For the cantilever retaining wall shown in Figure 10.20, calculate the width of the heel, b, required to ensure stability of the wall against overturning. In addition, determine the angle, θ, of the potential active shear plane with respect to horizontal.

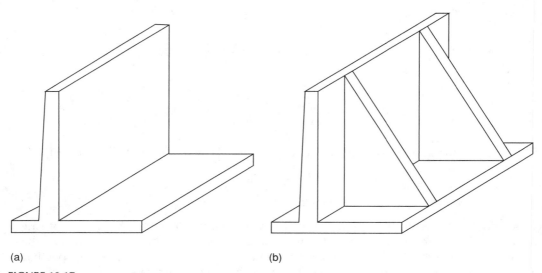

(a)   (b)

FIGURE 10.17
General view of (a) cantilever retaining wall and (b) counterfort wall.

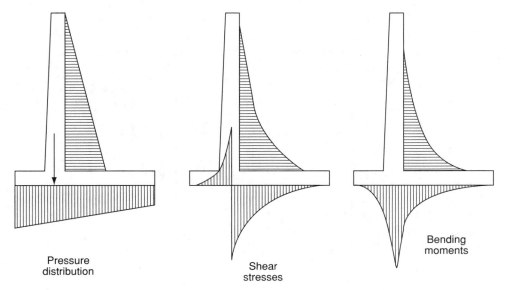

FIGURE 10.18
Schematic of shear and bending moment diagrams of cantilever wall.

Solution
In order to facilitate the calculation process, we divide the cantilever wall into sections. We then calculate the weight per unit width (W_i) and moment arm (x_i) for each block:

$$W_1 = 23.5 \times 0.5 \times 0.7 = 8.23 \text{ kN (per meter)}$$
$$W_2 = 23.5 \times 5 \times 0.5 = 58.75 \text{ kN}$$
$$W_3 = 23.5 \times 0.5 \times b = 11.75b \text{ kN}$$
$$W_4 = (17 \times 2.5 + 19 \times 2)b = 80.5b \text{ kN}$$

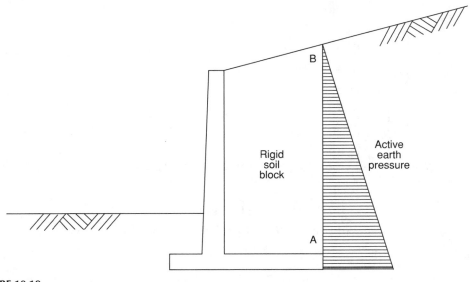

FIGURE 10.19
Rigid soil block assumption for design of cantilever retaining wall.

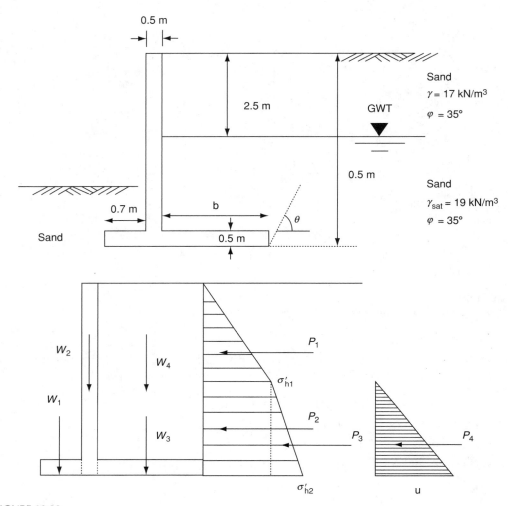

FIGURE 10.20
Illustration for Example 10.4.

$$x_1 = 0.35 \text{ m}$$
$$x_2 = 0.95 \text{ m}$$
$$x_3 = 1.20 + b/2$$
$$x_4 = 1.20 + b/2$$

We then calculate the active earth pressure and the water pressure on the wall. For lateral earth pressure calculations, we use $K_a = \tan^2(45 - 35/2) = 0.271$

$$\sigma'_{h1} = 17 \times 2.5 \times 0.271 = 11.52 \text{ kPa}$$
$$\sigma'_{h2} = (17 \times 2.5 + \{19 - 9.8\} \times 2.5) \times 0.271 = 17.75 \text{ kPa}$$
$$u = 9.8 \times 2.5 = 24.5 \text{ kPa}$$

The corresponding forces (per meter), P_1 to P_4, together with their moment arms, y_1 to y_4, are calculated as follows:

$$P_1 = 0.5 \times 11.52 \times 2.5 = 14.4 \text{ kN}$$
$$P_2 = 17.75 \times 2.5 = 44.38 \text{ kN}$$
$$P_3 = 0.5 \times (17.75 - 11.52) \times 2.5 = 7.79 \text{ kN}$$
$$P_4 = 0.5 \times 24.5 \times 2.5 = 30.63 \text{ kN}$$

$$y_1 = 2.5 + 2.5/3 = 3.33 \text{ m}$$
$$y_2 = 2.5/2 = 1.25 \text{ m}$$
$$y_3 = y_4 = 2.5/3 = 0.83 \text{ m}$$

The factor of safety against overturning is calculated from

$$\text{FS}_{\text{overturning}} = \frac{\sum_{i=1}^{4} W_i x_i}{\sum_{i=1}^{4} P_i y_i}$$

$$= \frac{8.23 \times 0.35 + 58.75 \times 0.95 + 11.75b \times (1.2 + b/2) + 80.5b \times (1.2 + b/2)}{14.4 \times 3.33 + 44.38 \times 1.25 + 7.79 \times 0.83 + 30.63 \times 0.83}$$

In order to ensure stability, the factor of safety must be at least equal to 1.5. Accordingly, we solve the equation above for b and obtain:

$$b = 1 \text{ m}$$

The angle, θ, that the potential active failure surface makes with respect to horizontal is simply equal to $45 + \varphi/2 = 45 + 35/2 = 62.5$.

10.6 MSE Walls

The design procedures for soils reinforced with horizontal metal strips originated in the mid-1960s in France (Vidal, 1966). The patented process, dubbed "reinforced earth," gave way to subsequent developments in soil reinforcement, especially with the advent of new reinforcement materials such as geosynthetics in the 1970s and the 1980s. By the mid-1990s, almost all newly constructed bridge abutment and retaining walls in the United States consisted of MSE structures. Compared to conventional gravity and cantilever retaining walls, MSE walls are more economical, easier to erect, and much more stable. Their performance under seismic conditions has also proven to be much more reliable due to their inherent ductility.

MSE walls are constructed by compacting the soil in layers separated by reinforcement strips or sheets (Figure 10.21). Typically, strip reinforcement consists of high-strength galvanized steel, while sheet reinforcement consists of geogrids or geotextiles, which are polymeric materials known as geosynthetics. Reinforcement strips are attached to facing units, and extend far enough into the backfill to ensure adequate pullout resistance. Although the facing units represent the finished wall surface, they actually have no structural function with respect to wall stability. Instead, MSE walls derive their stability from the internal stresses developing at the interface between the soil and the reinforcement elements. As such, MSE walls remain perfectly stable in the absence of facing units. The role of the facing units is to improve esthetics, protect the wall against vandalism, prevent local failure and erosion near the facing, and protect against ultraviolet degradation in the case of geosynthetic reinforcement.

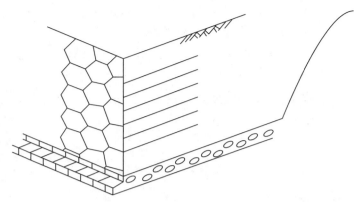

FIGURE 10.21
General view of MSE wall.

Originally, all external stability requirements (sliding, overturning, bearing capacity, and global stability) needed to be checked for MSE wall design. However, it has been found that, due to their monolithic nature, MSE walls are not prone to overturning. In addition, MSE walls must be designed to ensure internal stability, which includes checks against yielding and pullout of reinforcement. The design must also ensure adequate connection strength between reinforcement and facing in the case of timber or concrete panels.

10.6.1 Internal Stability Analysis and Design

The internal stability requirements for MSE walls dictate the extent of the reinforcement elements into the backfill, as well as their vertical and (if strips are used) horizontal spacing. Figure 10.22 represents a generic cross section of an MSE wall. Based on the existing or assumed vertical spacing, the vertical stresses are calculated at each reinforcement depth (z). The corresponding horizontal stress, $\sigma_{h,z}$, is then computed accordingly, assuming active earth pressure conditions. The horizontal earth pressure at depth z is calculated from:

$$\sigma_{h,z} = K_a \sigma_{v,z} \tag{10.18}$$

In the absence of any surcharge loading, the vertical stress is equal to the unit weight of the soil times the depth. Additional stresses resulting from surcharges at the surface may be calculated from a variety of methods such as elastic solutions and charts when applicable. The maximum tensile force in the reinforcement layer is calculated by multiplying the horizontal stress by the cross sectional "area of influence" of the reinforcement element. In the case of reinforcement strips, the area of influence is equal to $s_v \times s_h$, where s_v and s_h are the vertical and horizontal spacing between the reinforcement strips, respectively. In the case of geogrid and geotextile reinforcement, a unit width of the reinforcement is considered in lieu of the horizontal spacing, s_h. In this case, the calculation output is a force per unit length.

The factor of safety against yielding of the reinforcement is then calculated for each layer by dividing the yield strength of the reinforcement material by the maximum tensile strength:

$$\mathrm{FS}_{\text{yielding}} = \frac{F_{\max}}{\sigma_{h,z} \times s_v \times s_h} \geq 1.5 \tag{10.19}$$

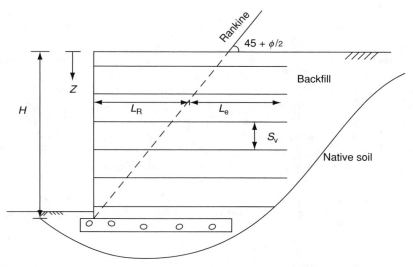

FIGURE 10.22
Cross section of MSE wall.

where F_{max} is the maximum design tensile resistance of the reinforcement element. In the case of galvanized steel, the yield strength may be used. However, in the case of geosynthetic reinforcement, the yield strength must be multiplied by a number of reduction factors to account for environmental conditions. As such, the maximum design strength of geosynthetic reinforcement is calculated from:

$$F_{max} = F_{yield} \times RF_{CR} \times RF_{ID} \times RF_{CD} \times RF_{BD} \tag{10.20}$$

where RF_{CR}, RF_{ID}, RF_{CD}, and RF_{BD} are reduction factors for creep deformation, installation damage, chemical degradation, and biological degradation, respectively. These values depend on the properties of the geosynthetic as well as the environmental conditions during operation and can vary within a very significant range. It is not uncommon for these factors to amount to an overall reduction factor of 10 or 20.

The second component of internal stability is the resistance to pullout, which dictates the extent of the reinforcement into the backfill. For design purposes, a potential Rankine-type failure wedge ($\theta = 45 + \varphi/2$) is considered to originate at the toe of the wall (Figure 10.22). The length of reinforcement within the Rankine wedge, L_R, is calculated from

$$L_R = (H - z) \tan (45 - \varphi/2) \tag{10.21}$$

Experimental evidence has shown that a Rankine wedge may not be representative of the actual potential failure surface, so more sophisticated design procedures may consider more realistic surfaces, such as curved or bilinear failure wedges. Since the failure wedge is assumed to be rigid, no internal deformations develop, and the length of reinforcement within this zone (L_R) does not contribute to resisting pullout. Instead, the effective length of reinforcement (L_e) is measured from the back end of the Rankine wedge. The factor of safety against pullout resistance is calculated by dividing the available pullout resistance by the maximum tensile force in the reinforcement for each reinforcement layer:

$$FS_{pullout} = \frac{2 \times w \times \sigma_{v,z} \times \tan \varphi_i \times L_e}{\sigma_{h,z} \times s_v \times s_h} \geq 1.5 \tag{10.22}$$

where w is the width of the reinforcement element and φ_i is the interface friction angle between the soil and the reinforcement. It is noted that a multiplier of 2 is included in the numerator to account for frictional stresses developing on both top and bottom faces of the embedded reinforcement. The total length, L_T, of the reinforcement for each layer is then calculated by adding the Rankine length, L_R, to the effective length, L_e.

For reinforcement elements distributed at uniform spacing, it is inevitable that design calculations will result in different required yield strength and length for each layer of reinforcement. However, from a constructability perspective, it is imperative to specify a constant set of values, corresponding to the most critical layer. As a result, the finished design ends up being overly conservative and extremely redundant in safety. In large projects where tall MSE walls are constructed, and when strict quality control measures are implemented in the field, it is possible to specify multiple sets of parameters over certain heights of the wall. For instance, it is not uncommon to use tighter vertical reinforcement spacing within the bottom half of a wall, where tensile forces are highest.

10.6.2 Reinforced Earth Walls

Reinforced earth walls are earth retaining structures that consist of steel strips connected to uniquely shaped concrete or metal facing panels. The most common facing design is the prefabricated concrete panel system shown in Figure 10.23, although other designs have also been used. Reinforcement elements consist of galvanized steel strips, approximately 0.1 m wide and 5 mm thick, with a patterned surface to enhance frictional interaction with the soil. Four strips are connected to each facing unit. Among the most critical issues concerning the response of these walls is corrosion of the steel strips, especially in marine environments. In such cases, the use of geosynthetic reinforcement may be warranted.

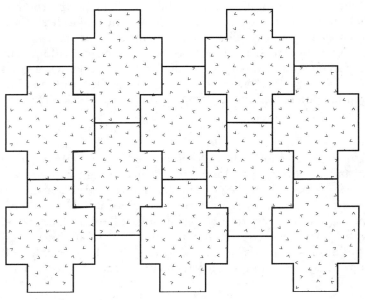

FIGURE 10.23
Reinforced earth wall facing panel system.

Design of reinforced earth walls and similar MSE systems starts by determining the reinforcement vertical and horizontal spacing. These values are typically predetermined from the geometry of the prefabricated concrete facing panels. Typical values of vertical and horizontal spacing are 0.75 and 0.5 m, respectively. A suitable reinforcement material is then chosen based on Equation (10.19), and the reinforcement length is determined. In addition, the connection at the facing must be able to sustain the maximum tensile forces in the reinforcement, although, in reality, the forces at the connection are much smaller.

Example 10.5
An MSE wall is reinforced with galvanized steel strips, spaced at 0.75 m vertically and 0.5 m horizontally. The strips are 0.10 m wide, and the yield strength of the galvanized steel is 240 MPa. The wall height is 9 m, and the backfill consists of select granular material with $\varphi = 35°$ and $\gamma = 18$ kN/m^3. The soil–steel interface friction angle is 25°. A surcharge of 200 kPa is applied at the top of the wall. Calculate the minimum required thickness of the steel strips and the total embedment length.

Solution
Since the wall height is 9 m, and the vertical spacing between the reinforcement layers is 0.75 m, the total number of reinforcement layers is 12, with the first layer embedded at 0.375 m from the top. Typically, it is advisable to perform all calculations in a tabulated (spreadsheet) format, with each row corresponding to a soil layer. In this particular example, because a uniform surcharge is applied at the top, the maximum horizontal stresses will develop at the bottom layer, where $z = 8.625$ m

$$\sigma_{h,z} = K_a(q + \gamma z) = \frac{1 - \sin 35°}{1 + \sin 35°} \times (200 + 18 \times 8.625) = 96.3 \text{ kPa}$$

From Equation (10.19) and assuming $FS_{yielding} = 1.5$, we determine F_{max}

$$F_{max} = 1.5 \times 96.3 \times 0.75 \times 0.5 = 54.2 \text{ kN}$$

The thickness of the steel strip is determined from the width and yield strength of the steel strips:

$$t = \frac{F_{max}}{\sigma_{yield} \times w} = \frac{54.2}{240,000 \times 0.1} = 2.3 \times 10^{-3} \text{ m} = 2.3 \text{ mm}$$

A minimum of 2 mm is typically added as a sacrificial thickness since corrosion is all but certain. Therefore, the total thickness is equal to 4.3 mm, which is rounded to the nearest practical thickness of 5 mm.

The effective length of reinforcement, L_e, is calculated from Equation (10.22), by assuming a factor of safety of 1.5. Since the ratio between $\sigma_{h,z}$ and $\sigma_{v,z}$ is equal to K_a, the value of L_e is independent of depth, and is calculated from

$$L_e = \frac{1.5 \times K_a \times s_v \times s_h}{2 \times w \times \tan \varphi_i} = \frac{1.5 \times 0.271 \times 0.75 \times 0.5}{2 \times 0.1 \times \tan 25°} = 1.63 \text{ m}$$

The maximum value of L_R (Rankine length) will occur at the top layer, where $z = 0.375$. From Equation (10.21), $L_R = (9 - 0.375) \tan(45 - 35/2) = 4.49$ m. The total length, L_T, is equal to

$$L_T = L_e + L_R = 1.63 + 4.49 = 6.12 \text{ m}$$

The value of L_T is rounded to the nearest practical length of 6.25 m.

10.6.3 Geogrid-Reinforced Walls

With the increased availability of high-strength geogrid materials in the 1990s, geogrid-reinforced walls were introduced as an alternative to metallic strip reinforcement. They provide increased interface area (since the coverage area can be continuous), better interlocking with the backfill (due to the geometry of the openings), resistance to corrosive environments, and lower cost. The most common type of geogrids used in earth reinforcement is the uniaxial type, owing to its high strength and stiffness in the main direction. Facing panel units may be connected to the geogrid using a steel bar interwoven into the grid (known as a Bodkin connector) or, more recently, through special plastic clamps that tie into a geogrid section embedded in the concrete panel.

Among the concerns associated with the use of geogrids in heavily loaded walls (such as bridge abutments) are the time-dependent stress relaxation (creep deformation), installation damage, and chemical degradation. It is, therefore, crucial to determine the design strength of the geogrid considering the various reduction factors described in Equation (10.20). In addition, it is extremely important to ensure that the geogrid is fully stretched during installation and compaction of the subsequent soil layer. Otherwise, significant deformation is needed before tensile stresses and interface friction is mobilized. A closely related problem that has been identified is the difficulty in keeping the facing elements plumb during installation, especially when close tolerance is needed in tall walls.

Design and construction procedures for geogrid-reinforced walls are almost identical to reinforced earth walls. One distinct exception is that the strength of the geogrid is expressed in terms of force per unit length, and the associated horizontal spacing, s_h, is taken as the unit length in all calculations. Another difference is that, because of the effective interlocking of the soil particles within the geogrid openings, the interface friction angle is usually equal to the internal friction angle of the soil.

10.6.4 Geotextile-Reinforced Walls

Unlike metallic and geogrid reinforcement, typical geotextile-reinforced wall designs do not require facing elements. Instead, the geotextile layer is wrapped around the compacted soil at the front to form the facing (Figure 10.24). The finished wall must be covered with shotcrete, bitumen, or Gunite to prevent ultraviolet radiation from reaching and damaging the geotextile. Such walls are usually constructed as temporary structures, or where aesthetics are not of prime importance. However, it is possible to cover the wall with a permanent "faux finish" that blends with the surrounding environment.

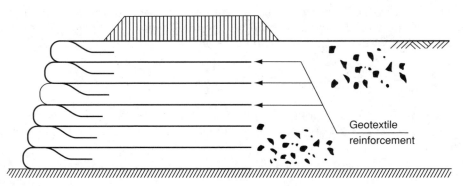

FIGURE 10.24
Geotextile-reinforced wall with wrapped-around facing.

The design procedures for geotextile-reinforced walls are also identical to those described earlier for steel and geogrid reinforcement. The interface friction angle between the soil and the geotextile sheet is typically equal to $(1/2)\varphi$ to $(2/3)\varphi$. In addition, the overlap length, L_o must be determined from the following equation:

$$L_o = \frac{S_v \times \sigma_{h,z}}{2\sigma_{v,z} \times \tan \varphi_i} \times FS_{overlap} \qquad (10.23)$$

The minimum acceptable overlap length is 1 m.

10.7 Sheet-Pile and Tie-Back Anchored Walls

Sheet-pile walls provide temporary or permanent support when excavations are to be carried out. They consist of steel, concrete, and sometimes timber sections, typically driven in the ground using percussion, vibration, or jetting. More recently, fiber-reinforced polymers (FRPs) have been used successfully in a number of projects where the sheet piles are driven to shallow depths. FRPs have the advantage of resisting a wide range of chemically aggressive environments. Typical cross sections of sheet piles (in plan view) are shown in Figure 10.25.

Once driven in the ground, excavation proceeds on one side, with the sheet pile providing the necessary earth support. For shallow depths (less than 6 m), cantilever-type sheet piles are adequate (Figure 10.26). In this case, the embedment depth of the sheet pile below the excavation level can deliver the moment required to resist the lateral earth pressure on the active side. For larger excavation depths, it becomes necessary to supplement the embedment resistance with a tie-rod anchor at a shallow depth. Such tie-rod anchors are often installed by excavating and re-compacting the soil. If multiple rows of anchors are required, a tie-back anchored retaining wall is constructed by driving the anchors and grouting them in place.

10.7.1 Cantilever Sheet Piles

A conceptual representation of the lateral earth pressure acting on a cantilever sheet pile is shown in Figure 10.27. Only active pressure is present on Side A, from the ground surface to the depth of excavation. Below the excavation depth, passive conditions are assumed to act on Side B of the sheet pile, while active conditions persist on Side A, up to point O, where a reversal of conditions occurs. Point O can be viewed roughly as the point of rotation of the sheet pile in the ground. Such rotation is necessary in order to achieve static equilibrium of the system. Below point O, active conditions develop on Side B while passive earth pressures are present on Side A.

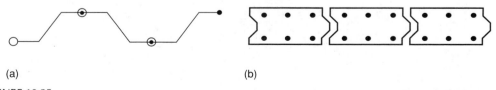

(a)   (b)

FIGURE 10.25
Typical cross sections of sheet-pile materials: (a) steel, and (b) concrete.

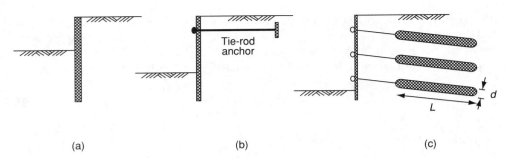

FIGURE 10.26
Typical sheet-pile support mechanisms: (a) cantilever sheet pile, (b) tie-rod anchored sheet pile, and (c) tie-back anchored wall.

Cantilever sheet-pile design typically involves the determination of the embedment depth, D, given other geometric constraints of the problem as well as soil properties. Therefore, the first step is to calculate the magnitude of the horizontal stresses σ_A, σ_B, and σ_C. The value of σ_A is readily calculated as the active earth pressure acting at depth H. The magnitudes of σ_O and σ_B must be calculated as a function of the embedment depth, D, and the depth to the rotation point, D_1, both of which are unknown. The value of σ_O is calculated assuming passive conditions on Side B and active conditions on Side A. Similarly, σ_B is calculated with passive earth pressure on Side A and active earth pressure on Side B. Two equilibrium conditions are to be satisfied: the sum of the horizontal forces and the sum of the moments in the system must be equal to zero. By solving both equilibrium equations, the two unknowns, D and D_1, can be determined.

As force and moment calculations become complex, it is often convenient to determine the value of σ_C, which is the hypothetical lateral earth pressure at depth D that corresponds to passive earth pressure on Side B and active earth pressure on Side A. As will be shown in the Example 10.6, the forces and their line of action are found easier through this procedure.

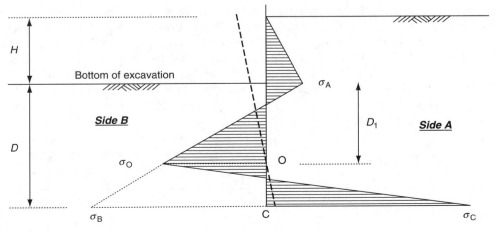

FIGURE 10.27
Conceptual representation of lateral earth pressure on a cantilever sheet-pile wall.

Example 10.6

For the sheet-pile wall shown in Figure 10.28, determine the minimum depth of excavation required to achieve equilibrium.

Solution

For the clay layer, K_a (from Equation 10.2) is equal to 0.406. The depth horizontal stress at point A is determined from Equation (10.10) and the critical depth z_c from Equation (10.11):

$$\sigma_A = 0.406 \times 16 \times 5 - 2 \times 20 \times \sqrt{0.406} = 7.0 \text{ kPa}$$

$$z_c = \frac{2 \times 20}{16\sqrt{0.406}} = 3.92 \text{ m}$$

The coefficients of active and passive lateral earth pressure, K_a and K_p, for the sand layer are equal to 0.333 and 3.0, respectively. The horizontal stress at the top of the sand layer (point B) is thus equal to

$$\sigma_B = 0.333 \times 16 \times 5 = 26.7 \text{ kPa}$$

The horizontal stress at point F is equal to the difference between passive pressure to the left side of the sheet pile and active pressure to the right:

$$\sigma_F = 3 \times 17 \times D - 0.333 \times (16 \times 5 + 17 \times D) = 45.34D - 26.7 \text{ kPa}$$

Similarly, the lateral earth pressure at G is equal to the difference between passive conditions on the right side of the sheet pile and active conditions on the left:

$$\sigma_G = 3 \times (16 \times 5 + 17 \times D) - 0.333 \times 17 \times D = 45.34D + 240 \text{ kPa}$$

It is possible to determine the lateral forces in the sheet piles by calculating the areas of triangles RNA, BRC, CEM, and MGQ. However, such calculations become cumbersome due to the fact that depths D and D_1 are unknown. Instead, it is possible to consider an equivalent set of forces, P_1 to P_4, such that the net earth pressure is the same:

1. P_1 is equal to the area of triangle RNA.
2. P_2 is equal to the area of rectangle RBJQ.

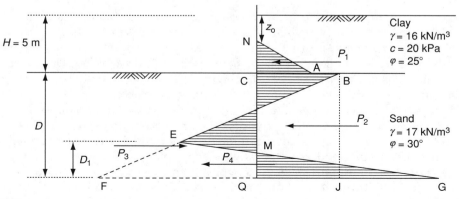

FIGURE 10.28
Illustration for Example 10.6.

3. P_3 is equal to the area of triangle BJF.

4. P_4 is equal to the area of triangle EFG.

We then calculate the forces, P_i, and their corresponding moment arms, y_i, from point R

$$P_1 = (1/2)(5.0 - 3.92)(7.0) = 3.78 \text{ kN}$$
$$P_2 = (D)(26.7) = 26.7D$$
$$P_3 = -(1/2)(D)(26.7 + 45.34D - 26.7) = -22.67D^2$$
$$P_4 = (1/2)(D_1)(45.34D - 26.7 + 45.34D + 240) = (D_1)(45.34D + 106.7)$$

$$y_1 = \tfrac{1}{3}(5.0 - 3.92) = 0.36 \text{ m}$$
$$y_2 = -\tfrac{1}{2}D$$
$$y_3 = -\tfrac{2}{3}D$$
$$y_4 = -(D - \tfrac{1}{3}D_1)$$

The next step is to satisfy the equations of equilibrium, in terms of horizontal forces and moments about point R

$$\sum_{i=1}^{4} P_i = 0$$

$$3.78 + 26.7D - 22.67D^2 + D_1(45.34D + 106.7) = 0$$

$$\sum_{i=1}^{4} P_i y_i = 0$$

$$(3.78)(0.36) - (26.7D)(\tfrac{1}{2}D) + (22.67D)(\tfrac{2}{3}D) - (D_1)(45.34D + 106.7)(D - \tfrac{1}{3}D_1) = 0$$

Solving the two equations above simultaneously, we obtain:

$$D = 2.1 \text{ m} \quad (\text{and } D_1 = 0.2 \text{ m})$$

The actual embedment depth is calculated by multiplying the theoretical depth by a factor of safety of 1.2. Therefore, the actual embedment depth is equal to $1.2D = 2.5$ m.

10.7.2 Anchored Sheet Piles

For large excavation depths the inclusion of an anchor tie rod is necessary in order to reduce the moment on the sheet-pile wall. Otherwise, unreasonably large cross sections may be needed in order to resist the moment. Anchored sheet piles may be analyzed using either the free-earth support or the fixed-earth support method. Under free-earth support conditions, the tip of the sheet pile (Figure 10.29) is assumed to be free to displace and rotate in the ground. Only active earth pressures develop on the tie-back side, while passive pressures act on the other side. No stress reversal of rotation points exist down the embedded depth of the sheet pile. In contrast, the tip of fixed-earth support sheet piles is assumed to be restricted from rotation. Stress reversal occurs down the embedded depth, and the sheet pile is analyzed as a statically indeterminate structure.

To design free-earth support sheet piles, typically the depth of the anchor tie rod needs to be known. Equilibrium conditions are checked in terms of horizontal forces and moments, and the force in the tie rod as well as the depth of embedment are calculated accordingly. Alternatively, the maximum allowable force in the tie rod may be given, with

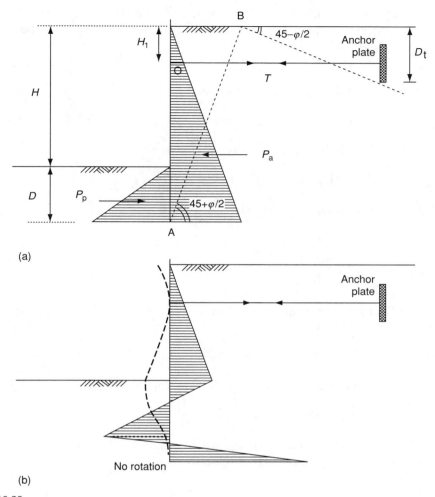

FIGURE 10.29
Conceptual representation of lateral earth pressure on tie-rod anchored sheet-pile walls: (a) free-earth support, and (b) fixed-earth support.

the depth of the tie rod and the embedment depth of the sheet pile as unknowns. It is usually convenient to sum the system moments about the connection of the tie rod with the sheet pile to eliminate the tensile force in the tie rod from the equation.

The necessary anchor resistance in the tie rod is supplied through an anchor plate or deadman located at the far end of the tie rod (Figure 10.29a). In order to ensure stability, the anchor plate must be located outside the active wedge behind the sheet pile, which is delineated by line AB from the tip. In addition, this active wedge must not interfere with the passive wedge through which the tension in the tie rod is mobilized, which is bounded by the ground surface and line BC. Therefore, the length of the tie rod, L_t is calculated from

$$L_t = (H + D) \tan (45° - \varphi/2) + D_t \tan (45° + \varphi/2) \qquad (10.24)$$

Example 10.7
For the sheet-pile wall shown in Figure 10.29(a), determine the depth of embedment, D, and the force in the tie rod. The soil on both sides of the sheet pile is well-graded sand,

with a unit weight $\gamma = 19$ kN/m^3 and an internal friction angle of 34°. The tie rods are spaced at 3 m horizontally, and are embedded at a depth $H_1 = 1$ m. The height of the excavation $H = 15$ m.

Solution
The active force, P_a, on the right side is calculated from

$$P_a = (1/2)K_a\gamma(H + D) = (1/2) \times 0.283 \times 19 \times (15 + D) = 40.328 + 2.689D$$

The passive force, P_p, on the right side is calculated from

$$P_p = (1/2)K_p\gamma D = (1/2) \times 3.537 \times 19 \times D = 33.60D$$

The sum of the moments about point O is

$$(40.328 + 2.689D)(\tfrac{2}{3}\{15 + D\} - 1) - (33.60D)(15 + \tfrac{2}{3}D) = 0$$

Solving for D, we obtain:

$D = 0.77$ m
The tension in the tie rods per meter width of the sheet-pile wall is then calculated from the equilibrium of the horizontal forces:

$$T \text{ (per meter)} = (40.328 + 2.689 \times 0.77) - (33.60 \times 0.77) = 16.53 \text{ kN/m}$$

The force in each tie rod is calculated by multiplying the result by the horizontal spacing:

$$T = 3 \times 16.53 = 50 \text{ kN}$$

The theoretical depth, D, is multiplied by a safety factor of 1.2, thereby giving a total embedment depth of 0.92 m, which is rounded to 1.0 m.

10.7.2.1 *Redundancy in Design*
It is important to note that, when the theoretical depth of embedment of the sheet pile is multiplied by a safety factor of 1.2, the resulting depth does not satisfy equilibrium conditions. In reality, when the depth of embedment is extended, full passive conditions do not develop. Therefore, if an analysis of the stability of an existing sheet pile is carried out, it would be incorrect to consider the earth pressure diagrams used in design. Instead, it is common practice to redesign the sheet-pile structure and then check the redundancy by calculating the factor of safety as

$$FS = \frac{\text{Actual embedment depth}}{\text{Theoretical embedment depth}} \qquad (10.25)$$

10.7.3 **Braced Excavations**

Excavations in urban environments are constrained by the lack of adjacent space for installing tie rods or ground anchors. Where cantilever sheet piles are impractical, it becomes imperative to provide support to the sheet piles through internal bracing and struts. Examples of typical braced excavations are shown in Figure 10.30. It is important to ensure that the bracing system is stiff enough to prevent or minimize adjacent ground

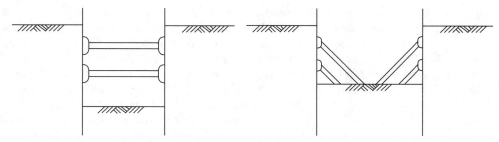

FIGURE 10.30
Examples of braced excavations.

movement, and strong enough to resist the earth pressure associated with such restricted deformations. Construction is usually initiated by driving the sheet piles or lateral support system, then excavating gradually from the ground level down. Rows of bracing or lateral support are installed as the excavation progresses down.

The earth pressure developing in the case of braced excavations is different from the theoretical linear increase with depth described earlier for conventional retaining walls. In braced excavations, the lateral earth pressure is dictated by the sequence of excavation, soil type, stiffness of the wall and struts, and movement allowed prior to installing the struts. Although accurate determination of the distribution of earth pressure in braced cuts is almost impossible, Terzaghi et al. (1996) provide approximate methods based on actual observations for use in strut design (Figure 10.31). In the case of sands, the envelope of the apparent earth pressure is constant and equal to $0.65K_a\gamma H$, where H is the depth of the excavation. In the case of soft/medium clays, and stiff fissured clays, the diagrams shown in Figure 10.31(b) and (c) are used, respectively. It is imperative, however, to monitor the development of forces in the struts and deformations along the depth of the braced cut during construction. Forces in the struts and deformations must be continuously adjusted to comply with design specifications.

10.7.4 Tie-Back Anchored Walls

In cases where adequate land adjacent to an excavated site is available, the use of tie-back anchors is preferred over braced cuts. Tie-back anchors do not interfere with the space available for construction equipment mobilization in the excavation, and can be used as

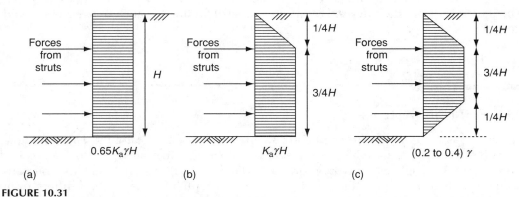

FIGURE 10.31
Approximate earth pressure diagrams for use in designing braced excavations: (a) sand, (b) soft to medium clay, and (c) stiff fissured clay.

permanent earth retaining systems. Construction proceeds by first driving soldier piles into the ground. Soldier piles are individual sections, typically steel I-beams that are driven in the ground at close spacing (1 or 2 m). Excavation is then carried out in stages, 1 to 2 m deep at a time, with timber lagging placed in between the soldier piles to prevent soil collapse. Tie-back anchors are subsequently installed at different levels, and locked in place to support the retaining structure (Figure 10.31c). The installation procedure consists of directionally drilling in the soil, typically at a 15° angle, then sliding a PVC or metal perforated casing into the hole. A steel anchor cable or rod is then inserted into the casing, and grout is injected to fill the casing around the cable and permeate through the casing perforations into the surrounding soil. The grouted zone must be fully located outside the active wedge $(45° + \varphi/2)$ measured from the bottom of the excavation. The tension at each level can be adjusted by tightening or loosening a lock nut at the connection between the tie-back anchor and the wall. Walls very large in height can be constructed using this technique. An alternative procedure for building tie-back anchored walls involves the construction of a concrete wall by slurry replacement, followed by a similar excavation and tie-back anchoring procedure, as described above.

When determining the maximum pullout resistance of a tie-back anchor, it is important to determine the soil type, and whether the structure is temporary or permanent. For permanent walls in clays, the use of effective stress analysis concepts is essential in calculating the long-term pullout capacity. On the other hand, total stress analysis (undrained conditions) may be adopted for short-term evaluation of tie-back anchor capacity. For tie-back anchors in sand, the maximum capacity, T_{ult}, is calculated from

$$T_{ult} = \pi d p_g L \tan \varphi \qquad (10.26)$$

where d is the diameter of the grouted bulb, p_g is the earth pressure on the grout, L is the grouted length, and φ is the friction angle of the soil. The value of p_g may be taken as $K_0 \sigma_v'$ for shallow depths, where arching is not expected in the soil. Alternatively, based on field measurements of typical grouted anchors, the value of $\pi d p_g$ may be taken to be equal to 500 kN/m for coarse granular materials, and 150 kN/m for fine sands. For undrained analysis in clays, T_{ult}, is calculated from

$$T_{ult} = 0.3\pi d L S_u \qquad (10.27)$$

where S_u is the undrained shear strength (undrained cohesion) of the clay.

The factor of safety for pullout resistance is typically taken to be 3 or 4 for long-term analysis and 2 for short-term analysis. Tie-back anchors are tested in the field, postconstruction, at 1.5 times their desired capacity. In case of failure, the anchor is regrouted and retested.

10.8   Soil Nail Systems

Soil nailing is a technique for stabilizing steep slopes and vertical cuts. The technique relies on driving soil nails, which are steel rods 25 to 50 mm in diameter, into a vertical or steep cut at a 15° angle as excavation proceeds (Figure 10.32). Alternatively, the soil nails may be installed through drilling and grouting. Conceptually, the method differs from tie-back anchored walls in that no prior driving or installation of an earth retaining wall (sheet pile, soldier piles, or slurry wall) is conducted. Instead, the exposed soil surface is

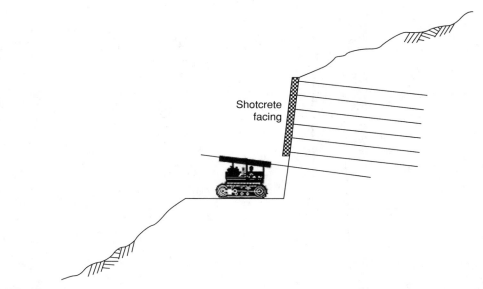

FIGURE 10.32
Construction of soil-nailed wall.

kept from caving in by installing a wire mesh on which the soil nails are connected through face plates. The technique works best in cohesive soils, since significant unraveling may occur in sandy soils. The wire mesh is then covered with shotcrete, and excavation is proceeded to the next level. Soil nail walls are typically used as temporary earth retaining systems, although they have been used successfully as permanent structures.

When analyzing soil nail systems, global stability is considered. Conventional methods for slope stability analysis are used, with the tension in the soil nails contributing to the stability of the slope. Typically, the method of slices is chosen, and the forces acting on each slice, including the tensile resistance of the soil nail are included. The global stability of the reinforced soil mass is then assessed, and a factor of safety is calculated. The length of the soil nails is determined accordingly, so as to satisfy equilibrium of the slope or vertical cut.

10.9 Drainage Considerations

Proper drainage of backfill materials is a crucial component of retaining wall design. In general, cohesive backfills are highly undesirable because of their poor drainage, loss of strength and increase in density upon wetting, and high coefficient of active earth pressure. This is particularly important in the case of MSE walls, where only select backfill material may be used. Acceptable soil types for such purpose are SW (well-graded sand), GW (well-graded gravel), and SP (poorly graded sand). The vast majority of current design codes prohibit the use of cohesive materials as backfill in MSE walls.

In addition, proper drainage provisions must be included in the design. This entails the inclusion of drains and filters in the cross section, such as those shown in Figure 10.33.

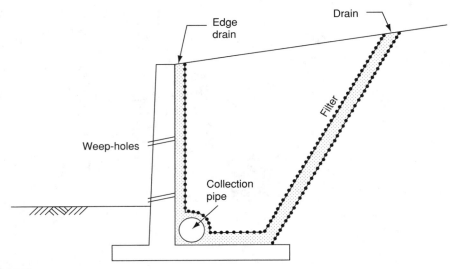

FIGURE 10.33
Drainage provisions in cantilever retaining wall.

While graded sand constituted the majority of filters and drains in the past, geosynthetics (geotextiles, geonets, and geocomposites) are used in almost all projects today. In order to select the proper filter material, the acceptable apparent opening size (O_{95}) of the geosynthetic must be first determined. Current guidelines require that O_{95} be smaller than $2.5D_{85}$, where D_{85} is the grain size corresponding to 85 percentile on the grain size distribution of the soil. Geonets selected for drainage must have a maximum flow rate that is greater than the anticipated flow rate in the structure.

10.10 Deformation Analysis

Recently, the recognition of the importance of limiting deformation levels in retaining walls has led to the emergence of deformation-based design approaches. Because all such methods require accurate estimation of the stress–strain behavior of soils, by itself a difficult task, deformation-based design is limited to large projects where, for instance, soil parameters can be determined accurately from extensive laboratory tests. Finite element analyses can then be performed to determine the anticipated levels of deformation under a given set of boundary conditions, and the wall design is modified accordingly to result in acceptable deformations.

   In projects where deformation is critical, such as bridge approaches and abutments, deformations are typically monitored during construction to ensure they remain within the acceptable range. In situations where tie-back anchored sheet-pile walls are constructed, the tension within the tie-backs may be adjusted at the connection between the facing and the tie-backs to prevent excessive deformations during construction. Levels of deformation required to achieve active and passive conditions during construction, for different soil types, are given in Table 10.1.

10.11 Performance under Seismic Loads

In seismically active areas, it is essential to consider the additional seismic forces acting on the earth retaining structure during design. For gravity and cantilever walls, the procedure most widely used in practice is the Mononobe–Okabe method, as modified by Seed and Whitman (1970). The method considers the stability of Coulomb-type wedge in a manner similar to that used in static design (Figure 10.34). However, an equivalent seismic load is assumed to act at the center of gravity of the wedge, with horizontal and vertical magnitudes equal to the mass of the wedge times the horizontal and vertical accelerations, respectively. The force resulting from the combination of static earth pressure and seismic loading is assumed to act at a height of $0.6H$ from the bottom of the wall. The active earth pressure resultant force on the wall, P_{AE}, is calculated from:

$$P_{AE} = (1/2)K_{AE}\gamma(1 - k_v)H^2 \tag{10.28}$$

$$K_{AE} = \frac{\cos^2{(\varphi - \psi - \theta_v)}}{\cos{\psi}\cos^2{\theta_v}\cos{(\psi + \theta_v + \delta)}\left(1 + \sqrt{\frac{\sin{(\varphi + \delta)}\sin{(\varphi - \psi - \beta)}}{\cos{(\beta - \theta_v)}\cos{(\psi + \theta_v + \delta)}}}\right)^2} \tag{10.29}$$

$$\psi = \frac{k_h}{1 - k_v} \tag{10.30}$$

where k_v is the vertical acceleration in gs, ψ is the seismic inertia angle calculated from Equation (10.30), θ_v is the inclination of the wall with respect to vertical, β is the slope angle of the ground surface, and δ is the soil–wall interface friction angle.

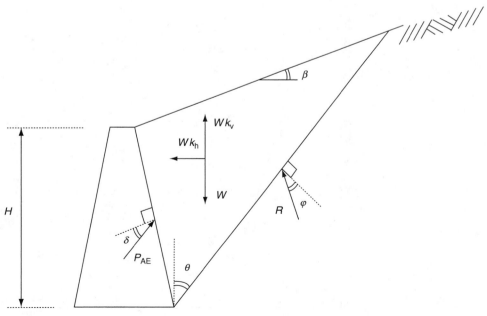

FIGURE 10.34
Mononobe–Okabe stability wedge under seismic loading.

Similarly, the force on the wall under passive earth pressure conditions is calculated from:

$$P_{PE} = (1/2)K_{PE}\gamma(1 - k_v)H^2 \tag{10.31}$$

$$K_{PE} = \frac{\cos^2(\varphi - \psi + \theta_v)}{\cos\psi \cos^2\theta_v \cos(\psi - \theta_v + \delta)\left(1 - \sqrt{\dfrac{\sin(\varphi + \delta)\sin(\varphi - \psi + \beta)}{\cos(\beta - \theta_v)\cos(\psi - \theta_v + \delta)}}\right)^2} \tag{10.32}$$

Field observations of postearthquake conditions of MSE walls indicate minimal damage to the structure. This is partly due to the high level of redundancy in design, but is mostly attributed to the ductility of the structure.

10.12 Additional Examples

Example 10.8

The concrete retaining wall shown in Figure 10.35 retains a granular backfill. The angle of friction between the backfill and concrete is 12°. Determine the resultant force on the wall due to the earth pressure, its direction, and the line of action.

Based on the figure, 4 m = 5 cm

Hence, the top width = 0.72 m and the bottom width = 2 m

$$\alpha = 180° - \tan[14/(2 - 0.72)] = 180° - 72° = 108°$$
$$\beta = 0$$
$$\delta = 12°$$
$$N = 30°$$

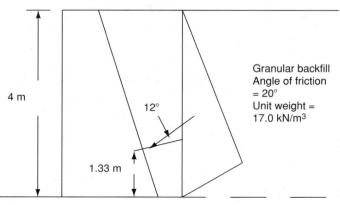

FIGURE 10.35
Illustration for Example 10.8.

Using Coulomb's active earth pressure coefficient (Equation 10.8)

$$K_A = \left(\frac{\sin(\alpha - \varphi)/\sin\alpha}{\sqrt{\sin(\alpha + \delta)} + \sqrt{\dfrac{\sin(\varphi + \delta)\sin(\varphi - \beta)}{\sin(\alpha - \beta)}}} \right)^2 = 0.455$$

The pressure diagram can be determined by the following equation:

$$\begin{aligned} \sigma_{h,a} &= K_a \gamma H - 2c\sqrt{K_a} \\ &= 0.455(17)z = 7.735z \quad \text{(see the plot in figure)} \end{aligned} \tag{10.10}$$

where P_A is the area of the pressure diagram equal to

$$\int_0^4 [7.735z]dz = 61.88 \text{ kN/m}$$

The line of action and direction are indicated in the figure.

Example 10.9

The steel sheet pile shown in Figure 10.36(a) is embedded in clayey ground and retains a granular backfill

(i) Plot the lateral earth pressure exerted on the sheet pile.

(ii) Plot the Mohr circles for points on the sheet pile at depths of 1 m and 6 m from the top.

(iii) Accurately indicate the failure surface

$$\gamma = \frac{(G_s + S_e)\gamma_w}{1 + e} = 11.8 \text{ kN/m}^3$$

where $G_s = 2.65$, $S_e = 0$, and $e = 1.2$

$$\gamma_s = \frac{(2.65 + 1 \times 1.2)}{1 + 1.2} 9.8 = 17.15 \text{ kN/m}^3$$

For sandy soil

$$K_a = \cos\alpha \frac{\cos\alpha - \sqrt{\cos^2\alpha - \cos^2\phi}}{\cos\alpha + \sqrt{\cos^2\alpha - \cos^2\phi}} = 0.5312, \quad \text{where } \alpha = 10°, \ \phi = 20°$$

or

$$K_a = \left[\frac{\sin(\alpha - \phi)/\sin\alpha}{\sqrt{\sin(\alpha + \delta)} + \sqrt{\dfrac{\sin(\alpha + \delta)\sin(\phi - \beta)}{\sin(\alpha - \beta)}}} \right]^2 \tag{10.8}$$

$$= 0.569 \approx 0.57$$

where

$$\begin{cases} \alpha = 90° \text{ vertical} \\ \phi = 20° \\ \delta = \text{surface friction} = 0 \\ \beta = \text{slope} = 10° \end{cases}$$

For clayey soil

$$K'_a = \frac{K_a}{\cos \alpha} = 1.0 \quad \text{or} \quad K_a = 1.0$$

$$K'_p = \frac{K_p}{\cos \alpha} = \frac{\cos \alpha + \sqrt{\cos^2 \alpha - \cos^2 \phi}}{\cos \alpha + \sqrt{\cos^2 \alpha - \cos^2 \phi}} = 1.0$$

or

$$K_p = \left[\frac{\sin (\alpha + \phi)/ \sin \alpha}{\sqrt{\sin (\alpha + \phi)} - \sqrt{\dfrac{\sin (\phi + \delta) \sin (\phi + \beta)}{\sin (\alpha - \beta)}}} \right]^2 = 1.0$$

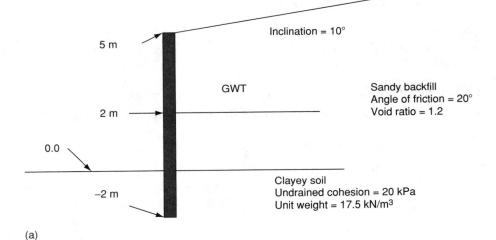

(a)

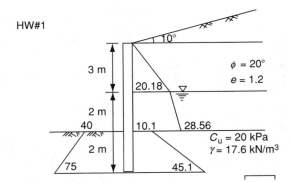

(b)

FIGURE 10.36
Continued

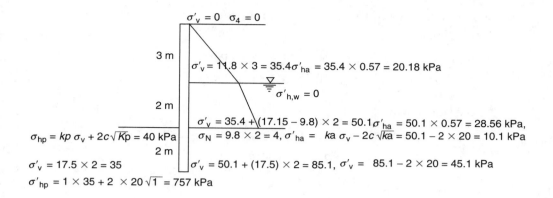

$\sigma'_v = 0$ $\sigma_4 = 0$

3 m

$\sigma'_v = 11.8 \times 3 = 35.4 \sigma'_{ha} = 35.4 \times 0.57 = 20.18$ kPa

$\sigma'_{h,w} = 0$

2 m

$\sigma'_v = 35.4 + (17.15 - 9.8) \times 2 = 50.1 \sigma'_{ha} = 50.1 \times 0.57 = 28.56$ kPa,

$\sigma_N = 9.8 \times 2 = 4, \sigma'_{ha} = \quad ka\, \sigma_v - 2c\, \sqrt{ka} = 50.1 - 2 \times 20 = 10.1$ kPa

$\sigma_{hp} = kp\, \sigma_v + 2c\sqrt{Kp} = 40$ kPa

2 m

$\sigma'_v = 17.5 \times 2 = 35$

$\sigma'_v = 50.1 + (17.5) \times 2 = 85.1, \sigma'_v = \quad 85.1 - 2 \times 20 = 45.1$ kPa

$\sigma'_{hp} = 1 \times 35 + 2 \times 20 \sqrt{1} = 757$ kPa

(c)

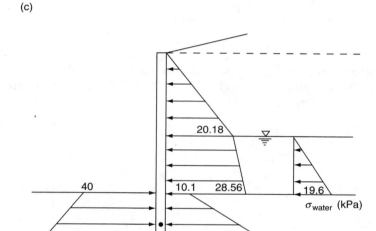

20.18

40

10.1   28.56

19.6

σ_{water} (kPa)

75

45.1

σ_{hp} (kPa) σ_{ha} (kPa)

(d)

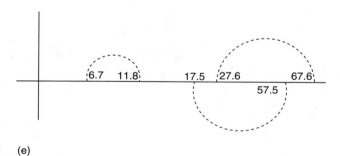

6.7   11.8   17.5   27.6   67.6

57.5

(e)

FIGURE 10.36

(a) Illustration for Example 10.9. (b) Earth pressure distribution on the sheet pile. (c) Earth pressure computations for Example 10.9. (d) Earth and water pressure distributions for Example 10.9. (e) Mohr circle plots for Example 10.9.

(ii) Stresses at depths of 1 m and 6 m from the top:

Active side at 1 m

$$\sigma_v = 11.8 \times 1 = 11.8 \text{ kPa}$$
$$\sigma_{ha} = 0.57 \times 11.8 = 6.7 \text{ kPa}$$

Active side at 6 m

$$\sigma_v = 11.8 \times (17.15 - 9.8) \times 2 + 17.15 \times 1 = 67.6 \text{ kPa}$$
$$\sigma_{ha} = 1 \times 67.6 - 2 \times 20\sqrt{1} = 27.6 \text{ kPa}$$

Passive side at 6 m

$$\sigma_v = 17.5 \times 1 = 17.5 \text{ kPa}$$
$$\sigma_{hp} = 1 \times 17.5 + 2 \times 20\sqrt{1} = 57.5 \text{ kPa}$$

Example 10.10
Estimate the passive pressure on the wall shown in Figure 10.37(a) by selecting a trial failure surface shown. The backfill supports a highway pavement of width 4.5 m, which imposes a distributed load of 40 kPa.
Horizontal, $\beta = 0°$

$$\text{Length} = \frac{5}{\sin 45°} = 7.07 \text{ m}$$

Soil area $= (1/2)(6.5)(5) = 16.25 \text{ m}^2$

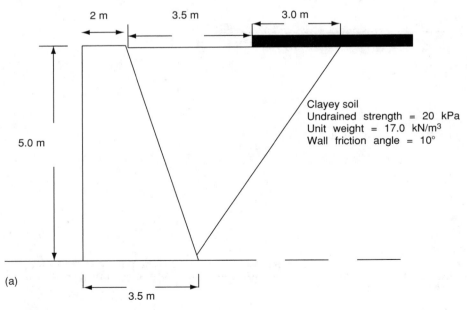

2 m   3.5 m   3.0 m

Clayey soil
Undrained strength = 20 kPa
Unit weight = 17.0 kN/m³
Wall friction angle = 10°

5.0 m

(a)

3.5 m

FIGURE 10.37
Continued

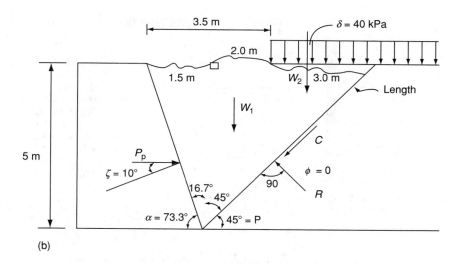

(b)

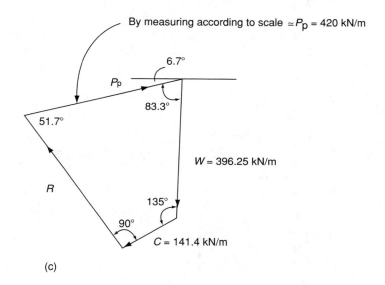

(c)

FIGURE 10.37
(a) Illustration for Example 10.10. (b) Free-body diagram for Example 10.10. (c) Force polygon for Example 10.10.

$W_1 = 16.25 \times 17 = 276.25$ kN/m
$W_2 = (40)(3) = 120$ kN/m
$W = W_1 + W_2 = 396.25$ kN/m
$C = S_u(\text{length}) = 20(7.07) = 141.4$ kN/m

For passive lateral force, the wall will be pushed inside, moves right.
By measuring according to scale,

$$\approx P_p = 420 \text{ kN/m}$$

Example 10.11

Investigate the stability of the concrete retaining wall shown in Figure 10.38(a) against sliding. Assume that the stem and the base are of width 0.5 m.

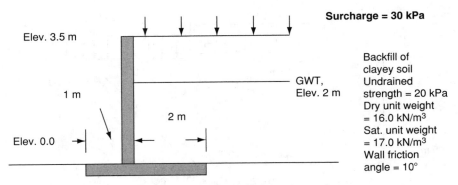

Surcharge = 30 kPa

Elev. 3.5 m

Backfill of
clayey soil
GWT,   Undrained
Elev. 2 m   strength = 20 kPa
            Dry unit weight
            = 16.0 kN/m³
            Sat. unit weight
            = 17.0 kN/m³
Wall friction
angle = 10°

1 m

2 m

Elev. 0.0

(a) Original ground of granular soil, Sat. unit weight = 16.5 kN/m³
Internal friction angle = 20°

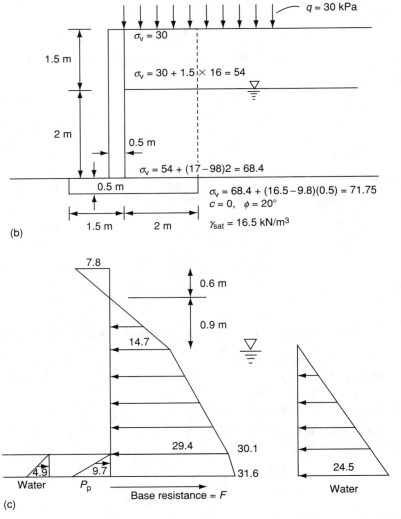

$q = 30$ kPa

1.5 m

$\sigma_v = 30$

$\sigma_v = 30 + 1.5 \times 16 = 54$

2 m

0.5 m

$\sigma_v = 54 + (17 - 9.8)2 = 68.4$

0.5 m

$\sigma_v = 68.4 + (16.5 - 9.8)(0.5) = 71.75$
$c = 0, \quad \phi = 20°$

1.5 m   2 m

$\gamma_{sat} = 16.5$ kN/m³

(b)

7.8

0.6 m

0.9 m

14.7

29.4   30.1

31.6

4.9   9.7
Water P_p

24.5

Base resistance = F

Water

(c)

FIGURE 10.38
(a) Illustration for Example 10.11. (b) Earth pressure calculations for Example 10.11. (c) Earth and water pressure distributions for Example 10.11.

$$K_a = \left[\frac{\sin(\alpha - \phi)/\sin\alpha}{\sqrt{\sin(\alpha + \delta)} + \sqrt{\dfrac{\sin(\alpha + \delta)\sin(\phi - \beta)}{\sin(\alpha - \beta)}}} \right]^2 = 1.02 \qquad (10.8)$$

where

$\alpha = 90°$

$\phi = 0°$

$\delta = 10°$

$\beta = 0°$

For foundation soil,

$$\left. \begin{array}{l} \beta = 0°; \quad c = 0; \quad \phi = 20° \\ \alpha = 90°; \quad \delta = 2/3; \quad \phi = 13.3° \end{array} \right\}$$

$$K_a = 0.44$$
$$K_p = 2.89$$

Active side:

$\sigma_{h,Dm} = 30 \times 1.02 - 2(20)\sqrt{1.02} = -9.8$ kPa

$\sigma_{h,1.5m} = 54 \times 1.02 - 2(20)\sqrt{1.02} = 14.7$ kPa

$\sigma_{h,3.5m_T} = 68.4 \times 1.02 - 2(20)\sqrt{1.02} = 29.4$ kPa

$\sigma_{h,3.5m_B} = 68.4 \times 0.44 = 30.1$ kPa

$\sigma_{h,4m} = 71.75 \times 0.44 = 31.6$ kPa

$\sigma_{water,4m} = 9.81 \times 2.5 = 24.5$ kPa

Passive side:

$\sigma_{h,0m} = 0$

$\sigma_{h,0.5m} = (16.5 - 9.8)(0.5)(2.89) = 9.7$ kPa

$\sigma_{water,0.5m} = 9.81(0.5) = 4.9$ kPa

$W = \text{surcharge} + \text{soil} + \text{stem} + \text{base}$

$\quad = 30 \times 2 + (16)(1.5 \times 2) + (17)(2 \times 2) + (0.5)(3.5)(23) + (0.5)(3.5)(23)$

$\quad = 60 + 116 + 80.5 = 256.5$ kN/m

$F = \text{base friction resistance} = W \tan\delta$

$\quad = 256.5 \, \tan\left(\dfrac{2}{3} \times 20\right)$

$\quad = 60.8$ kN/m

$$\text{Active force} = \frac{1}{2}(14.7)(0.9) + \frac{1}{2}(14.7 + 29.4)(2) + \frac{1}{2}(30.1 + 31.6)(0.5) + \frac{1}{2}(2.5)(24.5)$$

$$= 96.8 \text{ kN/m}$$

$$\text{Passive resistance} = \frac{1}{2}(0.5)(9.7) + \frac{1}{2}(0.5)(4.9) = 3.65 \text{ kN/m}$$

$$\text{FS} = \frac{3.65 + 60.8}{96.8} = 0.66 < 1.5$$

So, not OK. Hence, wall will slide.

Example 10.12
Check the stability of the gravity retaining wall shown in Figure 10.39(a) against over-turning and development of bottom tension.

(a) Clayey soil

$$Z = \frac{2C}{\gamma\sqrt{K_a}} = 2.25 \text{ m} \tag{10.11}$$

$$S_u = 20 \text{ kPa}; \quad \gamma = 17 \text{ kN/m}; \quad \delta = 10°$$

$$P_{ax} = (60.56)\cos 24 = 55.32 \text{ kN/m}$$
$$P_{ay} = (60.56)\sin 24 = 24.63 \text{ kN/m}$$

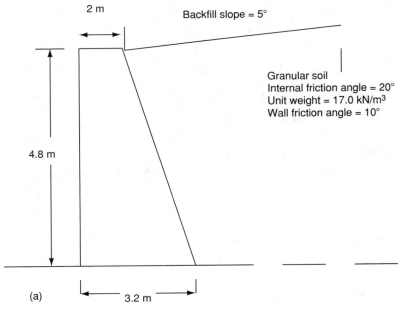

FIGURE 10.39
Continued

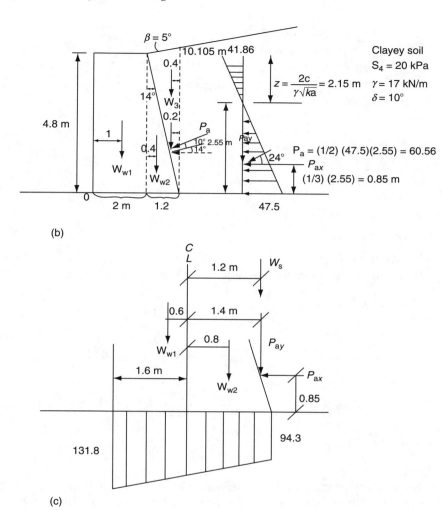

(b)

(c)

FIGURE 10.39
(a) Illustration for Example 10.12. (b) Earth pressure calculations for Example 10.12. (c) Bottom pressure distribution for Example 10.12.

$$K_a = \left[\frac{\sin(\alpha - \phi)/\sin\alpha}{\sqrt{\sin(\alpha + \delta)} + \sqrt{\frac{\sin(\alpha + \delta)\,\sin(\phi - \beta)}{\sin(\alpha - \beta)}}} \right]^2 = 1.095 \qquad (10.8)$$

Top, $\quad \sigma_h = \sigma_v K_a - 2c\sqrt{k_a}$

$\qquad\qquad = -2(20)\sqrt{1.095} = -41.86$ kPa

Bottom, $\quad \sigma_h = (17)(4.8)(1.095) - 2(20)\sqrt{1.095} = 47.5$ kPa

$\qquad w_{w_1} = (2)(4.8)(23) = 220.8$ kN/m $\qquad\qquad (10.10)$

$\qquad W_{w_2} = \left(\frac{1}{2}\right)(1.2)(4.8)(23) = 66.24$ kN/m

$\qquad W_s = \left(\frac{1}{2}\right)(1.2)(4.8 + 0.105)(17) = 50.03$ kN/m

$$\text{FS}_{\text{overturning}} = \frac{\sum \text{Resisting moments}}{\sum \text{Overturning moments}} \text{(taking moments at O)}$$

$$= \frac{(220.8)(1) + (66.24)(2 + 0.4) + (50.03)(3.2 - 0.4) + (24.6 \times 3)}{(55.3)(0.85)} \quad (10.13)$$

$$= \frac{593.66}{47} = 12.6 \gg 1.5$$

So, OK.

(b) Bottom tension

$$f = \frac{V}{A} \pm \frac{Mc}{I}$$

$$V = 220.8 + 66.24 + 50.03 + 24.6$$

$$= 361.67 \, \text{kN/m}$$

$$A = (3.2)(1) = 3.2 \, \text{m}^2/\text{m}$$

$$M = (220.8)(0.6) + (55.3)(0.85) - (66.24)(0.8) - (50.03)(1.2) - (24.6)(1.4) = 32.017 \, \text{kN m}$$

$$I = \frac{6h^3}{12} = \frac{(1)(3.2)^3}{12} = 2.73$$

$$\text{Max. stress} = F_{\text{max}} = \frac{361.67}{3.2} + \frac{(32)(1.6)}{2.73} = 131.8 \, \text{kPa}$$

$$\text{Min. stress} = F_{\text{min}} = \frac{361.67}{3.2} - \frac{(32)(1.6)}{2.73} = 94.3 \, \text{kPa}$$

There is no tension.

Example 10.13
Investigate the stability of the cantilever sheet-pile wall shown in Figure 10.40 using the free-earth support method. Assume that the depth embedment is 1.5 m.

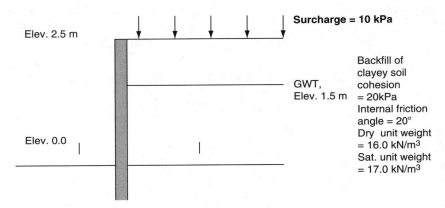

(a)   Original ground of clayey soil, Sat. unit weight = 16.5 kN/m³
          Undrained strength = 20 kPa

FIGURE 10.40
Continued

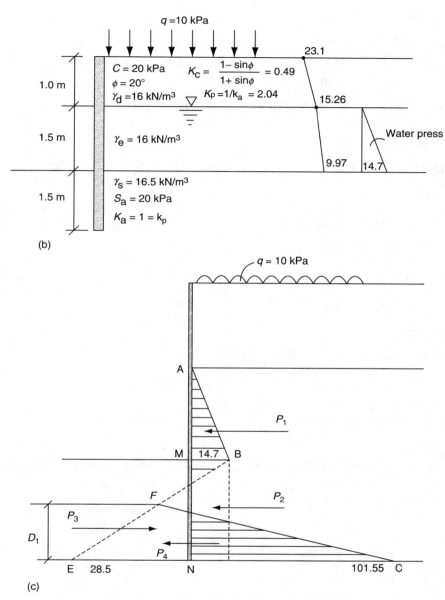

FIGURE 10.40
(a) Illustration for Example 10.13. (b) Earth pressure calculations for Example 10.13. (c) Earth pressure distribution for Example 10.13.

Right side:

$$\sigma_{h,0m} = (10)(0.49) - 2(20)\sqrt{0.49} = -23.1 \text{ kPa}$$

$$\sigma_{h,1m} = (10 + 16 \times 1)(0.49) - 2(20)\sqrt{0.49} = -15.26 \text{ kPa}$$

$$\sigma_{h,2.5m} = (10 + 16 + (17 - 9.8)(1.5))(0.49) - 2(20)\sqrt{0.49} = -9.97 \text{ kPa}$$

$$\sigma_{h,2.5m} = (36.8)(1) - 2(20)\sqrt{1} = -3.2 \text{ kPa}$$

$$\sigma_{h,4m} = (36.8 + (16.5 - 9.8)(1.5))(!) - 2(20)\sqrt{1} = +6.85 \text{ kPa}$$

$$\sigma_{h,w_{4m}} = (9.8)(3) = 29.4 \text{ kPa}$$

Left side:

$$\sigma_{p,0m} = 2(20)\sqrt{1} = 40 \text{ kPa}$$
$$\sigma_{p,1.5m} = (16.5 - 9.8)(1.5)(1) + 2(20)\sqrt{1} = 50.05 \text{ kPa}$$
$$\sigma_{a,0m} = -2(20)\sqrt{1} = -40 \text{ kPa}$$
$$\sigma_{a,1.5m} = (16.5 - 9.8)(1.5)(1) - 2(20)\sqrt{1} = 29.95 \text{ kPa}$$

From the figure,

Horizontal stress at E, $\sigma_E = (50.05 + 14.7) - (6.85 + 29.4) = 28.5 \text{ kPa}$

Horizontal stress at C, $\sigma_C = (86.85 + 29.4) - (14.7) = 101.55 \text{ kPa}$

$$P_1 = \frac{1}{2}(14.7)(1.5) = 11.02 \text{ kN/m}$$
$$P_2 = 14.7(1.5) = 22.05 \text{ kN/m}$$
$$P_3 = \frac{1}{2}(28.5 + 14.7)(1.5) = 32.4 \text{ kN/m}$$
$$P_4 = \frac{1}{2}(28.5 + 101.55)(D_1) = 65.02D_1$$
$$\sum F_x = 0$$

therefore, $11.02 + 22.05 - 32.4 + 65.02D_1 = 0$

$D_1 = \text{negative}$

Point of rotation does not exist.
Wall will move rightward, Unstable.

Example 10.14
Assuming that the depth of embedment is 1.2 m, estimate the magnitude of tensioning required for stability of the anchored sheet-pile wall shown in Figure 10.41(a). Comment on the depth of embedment provided.

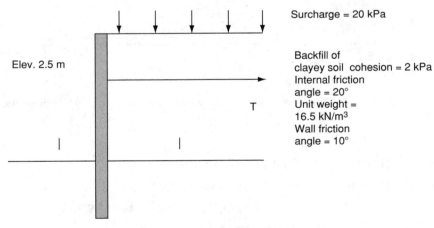

(a)   Original ground of clayey soil, Unit weight = 17.5 kN/m³
       Undrained strength = 10 kPa

FIGURE 10.41
Continued

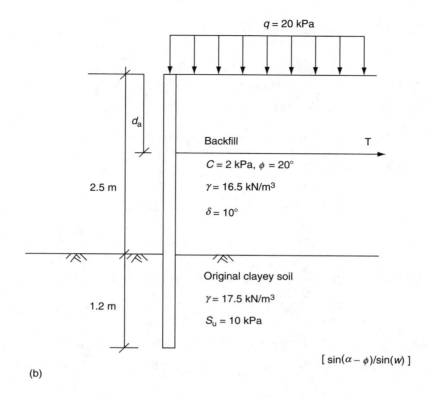

$q = 20$ kPa

d_a

Backfill   T

2.5 m

$C = 2$ kPa, $\phi = 20°$

$\gamma = 16.5$ kN/m³

$\delta = 10°$

Original clayey soil

$\gamma = 17.5$ kN/m³

1.2 m

$S_u = 10$ kPa

$[\sin(\alpha - \phi)/\sin(w)]$

(b)

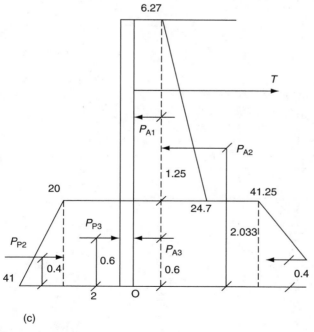

6.27

T

P_{A1}

P_{A2}

1.25

20

41.25

24.7

P_{P3}

2.033

P_{P2}

0.6

P_{A3}

0.4

0.6

41

2   O

0.4

(c)

FIGURE 10.41
(a) Illustration for Example 10.14. (b) Properties of different soil layers in Example 10.14. (c) Earth pressure distribution for Example 10.14.

(1) Backfill soil:

$$K_a = \left[\frac{\sin(\alpha - \phi)/\sin \alpha}{\sqrt{\sin(\alpha + \delta)} + \sqrt{\frac{\sin(\alpha + \delta)\sin(\phi - \beta)}{\sin(\alpha - \beta)}}} \right]^2$$

$$= 0.447$$

$$\phi = 20°; \quad \alpha = 90°; \quad \delta = 10°; \quad \beta = 0°$$

(2) Original clayey soil:

$$K_a = \tan^2\left(45 - \frac{\phi}{2}\right) = \tan^2(45 - 0) = 1$$

$$K_p = \frac{1}{K_a} = 1$$

$$\sigma_{ha,0} = \sigma_v K_a - 2C\sqrt{K_a} = (20)(0.447) - 2(2)(0.447) = 6.27 \text{ kPa}$$

$$\sigma_{ha,2.5m_T} = (20 + 16.5 \times 2.5)(0.447) - 2(2)\sqrt{0.447} = 24.7 \text{ kPa}$$

$$\sigma_{ha,2.5m_B} = (20 + 16.5 \times 2.5)(1) - 2(10)\sqrt{1} = 41.25 \text{ kPa}$$

$$\sigma_{ha,3.7m} = (20 + 16.5 \times 2.5 + 17.5 \times 1.2)(1) - 2(10)\sqrt{1} = 62.25 \text{ kPa}$$

$$\sigma_{hp,0m} = \sigma_v K_p + 2C\sqrt{K_p} = 0 + 2(10)\sqrt{1} = 20 \text{ kPa}$$

$$\sigma_{hp,1.2m} = (17.5 \times 1.2)(1) + 2(10)\sqrt{1} = 41 \text{ kPa}$$

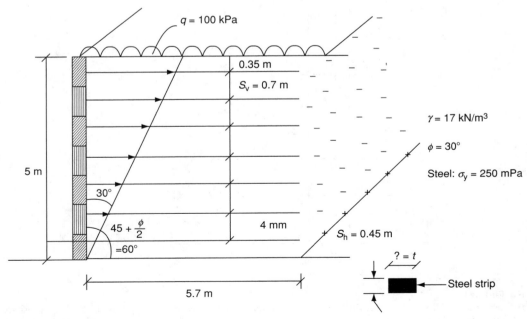

FIGURE 10.42
Illustration for Example 10.15.

$$P_{A_1} = (6.27)(2.5) = 15.68 \text{ kN/m}$$

$$P_{A_2} = \frac{1}{2}(24.7 - 6.27)(2.5) = 23.04 \text{ kN/m}$$

$$P_{A_3} = (41.25)(1.2) = 49.5 \text{ kN/m}$$

$$P_{A_4} = \frac{1}{2}(62.25 - 41.25)(1.2) = 12.6 \text{ kN/m}$$

$$P_{P_1} = (20)(1.2) = 24 \text{ kN/m}$$

$$P_{P_2} = \frac{1}{2}(41 - 20)(1.2) = 12.6 \text{ kN/m}$$

$$\sum F_x = 0$$
$$P_A = P_p + T$$
$$T = (24 + 12.6) + (15.68 + 23.04 + 9.5 + 12.6)$$
$$T = 64.22 \text{ kN/m}$$
$$\sum M_o = 0$$
$$T(3.7 - d_a) + (24)(0.6) + (12.6)(0.4) = (15.68)(1.25 + 1.2) + (23.04)(2.033)$$
$$+ (49.5)(0.6) + (12.6)(0.4)$$

therefore $\quad d_a = 2.13 \text{ m}$

Example 10.15

A 5 m MSE wall is reinforced with steel strips at vertical and horizontal spacing of 0.7 and 0.45 m, respectively. First draw the configuration of the strips and the wall. If the steel strips are 4 mm in thickness and 5.7 m in length determine the minimum width of the strips and the safety factor of the MSE wall against yielding under a surcharge of 100 kPa. Assume that the unit weight and internal friction are 17 kN/m^3 and 30°, respectively, and the yield strength of steel is 250 MPa.

$$\gamma = 17 \text{ kN/m}^3$$
$$\phi = 30°$$
$$\text{Steel}: \quad \sigma_y = 250 \text{ MPa} \qquad \qquad (10.2)$$
$$K_a = \frac{1 - \sin 30°}{1 + \sin 30°}$$
$$= 0.3333$$

Maximum horizontal stresses will develop at the bottom layer

$$h = 0.35 + 0.7 \times 6 = 4.55 \text{ m}$$

$$\sigma_{a,h} = K_a \sigma_v - 2C\sqrt{K_a}$$
$$= (0.333)(100 + 4.55 \times 17) = 59.06 \text{ kN/m}^2/\text{m}$$

Assuming $FS_{yielding} = 1.5$, $\quad F_{max} = 1.5 \times 59.06 \times 0.7 \times 0.45 = 27.91 \text{ kN}$

The width of steel stripes, $\quad w = \dfrac{F_{max}}{\sigma_y \times \text{thickness}} = \dfrac{27.9}{250{,}000 \times 0.004} = 0.0279 \text{ m} = 28 \text{ mm}$

For bottom layer, effective length $= 5.7 - (5 - 4.55)\tan 30 = 5.44\ m = L_e$

$$FS_{pullout} = \frac{2\omega \cdot \sigma_v \tan \delta L_e}{\sigma_a h S_v S_h}$$

$$= 1.5 = \frac{2\omega(100 + 4.55 \times 17)\tan(20)(5.44)}{(59.06)(0.35 + 0.45)(0.45)}$$

therefore $\omega = 0.045\ m \approx 45\ mm$

For top layer, effective length $= 5.7 - (5 - 0.35)\tan 30 = 3.02\ m$

$$\frac{2\omega(100 + 0.35 \times 17)\tan(20)(3.02)}{(0.333)(100 + 0.35 \times 17)(0.7)(0.45)} = 1.5$$

therefore $\omega = 0.0715\ m = 71.5\ mm$

therefore $F_{steel} = \sigma_y(w)(t) = 250,000(0.0715)(0.004) = 71.5\ kN$

$$FS_{yielding} = \frac{F_{steel}}{F_{bottom}} = \frac{71.5}{(59.06)(0.7)(0.45)} = 3.84 \geq 1.5$$

Glossary

Allowable stress design (ASD) — A design method in which a global factor of safety is calculated by comparing the resisting forces to the destabilizing forces.

Earth retaining structures — A broad class of civil engineering structures intended to support excavations or earth fills into vertical or nearly vertical geometries.

Geosynthetic reinforced soil (GRS) — A class of MSE materials that is specifically reinforced with geosynthetics.

Load and resistance factor design (LRFD) — A design method in which the load and resistance components are multiplied by factors related to the uncertainty associated with each component.

Mechanically stabilized earth (MSE) — A composite material made of soil and metal, polymer, or natural fiber inclusions that exhibits superior strength and stiffness characteristics compared to the soil.

References

Coulomb, C.A., 1776, Essai sure une application des règles de maximis et minimis à quelques problèmes de statique relatifs à l'architecture, *Mémoires de Mathématique et de Physique*, presented in 1773 at the Académie Royale des Sciences, 7:343–382.

Craig, R.F., 1987, *Soil Mechanics*, 4th edn, Van Nostrand Reinhold, New York.

Das, B.M., 1999, *Principles of Foundation Engineering*, 4th edn, Brooks Cole Publishing, Pacific Grove, CA.

Fang, H.Y., ed., 1991, *Foundation Engineering Handbook*, 2nd edn, Van Nostrand Reinhold, New York.

Holtz, R.D., Christopher, B.R., and Berg, R.R., 1997, *Geosynthetic Engineering*, BiTech Publishers, Richmond, BC, Canada.

Jáky, J. 1948, Earth pressure in silos, Proceedings of the Second International Conference on Soil Mechanics and Foundation Engineering, Rotterdam, Vol. 1, pp. 103–107.

Kulhawy, F.H. and Mayne, P.W., 1990, *Manual on Estimating Soil Properties for Foundation Design*, Electric Power Research Insitute, Palo Alto, CA.

Kulhawy, F.H., Jackson, C.S., and Mayne, P.W., 1989, First-order estimation of K_o in sands and clays, in: *Foundation Engineering: Current Principles and Practice*, Kulhawy, F.H., ed., ASCE, New York, pp. 121–134.

Lacasse, S. and Lunne, T., 1988, Calibration of dilatometer correlations, Proceedings of the First International Symposium on Penetration Testing — ISOPT-1, Orlando, Vol. 1, pp. 539–548.

Marchetti, S., 1980, *In-situ* tests by flat dilatometer, *Journal of the Geotechnical Engineering Division, ASCE*, 106(GT-3):299–321.

Rankine, W.J.M., 1857, On the stability of loose earth, *Philosophical Transactions of the Royal Society of London*, 147(Part 1):9–27.

Schmidt, B., 1966, Discussion of "Earth pressure at rest related to stress history," *Canadian Geotechnical Journal*, 3(4):239–242.

Seed, H.B. and Whitman, R.V., 1970, Design of earth retaining structures for dynamic loads, ASCE Specialty Conference on Lateral Stresses in the Ground and the Design of Earth-Retaining Structures, Cornell University.

Terzaghi, K., Peck, R.B., and Mesri, G., 1996, *Soil Mechanics in Engineering Practice*, 3rd edn, Wiley Interscience, New York.

Tomlinson, M.J., 2001, *Foundation Design and Construction*, 7th edn, Prentice Hall, Englewood Cliffs, NJ.

Vidal, H., 1966, Terre Armée, *Annales de l'Institue Technologique du Bâtiment et des Traveaux Publiques*, France, No. 228–229, pp. 888–938.

Withiam, J.L., Voytko, E.P., Barker, R.M., Duncan, J.M., Kelly, B.C., Musser, S.C., and Elias, V., 1998, Load and resistance factor design (LRFD) of highway bridge substructures, Publication No. FHWI HI-98-032, Federal Highway Administration, Washington, DC.

11

Stability Analysis and Design of Slopes

Manjriker Gunaratne

CONTENTS

11.1 Introduction

Construction of building foundations and highways on sloping ground or embankments can present instability problems due to potential shear failure. Therefore, geotechnical designers are often required to design stable embankments that would allow additional construction such as highways and buildings on top of them. On the other hand, instability could also result due to partial excavation of slopes during foundation construction. Furthermore, when one designs a structure in the vicinity of a slope, then safety considerations would naturally warrant a stability analysis of that slope. Hence, designers are often required to perform a ground stability analysis in addition to the foundation design. Stability analysis can be performed more effectively and accurately if the analyst comprehends the specific causes of potential slope failure under the given geological conditions.

The primary cause of slope instability due to possible shearing is the inadequate mobilization of shear strength to meet the shear stresses (τ) induced on any impending failure plane by the loading on the slope. Mathematically, the condition for instability can be expressed as in the following equation based on the Mohr–Coulomb criterion:

$$\tau > \sigma' \tan \phi + c \tag{11.1}$$

where σ is the normal stress on the potential failure plane.

One can identify the following factors that would trigger the above condition (Equation (11.1)):

1. Common factors that cause increased shear stresses in slopes:
    a. Static loads due to external buildings or highways
    b. Cyclic loads due to earthquakes
    c. Steepened slopes due to erosion or excavation
2. Common factors that cause reduction in shear strength of slopes:
    a. Increased pore pressures due to seepage and artesian conditions
    b. Loss of cementing materials
    c. Sudden loss of strength in sensitive clays

The limit equilibrium method is the most popular method adopted in slope-stability analysis. In this approach, it is assumed that the shear strength is mobilized simultaneously along the entire (predetermined) failure plane. Then the factor of safety for the predetermined failure wedge can be defined based on either the forces or moments as follows:

$$F = \frac{(\text{Force}/\text{Moment})_{\text{Stabilizing}}}{(\text{Force}/\text{Moment})_{\text{Destabilizing}}} \tag{11.2}$$

It must be noted that the stabilizing force or the stabilizing moment is the maximum force or the maximum moment that can be generated by the failing soil along the failure plane. Hence, these quantities can be determined by assuming that shear strength is mobilized along the entire failure surface. As discussed in Chapter 1, and employed in Equation (11.1), the shear strength of a soil is commonly determined by the Mohr–Coulomb shear strength criterion in foundation engineering.

On the other hand, the destabilizing force or moment active at a given instance can be determined in terms of the shear forces required to maintain the current state of equilib-

TABLE 11.1

Suggested Minimum Factors of Safety from FHWA

Condition	Recommended Minimum Factors of Safety (FS)
Highway embankment side slopes	1.25
Slopes affecting significant structures (e.g., bridge abutments, major retaining walls)	1.30

Source: From Federal Highway Administration, 1998, *Load and Resistance Factor Design (LRFD) for Highway Bridge Substructures*, Washington, DC. With permission.

rium. A number of common slope-stability analysis procedures will be outlined in the ensuing sections.

11.1.1 Required Minimum Factors of Safety

The minimum factors of safety as suggested by FHWA and AASHTO are given in Table 11.1 and Table 11.2.

11.2 Analysis of Finite Slopes with Plane Failures

Finite slopes such as natural embankments that are limited in extent can contain strata of relatively weak layers as shown in Figure 11.1(a) and (b) and Figure 11.2. Similar situations can also occur due to stratified deposits and interfaces between crusts (or shells) of dams that are typically granular soils and cores of dams made of impervious soils. By considering the weak layers to be planar surfaces (Figure 11.1a), one can perform simple stability analyses based on the limit equilibrium method.

TABLE 11.2

Required Minimum Factors of Safety from AASHTO

Condition	Required Minimum Factors of Safety (FS)	
	Detailed Exploration	Limited Exploration
Highway embankment slopes and retaining walls	1.3	1.5
Slopes supporting abutments or abutments above retaining walls	1.5	1.8

Source: From AASHTO, 1996, *Standard Specifications for Highway Bridges*, American Association for State Highway and Transportation Officials, Washington, DC. With permission.

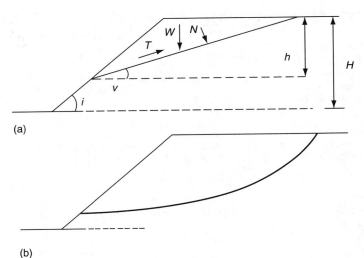

(a)

(b)

FIGURE 11.1
(a) Finite slope with a homogeneous failure plane. (b) Finite slope with a homogeneous nonplanar failure surface.

Case A. Homogeneous failure plane

Referring to Figure 11.1a, one can derive the following relations by considering force equilibrium parallel and perpendicular to the failure plane, respectively,

$$T = W\sin(v) \tag{11.3}$$

$$N = W\cos(v) \tag{11.4}$$

The weight of the failure mass can be expressed in terms of the unit weight of the failing soil mass as

$$W = \frac{1}{2}\gamma\frac{h}{\sin(i)}\sin(i - v)L \tag{11.5}$$

where

$$L = \frac{h}{\sin(v)}$$

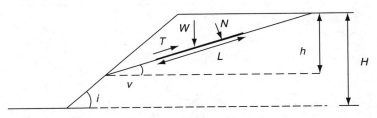

FIGURE 11.2
Finite slope with a nonhomogeneous failure plane.

The stabilizing force is determined by the available strength based on the Mohr–Coulomb criterion as

$$F_{\text{stab.}} = L[\sigma' \tan \phi + c] = L\left[\left(\frac{N}{L} - u\right) \tan \phi + c\right] \qquad (11.6)$$

where c and ϕ are the shear strength parameters of the weak soil layer (of length L) along the failure plane. It is also seen that under the current state of equilibrium, $F_{\text{destab.}}$ is equal and opposite to T.

Then, it follows from Equations (11.2) and (11.6) that

$$F = \frac{L[((N/L) - u) \tan \phi + c]}{T}$$

where u is the average pore pressure along the failure plane.

By using Equations (11.3)–(11.5), the factor of safety can be simplified to

$$F = \frac{\left[\left(\gamma \frac{h}{\sin (i)} \sin (i - v) \cos (v) - u\right) \tan \phi + c\right]}{\frac{1}{2} \gamma \frac{h}{\sin (i)} \sin (i - v) \sin(v)} \qquad (11.7)$$

A reasonable value of u representative of the pore pressures in the failure plane can be estimated by any of the methods outlined in Section 1.3.

Under "undrained" conditions ($\phi_u = 0$), Equation (11.7) simplifies to

$$F = \frac{c_u \sin(i)}{\gamma h \, \sin(i - v) \sin(v)} \qquad (11.8)$$

Case B. Nonhomogeneous failure plane
When the relatively weak stratum defines only a part of the potential failure plane, then destabilization of a slope occurs only if the stronger soil composing the failure mass allows the rest of the failure plane to form within itself, as shown in Figure 11.2. However, in such cases, it is difficult to obtain a closed-form solution for the safety factor without making a number of assumptions regarding the distribution of shear stresses on the entire failure plane comprising two different materials undergoing shear failure, i.e., weak stratum and the relatively stronger soil forming the rest of the failure surface.

Such assumptions are typically made in the "method of slices" described in Section 11.3. Hence, it is the method of slices that would be most suitable for analyzing the stability of slopes where conditions are nonuniform throughout the plane of failure or the failing soil mass.

11.3   Method of Slices

The method of slices is a numerical procedure that has been developed to handle stability analysis of slopes where conditions are nonhomogeneous within the soil mass making it impossible to deduce closed-form solutions. Some such nonhomogeneous conditions that are commonly encountered are as follows:

1. Irregularity of failure planes, i.e., failure planes cannot be defined by simple geometric shapes. This situation arises when relatively weak strata are randomly distributed within the slope or when the slope contains different soil types along the potential failure surface.
2. Presence of two different soil types within the failure wedge requiring the use of different soil properties in analysis.
3. Significant variation in the distribution of pore pressure along the failure plane, even under static groundwater conditions.
4. Irregularity of the slope geometry.
5. Significant variation in the buoyancy effects due to artesian conditions and seepage of groundwater.

The analysis requires the selection of a trial failure plane and discretization of the resulting failure wedge into a convenient number of slices as shown in Figure 11.3. The analyst is required to device the slicing in a manner that can incorporate any nonhomogeneity within the slope so that each resulting slice would be a homogeneous entity. Then, the stability of each slice can be analyzed separately using the limit equilibrium method and principles of statics, as done in Section 11.2.

The free-body diagram for each slice is illustrated in Figure 11.4 where it is seen that the side forces on the slices (X_i and Y_i) introduce additional unknowns into the analysis making it a statically indeterminate problem. Hence, the analyst needs to make simplifying assumptions to reduce the number of unknowns to facilitate a statically determinate solution. This flexibility has given rise to a variety of different analytical procedures, some of which will be outlined in this section. It is also realized that although the number of slices used in the analysis determines the accuracy of the solution, today's availability of superior computational devices and effective algorithms enable one to achieve solutions with reasonable accuracy for even the most complicated situations.

Without the need for any assumptions a simple expression can be derived for the safety factor by considering the equilibrium of the entire failure wedge as follows.

Under normal equilibrium conditions (Figure 11.3 and Figure 11.4), the destabilizing moment about the center of the trial failure surface is given by

$$M_{\text{destab.}} = R \sum_{1}^{n} W_i \sin(\alpha_i) = R \sum_{1}^{n} T_i$$

where n is the total number of slices.

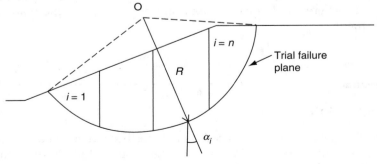

FIGURE 11.3
Illustration of the method of slices.

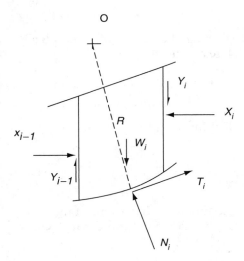

FIGURE 11.4
Free-body diagram for any slice *i*.

On the other hand, the stabilizing moment obtained from Equation (11.6) is based on the available strength and can be given as

$$M_{\text{stab.}} = R \sum_{1}^{n} [(N_i - ul_i)\tan \phi + cl_i]$$

where l_i is the arc length of slice *i* which can be expressed as

$$l_i = b_i \sec(\alpha_i)$$

Then, by employing Equation (11.2), the safety factor for the given slope can be expressed as

$$F = \frac{\sum_{1}^{n} [(N_i - ul_i) \tan \phi + cl_i]}{\sum_{1}^{n} W_i \sin(\alpha_i)} \tag{11.9a}$$

In order to suit relatively complex situations where nonhomogeneity with respect to soil properties and pore pressures prevails over the entire soil wedge, Equation (11.9a) can be rewritten as

$$F = \frac{\sum_{1}^{n} [(N_i - u_i l_i)\tan \phi_i + c_i l_i]}{\sum_{1}^{n} W_i \sin(\alpha_i)} \tag{11.9b}$$

A reasonable value of u can be estimated by any of the methods outlined in Section 1.3.2. It is the elimination of N_i that requires simplifying assumptions giving rise to several different approaches used in the method of slices.

11.3.1 Ordinary Method of Slices

Fellinius (1937) oversimplified the problem by neglecting the side forces completely. Then, based on the free-body diagram in Figure 11.4, one notes that

$$N_i = W_i \cos(\alpha_i)$$

Then, Equation (11.9b) reduces to the following explicit form:

$$F = \frac{\sum_1^n \left[(W_i \cos(\alpha_i) - u_i l_i) \tan \phi_i + c_i l_i \right]}{\sum_1^n W_i \sin(\alpha_i)} \tag{11.10}$$

Since the impending downslope movement of the slices involves distortions of the slices, shearing stresses do occur on sides. Therefore, the main limitation of the ordinary method of slices is the omission of the side forces.

11.3.2 Bishop's Simplified Method

Bishop (1955) assumed the side forces in all of the slices to be horizontal. This provides the following additional equations. For the vertical force equilibrium of each slice:

$$N_i \cos(\alpha_i) + T_i \sin(\alpha_i) = W_i \tag{11.11}$$

It is also assumed that the individual safety factor for each slice is equal to that of the entire failure wedge. Then, by using Equation (11.2) for each slice one obtains

$$F = \frac{(N_i - u_i l_i) \tan \phi_i + c_i l_i}{T_i} \tag{11.12}$$

Equations (11.11) and (11.12) can be used to solve for N_i in terms of F. Subsequent substitution for N_i in Equation (11.9b) yields an implicit expression for the safety factor as

$$F = \frac{\sum_1^n \left[\{ (W_i - u_i b_i) \tan \phi_i + c_i b_i \} \dfrac{\sec(\alpha_i)}{1 + \dfrac{\tan \phi_i \tan \alpha_i}{F}} \right]}{\sum_1^n W_i \sin(\alpha_i)} \tag{11.13}$$

Equation (11.13) can be solved on an iterative basis until convergence is achieved in terms of the assumed and computed safety factors.

Example 11.1

Investigate the stability of the embankment shown in Figure 11.5 (based on the trial failure surface drawn in Figure 11.5a), using

(i) the ordinary method of slices and (ii) Bishop's method of simple slices for:

Case 1. Dry embankment conditions

and using the ordinary method of slices only for (Table 11.3 and Table 11.4)

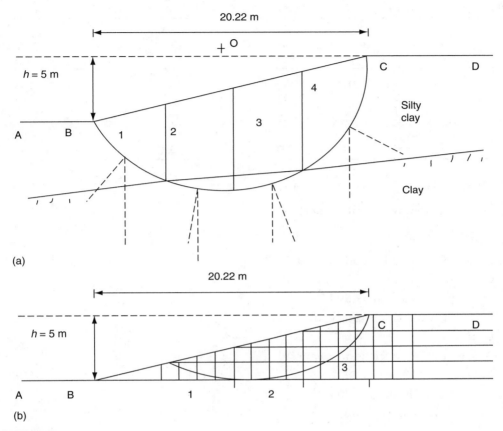

FIGURE 11.5
(a) Illustration for Example 11.1 (drawn to scale). (b) Investigation of slope failure under rapid drawdown.

Case 2. Immediately after compaction (assuming a pore pressure coefficient, $r_u = 0.4$)

Case 3. Under completely submerged conditions (groundwater table at the level CD)

Case 4. On sudden drawdown of the groundwater table to level AB for the trial failure surface drawn in Figure 11.5(b).

TABLE 11.3

Data for the Ordinary Slices Method (Example 11.1, Case 1)

Slice (*i*)	b_i (m)	h_i (m)	$W_i = \gamma b_i h_i$	α_i	$W_i \sin \alpha_i$	$l_i = b_i \sec \alpha_i$	$c_i l_i$	$W_i \cos \alpha_i \tan \Phi_i$
1	5.06	3.88	314.12	−40°	−201.91	6.61	99.15	138.93
2	5.06	6.3	581.44	−11°	−110.94	5.16	103.2	100.64
		+0.83						
3	5.06	6.93	657.40	21°	235.59	5.42	108.4	108.22
		+1.12						
4	5.06	5.82	471.2	60°	408.07	10.12	151.8	136.02
					$\sum = 330.81$		$\sum = 462.55$	$\sum = 483.81$

TABLE 11.4

Data for Bishop's Simplified Method (Example 11.1, Case 1); Assumed $F = 4.2$

Slice (i)	b_i (m)	h_i (m)	$W_i = \gamma b_i h_i$	α_i	$W_i \tan \phi_i$	$c_i b_i$	$\tan \phi_i \tan \alpha_i$	$\dfrac{\sec(\alpha_i)}{1 + \dfrac{\tan\phi_i \tan\alpha_i}{F}}$
1	5.06	3.88	314.12	$-40°$	181.36	75.9	-0.48	1.48
2	5.06	6.3	581.44	$-11°$	102.52	101.2	-0.034	1.03
		+0.83						
3	5.06	6.93	657.40	$21°$	115.92	101.2	0.068	1.05
		+1.12						
4	5.06	5.82	471.2	$60°$	272.05	75.9	1	1.62

Assume the following soil properties:

Silty clay
Angle of internal friction $= \phi = 30°$
Cohesion $= 15$ kPa
Undrained cohesion $= 20$ kPa
Dry unit weight $= 16.0$ kN/m^3
Clay
Angle of internal friction $= \phi = 10°$
Cohesion $= 20$ kPa
Undrained cohesion $= 30$ kPa
Dry unit weight $= 17.0$ kN/m^3

First, it is noted that slicing is done to separate the soil layers.

Solution
From Section 1.6, $\gamma_{dry} = G_s \gamma_w/(1 + e)$, and assuming G_s to be 2.65 and knowing that $\gamma_w = 9.8$ kN/m^3,
For silty clay, $e = 0.623$

Also $\gamma_{sat} = \gamma_w(G_s + e)/(1 + e)$
$\gamma_{sat} = 19.76$ kN/m^3
In addition, $\gamma_{sub} = 19.76 - 9.8 = 9.96$ kN/m^3

Similarly for clay,

$\gamma_{sat} = 20.37$ kN/m^3
$\gamma_{sub} = 20.37 - 9.8 = 10.57$ kN/m^3
Using Equation (11.10), $F = (483.81 + 462.55)/(330.81) = 2.86$
Using Equation (11.13), $F = [(181.36 + 75.9)1.48 + (102.52 + 101.2)1.03 + (115.92 + 101.2)1.05 + (272.05 + 75.9)1.62]/(330.81) = 4.18$

Case 2. Immediately after compaction
The Bishop and Morgenstern pore pressure coefficient, r_u, is defined as (Table 11.5)

TABLE 11.5

Pore Pressure Estimations for Example 11.1, Case 2

Slice (i)	h_i (m)	$W_i = \gamma h_i$	u_i (kPa)	$l_i = b_i \sec \alpha_i$	$u_i l_i$	$(W_i \cos \alpha_i - u_i l_i) \tan \phi_i + c_i l_i$
1	3.88	62.08	24.8	6.61	163.93	143.43
2	6.3	114.91	45.96	5.16	237.15	162.02
	+0.83					
3	6.93	129.92	51.96	5.42	281.62	166.96
	+1.12					
4	5.82	93.12	37.25	10.12	376.97	70.18
						$\sum = 542.59$

$$r_u = \frac{u}{\gamma h} \qquad (11.14)$$

Using Equation (11.10), $F = (542.59)/(330.81) = 1.64$

Case 3. Under completely submerged conditions
In this case, one can assume undrained conditions to be the most critical and both pore pressure and the surcharge water can be incorporated together in the analysis by considering the submerged weights of the slices (Table 11.6).
    Using Equation (11.10), $F = (652)/(301.57) = 2.16$.

Case 4. Rapid drawdown
In cohesive soils rapid drawdown conditions promote slope failures as the one shown in Figure 11.5(b) due to the transient seepage condition developing at the face of the slope as shown in Figure 11.5(b). Then, the safety factor for the trial base failure surface shown in Figure 11.5(b) can be computed as illustrated in Table 11.7. It must be noted that the pore pressure values have been computed using the flownet principles discussed in Section 13.2 Since pore pressures are separately computed, the weights of slices are evaluated based on the saturated unit weight of $19.8\,\text{kN/m}^3$. It is also reasonable to assume that drained conditions occur in a transient flow regime. Hence, an effective stress analysis is performed using the evaluated pore pressures. From Equation (11.10), $F = 1.1$.
    The method of slices can be employed to perform stability analysis of slopes under steady-state seepage conditions as well. In such cases, one can conveniently predict the pore pressures using Bernoulli's equation (Section 13.2, Equation 13.1)

TABLE 11.6

Pore Pressure Estimations for Example 11.1, Case 3

Slice (i)	b_i (m)	h_i (m)	$W_i = \gamma_{sub} b_i h_i$	α_i	$W_i \sin \alpha_i$	$l_i = b_i \sec \alpha_i$	$(C_u)_i l_i$	$W_i \cos \alpha_i \tan \Phi_i$
1	5.06	3.88	286.8	$-40°$	-184.35	6.61	132.2	0
2	5.06	6.3	531.8	$-11°$	-101.47	5.16	154.8	0
		+0.83						
3	5.06	6.93	600.67	$21°$	215.26	5.42	162.6	0
		+1.12						
4	5.06	5.82	429.7	$60°$	372.13	10.12	202.4	0
					$\sum = 301.57$	$\sum = 652$		

TABLE 11.7

Pore Pressure Estimations for Example 11.1, Case 4

Slice (i)	α_i	b_i	l_i	u_i (kPa)	$u_i b_i$	$u_i l_i$	$(W_i \cos \alpha_i - u_i l_i) \tan \phi_i + c_i l_i$	$W_i \sin \alpha_i$
1	10°	5.1	5.2	13.2	66.7	68.6	21.6 + 78	26.3
2	15°	5.1	5.3	26.4	133.5	139.9	40.94 + 80	78.4
3	50°	5.1	7.9	41.5	209.8	327.9	−53 + 119	154.7

11.4 Slope-Stability Analysis Using the Stability Number Method

In the stability number method (Taylor, 1948), the limit equilibrium computations are based on an assumed linear or circular rupture surface. Then the safety factor, F, is defined as the ratio of maximum shear strength that can be mobilized to the shear strength required on the assumed failure surface to maintain the slope in equilibrium. Based on the Mohr–Coulomb criterion, F can be written by

$$F = \frac{c + \sigma_n \tan \phi}{c_d + \sigma_n \tan \phi_d} \tag{11.15}$$

where c and ϕ are the strength parameters and the subscript "d" indicates the strength parameters required for equilibrium or the developed strength. σ_n is the average normal stress on the failure surface.

The individual safety factors with respect to cohesion and friction can be also defined, respectively, as follows:

$$F_c = \frac{c}{c_d} \tag{11.16}$$

$$F_\phi = \frac{\tan \phi}{\tan \phi_d} \tag{11.17}$$

Observation of Equations (11.15)–(11.17) shows that the actual (true) safety factor is obtained under the following condition:

$$F_c = F_\phi = F \tag{11.18}$$

11.4.1 Stability Analysis of a Homogeneous Slope Based on an Assumed Failure Surface

The following procedure is typically followed in obtaining the true safety factor guided by Equations (11.16)–(11.18):

Step 1. Assume a reasonable F_ϕ for the assumed failure surface.

Step 2. Use Equation (11.17) to estimate ϕ_d knowing the available friction.

Step 3. Knowing the developed frictional strength ($\tan \phi_d$), perform a static equilibrium analysis of the failure mass to determine the developed cohesive strength c_d. This is conventionally designated as a nondimensional cohesive strength or the stability number for the given failure mass using γ (unit weight) and H (height of slope).

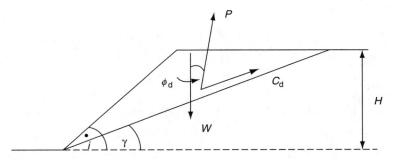

FIGURE 11.6
Equilibrium of failure wedge on a planar failure surface.

$$\frac{c_d}{\gamma H} = \frac{c}{F_c \gamma H} \tag{11.19}$$

If the failure mass has a simple geometric shape, the stability number can be derived based on principles of statics. For example, the following stability number expressions are available for planar and circular failure surfaces shown in Figure 11.6 and Figure 11.7, respectively.

11.4.1.1 Plane Failure Surface

In terms of the notation in Figure 11.6, the stability number can be expressed as follows:

$$\frac{c}{F_c \gamma H} = \frac{1}{2} \csc(i) \frac{\sin(i - \nu) \sin(\nu - \phi_d)}{\cos(\phi_d)} \tag{11.20}$$

11.4.1.2 Circular Failure Surface

For assumed circular trial failure surfaces, the stability numbers can be expressed in Equations (11.21a) and (11.21b), in terms of the notation in Figure 11.7 for toe failure (Taylor, 1937)

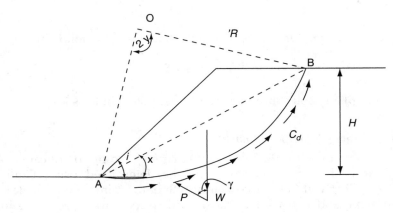

FIGURE 11.7
Equilibrium of failure wedge on a circular failure surface.

Toe failure

$$\frac{c}{F_c \gamma H} = \frac{\frac{1}{2} \csc^2(x)[y \csc^2(y) - \cot(y)] + \cot(x) - \cot(i)}{2[\cot(x) \cot(\nu) + 1]}$$ (11.21a)

Base failure

$$\frac{c}{F_c \gamma H} = \frac{\frac{1}{2} \csc^2(x)[y \csc^2(y) - \cot(y)] + \cot(x) - \cot(I) - 2\eta}{2[\cot(x) \cot(\nu) + 1]}$$ (11.21b)

where i is the slope angle, H is the height of the slope, x, y, and ν are angles shown in Figure 11.7, and η is the ratio of the distance between the toe of the slope and point A and the slope height H (for base failure).

Figure 11.7 also shows the forces that ensure the equilibrium of the assumed failure wedge. These are the weight, W, the design cohesion, C_d, and the resultant of the normal and frictional forces, P. The force P must be tangent to a circle of radius $R \sin \phi_d$ and centered at o (Figure 11.14). This circle is known as the friction circle.

> *Step 4.* Knowing the maximum available cohesion, c, estimate F_c from Equations (11.21)
>
> *Step 5.* If F_c and F_ϕ are not equal, repeat the procedure from Steps 1 to 4 for different F_ϕ values until the condition of $F_c = F_\phi$ is satisfied. Once it is satisfied, this F_c (or F_ϕ) is the true safety factor FS (Equation (11.18)).

11.4.2 Stability Analysis of a Homogeneous Slope Based on the Critical Failure Surface

If the ultimate goal is to find the failure surface with the minimum safety factor, i.e., the critical failure surface, then the above procedure has to be repeated for a number of different trial failure surfaces.

11.4.2.1 Closed-Form Solution

For a planar critical surface, a closed-form solution for minimum safety factor can be obtained using Equation (11.20) in terms of ϕ as

$$F = \frac{4c}{\gamma H} \frac{\sin(i) \cos(\phi)}{[1 - \cos(i - \phi)]}$$ (11.22a)

Then the corresponding failure plane would be defined by the following inclination (ν):

$$\nu = \tfrac{1}{2}(i + \phi)$$ (11.22b)

The above failure plane would provide the highest potential for sliding.

11.4.2.2 Use of Taylor's Stability Charts

The stability number $c_d / \gamma H$ (Equation (11.19)), corresponding to critical circular failure surfaces with respect to a given slope, can be determined using Figure 11.8 and Figure 11.9. Accordingly, if the analyst is in search of the minimum possible safety factor for the given slope and not a safety factor with respect to a specific failure surface, Figure 11.8 and Figure 11.9 will immensely reduce the volume of computations in Step 3 of the above iterative procedure.

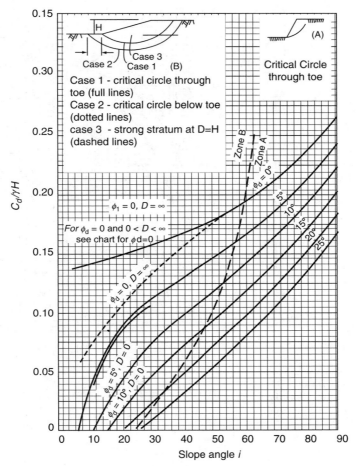

FIGURE 11.8

Stability chart for soils with friction angle. (From Taylor, D.W., 1948, *Fundamentals of Soil Mechanics*, John Wiley, New York. With permission.)

Taylor's simplified method of stability analysis will be illustrated in Examples 11.2–11.4.

Example 11.2

With respect to the slope shown in Figure 11.10, estimate the safety factor corresponding to the trial failure plane BD. Assume the following soil properties:

Angle of friction $= 20°$

Cohesion $= 15\,\text{kPa}$

Dry unit weight $= 17.5\,\text{kN/m}^3$

Solution
Trial 1
Assume an F_ϕ of 1.5. Then, from Equation (11.17), $\tan\phi_\text{d} = \tan 20°/1.5° = 0.243$ and $\phi_\text{d} = 13.6°$.

By applying Equation (11.20),

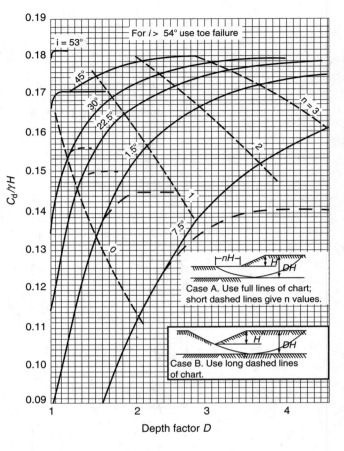

FIGURE 11.9
Stability chart for soils with zero-friction angle (From Taylor, D.W., 1948, *Fundamentals of Soil Mechanics*, John Wiley, New York. With permission.)

$$\frac{c}{F_c \gamma H} = \frac{1}{2} \csc (31) \frac{\sin (31 - 20) \sin (20 - 13.6)}{\cos (13.6)} = 0.0056$$

$$F_c = 18.66$$

Trial 2
For an assumed F_ϕ of 5, a similar procedure produces $F_c = 7.77$.

Trial 3
Finally, for an assumed F_ϕ of 6.2, one obtains $F_c = 1.97$.

Then the results of the above iterative procedure can be plotted in Figure 11.11, from which it is seen that the true factor of safety $F (= F_c = F_\phi) = 5.5$.

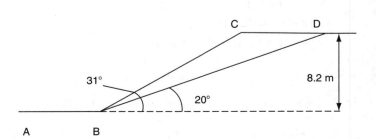

FIGURE 11.10
Illustration for Example 11.2.

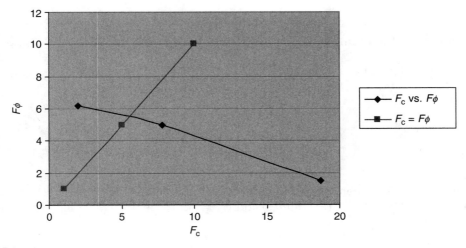

FIGURE 11.11
Plot of F_c vs. F_ϕ for Example 11.2.

Example 11.3

With respect to the slope shown in Figure 11.12, estimate the minimum safety factor corresponding to a critical failure plane passing through the toe. Assume the following soil properties:

Angle of friction $= 20°$

Cohesion $= 15\,\text{kPa}$

Dry unit weight $= 17.5\,\text{kN/m}^3$

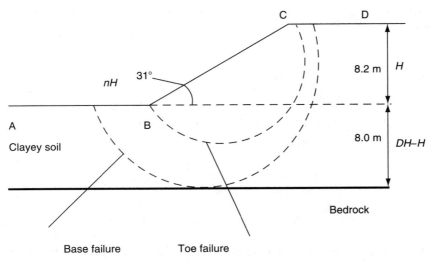

FIGURE 11.12
Illustration for Examples 11.3 and 11.4

Solution

Trial 1

Assume an F_ϕ of 1.5. Then, from Equation (11.17) $\tan \phi_d = \tan 20°/1.5° = 0.243$ and $\phi_d = 13.6°$.

From Figure 11.8, $D = 0$ (toe failure), $i = 31°$, and $\phi_d = 13.6°$ yield

$$\frac{c_d}{\gamma H} = \frac{c}{F_c \gamma H} = 0.05$$

Thus, $F_c = 2.09$.

Trial 2

Assume an F_ϕ of 1.7. Then, from Equation (11.17), $\phi_d = 12.1°$.

From Figure 11.8, $D = 0$ (toe failure) and $\phi_d = 12.1°$ yield

$$\frac{c_d}{\gamma H} = \frac{c}{F_c \gamma H} = 0.065$$

Thus, $F_c = 1.537$.

Trial 3

Assume an F_ϕ of 1.6. Then, from Equation (11.17), $\phi_d = 12.8°$.

From Figure 11.8, $D = 0$ (toe failure) and $\phi_d = 12.82°$ yield

$$\frac{c_d}{\gamma H} = \frac{c}{F_c \gamma H} = 0.068$$

Thus, $F_c = 1.608$.

Hence, considering a toe failure, the true minimum safety factor for the slope in Figure 11.12 is about 1.6.

Example 11.4

With respect to the slope shown in Figure 11.12, estimate the minimum safety factor corresponding to a critical failure plane that touches the bedrock. Assume the following soil properties:

Undrained cohesion $= 25\,\text{kPa}$

Dry unit weight $= 17.5\,\text{kN/m}^3$

Solution

From Figure 11.9, $D = 16.2/8.2 = 1.975$ (base failure) and $i = 31°$ yield

$$\frac{c_d}{\gamma H} = \frac{c}{F_c \gamma H} = 0.173$$

Thus, $F = F_c = 1.007$.

Figure 11.9 also shows that the critical failure surface intersects the base at a distance of nH, where $n = 1.3$.

Hence, the critical failure surface passes at a distance of 10.67 m from the toe producing a safety factor of 1.007.

11.5 Stabilization of Slopes with Piles

The use of piles as a restraining element has been applied successfully in the past and proven to be an effective solution, since piles can often be installed without disturbing the equilibrium of the slope. Piles used to stabilize slopes are in a passive state and the lateral forces acting on the piles are dependent on the soil movements that are in turn affected by the presence of piles. Due to their relatively low cost and the insignificant axial strength and length demand in this particular application, timber piles are ideal for stabilization of slopes.

11.5.1 Lateral Earth Pressure on Piles

Poulos (1973) first suggested a method to determine the lateral forces on piles. Ito and Matsui (1975) proposed a different theoretical approach to analyze the growth mechanism of lateral forces acting on stabilizing piles assuming that soil is forced to squeeze between the piles. This condition is applicable to relatively small gaps between piles. Then, the passive force on the pile per unit length (z) (Figure 11.13) can be computed by the following equation based on Ito et al. (1975):

$$q = Ac\left(\frac{1}{N_\phi \tan\phi}\left\{\exp\left[\frac{D_1 - D_2}{D_2}N_\phi \tan\phi \tan\left(\frac{\pi}{8}+\frac{\phi}{4}\right)\right] - 2N_\phi^{1/2}\tan\phi - 1\right\} + \frac{2\tan\phi + 2N_\phi^{1/2} + N_\phi^{-1/2}}{N_\phi^{1/2}\tan\phi + N_\phi - 1}\right)$$

$$- c\left(D_1\frac{2\tan\phi + 2N_\phi^{1/2} + N_\phi^{-1/2}}{N_\phi^{1/2}\tan\phi + N_\phi - 1} - 2D_2N_\phi^{-1/2}\right) + \frac{\gamma z}{N_\phi}\left\{A\exp\left[\frac{D_1 - D_2}{D_2}N_\phi \tan\phi \tan\left(\frac{\pi}{8}+\frac{\phi}{4}\right)\right] - D_2\right\}$$

$$(11.23)$$

where

$$A = D_1\left(\frac{D_1}{D_2}\right)^{N_\phi^{1/2}\tan\phi + N_\phi - 1}$$

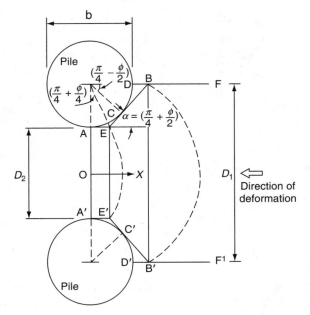

FIGURE 11.13
Plastically deforming ground around stabilizing piles. (From Hassiotis, S., Chameau, J.L., and Gunaratne, M., 1997, *Journal of Geotechnical and Geoenvironmental Engineering, ASCE*, 123(4). With permission.)

$$N_\phi = \tan^2\left(\frac{\pi}{4} + \frac{\phi}{2}\right)$$

D_1 is the pile spacing, D_2 is the opening between the piles, c is cohesion and ϕ is the friction angle.

Under undrained conditions ($\phi_u = 0$), Equation (11.23) reduces to

$$q = c\left(D_1\left\{3\log\frac{D_1}{D_2} + \frac{D_1 - D_2}{D_2}\tan\frac{\pi}{8}\right\} - 2(D_1 - D_2)\right) + \gamma z(D_1 - D_2) \qquad (11.24)$$

where γz is the overburden stress.

11.5.2 Analysis Using the Friction-Circle Method

Modified stability number. For a slope of inclination i and height H (Figure 11.14), the stability number can be expressed as in the following equations (Hassiotis et al., 1997).

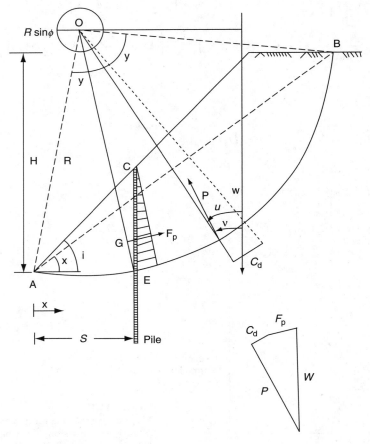

FIGURE 11.14
Forces on a slope reinforced with piles. (From Hassiotis, S., Chameau, J.L., and Gunaratne, M., 1997, *Journal of Geotechnical and Geoenvironmental Engineering*, ASCE, 123(4). With permission.)

For toe failure

$$\frac{C}{F_c \gamma H} = \frac{E - \dfrac{12F_P}{\gamma H^3} \left[\dfrac{\csc{(CEO)}}{\sin \nu} \dfrac{H}{2} \csc{(x)} \csc{(y)} \sin \phi + OG \right]}{6 \csc^2{(x)} \csc{(y)} \sin \phi \left[\dfrac{\cos{(x)}}{\sin \nu} + \csc{(u - v)} \cos{(x - u)} \right]}$$

(11.25a)

For base failure

$$\frac{C}{F_c \gamma H} = \frac{E + 6\eta^2 - 6\eta \csc{(x)} \csc{(y)} \sin \phi - \dfrac{12F_P}{\gamma H^3} \left[\dfrac{\csc{(CEO)}}{\sin \nu} \dfrac{H}{2} \csc{(x)} \csc{(y)} \sin \phi + OG \right]}{6 \csc^2{(x)} \csc{(y)} \sin \phi \left[\dfrac{\cos{(x)}}{\sin \nu} + \csc{(u - v)} \cos{(x - u)} \right]}$$

(11.25b)

where

$E = 1 - 2 \cot^2{(i)} + 3 \cot{(i)} \cot{(x)} - 3 \cot{(i)} \cot{(y)} + 3 \cot{(x)} \cot{(y)}$

F_P = Passive pressure on the pile row per unit width of slope, computed by integrating Equation (11.23) along CE (Figure 11.14)

$CEO, x, y,$ and v are angles shown in Figure 11.14

OG = Moment arm of the lateral force F_P around the center of the friction circle (Figure 11.14)

11.5.3 Design Methodology

The following steps have been proposed (Hassiotis et al., 1997) to design the stabilizing piles on a given slope by selecting an appropriate pile spacing (D_1) for a given pile size ($D_1 - D_2$):

Step 1. Dictated by the site conditions or on an arbitrary basis, select an appropriate range of the horizontal spacing, *s*, between the row of piles and the toe of the slope. Select a number of trial design values of *s*, within this range.

Step 2. Select a number of trial pile spacing (D_1) values. It must be noted that since the pile size ($D_1 - D_2$) is fixed, any selected pile spacing (D_1) fixes the D_2/D_1 ratio. As pointed out previously, Equation (11.23) is only applicable for relatively small ratios of D_2/D_1.

Step 3. Select a trial toe failure surface or trial base failure surface. Then, numerically integrate *q* in Equation (11.23) along the depth CE (Figure 11.14), and divide *q* by the pile spacing D_1 to obtain the passive force on the slope exerted by the piles per unit width of the slope.

Step 4. From Figure 11.14, also obtain the distance OG and the angles $CEO, x, y,$ and to compute the safety factor from Equation (11.25a) or (11.25b), whichever is applicable.

Step 5. Repeat steps 3 and 4 for different trial failure surfaces, and record the conditions producing the minimum safety factor for given *s* and D_1.

Step 6. Repeat steps 1 to 5 to obtain the minimum safety factors for different combinations of *s* and D_1 and plot the results as shown in Figure 11.15. Use Figure 11.15 to select the optimum design parameters for the desired safety factor.

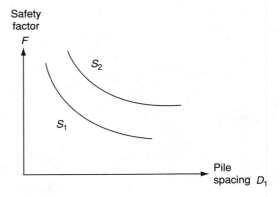

FIGURE 11.15
Illustration of the selection of design parameters.

Step 7. Assuming an infinite pile embedment, the horizontal deflection, the shear force, and the bending moment along the length of the pile can be determined assuming the pile to be an infinite beam embedded in an elastic foundation. This design methodology is discussed in detail in Chapter 8.

Step 8. Finally, the optimum embedment can be decided upon by selecting the length of the pile above the region where the horizontal deflection, the shear force, and the bending moment approach zero for all practical purposes.

11.6 Reinforcement of Slopes with Geotextiles and Geogrids

11.6.1 Reinforcement with Geotextiles

In designing slopes with geotextiles, the usual geotechnical engineering approach (Section 11.3) to the slope-stability problem is extended to include the tensile force, T, provided by the geotextile (Figure 11.16). The resulting equation for the safety factor on a circular failure surface (Koerner, 1998) is given in the following equation:

$$F = \frac{\sum_{i=1}^{n} (\bar{N}_i \tan \bar{\phi} + \bar{c}\Delta l_i)R + \sum_{i=1}^{m} T_i y_i}{\sum_{i=1}^{n} (W_i \sin \alpha_i)R} \qquad (11.26)$$

where $N_i = W_i \cos \alpha_i$, W_i is the weight of the slice, α_i is the angle of intersection of horizontal to tangent at the center of the slice, Δl_i is the arc length of the slice, R is the radius of the failure circle, ϕ is the effective angle of shearing resistance, c is the cohesion, T_i is the allowable geotextile tensile strength, y_i is the moment arm for the geotextile, n is the number of slices, m is the number of geotextile layers, $\bar{N} = N_i - u_i\Delta x_i$, in which $u_i = h_i\gamma_w$ = pore water pressure, h_i is the height of water above the base of the circle, Δx_i is the width of the slice, and γ_w is the unit weight of water.

For saturated fine-grained cohesive soils with $\phi = 0$, Equation (11.26) simplifies to

$$F = \frac{C_u LR + \sum_{i=1}^{m} T_i y_i}{WX} \qquad (11.27)$$

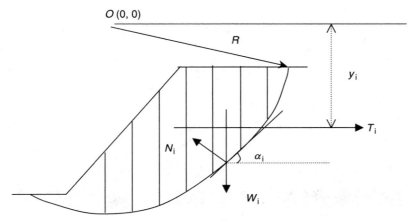

FIGURE 11.16
Illustration for stability analysis of geotextile reinforced slope.

where C_u is the undrained cohesion, L is the length of the failure arc, W is the weight of the failure zone, and X is the moment arm of the failure arc about the center of the failure circle.

11.6.2 Geogrid Reinforced Slope

Koerner (1998) also provides the following expression for the safety factor of a slope analyzed using limit equilibrium methods using a circular arc failure plane (Figure 11.17)

$$F = \frac{M_{stab} + \sum_{i=1}^{m} T_i y_i}{M_{destab}} \tag{11.28}$$

where M_{stab} are moments resisting failure due to the shear strength of the soil, M_{destab} are moments resisting failure due to gravity, seepage, live and other disturbing loads, T_i is the allowable geogrid tensile strength, y_i is the appropriate moment arm, and m is the number of separate reinforcement layers. Expressions for M_{stab} and M_{destab} are found in Section 11.3.

11.7 Reliability-Based Slope Design

For a homogeneous slope made of a soil possessing shear strength parameters of c and ϕ, the safety factor expression corresponding to the ordinary method of slices (Equation (11.10)) can be rewritten as

$$F = \frac{\sum [W_i \cos(\alpha_i) - u_i l_i]}{\sum W_i \sin(\alpha_i)} \tan\phi + \frac{\sum l_i}{\sum W_i \sin(\alpha_i)} c \tag{11.29}$$

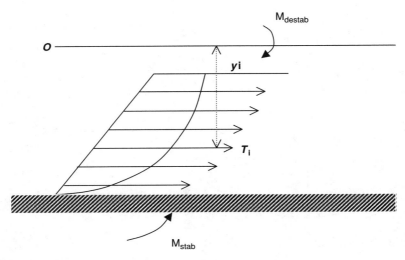

FIGURE 11.17
Illustration for stability analysis of geogrid reinforced slope.

Spatial variation of soil strength properties within a given embankment often makes it a difficult to estimate the soil properties required for stability analysis. If laboratory testing such as triaxial tests (Section 1.4) is conducted based on sampling done in one location of the embankment, to determine soil properties applicable to the entire embankment, one would have to account for the random spatial variation of soil properties prior to their use in stability analysis for the embankment. In the wake of possible variations in soil properties within the considered slope, the safety factor, F, would inherit a random uncertainty. Under this scenario, the best option available for the analyst is to consider the safety factor, F, as a randomly distributed variable. Then the statistical properties of the distribution of F can be estimated based on the statistics of the spatial variation of strength properties (c and ϕ) and the Taylor series approach for manipulation of statistical parameters (Harr, 1977, 1987, Rosenblueth, 1975).

Mean safety factor
Based on the Taylor series approach, and Equation (11.29) the mean safety factor for a slope consisting of two soil types can be evaluated by the following expression:

$$\overline{F} = \frac{\sum [W_i \cos(\alpha_i) - u_i l_i]}{\sum_{i=1}^{n} W_i \sin(\alpha_i)} \overline{\tan \phi_j} + \frac{\sum [W_i \cos(\alpha_i) - u_i l_i]}{\sum_{i=n+1}^{n+m} W_i \sin(\alpha_i)} \overline{\tan \phi_k}$$
$$+ \frac{\sum l_i}{\sum_{i=1}^{n} W_i \sin(\alpha_i)} \overline{c}_j + \frac{\sum l_i}{\sum_{i=n+1}^{n+m} W_i \sin(\alpha_i)} \overline{c}_k \tag{11.30}$$

where $[\overline{\tan \phi_j}, \overline{c}_j]$ and $[\overline{\tan \phi_k}, \overline{c}_k]$ are the mean values of the soil properties of two layers that make up n and m slices, respectively, of the designed slope. Generally, if only single estimates of the strength properties are available, then those are regarded as estimates themselves of the mean values of the respective soil properties.

Standard deviation of the safety factor
Taylor Series method. Using the Taylor series approach, and Equation (11.29) an approximate estimate of the standard deviation of the safety factor can be obtained as follows for a slope made up of two distinct soil types:

$$s_F^2 = \left[\frac{\sum [W_i \cos (\alpha_i) - u_i l_i]}{\sum_{i=1}^{n} W_i \sin (\alpha_i)} \right]^2 s_{\tan \phi_j}^2 + \left[\frac{\sum [W_i \cos (\alpha_i) - u_i l_i]}{\sum_{i=1}^{n} W_i \sin (\alpha_i)} \right]^2 s_{\tan \phi_k}^2$$
$$+ \left[\frac{\sum l_i}{\sum_{i=n+1}^{n+m} W_i \sin (\alpha_i)} \right]^2 s_{c_j}^2 + \left[\frac{\sum l_i}{\sum_{i=n+1}^{n+m} W_i \sin (\alpha_i)} \right]^2 s_{c_k}^2 \qquad (11.31)$$

where $[s_{\tan \phi_j}, s_{c_j}]$ and $[s_{\tan \phi_k}, s_{c_k}]$ are the standard deviation values of the soil properties of the two layers that make up n and m slices, respectively. If test data are insufficient to estimate the strength properties, then one can use typical standard deviations of soil properties within a given site (Table 11.8).

If the safety factor, F, can be considered to be a random variable, then the issue of slope stability becomes one in which the degree of instability is expressed by the probability of failure, p_f, estimated as follows:

$$p_f = \text{Prob}(F \leq 1) \qquad (11.32a)$$

Alternatively, the reliability of the slope design is expressed as

$$R = \text{Prob}(F > 1) = 1 - p_f \qquad (11.32b)$$

As for the distribution of the safety factor, the most popular choices are (1) normal or Gaussian distribution and (2) the lognormal distribution.

11.7.1 Reliability Estimates with Random Loads

If the resistance (S) and the loads (L) are both affected by random uncertainty, then in order to investigate the probability of failure of the system, a joint probability distribution must be defined for the random variable ($S - L$). If this is denoted by $f_{S-L}(s, l)$, then the probability of failure is evaluated based on the following condition:

$$p_f = p[(s - L) < 0] \qquad (11.33a)$$

By assuming an appropriate distribution for $f_{S-L}(s, l)$, p_f can be conveniently evaluated.

Alternatively, the following convolution method can also be used to estimate the probability of failure in cases where both the resistance and the loads have to be considered as random variables:

$$p_f = \int_{-\infty}^{\infty} F_S(l) f_L(l) \, dl \qquad (11.33b)$$

TABLE 11.8

Reported Coefficients of Variation of Soil Properties

Soil Property	Coefficient of Variation
Unit weight	3–8%
Effective angle of friction, Φ	2–21%
Undrained strength, s_u	13–49%

Source: From Harr, M.E., 1987, *Reliability-Based Design in Civil Engineering*, McGraw-Hill, New York. With permission.

where $F_S(l)$ is the cumulative distribution of the resistance in the load domain and $f_L(l)$ is the probability density function of the load.

Example 11.5

If the strength properties of Example 11.1 were determined using CPT tests, the interpretation of which resulted in the data shown in Table 11.9, estimate the reliability of the slope assuming that the safety factor is normally distributed.

Mean and standard deviation values of soil properties in Table 11.9 are computed by the following expressions:

$$\bar{x} = \frac{\sum x_i}{n} \tag{11.34a}$$

$$s_x^2 = \frac{\sum (x_i - \bar{x})^2}{n - 1} \tag{11.34b}$$

Applying Equation (11.30) for the computed values in Table 11.9,
$\bar{F} = [(476.23)/(330.81)](0.57) + [(1184.5)/(330.81)](0.177) + [16.73/330.81](15) + [10.58/330.81](20) = 946.36/330.81$
$\bar{F} = 2.86$.

Then, Equation (11.31) and the information in Table 11.9 also produce,
$s_F^2 = [(476.23)/(330.81)]^2(0.072)^2 + [(1184.5)/(330.81)]^2(0.062)^2 + [16.73/330.81]^2(6.56)^2 + [10.58/330.81]^2(6.85)^2 = 0.0107 + 0.04928 + 0.11 + 0.048 = 0.2181$
Hence $s_F = 0.467$.

From Table 11.10, the standard normal value (z) corresponding to $F = 1$ is found as (Figure 11.18)

$$z = \frac{1 - \bar{F}}{s_F} = \frac{(1 - 2.86)}{0.467} = -3.98$$

In Figure 11.18, area $\psi(z) = 0.499966$
Therefore, $p_f = \text{Prob}(F \leq 1) = 3.5 \times 10^{-5}$.

11.7.1.1 Use of the Lognormal Distribution

There are two obvious drawbacks in the normal (Gaussian) distribution:

1. It assumes that the random variable takes a range of values between $-\infty$ and ∞, which is unrealistic for positive variables such as the safety factor.

TABLE 11.9

Soil Properties Inferred from *In Situ* CPT Tests (for Example 11.5)

Test	Silty Clay			Clay		
	c (kPa)	Φ	tan Φ	c (kPa)	Φ	tan Φ
1	9.0	31°	0.600	26	12°	0.213
2	14.0	32°	0.623	12.5	12°	0.213
3	22.0	26°	0.488	21.5	6°	0.105
Mean	15		0.570	20		0.177
Std. dev.	6.56		0.072	6.85		0.062

TABLE 11.10

Area under the Standard Normal distribution ($P(z)$)

z	0	0.01	0.02	0.03	0.04	0.05	0.06	0.07	0.08	0.09
0	0	0.003989	0.007978	0.011966	0.015953	0.019938	0.023922	0.027902	0.031881	0.035855
0.1	0.039827	0.043794	0.047757	0.051715	0.055669	0.059616	0.063558	0.067493	0.071422	0.075344
0.2	0.079258	0.083164	0.087062	0.090952	0.094832	0.098704	0.102566	0.106417	0.110258	0.114089
0.3	0.117908	0.121716	0.125513	0.129297	0.133068	0.136827	0.140573	0.144305	0.148024	0.151728
0.4	0.155418	0.159093	0.162753	0.166398	0.170027	0.17364	0.177237	0.180818	0.184382	0.187928
0.5	0.191458	0.194969	0.198463	0.201939	0.205396	0.208835	0.212255	0.215656	0.219037	0.222399
0.6	0.225741	0.229063	0.232365	0.235647	0.238908	0.242148	0.245367	0.248565	0.251742	0.254897
0.7	0.25803	0.261142	0.264231	0.267298	0.270344	0.273366	0.276366	0.279343	0.282298	0.285229
0.8	0.288138	0.291023	0.293885	0.296724	0.299539	0.30233	0.305098	0.307843	0.310563	0.31326
0.9	0.315932	0.318581	0.321206	0.323807	0.326384	0.328936	0.331465	0.333969	0.336449	0.338905
1	0.341337	0.343744	0.346128	0.348487	0.350822	0.353133	0.35542	0.357682	0.359921	0.362135
1.1	0.364326	0.366492	0.368635	0.370754	0.372849	0.37492	0.376967	0.378991	0.380991	0.382968
1.2	0.384922	0.386852	0.388759	0.390643	0.392504	0.394342	0.396157	0.397949	0.399719	0.401466
1.3	0.403191	0.404893	0.406574	0.408232	0.409869	0.411483	0.413076	0.414648	0.416198	0.417727
1.4	0.419234	0.420721	0.422187	0.423633	0.425057	0.426462	0.427846	0.42921	0.430554	0.431879
1.5	0.433184	0.434469	0.435735	0.436983	0.438211	0.43942	0.440611	0.441783	0.442937	0.444074
1.6	0.445192	0.446292	0.447375	0.44844	0.449488	0.450519	0.451534	0.452531	0.453512	0.454477
1.7	0.455425	0.456358	0.457275	0.458176	0.459061	0.459932	0.460787	0.461627	0.462453	0.463264
1.8	0.464061	0.464843	0.465611	0.466366	0.467107	0.467834	0.468548	0.469249	0.469937	0.470612
1.9	0.471274	0.471924	0.472562	0.473188	0.473801	0.474403	0.474993	0.475572	0.476139	0.476695
2	0.477241	0.477775	0.478299	0.478813	0.479316	0.479809	0.480292	0.480765	0.481228	0.481682
2.1	0.482127	0.482562	0.482988	0.483405	0.483814	0.484213	0.484605	0.484988	0.485362	0.485729
2.2	0.486088	0.486438	0.486782	0.487117	0.487446	0.487767	0.48808	0.488387	0.488687	0.48898
2.3	0.489267	0.489547	0.489821	0.490088	0.490349	0.490604	0.490854	0.491097	0.491335	0.491567
2.4	0.491794	0.492015	0.492231	0.492442	0.492648	0.492848	0.493044	0.493236	0.493422	0.493604
2.5	0.493782	0.493955	0.494123	0.494288	0.494449	0.494605	0.494758	0.494906	0.495051	0.495192
2.6	0.49533	0.495464	0.495595	0.495722	0.495846	0.495967	0.496084	0.496199	0.49631	0.496419
2.7	0.496524	0.496627	0.496727	0.496825	0.496919	0.497012	0.497101	0.497188	0.497273	0.497356
2.8	0.497436	0.497514	0.49759	0.497664	0.497736	0.497805	0.497873	0.497939	0.498003	0.498065
2.9	0.498126	0.498184	0.498241	0.498297	0.49835	0.498402	0.498453	0.498502	0.49855	0.498596
3	0.498641	0.498685	0.498728	0.498769	0.498808	0.498847	0.498885	0.498921	0.498956	0.498991
3.1	0.499024	0.499056	0.499087	0.499117	0.499147	0.499175	0.499203	0.499229	0.499255	0.49928
3.2	0.499304	0.499328	0.49935	0.499372	0.499394	0.499414	0.499434	0.499454	0.499472	0.49949
3.3	0.499508	0.499525	0.499541	0.499557	0.499573	0.499587	0.499602	0.499616	0.499629	0.499642
3.4	0.499654	0.499667	0.499678	0.49969	0.499701	0.499711	0.499721	0.499731	0.499741	0.49975
3.5	0.499759	0.499767	0.499776	0.499784	0.499791	0.499799	0.499806	0.499813	0.49982	0.499826
3.6	0.499832	0.499838	0.499844	0.49985	0.499855	0.49986	0.499865	0.49987	0.499875	0.499879
3.7	0.499884	0.499888	0.499892	0.499896	0.499899	0.499903	0.499906	0.49991	0.499913	0.499916
3.8	0.499919	0.499922	0.499925	0.499927	0.49993	0.499932	0.499935	0.499937	0.499939	0.499941
3.9	0.499943	0.499945	0.499947	0.499949	0.499951	0.499952	0.499954	0.499956	0.499957	0.499958
4	0.49996	0.499961	0.499962	0.499964	0.499965	0.499966	0.499967	0.499968	0.499969	0.49997

Source: From Huang, Y.H., 2004, *Pavement Analysis and Design*, Pearson Education Inc., Upper Saddle River, NJ. With permission.

2. It is inappropriate for representing variables random variables that can vary within a large range of values.

Hence, it is also common to use the lognormal distribution as an alternative distribution. If X is a lognormally distributed random variable, then log X will be normally distributed. By knowing the following relationship for the mean and the standard deviation of log X in terms of the corresponding statistics of X, the normal distribution procedure exemplified in Example 11.5 with Table 11.10 and Figure 11.18 can be

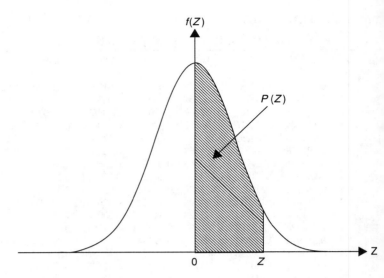

FIGURE 11.18
Standard normal distribution.

conveniently used to evaluate the probabilities associated with the lognormal distribution

$$z = \frac{\log x - \log \mu}{s_{\log x}} \tag{11.35a}$$

$$s_{\log x} = \frac{0.4343}{\mu} s \tag{11.35b}$$

where μ and s are the mean and the standard deviation of X, respectively.

11.8 Slope Instability and Landslides

Landslide studies have broadly identified factors such as (1) bedrock geology, (2) hydrology and drainage, (3) surface overburden, (4) slope range, (5) land use, and (6) landform as major causative factors of landslides. These factors contribute in different degrees to initiate a landslide, and in many occasions their impact is compounded. Therefore, the integration of the effect of such factors to evaluate the overall potential instability depends on how well their independent contribution is quantified.

11.8.1 Factors Causing Landslides

11.8.1.1 Bedrock Geology

Highly jointed rocks such as quartzite and Charnockite contribute more toward slope failure mainly because of their vulnerability to be broken into slabs or large blocks of rocks. On the contrary, rocks such as marble that remain as massive rocks contribute less toward slope failures, even though they are highly susceptible to weathering. Moreover, quartzitic rocks, when decomposed, adopt a clayey composition, and, therefore, landslides are more abundant in the residual soils of these rocks.

11.8.1.2 Hydrology

Since infiltration of rainwater contributes to triggering of most landslides. The following secondary attributes determine the overall impact of hydrology:

(1) Relief amplitude: difference between the highest and lowest points in a drainage basin.

(2) Hydrological basin form factor: the map unit or basin form factor is defined by the ratio of basin area−(basin length)^2. Basin length is the distance between the lowest discharge point and the most distant point on the crest line.

(3) Watershed basin: the area of the watershed basin governs the quantity of water flow.

(4) Drainage density: the total length of stream channels per unit area of a basin.

(5) Proximity to water bodies.

11.8.1.3 Surface Overburden

Most surface deposits involve either transported (colluvial, alluvial, etc.) or residual soil. On occasions, transported soils are present over the residual soil on slopes; hence, the impact of these two materials on slope instability cannot be distinguished. However, in residual soils, relict structure, relict joints, and relict bedding planes are usually preserved and have a control on slope failures. Hence, a given thickness of transported soils tends to be more unstable than the same thickness of residual soil. The extent and depth of the overburden deposits must be estimated visually by examining cuts and exposures. Drilling or detailed surveys would also be helpful in this task.

11.8.1.4 Slope Range

Since gravity is the driving force of unstable overburden, the steepness of a slope is directly proportional to that slope's potential for failure.

11.8.1.5 Land Use

Improper land uses such as clear cutting, de-rooting, quarrying, and mining impart a significant impact on slope failures by facilitating surface erosion and changing the internal stresses in the soil of a slope. Therefore, the contribution of land use depends on how well the land is managed or maintained.

11.8.1.6 Landform

Geomorphology or the shapes of slopes also affect the slopes' stability. A straight or complex slope with higher relief is more susceptible to failure than a terraced slope. A rough or broken rocky slope with gullies is more vulnerable to failure than an undulating to rolling slope.

11.8.2 Impact of Rainfall on Slope Instability

Rainfall presents an added hazard with respect to landslides (Okada, 1999). This is mainly due to the reduction in effective stresses in the slope due to pore pressure increases and partly due to possible erosion. Evaluation of changing effective stress conditions due to pore water pressure variations is rather a complex problem, especially if they are addressed within the impacts of the vegetation of the slope and the nature of its subsoil. According to Bhandari and Thayalan (1994), the best approach to evaluate the impact of ground water conditions is to measure the *in situ* suction and the pore water pressure. However, the cost of instrumentation that can accurately measure these parameters overrides the incorporation of these parameters into any evaluations, specifically at a large scale. Hence, the current perception is to consider rainfall intensity as an external triggering factor in landslide zonation.

Research conducted by Chen et al. (2004) has also shown that matric suction contributes to the dilative or contractive behavior of unsaturated soils. Further, the *in situ* hydrologic response to rainstorms indicates that soil suction in largely unsaturated soils is reduced by rainfall infiltration, which often becomes the triggering factor in initiating slope instability.

11.9   Investigation of Slope Failures

Historically unstable slopes continue to present slope hazards. Geotechnical investigators often study well-defined slope failure histories in order to learn useful information regarding the mechanisms of slope failure. Information sought in this regard include the development of slip surfaces and mobilization of shear strength along the slip surface. This also enables analysts to determine the levels of safety factors that assure stability. Stark and Eid (1997) studied a number of first-time slides to understand the mobilization of shear strength in stiff fissured clay where significant differences were observed in the peak and the residual shear strengths (Figure 11.19). The above study concluded that the peak shear strength of the soil mass must be used to locate the critical slip surface in slopes that have not undergone previous sliding. On the other hand, in slopes that have undergone previous sliding, the critical slip surface is usually well defined and the shear strength mobilized is an average between the peak shear (for high plasticity clays) and the residual shear.

11.10   Approximate Three-Dimensional Slope-Stability Analysis

Although the limit equilibrium methods discussed in the previous sections are strictly based on two-dimensional analysis, investigation of almost every slope failure reveals

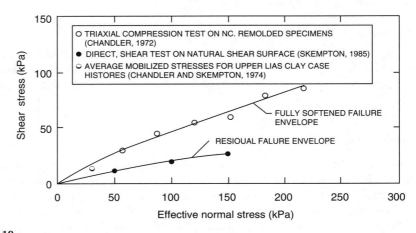

FIGURE 11.19
Drained shear strength failure envelopes for upper Lias clay. (From Stark, T. and Eid, H., 1997, *Journal of Geotechnical and Geoenvironmental Engineering, ASCE*, 123(4). With permission.)

that the deformations along a direction parallel to the slope vary significantly. Hence, actual slope failures cannot be accurately modeled by two-dimensional conditions and discrepancies are often observed between analytical predictions and field observations. On the other hand, three-dimensional analysis of slope stability presents extreme analytical and numerical difficulties mainly because, to be meaningful, the analysis has to be performed on a site-specific basis. Therefore, in order to address these issues, analysts usually resort to psuedo-three-dimensional techniques that are mostly based on generalizing two-dimensional formulations to three dimensions (Lam et al., 1993, Michalowski, 1989). Researchers have introduced alternative methods of three-dimensional analysis, one of which is the resistance-weighted procedure presented by Loehr et al. (2004). Although the resistance-weighted procedure is approximate, it serves as a simple means for estimating the magnitude of three-dimensional effects when a more rigorous three-dimensional procedure is not available.

Performing a quasi-three-dimensional analysis using the resistance-weighted procedure involves the following steps:

1. Assume a potential three-dimensional slip surface and direction of sliding.
2. Divide the soil mass above the slip surface into a number of two-dimensional, vertical cross sections aligned with the direction of sliding. Approximately 20 cross sections are sufficient to accurately represent the stability of most sliding bodies.
3. Determine the geometry of the "mean cross section" midway between each cross section created in step 2.
4. Analyze each "mean cross section" using a conventional two-dimensional slope stability analysis procedure to compute the factor of safety, F_2, and total mobilized shear force, T_1, for each section. A complete (force and moment) equilibrium procedure, such as the method of slices (Section 11.3), is typically chosen for this purpose to produce an accurate set of forces.
5. Compute the three-dimensional factor of safety from Equation (11.36) using the factors of safety, F_2, and mobilized shear forces, T_1, computed in step 4.
6. Repeat the analyses for other assumed three-dimensional slip surfaces and directions of sliding to establish the most critical three-dimensional slip surface and direction of sliding

$$F_3 = \frac{\sum_1^n F_{2-i} T_{t,i} (ds/dx)_i}{\sum_{i=1}^n T_{t,i}} \tag{11.36}$$

Where s is measured along the slip surface perpendicular to the 2-D cross-section and x is measured perpendicular to the 2-D cross section. Although the above method is approximate, results of comparative analyses presented show that the method produces three-dimensional factors of safety that are in close agreement with factors of safety computed using more rigorous three-dimensional slope-stability analysis procedures.

11.11   Additional Examples

Example 11.6
Figure 11.20 shows the typical cross section of an earthen dam built on an impervious foundation. This Figure has been drawn to a scale of 1 in. = 3 m.

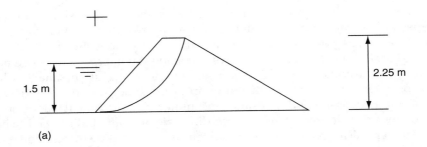

(a)

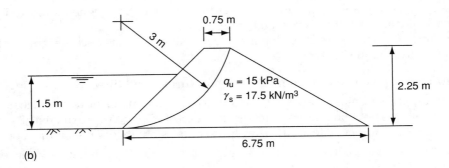

(b)

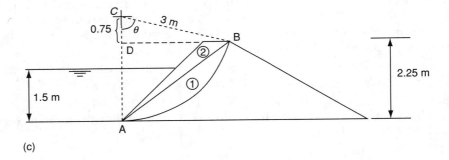

(c)

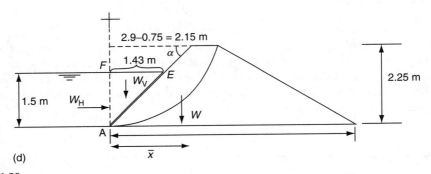

(d)

FIGURE 11.20
(a) Illustration for Example 11.6 (scale 1 in. = 3 m). (b) Problem configuration. (c) Aid for computation of the weight. (d) Computation of hydrostatic forces.

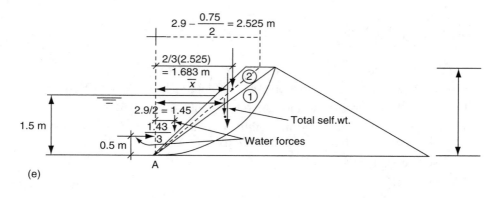

(e)

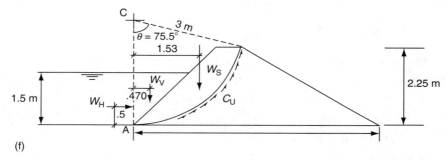

(f)

FIGURE 11.20
(e) Computation of moments. (f) Free-body diagram for Example 11.6.

The dam mostly consists of a clay with an unconfined compression strength (q_u) of 15 kPa and a saturated unit weight of 17.5 kN/m³. In order to investigate the stability of the upstream slope, one trial circular failure surface is selected as shown in the Figure:

(a) Evaluate the weight of the trial failure mass by separating it to a triangle and a circle segment. Assume that the entire dam is saturated.

(b) Evaluate the horizontal and vertical components of the water pressure force on the upstream embankment due to the reservoir.

(c) Clearly indicate the points of action of the above forces (due to the self-weight and water pressure).

(d) Evaluate the safety factor for the trail failure surface.

(a) $\theta = \cos^{-1}(0.75/(1'' \times 3)) = 75.5°$

$BD = 0.75 \tan \theta = 0.75 \tan 75.5 = 2.9\,\text{m}$

Circle segment area ($\theta = 75.5\,°$) $= \Pi(3)^2 \times (75.5/360) = 5.93\,\text{m}^2$

Triangle ABC area $= \frac{1}{2}(3)(2.9) = 4.35\,\text{m}^2$

Arc area $= (1) = 5.93 - 4.35 = 1.58\,\text{m}^2$

Triangle (2) area $= \frac{1}{2}(0.75)(2.25) = 0.84\,\text{m}^2$

Therefore, total area (1) + (2) $= 1.58 + 0.84 = 2.42\,\text{m}^2$

Weight of trial failure mass $= 2.42 \times 1 \times 17.5 = 42.35\,\text{kN/m}$

(b) $\alpha = \tan^{-1}(2.25/2.15) = 46.3°$

$EF = \frac{AF}{\tan 46.30} = 1.43$ m

$AE = \sqrt{1.5^2 + 1.43^2} = 2.07$ m

Total water force on $AE = \frac{1}{2}(9.8)(1.5)(2.07) = 15.2$ kN/m

$W_H = $ Horizontal water force on $AE = \frac{1}{2}(9.8)(1.5)(1.5) = 11.02$ kN/m

$W_V = $ Vertical water force on $AE = \frac{1}{2}(9.8)(1.5)(1.43) = 10.51$ kN/m

(c) Self-weight at (1) $=$ area $\times$ unit weight $= 1.58 \times 17.5 = 27.65$ kN/m

Self-weight at (2) $= 0.84 \times 17.5 = 14.7$ kN/m

Total self-weight $= 42.35$ kN/m

Taking moment at A

$(42.35)\bar{x} = (27.65)(1.45) + (14.7)(1.683)$

Therefore, $\bar{x} = \dfrac{64.83}{42.35} = 1.53$ m

Failure curve length $= 2\pi(3)(75.5/360) = 3.95$ m

Stabilizing moment $= (C_u)(3.95)(3) + (W_H)(3 - 0.5)$

$= (7.5)(3.95)(3) + (11.02)(2.5) = 116.42$ kN m/m

Destabilizing moment $= (W_s)(1.53) + (W_V)(0.476)$

$= (42.35)(1.53) + (10.51)(0.476)$

$= 69.8$ kN m/m

$\text{FOS} = \dfrac{\text{Stabilizing moment}}{\text{Destabilizing moment}} = \dfrac{116.42}{69.8} = 1.66$

Example 11.7

Estimate the maximum load that can be imposed on the building foundation in order to ensure the stability of the sandy embankment shown in Figure 11.21. The unit weight of the embankment sand is 16.5 kN/m^3 and the shear strength parameters of the clay layer are 10 kPa and 15°.

Self $-$ weight of soil above failure line $= \frac{1}{2}(1.75)(1.3125)(16.5)$

$= 18.95$ kN/m

Total weight $=$ self-weight $+$ foundation weight

$W = (18.95 + W_f)$kN/m

force $= (C + \sigma \tan \phi)$ (failure length x1)

Destabilizing force $= W \sin \theta_1 = (18.95 + W_f) \sin 25$

Stabilizing force $= (C + \sigma \tan \phi)$ (failure length $\times$ 1)

$$= \left(10 + \frac{W \cos \theta_1}{\text{Length} \times 1} \tan 15\right)(\sqrt{2.8125^2 + 1.3124^2})$$

$$= \left(10 + \left(\frac{18.95 + w_f}{3.103 \times 1}\right) \cos 25 \tan 15\right)(3.103)$$

(11.6)

$$= 35.63 + 0.2428w_f$$

$\text{FOS} = 2.5 = \dfrac{\text{Stabilizing force}}{\text{Destabilizing force}} = \dfrac{35.63 + 0.2428W_f}{(18.95 + W_f)\sin 25}$

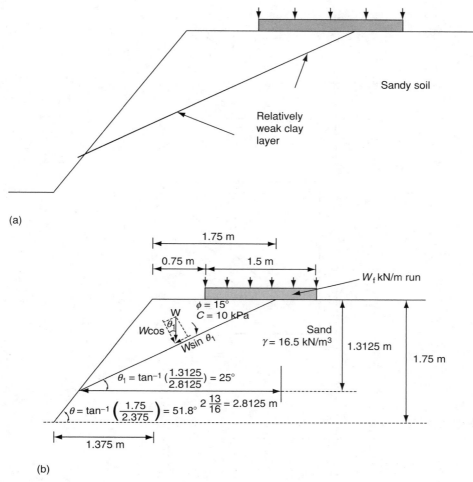

FIGURE 11.21
(a) Illustration for Example 11.7 (scale 1 in. = 1 m). (b) Free-body diagram for Example 11.7.

Therefore, $20.02 + m1.056W_f = 35.63 + 0.2428W_f$
Therefore, $W_f = 19.2$ kN/m.

Example 11.8
Figure 11.22(a) shows the cross section of a highway embankment close to a waterway. It is constructed by compacting a fill of clayey material to have the following properties:

Drained strength: Cohesion $= 10$ kPa, angle of internal friction $= 12°$
Undrained strength $= 7.5$ kPa
Saturated unit weight $= 17.5$ kN/m^3
Dry unit weight $= 16.5$ kN/m^3

i. Use Taylor's stability charts to investigate the stability of the embankment in the long term (i.e., long time after it has been built) with respect to toe failure. Assume that the embankment is dry.

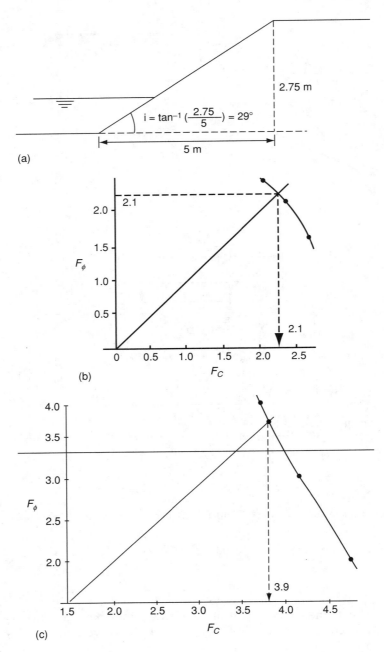

FIGURE 11.22
(a) Illustration for Example 11.8. (b) Relationship between safety factors (dry embankment). (c) Relationship between safety factors (wet embankment).

    ii. Use Taylor's stability charts to investigate the stability of the embankment in the long term (i.e., long time after it has been built) with respect to toe failure. Assume that the embankment is completely submerged by the rising water level. What conclusions can you reach by comparing your responses to parts (i) and (ii)?

iii. Use Taylor's stability charts to investigate the stability of the embankment in the short term (i.e., immediately after it has been built) with respect to toe failure. How far below the toe would the lowest point of the critical circle be?

(i) Taylor's stability chart method (Section 11.4.2.2):

$$F_\phi = \frac{\tan \phi}{\tan \phi_d} \quad \text{Therefore,} \quad \phi_d = \tan^{-1}\left[\frac{\tan \phi}{F_\phi}\right] \quad \{\phi = 12°, i = 29°\}$$

SN = stability number

Using $C = 10$ kPa, $\gamma_d = 16.5$ kN/m^3, $H = 2.75$ m (Table 11.11)
So, from the plot, $F_\phi = F_c = 2.1$.

(ii) Taylor's stability chart method:
Toe failure (Case 1 of Figure 11.8)
Long-term condition (drained) $C = 10$ kPa, $\phi = 12°$ (Table 11.12)
Completely submerged, $\gamma_{sub} = \gamma_{sat} - 9.81 = 17.5 - 9.8 = 7.7$ kN/m^3
From the plot area, $F_\phi = F_c = 3.9$.
Higher stability due to rising water level.

(iii) Taylor's stability chart method:
Toe failure (Case A, Figure 11.9, $n = 0$)
Short-term (undrained condition, $\phi = 0$, $C_u = 7.5$ kPa)
Saturated unit weight, $\gamma_s = 17.5$ kN/m^3
From Figure 11.9 (for zero-friction angle), for $n = 0$, $i = 29°$

$$\frac{C_d}{\gamma H} = 0.152 \quad \text{and} \quad D = 1.23$$

TABLE 11.11

Worksheet for Example 11.8

F_ϕ	ϕ_d	SN $= \frac{C_d}{\gamma H} = \frac{C}{F_c \gamma H}$	$F_c = \frac{C}{SN \gamma H}$
1.5	8.06	0.085	2.6
2.0	6.06	0.100	2.2
2.5	4.86	0.11	2.0

TABLE 11.12

Worksheet for Example 11.8

F_ϕ	$\phi_d = \tan^{-1}\left[\frac{\tan\phi}{F_\phi}\right]$	SN $= \frac{C}{F_c \gamma_{sub} H}$	$F_c = \frac{C}{SN \gamma_{sub} H}$	$\begin{bmatrix} C = 10 \text{ kPa}, \gamma_{sub} = 7.7 \text{ kN/m}^3 \\ H = 2.75 \text{ m} \end{bmatrix}$
2.0	6.06	0.100		4.7
4.0	3.04	0.125		3.8
3.0	4.05	0.115		4.1

$$\frac{C_d}{\gamma H} = \frac{c}{F_c \gamma_s H} = 0.152$$

$$\text{Therefore,} \quad F_c = \frac{7.5}{0.152 \times 17.5 \times 2.75} = 1.03$$

Depth factor, $D = 1.23$

Therefore, $\quad DH - H = 0.23 \times 2.75 = 0.63$ m

So, the lowest point of the critical circle would be 0.63 m below toe.

Example 11.9

A highway embankment of 3 m (Figure 11.23) is designed in a densely compacted natural sand at a slope of 25°. It is known that a clay layer passes through the entire embankment starting from its toe and emerging from the top at a horizontal distance of 2 m from the top edge of the embankment. The two soil types have the following properties:

Clay:

Cohesion $= 10$ kPa, angle of internal friction $= 12°$

Unit weight $= 17.0$ kN/m^3

Sand:

Angle of internal friction $= 30°$

Unit weight $= 18.0$ kN/m^3

   i. Use the Stability number method to estimate the safety factor of the embankment with respect to possible failure along the clay layer.
  ii. Compare your answer against the estimate given by Equation (11.7) and comment on it.

Using the stability number method (i.e., Equation (11.20)) and $F_\phi = \tan\phi / \tan\phi_d$

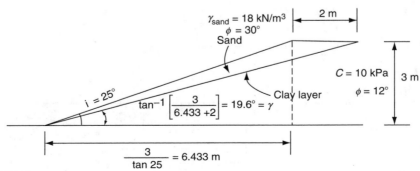

FIGURE 11.23

Illustration for Example 11.9.

$$\frac{c}{F_c \gamma H} = \frac{1}{2} \csc (i) \frac{\sin (i - \nu) \sin (\nu - \phi_d)}{\cos (\phi_d)}$$

$$\left[i = 25°, \; \nu = 19.6°, \; \phi_d = \tan^{-1} \left(\frac{\tan \phi}{F_\phi} \right) \right]$$

$C = 10 \text{kPa}$

$\phi = 12°$

$\gamma = 18 \text{ kN/m}^3$

So, $F = F_\phi = F_c = 5.55$. (Table 11.13)

(ii) Equation (11.7):

$$F = \frac{\left[\left(\gamma \dfrac{h}{\sin i} \dfrac{\sin (i - \nu) \cos (\nu)}{2} - u \right) \tan \phi + c \right]}{\gamma \dfrac{h}{\sin (i)} \dfrac{\sin (i - \nu) \sin (\nu)}{2}}$$

$$\left. \begin{array}{l} u = \text{porepressure} = 0 \\ i = 25° \\ \nu = 19.6° \\ \gamma = 18 \text{ kN/m}^3 \\ \phi = 12° \\ C = 10 \text{ kPa} \\ h = 3 \text{ m} \end{array} \right\} F = \frac{11.2}{2.01} = 5.57$$

The two values of FOS are quite close.

Example 11.10

Figure 11.24(a) shows the cross section of a completely inundated dike. The dike is made up of material that has the following properties:

Cohesion $= 10 \text{ kPa}$
Angle of internal friction $= 12°$
Saturated unit weight $= 17.5 \text{ kN/m}^3$
Water content $= 15\%$

Use the ordinary slices method to investigate the stability of the trial failure mass shown in the Figure:

TABLE 11.13

Worksheet for Example 11.9

F_ϕ	ϕ_d	SN	$F_c = \frac{c}{SN \gamma H}$
2.0	6.06	0.0262	7.07
5.0	2.43	0.0329	5.63
6.0	2.03	0.0336	5.51
5.55	2.19	0.0333	5.55

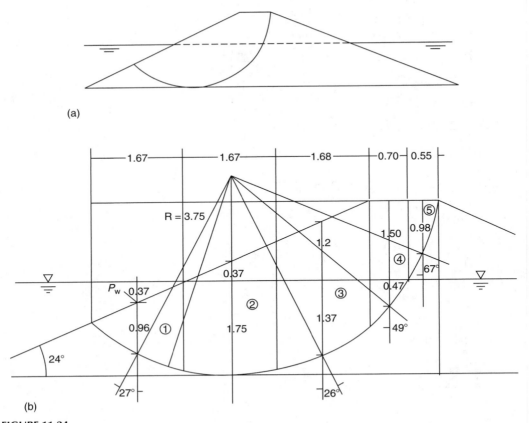

FIGURE 11.24
(a) Illustration for Example 11.10 (scale: 1 in. = 4 m). (b) Configuration and information on slices for Example 11.10.

Cohesion = 10 kPa

$\phi = 12°$ $\gamma_s = \dfrac{\gamma_s}{1 + \omega_c} = \dfrac{17.5}{1 + 0.15} = 15.22 \text{ kN/m}^2$

$\gamma_s = 17.5 \text{ kN/m}^2$ $\gamma_{sub} = 17.5 - 9.81 = 7.7 \text{ kN/m}^2$

$\omega_c = 15\%$

Using the ordinary slice method (Equation 11.10) and Table 11.14:

$$FOS = \frac{\sum [W_i \cos \alpha_i \, \tan \phi + c_i l_i]}{\sum W_i \sin \alpha_i}$$

$$= \frac{99.91 \, (\tan 12) + 78.87}{36.24} = 2.76$$

Example 11.11

Figure 11.25 shows the cross section of a highway embankment close to an interchange. The highway pavement, which is 0.61 m (2 ft) thick, is constructed out of 0.101 m (4 in.) of asphalt and compacted base. The unit weights of asphalt, base soil, and the embankment

TABLE 11.14

Worksheet for Example 11.10

Slice	Width b_i (m)	Height h_i (m)	Weight W_i (kN/m)	α_i	$W_i \sin \alpha_i$	$W_i \cos \alpha_i$	$c_i b_i \sec \alpha_i (l_i)$
1	1.67	0.96	12.34	$-27°$	-5.6	11.0	18.74
2	1.67	1.75	22.5	0	0	31.9	16.7
		0.37	$\sum = \dfrac{9.4}{31.9}$				
3	1.68	1.37	17.72	26°	20.32	41.67	18.69
		1.12	$\sum = \dfrac{28.64}{46.36}$				
4	0.70	0.47	2.53	67°	13.97	12.14	10.67
		1.50	$\sum = \dfrac{15.98}{18.51}$				
5	0.55	0.98	8.2		7.55	3.2	14.07
					$\sum = 36.24$	$\sum = 99.91$	$\sum = 78.87$

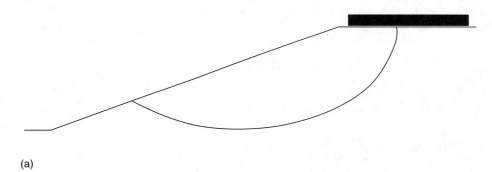

(a)

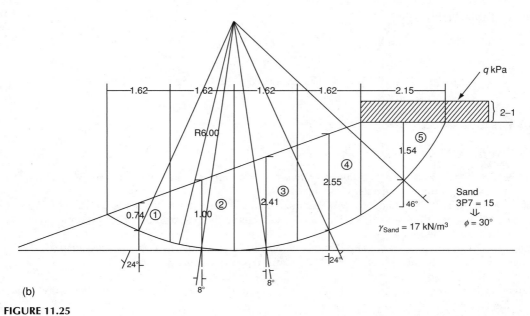

(b)

FIGURE 11.25

(a) Illustration for Example 11.11 (scale: 1 in. = 3 m). (b) Configuration and information on slices for Example 11.11.

TABLE 11.15

Worksheet for Example 11.11

Slice	Width b_i (m)	Height h_i (m)	Weight W_i (kN/m)	α_i	W_i $\sin \alpha_i$	$\tan \Phi_i$ $\tan \alpha_i$	$\sec \alpha_i$	For $F = 2.0$	For $F = 2.2$	For $F = 2.15$
1	1.62	0.74	20.38	−24	−8.29	−0.257	1.09	14.8	14.58	14.63
2	1.62	1.80	49.57	−8	−6.9	−0.081	1.009	30.12	30.00	30.03
3	1.62	2.41	66.37	8	9.23	0.081	1.009	37.19	37.32	37.29
4	1.62	2.55	70.23	24	28.56	0.257	1.095	39.32	39.74	39.64
5	2.15	1.54	56.28	46	58.16	0.5980	1.44	51.74	52.84	52.58
			± 24.57		$\sum = 80.76$			$\sum = 173.16$	$\sum = 174.5$	$\sum = 174.18$
			$\sum = 80.85$							
			$\sum = 287.4$							

soil are 22.5, 18.0, and 17.0 kN/m^3, respectively. Assuming that the embankment is made up of a sandy material compacted to an average SPT value of 15, use Bishop's method to investigate the stability of the failure surface indicated in the Figure.

$$\gamma_{asphalt} = 22.5 \text{ kN/m}^3 \quad (0.101 \text{ m thick})$$

$$\gamma_{base} = 18.0 \text{ kN/m}^3 \quad (0.509 \text{ m thick})$$

$$\gamma_{sand} = 17.0 \text{ kN/m}^3$$

$$q = 0.101 \times 22.5 + 0.509 \times 18 = 11.43 \text{ kN/m}^2$$

$$F = \frac{\sum \left[(w_i \tan \phi) \dfrac{\sec \alpha_i}{1 + \frac{\tan \phi \tan \alpha_i}{F}} \right]}{\sum w_i \sin \alpha_i} \tag{11.13}$$

using Table 11.15

Assume $F = 2.0$, $F = 173.16/80.76 = 2.14$

Assume $F = 2.2$, $F = 174.5/80.76 = 2.16$

Assume $F = 2.15$, $F = 174.18/80.76 = 2.15$ Satisfactory.

References

AASHTO, 1996, *Standard Specifications for Highway Bridges*, American Association for State Highway and Transportation Officials, Washington, DC.

Bhandari, R.K. and Thayalan, N., 1994, Landslides and other mass movements including failure of cuttings in residual soils of Sri Lanka, Proceedings of the National Symposium on Landslides, Colombo, Sri Lanka, 1994.

Bishop, A.W., 1955, The use of slip circle in stability analysis of earth slopes, *Geotechnique*, 5: 7–17.

Chen, H., Lee, C.F., and Law, K.T., 2004, Causative mechanisms of rainfall-induced fill slope failures, *Journal of Geotechnical and Geoenvironmental Engineering*, ASCE, 130(6): 593–602.

Chandler R.J., 1972, Lias clays, weathering process and their effect on shear strength, *Geotechnique*, 22(3): 403–431.

Chandler, R.J. and Skempton, A.W., 1974, Design of permanent cutting slopes in stiff fissured clay, *Geotechnique*, 24(4): 457–466.

Federal Highway Administration, 1998, *Load and Resistance Factor Design (LRFD) for Highway Bridge Substructures*, Washington, DC.

Fellinius, W., 1937, *Erdstatiche Berechnungen*, revised edition, W. Ernst u. Sons, Berlin, Germany.

Harr, M.E., 1977, *Mechanics of Particulate Media*, McGraw-Hill, New York.

Harr, M.E., 1987, *Reliability-Based Design in Civil Engineering*, McGraw-Hill, New York.

Hassiotis, S., Chameau, J.L., and Gunaratne, M., 1997, Design method for stabilization of slopes with piles, *Journal of Geotechnical and Geoenvironmental Engineering, ASCE*, 123(4): 314–323.

Huang, Y.H., 2004, *Pavement Analysis and Design*, Pearson Education Inc., Upper Saddle River, NJ.

Ito, T. and Matsui, T., 1975, Methods to estimate lateral force acting on stabilizing piles, *Soils and Foundation*, 15: 43–59.

Koerner, R., 1998, *Designing with Geosynthetics*, Prentice Hall, Englewood Cliffs, NJ.

Lam, L. and Fredlund, D.G., 1993, A general limit equilibrium model for three-dimensional slope stability analysis, *Canadian Geotechnical Journal*, 30(6): 905–919.

Loehr, J.E., McCoy, B.F., and Wright, S.G., 2004, Three-dimensional slope stability analysis method for general sliding bodies, *Journal of Geotechnical and Geoenvironmental Engineering, ASCE*, 130(6): 551–560.

Okada, K., 1999, Risk assessment and zonation on areas susceptible to rain induced slope failures in railway embankments, *Proceedings of the Workshop on Rain Induced Slope Failures*, Colombo, Sri Lanka.

Poulos, H.G., 1973, Analysis of piles in soils undergoing lateral movement, *Journal of Soil Mechanics and Foundations Division, ASCE*, 99: 391–406.

Rosenblueth, E., 1975, Point estimates of probability moments, *Proceedings of the National Academy of Science of the USA*, 72(10): 3812–3814.

Skempton, A.W., 1985, Residual strength of clays in landslides, folded strata and the laboratory, *Geotechnique*, 35(1): 3–8.

Stark, T. and Eid, H., 1997, Slope stability analysis in stiff fissured clays, *Journal of Geotechnical and Geoenvironmental Engineering, ASCE*, 123(4): 335–343.

Taylor, D.W., 1937, Stability of earth slopes, *Journal of the Boston Society of Civil Engineers*, 24(3): 337–386.

Taylor, D.W., 1948, *Fundamentals of Soil Mechanics*, John Wiley, New York.

12

Methods of Soft Ground Improvement

James D. Hussin

CONTENTS

12.1 Introduction

When a suitable foundation has to be designed for a superstructure, the foundation engineer typically follows a decision-making process in selecting the optimum type of foundation. The flowchart shown in Figure 12.1 illustrates the important steps of that decision process, which is based on the principle that cost-effective alternatives must be sought first before considering relatively costly foundation alternatives. It is seen that, in keeping with the decision sequence advocated in Figure 12.1, one must consider applicable site specific techniques for improvement of soft ground conditions, before resorting to deep foundations.

This chapter gives an overview of techniques that are commonly used by specialty contractors in the United States to improve the performance of the ground *in situ*. Not included are less specialized methods of ground improvement such as surface compaction with vibratory rollers or sheep foot type compactors, or methods that involve the placement of geotextile or geogrid materials in soil fill as it is placed. The techniques are divided into three categories:

1. *Compaction* — techniques that typically are used to compact or densify soil *in situ*.
2. *Reinforcement* — techniques that typically construct a reinforcing element within the soil mass without necessarily changing the soil properties. The performance of the soil mass is improved by the inclusion of the reinforcing elements.
3. *Fixation* — techniques that fix or bind the soil particles together thereby increasing the soil's strength and decreasing its compressibility and permeability.

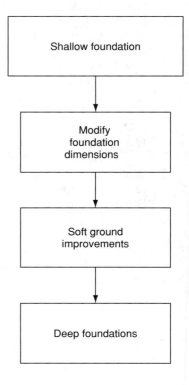

FIGURE 12.1
Decision process involved with selection of foundation type.

Techniques have been placed in the category in which they are most commonly used even though several of the techniques could fall into more than one of the categories. As each technique is addressed, the expected performance in different soil types is presented. An overview of the design methodology for each technique is also presented as are methods of performing quality assurance and quality control (QA/QC). Several *in situ* techniques of soil improvement exist that are not commonly used. These techniques are briefly described at the end of each category.

This chapter is intended to give the reader a general understanding of each of the techniques, how each improves the soil performance, and an overview of how each is analyzed. The purpose is neither to present all the nuances of each technique nor to be a detailed design manual. Indeed, entire books have been written on each technique separately. In addition, this chapter does not address all the safety issues associated with each technique. Many of these techniques have inherent dangers associated with them and should only be performed by trained and experienced specialty contractors with documented safety records.

12.2 Compaction

12.2.1 Dynamic Compaction

Dynamic compaction (DC), also known as dynamic deep compaction, was advanced in the mid-1960s by Luis Menard, although there are reports of the procedure being performed over 1000 years ago. The process involves dropping a heavy weight on the surface of the ground to compact soils to depths as great as 40 ft or 12.5 m (Figure 12.2). The

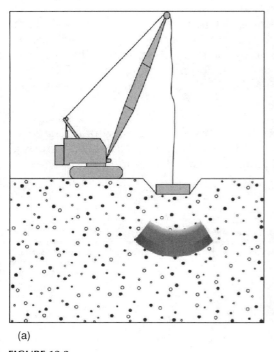

(a)   (b)

FIGURE 12.2
Deep dynamic compaction: (a) schematic, (b) field implementation. (From Hayward Baker Inc. With permission.)

method is used to reduce foundation settlements, reduce seismic subsidence and liquefaction potential, permit construction on fills, densify garbage dumps, improve mine
spoils, and reduce settlements in collapsible soils.

Applicable soil types: Dynamic compaction is most effective in permeable, granular soils.
Cohesive soils tend to absorb the energy and limit the technique's effectiveness. The
expected improvement achieved in specific soil types is shown in Table 12.1. The ground
water table should be at least 6 ft below the working surface for the process to be effective.
In organic soils, dynamic compaction has been used to construct sand or stone columns
by repeatedly filling the crater with sand or stone and driving the column through the
organic layer.

Equipment: Typically a cycle duty crane is used to drop the weight, although specially
built rigs have been constructed. Since standard cranes are typically not designed for the
high cycle, dynamic loading, the cranes must be in good condition and carefully maintained and inspected during performance of the work to maintain a safe working environment. The crane is typically rigged with sufficient boom to drop the weight from
heights of 50 to 100 ft (15.4 to 30.8 m), with a single line to allow the weight to nearly
"free fall," maximizing the energy of the weight striking the ground. The weight to be
dropped must be below the safe single line capacity of the crane and cable. Typically
weights range from 10 to 30 tons (90 to 270 kN) and are constructed of steel to withstand
the repetitive dynamic forces.

Procedure: The procedure involves repetitively lifting and dropping a weight on the
ground surface. The layout of the primary drop locations is typically on a 10 to 20 ft (3.1
to 6.2 m) grid with a secondary pass located at the midpoints of the primary pass. Once
the crater depth has reached about 3 to 4 ft (about 1 m), the crater is filled with granular
material before additional drops are performed at that location.

The process produces large vibrations in the soil which can have adverse effects on
nearby existing structures. It is important to review the nearby adjacent facilities for
vibration sensitivity and to document their preexisting condition, especially structures
within 500 ft (154 m) of planned drop locations. Vibration monitoring during DC is also
prudent. Extreme care and careful monitoring should be used if treatment is planned
within 200 ft (61.5 m) of an existing structure.

Materials: The craters resulting from the procedure are typically filled with a clean, free
draining granular soil. A sand backfill can be used when treating sandy soils. A crushed
stone backfill is typically used when treating finer-grained soils or landfills.

TABLE 12.1

Expected Improvement and Required Energy with Dynamic Compaction

Soil Description	Expected Improvement	Typical Energy Required (tons ft/cf)[a]
Gravel and sand <10% silt, no clay	Excellent	2–2.5
Sand with 10–80% silt and <20% clay, pI < 8	Moderate if dry; minimal if moist	2.5–3.5
Finer-grained soil with pI > 8	Not applicable	—
Landfill	Excellent	6–11

[a]Energy = (drop height × weight × number of drops)/soil volume to be compacted; 1 ton ft/ft^3 = 94.1 kJ/m^3.

Design: The design will begin with an analysis of the planned construction with the existing subsurface conditions (bearing capacity, settlement, liquefaction, etc.). Then the same analysis is performed with the improved soil parameters (i.e., SPT N value, etc.) to determine the minimum values necessary to provide the required performance. Finally, the vertical and lateral extent of improved soil necessary to provide the required performance is determined.

The depth of influence is related to the square root of the energy from a single drop (weight times the height of the drop) applied to the ground surface. The following correlation was developed by Dr Robert Lucas based on field data:

$$D = k(W \times H)^{1/2} \tag{12.1}$$

where D is the maximum influence depth in meters beneath the ground surface, W is the weight in metric tons (9 kN) of the object being dropped, and H is the drop height in meters above the ground surface. The constant k varies with soil type and is between 0.3 and 0.7, with lower values for finer-grained soils.

Although this formula predicts the maximum depth of improvement, the majority of the improvement occurs in the upper two-thirds of this depth with the improvement tapering off to zero in the bottom third. Repeated blows at the same location increases the degree of improvement achieved within this zone. However, the amount of improvement achieved decreases with each drop eventually resulting in a point of diminishing returns. The expected range of unit energy required to achieve this point is presented in Table 12.1.

Treatment of landfills is effective in reducing voids; however, it has little effect on future decomposition of biodegradable components. Therefore treatment of landfills is typically restricted to planned roadway and pavement areas, and not for structures. After completion of dynamic compaction, the soils within 3 to 4 ft (1 m) of the surface are loose. The surface soils are compacted with a low energy "ironing pass," which typically consists of dropping the same weight a couple of times from a height of 10 to 15 ft (3.0 to 4.5 m) over the entire surface area.

Quality control and quality assurance: In most applications, penetration testing is performed to measure the improvement achieved. In landfills or construction debris, penetration testing is difficult and shear wave velocity tests or large scale load tests with fill mounds can be performed. A test area can be treated at the beginning of the program to measure the improvement achieved and to make adjustments if required. The depth of the craters can also be measured to detect "soft" areas of the site requiring additional treatment. The decrease in penetration with additional drops gives an indication when sufficient improvement is achieved.

12.2.2 Vibro Compaction

Vibro compaction (VC), also known as Vibroflotation™ was developed in the 1930s in Europe. The process involves the use of a down-hole vibrator (vibroflot), which is lowered into the ground to compact the soils at depth (Figure 12.3). The method is used to increase bearing capacity, reduce foundation settlements, reduce seismic subsidence and liquefaction potential, and permit construction on loose granular fills.

Applicable soil types: The VC process is most effective in free draining granular soils. The expected improvement achieved in specific soil types is shown in Table 12.2. The typical spacing is based on a 165-horsepower (HP) (124 kW) vibrator. Although most effective below the groundwater table, VC is also effective above.

TABLE 12.2

Expected Improvement and Typical Probe Spacing with Vibro Compaction

Soil Description	Expected Improvement	Typical Probe Spacing (ft)[a]
Well-graded sand <5% silt, no clay	Excellent	9–11
Uniform fine to medium sand with <5% silt and no clay	Good	7.5–9
Silty sand with 5–15% silt, no clay	Moderate	6–7.5
Sand/silts, >15% silt	Not applicable[b]	—
Clays and garbage	Not applicable	—

[a]Probe spacing to achieve 70% relative density with 165 HP vibroflot, higher densities require closer spacing (1 ft = 0.308 m).
[b]Limited improvement in silts can be achieved with large displacements and stone backfill.

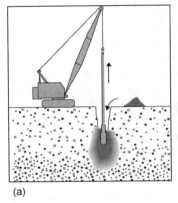

(a)   (b)

FIGURE 12.3
Vibroflotation: (a) schematic, (b) field implementation. (From Hayward Baker Inc. With permission.)

Equipment: The vibroflot consists of a cylindrical steel shell with and an interior electric or hydraulic motor which spins an eccentric weight (Figure 12.4). Common vibrator dimensions are approximately 10 ft (3.1 m) in length and 1.5 ft (0.5 m) in diameter. The vibration is in the horizontal direction and the source is located near the bottom of the probe, maximizing the effect on the surrounding soils. Vibrators vary in power from about 50 to over 300 HP (37.7 to 226 kW). Typically, the vibroflot is hung from a standard crane, although purpose built machines do exist. Extension tubes are bolted to the top of the vibrator so that the vibrator can be lowered to the necessary treatment depth.

Electric vibrators typically have a remote ammeter, which displays the amperage being drawn by the electric motor. The amperage will typically increase as the surrounding soils densify.

Procedure: The vibrator is lowered into the ground, assisted by its weight, vibration, and typically water jets in its tip. If difficult penetration is encountered, predrilling through the firm soils may also be performed. The compaction starts at the bottom of the treatment depth. The vibrator is then either raised at a certain rate or repeatedly raised and lowered as it is extracted (Figure 12.5). The surrounding granular soils rearranged into a denser configuration, achieving relative densities of 70 to 85%. Treatment as deep as 120 ft (37 m) has been performed.

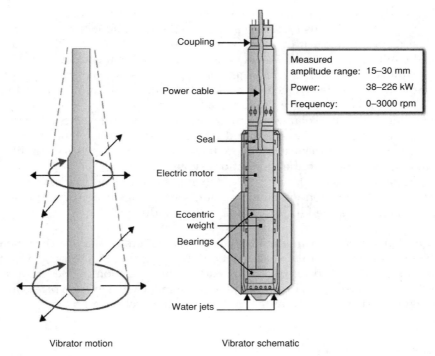

Measured	
amplitude range:	15–30 mm
Power:	38–226 kW
Frequency:	0–3000 rpm

Coupling

Power cable

Seal

Electric motor

Eccentric weight

Bearings

Water jets

Vibrator motion

Vibrator schematic

FIGURE 12.4
Electric vibroflot cross section. (From Hayward Baker Inc. With permission.)

Sand added around the vibrator at the ground surface falls around the vibrator to its tip to compensate for the volume reduction during densification. If no sand is added, the *in situ* sands will fall, resulting in a depression at the ground surface. Loose sand will experience a 5 to 15% volume reduction during densification. Coarser backfill, up to gravel size, improves the effectiveness of the technique, especially in silty soils. The technique does not densify the sands within 2 to 3 ft (0.6 to 0.9 m) of the ground surface. If necessary, this is accomplished with a steel drum vibratory roller.

Materials: Backfill usually consists of sand with less than 10% silt and no clay, although gravel size backfill can also be used. A coarser backfill facilitates production and densification.

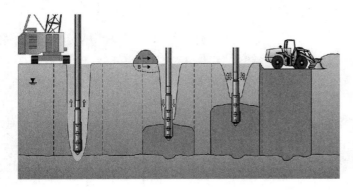

FIGURE 12.5
Vibro compaction process. (From Hayward Baker Inc. With Permission.)

Design: The design will begin with an analysis of the planned construction with the existing subsurface conditions (bearing capacity, settlement, liquefaction, etc.). Then the same analysis is performed with the improved soil parameters (i.e., SPT N value, etc.) to determine the minimum soil parameters necessary to provide the required performance. And finally, the vertical and lateral extent of improved soil necessary to provide the required performance is determined. In the case of settlement improvement for spread footings, it is common to improve the sands beneath the planned footings to a depth of twice the footing width for isolated column footings and four times the footing width for wall footings. Area treatments are required where an area load is planned or in seismic applications. For treatment beneath shallow foundations for nonseismic conditions, it is common to treat only beneath the foundations (Figure 12.6).

The degree of improvement achievable depends on the energy of the vibrator, the spacing of the vibrator penetrations, the amount of time spent densifying the soil, and the quantity of backfill added (or *in situ* soil volume reduction).

Quality control and quality assurance: Production parameters should be documented for each probe location, such as depth, compaction time, amperage increases, and estimated volume of backfill added. If no backfill is added, the reduction in the ground surface elevation should be recorded. The degree of improvement achieved is typically measured with penetration tests performed at the midpoint of the probe pattern.

12.2.3 Compaction Grouting

Compaction grouting, one of the few US born ground improvement techniques, was developed by Ed Graf and Jim Warner in California in the 1950s. This technique densifies soils by the injection of a low mobility, low slump mortar grout. The grout bulb expands as additional grout is injected, compacting the surrounding soils through compression. Besides the improvement in the surrounding soils, the soil mass is reinforced by the resulting grout column, further reducing settlement and increasing shear strength. The method is used to reduce foundation settlements, reduce seismic subsidence and liquefaction potential, permit construction on loose granular fills, reduce settlements in collapsible soils, and reduce sinkhole potential or stabilize existing sinkholes in karst regions.

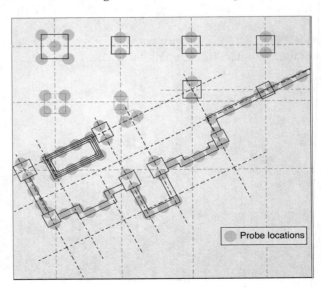

FIGURE 12.6
Typical vibro compaction layout for nonseismic treatment beneath foundations. (From Hayward Baker Inc. With permission.)

Applicable soil types: Compaction grouting is most effective in free draining granular soils and low sensitivity soils. The expected improvement achieved in specific soil types is shown in Table 12.3. The depth of the groundwater table is not important as long as the soils are free draining.

Equipment: Three primary pieces of equipment are required to perform compaction grouting, one to batch the grout, one to pump the grout, and one to install the injection pipe. In some applications, ready-mix grout is used eliminating the need for on-site batching. The injection pipe is typically installed with a drill rig or is driven into the ground. It is important that the injection pipe is in tight contact with the surrounding soils. Otherwise the grout might either flow around the pipe to the ground surface or the grout pressure might jack the pipe out of the ground. Augering or excessive flushing could result in a loose fit. The pump must be capable of injecting a low slump mortar grout under high pressure. A piston pump capable of achieving a pumping pressure of up to 1000 psi (6.9 MPa) is often required (Figure 12.7).

Procedure: Compaction grouting is typically started at the bottom of the zone to be treated and precedes upward (Figure 12.8). The treatment does not have to be continued to the ground surface and can be terminated at any depth. The technique is very effective in targeting isolated zones at depth. It is generally difficult to achieve significant improvement within about 8 ft (2.5 m) of the ground surface. Some shallow improvement can be accomplished using the slower and more costly top down procedure. In this procedure, grout is first pumped at the top of the treatment zone. After the grout sets up, the pipe is

TABLE 12.3

Expected Improvement with Compaction Grouting

Soil Description	Densification	Reinforcement
Gravel and sand <10% silt, no clay	Excellent	Very good
Sand with between 10 and 20% silt and <2% clay	Moderate	Very good
Finer-grained soil, nonplastic	Minimal	Excellent
Plastic soil	Not applicable	Excellent

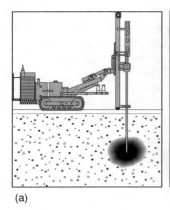
(a)

(b)

FIGURE 12.7
Compaction grout process: (a) schematic, (b) field implementation. (From Hayward Baker Inc. With permission.)

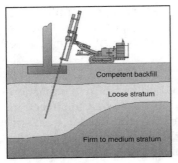
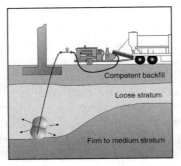
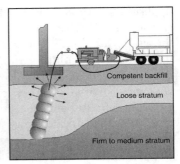

FIGURE 12.8
Compaction grouting process. (From Hayward Baker Inc. With permission.)

drilled to the underside of the grout and additional grout is injected. This procedure is repeated until the bottom of the treatment zone is grouted. The grout injection rate is generally in the range of 3 to 6 ft^3/min (0.087 to 0.175 m^3/min), depending on the soils being treated. If the injection rate is too fast, excess pore pressures or fracturing of the soil can occur, reducing the effectiveness of the process.

Materials: Generally, the compaction grout consists of Portland cement, sand, and water. Additional fine-grained materials can be added to the mix, such as natural fine-grained soils, fly ash, or bentonite (in small quantities). The grout strength is generally not critical for soil improvement, and if this is the case, cement has been omitted and the sand replaced with naturally occurring silty sand. A minimum strength may be required if the grout columns or mass are designed to carry a load.

Design: The design will begin with an analysis of the planned construction with the existing subsurface conditions (bearing capacity, settlement, liquefaction, etc.). Then the same analysis is performed with the improved soil parameters (i.e., SPT *N* value, etc.) to determine the minimum parameters necessary to provide the required performance. Finally, the vertical and lateral extent of improved soil necessary to provide the required performance is determined. In the case of settlement improvement for spread footings, it is common to improve the sands beneath the planned footings to a depth of twice the footing width for isolated column footings and four times the footing width for wall footings. A conservative analysis of the post-treatment performance only considers the improved soil and does not take into account the grout elements. The grout elements are typically columns. A simplified method of accounting for the grout columns is to take a weighted average of the parameters of the improved soil and grout. The grout columns can also be designed using a standard displacement pile methodology.

   The degree of improvement achievable depends on the soil (soil gradation, percent fines, percent clay fines, and moisture content) as well as the spacing and percent displacement (the volume of grout injected divided by volume of soil being treated).

Quality control and quality assurance: Depending on the grout requirements, grout slump and strength is often specified. Slump testing and sampling for unconfined compressive strength testing is performed during production. The production parameters should also be monitored and documented, such as pumping rate, quantities, pressures, ground heave, and injection depths. Postgrouting penetration testing can be performed between injection locations to verify the improvement of granular soil.

12.2.4 Surcharging with Prefabricated Vertical Drains

Surcharging consists of placing a temporary load (generally soil fill) on sites to pre-consolidate the soil prior to constructing the planned structure (Figure 12.9). The process improves the soil by compressing the soil, increasing its stiffness and shear strength. In partially or fully saturated soils, prefabricated vertical drains (PVDs) can be placed prior to surcharge placement to accelerate the drainage, reducing the required surcharge time.

Applicable soil types: Preloading is best suited for soft, fine-grained soils. Soft soils are generally easy to penetrate with PVDs and layers of stiff soil may require predrilling.

Equipment: Generally, a surcharge consists of a soil embankment and is placed with standard earthmoving equipment (trucks, dozers, etc). Often the site surface is soft and wet, requiring low ground pressure equipment.

The PVDs are installed with a mast mounted on a backhoe or crane, often with low ground pressure tracks. A predrilling rig may be required if stiff layers must be penetrated.

Procedure: Fill soil is typically delivered to the area to be surcharged with dump trucks. Dozers are then used to push the soil into a mound. The height of the mound depends on the required pressure to achieve the required improvement.

The PVDs typically are in 1000 ft (308 m) rolls and are fed into a steel rectangular tube (mandrel) from the top. The mandrel is pushed, vibrated, driven or jetted vertically into the ground with a mast mounted on a backhoe or crane. An anchor plate or bar attached to the bottom of the PVD holds it in place in the soil as the mandrel is extracted. The PVD is then cut off slightly above the ground surface and another anchor is attached. The mandrel is moved to the next location and the process is repeated. If obstructions are encountered during installation, the wick drain location can be slightly offset.

In very soft sites, piezometers and inclinometers, as well as staged loading, may be required to avoid the fill being placed too quickly, causing a bearing capacity or slope stability failure. If stiff layers must be penetrated, predrilling may be required.

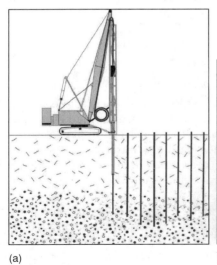

(a)   (b)

FIGURE 12.9
Surcharging with prefabricated vertical drains: (a) schematic, (b) field implementation. (From Hayward Baker Inc. With permission.)

Settlement plates are placed in the surcharge. The elevation of these plates is measured to determine when the design settlement has occurred.

Materials: The first layer of surcharge generally consists of a drainage material to drain the water displaced from the ground during compression. Since surcharge soils are generally temporary in nature, their composition and degree of compaction are generally not critical. If the site settlement will result in some of the surcharge soil settling below finish grade, this height of fill is initially placed as compacted structural fill, to avoid having to excavate and replace it at the end of the surcharge program.

The PVD is composed of a 4-in. (10 cm) wide strip of corrugated or knobbed plastic wrapped in a woven filter fabric. The fabric is designed to remain permeable to allow the ground water to flow through it but not the soil.

Design: Generally, a surcharge program is considered when the site is underlain by soft fine-grained soils which will experience excessive settlement under the load of the planned structure. Using consolidation test data, a surcharge load and duration is selected to preconsolidate the soils sufficiently such that when the surcharge load is removed and the planned structure is constructed, the remaining settlement is acceptable.

PVDs are selected if the required surcharge time is excessive for the project. The time required for the surcharge settlement to occur depends on the time it takes for the excess pore water pressure to dissipate. This is dictated by the soils permeability and the square of the distance the water has to travel to get to a permeable layer. The PVDs accelerate the drainage by shortening the drainage distance. The spacing of the PVDs are designed to reduce the consolidation time to an acceptable duration. The closer the drains are installed (typically 3 to 6 ft on center) the shorter the surcharge program is in duration.

Quality control and quality assurance: The height and unit weight of the surcharge should be documented to assure that the design pressure is being applied. The PVD manufacturer's specifications should be reviewed to confirm that the selected PVD is suitable for the application. During installation, the location, depth, and verticality are important to monitor and record. The settlement monitoring program is critical so that the completion of the surcharge program can be determined.

12.2.5   Infrequently-Used Compaction Techniques

12.2.5.1   *Blast-Densification and Vacuum-Induced Consolidation*

Blast-densification densifies sands with underground explosives. The technique was first used in the 1930s in the former Soviet Union and in New Hampshire. The below grade explosion causes volumetric strains and shearing which rearranges of soil particles into a denser configuration. The soils are liquefied and then become denser as the pore pressures dissipate. Soils as deep as 130 ft (40 m) have been treated. A limited number of projects have been performed and generally only for remote location where the blast-induced vibrations are not a concern.

Vacuum-induced consolidation (VIC) uses atmospheric pressure to apply a temporary surcharge load. The concept of VIC was introduced in the 1950s; however, the first practical project was performed in 1980 in China. Following that, a number of small projects have been performed, but few outside China. A porous layer of sand or gravel is placed over the site and it is covered with an air tight membrane, sealed into the clay below the ground surface. The air is then pumped out of the porous layer, producing a pressure difference of 0.6 to 0.7 atm, equivalent to about 15 ft (4.6 m) of fill. The process can be accelerated by the use of PVDs. The process eliminates the need for surcharge fill and avoids shear failure in the soft soil; however, any sand seams within the compressible layer can make it difficult to maintain the vacuum.

12.3 Reinforcement

12.3.1 Stone Columns

Stone columns refer to columns of compacted, gravel size stone particles constructed vertically in the ground to improve the performance of soft or loose soils. The stone can be compacted with impact methods, such as with a falling weight or an impact compactor or with a vibroflot, the more common method. The method is used to increase bearing capacity (up to 5 to 10 ksf or 240 to 480 kPa), reduce foundation settlements, improve slope stability, reduce seismic subsidence, reduce lateral spreading and liquefaction potential, permit construction on loose/soft fills, and precollapse sinkholes prior to construction in karst regions.

Applicable soil types: Stone columns improve the performance of soils in two ways, densification of surrounding granular soil and reinforcement of the soil with a stiffer, higher shear strength column. The expected improvement achieved in specific soil types is shown in Table 12.4. The depth of the ground water is generally not critical.

Procedure: The column construction starts at the bottom of the treatment depth and proceeds to the surface. The vibrator penetrates into the ground, assisted by its weight, vibration, and typically water jets in its tip, the wet top feed method (Figure 12.10 and Figure 12.11a). If difficult penetration is encountered, predrilling through the firm soils may also be performed. A front end loader places stone around the vibroflot at the ground surface and the stone falls to the tip of the vibroflot through the flushing water around the exterior of the vibroflot. The vibrator is then raised a couple of feet and the stone falls around the vibroflot to the tip, filling the cavity formed as the vibroflot is raised. The vibroflot is then repeatedly raised and lowered as it is extracted, compacting and displacing the stone in 2 to 3 ft (0.75 to 0.9 m) lifts. The flushing water is usually directed to a settlement pond where the suspended soil fines are allowed to settle.

If the dry bottom feed procedure is selected, the vibroflot penetrates into the ground, assisted by its weight and vibrations alone (Figure 12.11b). Again, predrilling may be used if necessary or desired. The remaining procedure is then similar except that the stone is feed to the tip of the vibroflot though the tremie pipe. Treatment depth as deep as 100 ft (30 m) has been achieved.

TABLE 12.4

Expected Densification and Reinforcement Achieved with Stone Columns

Soil Description	Densification	Reinforcement
Gravel and sand <10% silt, no clay	Excellent	Very good
Sand with between 10 and 20% silt and <2% clay	Very good	Very good
Sand with >20% silt and nonplastic silt	Marginal (with large displacements)	Excellent
Clays	Not applicable	Excellent

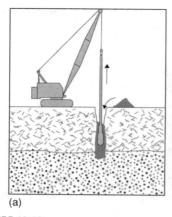 

(a)   (b)

FIGURE 12.10
Installation of stone columns: (a) schematic, (b) field implementation. (From Hayward Baker Inc. With permission.)

Equipment: When jetting water is used to advance the vibroflot, the equipment and setup is similar to VC. If jetting water is not desired for a particular project, the dry bottom feed process can be used (Figure 12.11b). A tremie pipe, through which stone is fed to the tip of the vibroflot, is fastened to the side of the vibroflot. A stone skip is filled with stone on the ground with a front end loader and a separate cable raises the skip to a chamber at the top of the tremie pipe.

A specific application is referred to as vibro piers. The process refers to short, closely spaced stone columns designed to create a stiff block to increase bearing capacity and reduce settlement to acceptable values. Vibro piers are typically constructed in cohesive soils in which a full depth predrill hole will stay open. The stone is compacted in 1 to 2 ft (0.4 to 0.8 m) lifts, each of which is rammed and compacted with the vibroflot.

Materials: The stone is typically a graded crushed hard rock, although natural gravels and pebbles have been used. The greater the friction angle of the stone, the greater the modulus and shear strength of the column.

Design: Several methods of analysis are available. For static analysis, one method consists of calculating weighted averages of the stone column and soil properties (cohesion, friction angle, etc.). The weighted averages are then used in standard geotechnical methods of analysis (bearing capacity, settlement, etc.). Another method developed by Dr Hans Priebe, involves calculating the post-treatment settlement by dividing the untreated settlement by an improvement factor (Figure 12.12). In static applications, the treatment limits are typically equal to the foundation limits.

For liquefaction analysis, stone column benefits include densification of surrounding granular soils, reduction in the cyclic stress in the soil because of the inclusion of the stiffer stone columns, and drainage of the excess pore pressure. A method of evaluation for all three of these benefits was presented by Dr Juan Baez. Dr Priebe has also presented a variation of his static method for this application. In liquefaction applications, the treatment generally covers the structure footprint and extends laterally outside the areas to be protected, a distance equal to two-thirds of the thickness of the liquefiable zone.

This is necessary to avoid surrounding untreated soils from adversely affecting the treated area beneath the foundation.

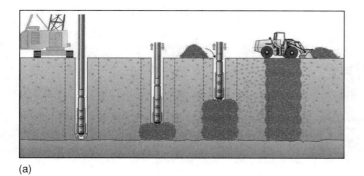

(a)

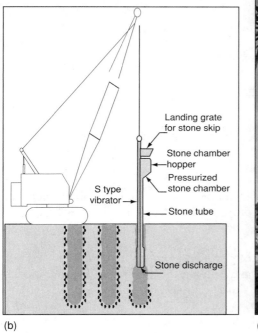

(b)

(c)

FIGURE 12.11
Stone column construction: (a) wet top feed method, (b) schematic, and (c) field implementation of dry bottom feed method. (From Hayward Baker Inc. With permission.)

Quality control and quality assurance: During production, important parameters to monitor and document include location, depth, ammeter increases (see Section 12.2.2), and quantity of stone backfill used. Post-treatment penetration testing can be performed to measure the improvement achieved in granular soils. Full-scale load tests are becoming common with test footings measuring as large as 10 ft square (3.1 m) and loaded to 150% of the design load (Figure 12.13).

12.3.2 Vibro Concrete Columns

Vibro concrete columns (VCCs) involve constructing concrete columns *in situ* using a bottom feed vibroflot (Figure 12.14). The method will densify granular soils and transfer

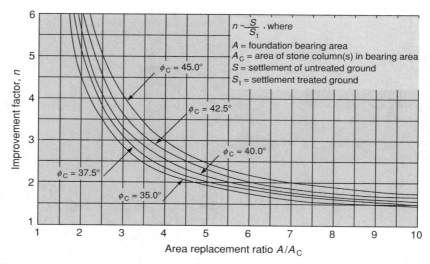

FIGURE 12.12
Chart to estimate improvement factor with stone columns.

loads through soft cohesive and organic soils. The method is used to reduce foundation settlements, to increase bearing capacity, to increase slope stability, and as an alternative to piling.

Applicable soil types: VCCs are best suited to transfer area loads, such as embankments and tanks, through soft and/or organic layers to an underlying granular layer. The depth of the groundwater table is not critical.

Equipment: The equipment is similar to the bottom feed stone column setup. A concrete hose connects a concrete pump to the top of the tremie pipe. Since verticality is important, the vibroflot is often mounted in a set of leads or a spotter.

Procedure: The vibroflot is lowered or pushed through the soft soil until it penetrates into the bearing stratum. Concrete is then pumped as the vibroflot is repeatedly raised and lowered about 2 ft (0.75 m) to create an expanded base and densifying surrounding granular soils. The concrete is pumped as the vibroflot is raised to the surface. At the

FIGURE 12.13
Full-scale load test (10 ft or 3.1 m², loaded to 15 ksf or 719 kPa). (From Hayward Baker Inc. With permission.)

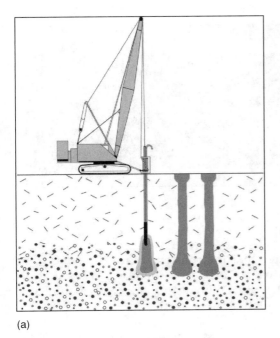

(a)   (b)

FIGURE 12.14
Installation of vibro concrete columns: (a) schematic, (b) field implementation. (From Hayward Baker Inc. With permission.)

ground surface, the vibroflot is again raised and lowered several times to form an expanded top. Most VCC applications are less than 40 ft (12.3 m) in depth.

Materials: Concrete or cement mortar grout is typically used. The mix design depends on the requirements of the application.

Design: The analysis and design of VCCs are essentially the same as would be performed for an expanded base pile except that the improved soil parameters are used.

Quality control and quality assurance: During production, important parameters to monitor and document include location, depth, verticality, injection pressure and quantity, and concrete quality. It is very important to monitor the pumping and extraction rates to verify that the grout pumping rate matches or slightly exceeds the rate at which the void is created as the vibroflot is extracted. VCCs can be load tested in accordance with ASTM D 1143.

12.3.3 Soil Nailing

Soil nailing is an *in situ* technique for reinforcing, stabilizing, and retaining excavations and deep cuts through the introduction of relatively small, closely spaced inclusions (usually steel bars) into a soil mass, the face of which is then locally stabilized (Figure 12.15). The technique has been used for four decades in Europe and more recently in the United States. A zone of reinforced ground results that functions as a soil retention system. Soil nailing is used for temporary or permanent excavation support/retaining walls, stabilization of tunnel portals, stabilization of slopes, and repairing retaining walls.

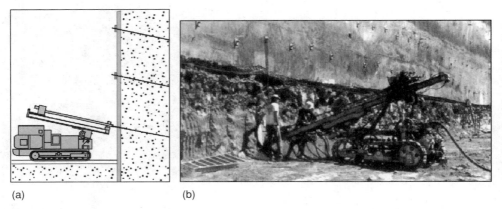

FIGURE 12.15
Soil nailing: (a) schematic, (b) field implementation. (From Hayward Baker Inc. With permission.)

Applicable soil types: The procedure requires that the soil temporarily stand in a near vertical face until a row of nails and facing are installed. Therefore, cohesive soil or weathered rock is best suited for this technique. Soil nails are not easily performed in cohesionless granular soils, soft plastic clays, or organics/peats.

Equipment: The technique requires some piece of earth moving equipment (such as a dozer or backhoe) to excavate the soil, a drill rig to install the nails, a grout mixer and pump (for grouted nails), and a shotcrete mixer and pump (if the face is to be stabilized with shotcrete).

Procedure: The procedure for constructing a soil nail excavation support wall is a top down method (Figure 12.16). A piece of earth moving equipment (such as a dozer or backhoe) excavates the soil in incremental depths, typically 3 to 6 ft (1 to 2 m). Then a drill rig typically is used to drill and grout the nails in place, typically on 3 to 6 ft (1 to 2 m) centers. After each row of nails is installed, the excavated face is stabilized, typically by fastening a welded wire mesh to the nails and then placing shotcrete.

Materials: Soil nails are typically steel reinforcing bars but may consist of steel tubing, steel angles, or high-strength fiber rods. Grouted nails are usually installed with a Portland

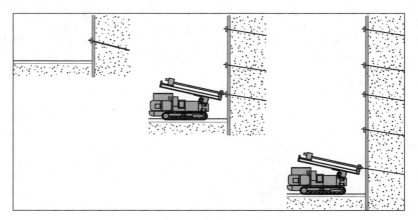

FIGURE 12.16
Soil nailing process. (From Hayward Baker Inc. With permission.)

cement grout slurry. The facing can be prefabricated concrete or steel panels, but is usually shotcrete, reinforced with welded wire mesh, rebar or steel or polyester fibers.

Design: Soil nails are designed to give a soil mass an apparent cohesion by transferring of resisting tensile forces generated in the inclusions into the ground. Frictional interaction between the ground and the steel inclusions restrain the ground movement. The main engineering concern is to ensure that the ground–inclusion interaction is effectively mobilized to restrain ground displacements and can secure the structural stability with an appropriate factor of safety. There are two main categories of design methods:

1. Limit equilibrium design methods
2. Working stress design methods.

Many software design programs are available including one developed in 1991 by CALTRANS called Snail.

Soil nail walls are generally not designed to withstand fluid pressures. Therefore, drainage systems are incorporated into the wall, such as geotextile facing, or drilled in place relief wells and slotted plastic collection piping. Surface drainage control above and behind the retaining wall is also critical.

Extreme care should be exercised when an existing structure is adjacent to the top of a soil nail wall. The soil nail reinforced mass tends to deflect slightly as the mass stabilizes under the load. This movement may cause damage to the adjacent structure.

Quality control and quality assurance: The location and lengths of the nails are important to monitor and document. In addition, the grout used in the installation of grouted nails can be sampled and tested to confirm that it exceeds the design strength. Tension tests can also be performed on test nails to confirm that the design bond is achieved.

12.3.4 Micropiles

Micropiles, also known as minipiles and pin piles, are used in almost any type of ground to transfer structural load to competent bearing strata (Figure 12.17). Micropiles were originally small diameter (2 to 4 in., or 5 to 10 cm), low-capacity piles. However, advances

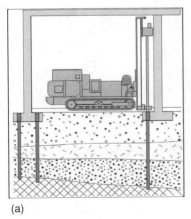

(a) (b)

FIGURE 12.17
Micropiling: (a) schematic, (b) field implementation. (From Hayward Baker Inc. With permission.)

in drilling equipment have resulted in design load capacities in excess of 300 tons (2.7 MN) and diameters in excess of 10 in. (25 cm). Micropiles are often installed in restricted access and limited headroom situations. Micropiles can be used for a wide range of applications; however, the most common applications are underpinning existing foundations or new foundations in limited headroom and tight access locations.

Applicable soil types: Since micropiles can be installed with drilling equipment and can be combined with different grouting techniques to create the bearing element, they can be used in nearly any subsurface soil or rock. Their capacity will depend on the bearing soil or rock.

Equipment: The micropile shaft is usually driven or drilled into place. Therefore, a drill rig or small pile driving hammer on a base unit is required. The pipe is filled with a cement grout so the appropriate grout mixing and pumping equipment is required. If the bearing element is to be created with compaction grout or jet grout, the appropriate grouting equipment is also required.

Procedure: The micropile shaft is usually either driven or drilled into place. Unless the desired pile capacity can be achieved in end bearing and side friction along the pipe, some type of bearing element must be created (Figure 12.18). If the tip is underlain by rock, this could consist of drilling a rock socket, filling the socket with grout and placing a full-length, high-strength threaded bar. If the lower portion of the pipe is surrounded or underlain by soil, compaction grouting or jet grouting can be performed below the bottom of the pipe. Also, the pipe can be filled with grout which is pressurized as the pipe is partially extracted to create a bond zone. The connection of the pipe to the existing or planned foundation must then be constructed.

Materials: The micropile typically consists of a steel rod or pipe. Portland cement grout is often used to create the bond zone and fill the pipe. A full length steel threaded bar is also common, composed of grade 40 to 150 ksi steel. In some instances, the micropile only consists of a reinforced, grout column.

Design: The design of the micropile is divided into three components: the connection with the existing or planned structure, the pile shaft which transfers the load to the bearing zone, and the bearing element which transfers the load to the soil or rock bearing layer.

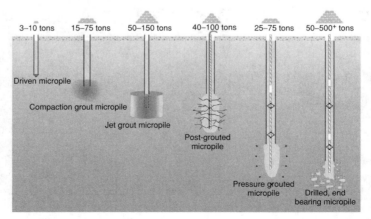

FIGURE 12.18
Sample of micropile bearing elements. (From Hayward Baker Inc. With permission.)

A standard structural analysis is used to design the pile section. If a grouted friction socket is planned, Table 12.5 can be used to estimate the sockets diameter and length. Bond lengths in excess of 30 ft (9.2 m) do not increase the piles capacity.

Quality control and quality assurance: During the construction of the micropile, the drilling penetration rate can be monitored as an indication of the stratum being drilled. Grout should be sampled for subsequent compressive strength testing. The piles verticality and length should also be monitored and documented.

A test pile is constructed at the beginning of the work and load tested to 200% of the design load in accordance with the standard specification ASTM D 1143 (Figure 12.19).

12.3.5 Fracture Grouting

Fracture grouting, also known as compensation grouting, is the use of a grout slurry to hydro-fracture and inject the soil between the foundation to be controlled and the process causing the settlement (Figure 12.20). Grout slurry is forced into soil fractures, thereby causing an expansion to take place counteracting the settlement that occurs or producing a controlled heave of the foundation. Multiple, discrete injections at multiple elevations can create a reinforced zone. The process is used to reduce or eliminate previous settlements, or to prevent the settlement of structures as underlying tunneling is performed.

A variation of fracture grouting is injection systems for expansive soils. The technique reduces the post-treatment expansive tendencies of the soil by either raising the soils' moisture content, filling the desiccation patterns in the clay or chemically treating the clay to reduce its affinity to water.

Applicable soil types: Since the soil is fractured, the technique can be performed in any soil type.

Equipment: For fracture grouting, the equipment consists of a drill rig to install the sleeve port pipes, grout injection tubing with packers, grout mixer, and a high-pressure grout

TABLE 12.5

Estimated Soil and Rock Bond Values for Micropiles

Soil/Rock Description	SPT N value (blows/ft)	Grout Bond with Soil/Rock (ksf)[a]
Nonpressure grouted		
Silty clay	3–6	0.5–1.0
Sandy clay	3–6	0.7–1.0
Medium clay	4–8	0.75–1.25
Firm clay or stiffer	>8	1.0–1.5
Sands	10–30	2–4
Soft shales		5–15
Slate and hard shales		15–28
Sandstones		15–35
Soft limestone		15–33
Hard limestone		20–35
Pressure grouted		
Medium dense sand		3.5–6.5
Dense sand		5.5–8.5
Very dense sand		8–12

[a] Design values, 1 ft = 0.308 m, 1 ksf = 47.9 kPa.

FIGURE 12.19
Micropile load test. (From Hayward Baker Inc. With permission.)

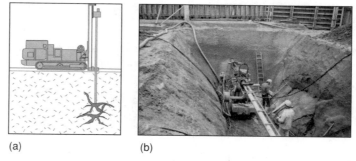

(a)   (b)

FIGURE 12.20
Fracture grouting: (a) schematic, (b) field implementation. (From Hayward Baker Inc. With permission.)

pump. A sleeve port pipe is a steel or PVC pipe with openings at regular intervals along its length to permit grout injection at multiple locations along the pipes length. Also a precise real-time level surveying system is often required to measure the movements of the structure or the ground surface.

For injection of expansive soils, the equipment generally consists of a track mounted rig that pushes multiple injection pipes into the ground at the same time (Figure 12.21). A mixing plant, storage tank and pump prepare, store, and deliver the solution to be injected.

Procedure: For fracture grouting beneath existing structures, large diameter shafts (10 to 15 ft, or 3 to 4.6 m, in diameter) or pits are constructed adjacent to the exterior of the structure to be controlled. From these shafts, a drill rig installs the sleeve port pipes horizontally beneath the structure. Then a grout injection tube is inserted into the sleeve port pipe. Packers on the injection tube are inflated on either side of an individual port and grout is injected. The packers are then deflated, the injection tube moved to another port, and the process repeated as necessary to achieve either the desired heave or

FIGURE 12.21
Injection rig for treatment of expansive soils. (From Hayward Baker Inc. With permission.)

prevent settlement. A level surveying system provides information on the response of the ground and overlying structure which is used to determine the location and quantity of the grout to be injected.

For injection of expansive soils, multiple injection rods are typically pushed into the ground to the desired treatment depth (typically 7 to 12 ft, or 2.2 to 3.7 m) and then an aqueous solution is injected as the rods are extracted.

Materials: For fracture grouting beneath structures, the grout typically consists of Portland cement and water.

For injection of expansive soils, the following solutions have been used:

Water — used to swell expansive clays as much as possible prior to construction.

Lime and fly ash — used to fill the desiccation pattern of cracks, reducing the avenues of moisture change.

Potassium chloride and ammonium lignosulfonate — used to chemically treat the clay and reduce its affinity for water.

Design: For fracture grouting beneath a structure, the design involves identifying the strata which has or will result in settlement, and placing the injection pipes between the shallowest stratum and the structure. For injection of expansive soils, the design includes identifying the lateral and vertical extents of the soils requiring treatment.

Quality control and quality assurance: For fracture grouting beneath existing structures, it is critical to know where all the injection ports are located, both horizontally and vertically. The monitoring of the overlying structure is then critical so that the affected portion of the structure is accurately identified and the injection is performed in the correct ports.

For injection of expansive soil, acceptance is typically based on increasing the *in situ* moisture content to the plastic limit plus 2 to 3 moisture points, reducing pocket penetrometer readings to 3 tsf (288 kPa) or less, and reducing the average swell to 1% or less within the treatment zone.

12.3.6 Infrequently-Used Reinforcement Techniques

12.3.6.1 *Fibers and Biotechnical*

Fiber reinforcement consists of mixing discrete, randomly oriented fibers in soil to assist the soil in tension. The use of fibers in soil dates back to ancient time but renewed interest was generated in the 1960s. Laboratory testing and computer modeling have been performed; however, field testing and evaluation lag behind. There are currently no standard guidelines on field mixing, placement and compaction of fiber-reinforced soil composites.

Biotechnical reinforcement involves the use of live vegetation to strengthen soils. This technique is typically used to stabilize slopes against erosion and shallow mass movements. The practice has been widely used in the United States since the 1930s. Recent applications have combined inert construction materials with living vegetation for slope protection and erosion control. Research has been sponsored by the National Science Foundation to advance the practice.

12.4 Fixation

12.4.1 Permeation Grouting

Permeation grouting is the injection of a grout into a highly permeable, granular soil to saturate and cement the particles together. The process is generally used to create a structural, load carrying mass, a stabilized soil zone for tunneling, and water cutoff barrier (Figure 12.22).

Applicable soil types: The permeability requirement restricts the applicable soils to sands and gravels, with less than 18% silt and 2% clay. The depth of the groundwater table is not critical in free-draining soils, since the water will be displaced as the grout is injected. Loose sands will have reduced strengths when grouted compared to sands with SPT N values of 10 or greater.

Equipment: The mixing plant and grout pump vary depending on the type of grout used. Drill rigs typically install the grout injection pipe. The rigs can vary from very small to very large, depending on the project requirements. When the geometry of the grouted mass is critical, sleeve port pipes will be used.

Procedure: The grout can be mixed in batches (cementacious slurries) or stream mixed (silicates and other chemical grouts). Batch mixing involves mixing a selected volume of grout, possibly 1 yard^3 or 0.79 m^3, and then injecting it before the next batch is mixed. The amount batched depends on the speed of injection and amount of time the specific grout

(a)

(b)

FIGURE 12.22
Permeation grouting: (a) schematic, (b) field implementation. (From Hayward Baker Inc. With permission.)

can be held and still be usable. Steam mixing involves storing the grout components in several tanks and then pumping them through separate hoses that combine before the grout reaches the injection pipe. If the geometry of the grouted mass is not important, the grout can be pumped through and out the bottom of the injection pipe. The pipe is then raised in increments, 1 to 3 ft (0.3 to 0.9 m), as the specified volume is injected at each interval.

A sleeve port pipe is used when the grouted geometry is important, such as excavation support walls. A sleeve port pipe is a steel or PVC pipe with holes, or ports, located at regular intervals, possibly 1 to 3 ft (0.3 to 0.9 m), along its length. A thin rubber membrane is placed over each port. The rig drills a hole in the soil, fills it with a weak, brittle, Portland cement grout, and inserts the sleeve port pipe. After the weak grout has hardened, a grout injection pipe with two packers is inserted into the sleeve port pipe allowing the grout to be injected through one port at a time (Figure 12.23). The injection pipe is then raised or lowered to another port and the process repeated in a sequence that includes primary, secondary, and tertiary injections.

Materials: The type of grout used depends on the application and soil grain size. For structural applications in gravel, Portland cement and water can be used. However, the particle size of the Portland cement is too large for sands. A finely ground Portland cement is available for use in course to medium sands. In fine, medium, and coarse sand, chemical grout can be used. The most common chemical grout used for structural applications is sodium silicate. Other chemical grouts are acrylates and polyurethanes.

Design: Generally, unconfined compressive strength and permeability are the design parameters. Sands grouted with sodium silicate can achieve a permeability of 1×10^{-5} cm/sec and an unconfined compressive strength of 50 to 300 psi (0.345 kPa to 2.07 MPa), although consistently achieving values in the field greater than 100 psi (0.69 MPa) is difficult. A standard analysis is performed assuming that the grouted soil is a mass with the design parameters. For excavation support walls, the mass is analyzed as a gravity structure, calculating the shear, sliding and overturning of the mass, as well as the global stability of the system.

Quality control and quality assurance: The mix design of the grouted soil can be estimated in the lab by compacting the soil to be grouted in a cylinder or cube molds at about the same

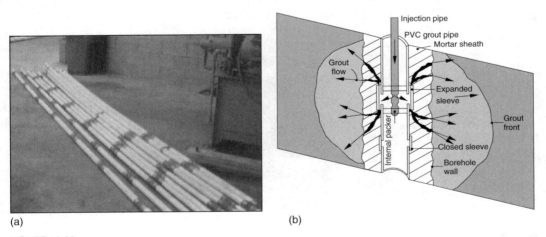

(a)   (b)

FIGURE 12.23
(a) Sleeve port pipes and (b) cross section of grout injection through a port. (From Hayward Baker Inc. With permission.)

density as exists *in situ* and then saturating the soil with the grout. Laboratory permeability or unconfined compressive strength tests can be performed after a specified cure time, such as 3, 7, 14, and 28 days. During production, the grout volume and pressure should be monitored and documented. The grouted soil can also be cored and tested after grouting.

12.4.2 Jet Grouting

Jet grouting (Figure 12.24) was conceived in the mid-1970s and introduced in the United States in the 1980s. The technique hydraulically mixes soil with grout to create *in situ* geometries of soilcrete. Jet grouting offers an alternative to conventional grouting, chemical grouting, slurry trenching, underpinning, or the use of compressed air or freezing in tunneling. A common application is underpinning and excavation support of an existing structure prior to performing an adjacent excavation for a new, deeper structure.

Super jet grouting is a modification to the system allowing creation of large diameters (11 to 16 ft, or 3.4 to 4.9 m) and is efficient in creating excavation bottom seals and treatment of specific soil strata at depth.

Applicable soil types: Jet grouting is effective across the widest range of soils. Because it is an erosion-based system, soil erodibility plays a major role in predicting geometry, quality, and production. Granular soils are the most erodible and plastic clays the least. Since the soil is a component of the final mix, the soil also affects the soilcrete strength (Figure 12.25). Organic soils are problematic and can be the cause for low strengths unless partially removed by an initial erosion pass before grouting. Flowing water can also be a problem.

Equipment: An on-site batch plant is required to mix the grout as needed. Pumps are also required to pump the grout and sometimes water and air to the drill rig. The drill rig is necessary to flush the jet grout monitor into the ground. Compact drills are capable of low headroom and tight access work. Pumps may also be required to remove the soilcrete waste.

Procedure: Jet grout is a bottom-up process (Figure 12.26). The drill flushes the monitor to the bottom of the treatment zone. The erosion and grout jets are then initiated as the monitor is rotated and extracted to form the soilcrete column. Varying geometries can be formed. Rotating the monitor through only a portion of a circle will create a portion of a column. Extracting the monitor without rotating it will create a panel. Treatment depths greater than 60 ft (18.5 m) require special precautions.

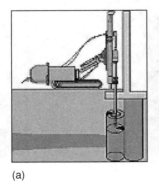
(a)

(b)

FIGURE 12.24
Jet grouting: (a) schematic, (b) field implementation. (From Hayward Baker Inc. With permission.)

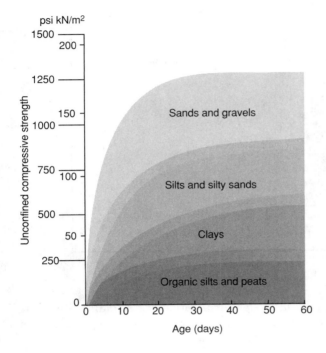

FIGURE 12.25
Range of soilcrete strengths based on soil type. (From Hayward Baker Inc. With permission.)

There are three traditional jet grout systems (Figure 12.27). Selection of the most appropriate system is determined by the *in situ* soil, the application, and the required strength of the soilcrete. The three systems are single, double, and triple fluid.

The single-fluid system uses only a high-velocity cement slurry grout to erode and mix the soil. This system is most effective in cohesionless soil and is generally not an appropriate underpinning technique because of the risk of pressurizing and heaving the ground.

The double-fluid system surrounds the high-velocity cement slurry jet with an air jet. The shroud of air increases the erosion efficiency. Soilcrete columns with diameters over 3 ft (0.9 m) can be achieved in medium to dense soils, and more than 6 ft (1.8 m) in loose soils. The double-fluid system is more effective in cohesive soils than the single-fluid system.

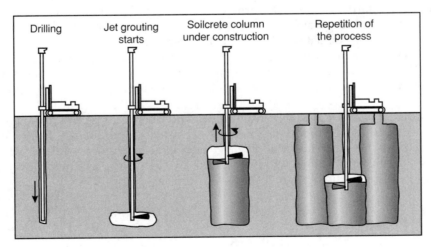

FIGURE 12.26
Jet grout process. (From Hayward Baker Inc. With permission.)

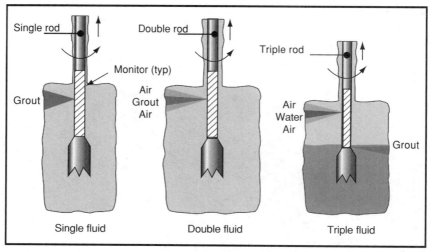

FIGURE 12.27
Single-, double-, and triple-jet grout systems. (From Hayward Baker Inc. With permission.)

The triple-fluid system uses a high-velocity water jet surrounded by an air jet to erode the soil. A lower jet injects the cement slurry at a reduced pressure. Separating the erosion process from the grouting process results in higher quality soilcrete and is the most effective system in cohesive soils.

Since material is pumped into the ground and mixed with the soil, the final mixed product has a larger volume than the original *in situ* soil. Therefore, as the mixing is performed, the excess soilcrete exits to the ground surface through the annulus around the drill steel. This waste material must be pumped or directed to an onsite retention area or trucked off-site. Since the waste contains cement, the waste sets up overnight and can be handled as a solid the following day.

Materials: Portland cement and water are generally the only two components, although additives can be utilized.

Design: Generally, either unconfined compressive strength or permeability is the design parameter. A standard analysis is performed to determine the required soilcrete geometry necessary based on the parameters achievable in the soil to be mixed. For excavation support walls, the mass must resist the surcharge, soil and water pressure imposed after excavation. This may include analysis of shear, sliding and overturning, as well as the global stability of the system. For underpinning applications, a standard bearing capacity and settlement analysis is performed as would be done for any cast in place pier.

Quality control and quality assurance: Monitoring and documenting the production parameters and procedures is important to assure consistency and quality. Test cylinders or cubes made from the waste material give a conservative assessment of the *in situ* characteristics. Wet sampling of the soilcrete *in situ* can also be performed although it is problematic. Coring of the hardened soilcrete is typical.

12.4.3 Soil Mixing

Soil mixing mechanically mixes soil with a binder to create *in situ* geometries of cemented soil. Mixing with a cement slurry was originally developed for environmental applications; however, advancements have reduced the costs to where the process is used for many general civil works, such as *in situ* walls, excavation support, port development on soft sites, tunneling support, and foundation support. Mixing with dry lime and cement was developed in the Scandinavian countries to treat very wet and soft marine clays.

Applicable soil types: The system is most applicable in soft soils. Boulders and other obstructions can be a problem. Cohesionless soils are easier to mix than cohesive soils. The ease of mixing cohesive soils varies inversely with plasticity and proportionally with moisture content. The system is most commonly used in soft cohesive soils as other soils can often be treated more economically with other technologies. Organic soils are problematic and generally require much larger cement content. The quality achieved with soil mixing is slightly lesser than that achieved with jet grouting in the same soils, with unconfined compressive strengths between 10 and 500 psi (0.69 to 3.45 MPa), and permeabilities as low as 1×10^{-7} cm/sec, depending on the soil type and binder content.

Equipment: A high-volume batching system is required to maintain productivity and economics. The components consist of an accurately controlled mixer, temporary storage, and high-volume pumps.

A drilling system is required to turn the mixing tool in the ground. The system varies from conventional hydraulic drill heads to dual-motor, crane-mounted turntables with torque requirements ranging from 30,000 to 300,000 ft lb (41 to 411 kJ). Multiaxis, electrically powered drill heads are also used, primarily for walling applications.

The mixing tool is generally a combination of partial flighting, mix blades, injection ports and nozzles, and shear blades. It can be a single- or multiple-axis tool (Figure 12.28). Tool designs vary with soil types and are often custom-built for specific projects (Figure 12.29). The diameter of the tool can vary from 1.5 to 12 ft (0.46 to 3.7 m).

Procedure: The binder is injected as the tool is advanced down to assist in penetration and to take advantage of this initial mixing. The soil and binder are mixed a second time as the tool is extracted. The rate of penetration and extraction is controlled to achieve adequate mixing. Single columns or integrated walls are created as the augers are worked in overlapping configurations. Treatment depths as great as 100 ft (31 m) have been achieved.

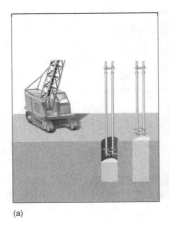
(a)

(b)

FIGURE 12.28
Soil Mixing: (a) Schematic, (b) Field implementation.

FIGURE 12.29
Example of soil mixing tools. (From Hayward Baker Inc. With permission.)

Materials: For wet soil mixing, the binder is delivered in a slurry form. Slurry volumes range from 20 to 40% of the soul volume being mixed. Common binders are Portland cement, fly ash, ground blast furnace slag, and additives. For dry soil mixing, the same materials (also line) are pumped dry using compressed air. Preproduction laboratory testing is used to determine mix energy and grout proportions.

Design: As with jet grouting, unconfined compressive strength and permeability are generally the design parameters. A standard analysis is performed to determine the required geometry based on the parameters achievable in the soil to be mixed. For excavation support walls, the mass can be designed as a standard excavation wall, or a thicker mass can be created and analyzed as a gravity structure, calculating the mass' shear, sliding and overturning, as well as the global stability of the system. When used as structural load bearing columns, a standard bearing capacity and settlement analysis is performed as would be for any cast in place pier. Anchored retention using steel reinforcement is common for support walls.

12.4.3.1 Dry Soil Mixing

Dry soil mixing (Figure 12.30) is a low-vibration, quiet, clean form of ground treatment technique that is often used in very soft and wet soil conditions and has the advantage of producing very little spoil. The high speed rotating mixing tool is advanced to the maximum depth, "disturbing" the soil on the way down. The dry binder is then pumped with air through the hollow stem as the tool is rotated on extraction. It is very effective in soft clays and peats. Soils with moisture content, greater than 60% are most economically

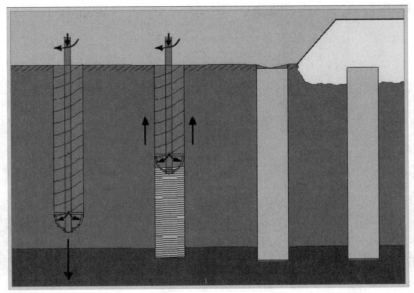

FIGURE 12.30
Illustration of dry soil mixing technique. (From Hayward Baker Inc. With permission.)

treated. This process uses cementacious binders to create bond among soil particles and thus increases the shear strength and reduces the compressibility of weak soils.

The most commonly used binding agents are cement, lime, gypsum, or slag. Generally, the improvement in shear strength and compressibility increases with the binder dosage. By using innovative mixtures of different binders engineers usually achieve improved results. It is known that strength gains are optimum for inorganic soils. It is realized that the strength gain would decrease with increasing organic and water content. The binder content varies from about $5\,lb/ft^3$ for soft inorganic clays to about $18\,lb/ft^3$ for peats with a high organic content.

12.4.3.2 Wet Soil Mixing

Wet soil mixing (Figure 12.31) is a similar technique except that a slurry binder is used making it more applicable with dryer soils (moisture contents less than 60%). The grout slurry is pumped through the hollow stem to the trailing edge of the mixing blades both during penetration and extraction. Depending on the *in situ* soils, the volume of grout slurry necessary varies from 20 to 40% of the soil volume. The technique produces a similar amount of spoil (20 to 40%) which is essentially excess mixed soil which, after setting up, can often be used as structural fill. The grout slurry can be composed of Portland cement, fly ash, and ground granulated blast furnace slag.

Quality control and quality assurance: Preproduction laboratory testing is often performed to prescribe the mixing energy and binder components and proportions. During production, it is necessary to monitor and document parameters such as mixing depth, mixing time, grout mix details, grout injection rates, volumes and pressures, tool rotation, penetration, and withdrawal rates.

Test cylinders or cubes can be cast from wet samples, but are problematic. The hardened columns can also be cored. In weaker mixes, penetration tests can be performed.

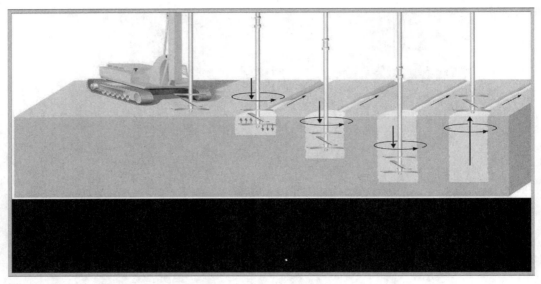

FIGURE 12.31
Illustration of wet soil mixing technique.

12.4.4 Infrequently-Used Fixation Techniques

12.4.4.1 Freezing and Vitrification

Ground freezing involves lowering the temperature of the ground until the moisture in the pore spaces freezes. The frozen moisture acts to "cement" the soil particles together. The first use of this technique was in 1862 in South Wales. The process typically involves placing double walled pipes in the zone to be frozen. A closed circuit is formed through which a coolant is circulated. A refrigeration plant is used to maintain the coolant's temperature. Since ice is very strong in compression, the technique has been most commonly used to create cylindrical retaining structures around planned circular excavations.

Vitrification is a process of passing electricity through graphite electrodes to melt soils *in situ*. Electrical plasma arcs have also been used and are capable of creating temperatures in excess of 4000°C. The soil becomes magma, and after several days of cooling it hardens into an artificial igneous rock. Although laboratory testing is ongoing, the electrical usage of the process to date appears to make it uneconomical. It is possible that the process could find application in the field of environmental cleanup.

12.5 Other Innovative Soft-Ground Improvements Techniques

12.5.1 Rammed Aggregate Piers

Rammed aggregate piers (RAPs) are a type of stone column as presented in Section 12.3.1. Aggregate columns installed by compacting successive lifts of aggregate material in a preaugered hold (Figure 12.32). The predrilled holes, which typically have diameters of 24 to 36 in. (0.6 to 1.2 m), can extend up to about 20 ft. As seen in Figure 12.33, aggregate is compacted in lifts with a beveled tamper to create passive soil pressure conditions both at the bottom and the sides of the piers. RAPs are generally restricted to cohesive soils in

FIGURE 12.32
Installation of rammed aggregate piers, a type of stone column. (From Geopiers Foundation Co. With permission.).

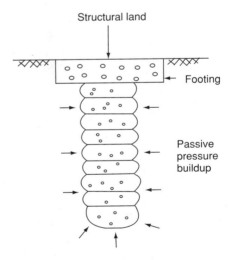

FIGURE 12.33
Schematic diagram of a rammed aggregate pier.

which a predrill hole will stay open. Although constructed differently than store columns or vibro piers (Section 12.3.1) all provide similar improvement to cohesive soils. The vertical tamping used to construct RAPs results in minimal densification in adjacent granular soils compared to vibratory probe construction.

RAPs can be used in some of the following stone column applications that are outlined below:

1. Support shallow footings in soft ground.
2. Reinforces soils to reduce earthquake-induced settlements, however, does not densify sands against liquefaction.
3. Increase drainage and consequently expedite long-term settlement in saturated fine-grained soils.
4. Increase global stability and bearing capacity of retaining walls in soft ground.
5. Improve stability of slopes if RAPs can be installed to intersect potential shear failure planes.
6. Reduce the need for steel reinforcements when RAPs are installed below concrete mat or raft foundations.

12.5.2 Reinforced Soil Foundations

Bearing capacity of foundation soils can be improved using geogrids and geosythetics placed as a continuous single layer, closely spaced continuous mutilayer set or mattress consisting of three-dimensional interconnected cells. Although standards on design of footings on reinforced soils are currently unavailable, Koerner (1998) provides some numerical guidelines on the extent of the improvement of bearing capacity and reduction of settlement. Figure 12.34(a) and (b) shows the results of laboratory tests where geotextiles were used to improve the bearing capacity of loose sands and saturated clay, respectively.

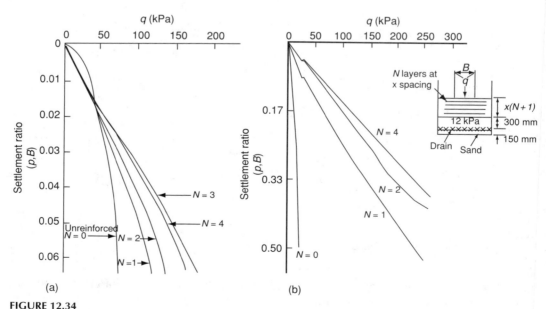

FIGURE 12.34
Improvement of soil bearing capacity with geotextiles: (a) loose sand, (b) saturated clay. (From Koerner, R., 1994. *Designing with Geosynthetics*, Prentice Hall, Englewood Cliffs, NJ. With permission.)

Figure 12.35 also shows the general approximations that the author has drawn from the results of large laboratory tests (Milligan and Love, 1984), which shows the improvement of settlement properties of saturated clay reinforced with geogrids.

A large number of load tests have been conducted in the test pits at the Turner-Fairbank Highway Research Center (TFHRC) in Alaska, USA, to evaluate the effects of single and multiple layer of reinforcement placed below shallow spread footings (FHWA, 2001). In this test program, two different geosynthetics were evaluated; a stiff biaxial geogrid and a geocell. Parameters of the testing program include: number of reinforcement layers; spacing between reinforcement layers; depth to the first reinforcement layer; plan area of the reinforcement; type of reinforcement; and soil density. Test results indicated that the use of geosynthetic reinforced soil foundations may increase the ultimate bearing capacity of shallow spread footings by a factor of 2.5 (FHWA, 2001).

12.5.2.1 *Mechanisms of Bearing Capacity Failure in Reinforced Soils*

In spite of the known favorable influence of geotextiles and geogrids on soil bearing capacity, the foundation designer needs to be aware of a number of mechanisms of bearing capacity failure even with reinforcements. These are discussed in Koerner (1998) as seen in Figure 12.36(a)–(d). Figure 12.36(a) shows the lack of reinforcement in the foundation influence zone while Figure 12.36(b) illustrates insufficient embedment of geotextiles or geogrids. Bearing capacity failures leading to inadequate tensile strength and excessive creep (long-term deformation) of reinforcements is shown in Figure 12.36(c) and (d), respectively.

These are discussed in Koerner (1998) as situations arising from;

- the lack of reinforcement in the foundation influence zone while Fig. 12.37
- insufficient embedment of geotextiles or geogrids.
- bearing capacity failures leading to inadequate tensile strength, and
- excessive creep (long-term deformation) of reinforcements

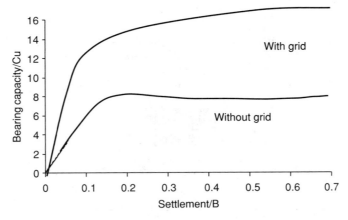

FIGURE 12.35
Improvement of settlement properties of saturated clays.

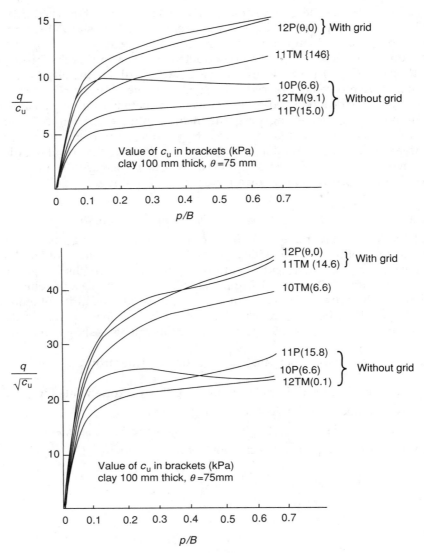

FIGURE 12.36
Improvement of settlement properties in saturated clay with geogrids. (From Koerner, R., 1994. *Designing with Geosynthetics*, Prentice Hall, Englewood Clifts, NJ. With permission.)

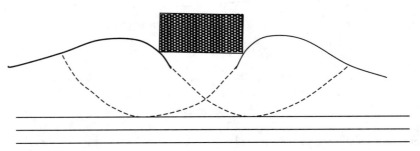

FIGURE 12.37
Lack of reinforcement in the foundation influence zone.

References

Baez, J.I., Martin, C.R., 1992, Quantitative evaluation of stone column techniques for earthquake liquefaction mitigation, Earthquake Engineering: Tenth World Conference 1992, Balkema, Rotterdam, Swets & Keitlinger, NL.

Federal Highway Administration (FHWA), 2001, Performance Tests for Geosynthetic-Reinforced Soil Including Effects of Preloading, FHWA-RD-01-018, June.

Koerner, R., 1994. *Designing with Geosynthetics*, Prentice Hall, Englewood Cliffs, NJ.

Milligan, G.W.E., Love, J.P., 1984, *Model testing of geogrids under aggregate layer in soft ground*, Proceedings of the Symposium on polymer reinforcement in Civil Engineering, London, ICE, 1984, pp. 128–138.

Priebe, H.J., 1995, *The Design of Vibro Replacement*, Ground Engineering, December.

Schaefer, V., Abramson, L., Drumbeller, J., Hussin, J., and Sharp, K., 1997, Ground improvement, ground reinforcement, ground treatment developments 1987–1997, Geotechnical Special Publication No. 69, American Society of Civil Engineers, NY.

13

Impact of Groundwater on the Design of Earthen Structures

Manjriker Gunaratne

CONTENTS

13.1 Groundwater and Seepage

Seeping groundwater has a major impact on the design of earthen structures. Stability analysis of soil slopes in groundwater flow regimes requires the knowledge of seepage forces. Furthermore, water-retaining structures are often built in groundwater flow regimes caused by differential hydraulic heads. Therefore, an analysis of groundwater seepage is essential in the design of water-retaining structures when estimating the uplift forces.

Foundation engineers employ a variety of approaches to understand the effects of groundwater on structures. They can be basically classified as:

1. Graphical approaches based on flow nets.
2. Numerical approaches based on the finite difference or the finite element method.
3. Analytical approaches based on mathematical transformations.

13.2 Graphical Solution to Groundwater Problems: Flow Nets

The most common and the simplest means of seepage analysis is by the method of flow nets. In this method, two orthogonal families of equipotential and flow lines are sketched in the flow domain (Figure 13.1) using the basic concepts defining the two families. A flow line is an identified or a visualized flow conduit boundary in the flow domain. An equipotential line, on the other hand, is an imaginary line possessing the same total energy head (energy per unit weight).

Rules Governing the Construction of a Flow net

1. Equipotential lines do not intersect each other.
2. Flow lines do not intersect each other.
3. Equipotential lines and flow lines form two orthogonal families.
4. In order to ensure equal flow in the drawn flow conduits and equal head drop between adjacent equipotential lines, individual flow elements formed by adjacent equipotential lines and flow lines must bear the same height–width ratio (this is typically selected as 1.0 for ease of plotting).

Useful guidelines regarding the plotting of reasonably accurate flow nets for different flow situations are found in Cedergreen (1989).

With seepage velocities generally relatively low, the pressure (p) exerted by seeping water and the potential energy contributes to the total hydraulic head (energy per unit weight) of water as

$$h = \frac{p}{\gamma_w} + z \tag{13.1}$$

The quantity of groundwater flow at any location in a porous medium such as soil can be expressed by D'Arcy's law as

$$q = kiA \tag{13.2a}$$

where k is the coefficient of permeability (or hydraulic conductivity) at that location while i, the hydraulic gradient, can be expressed by

$$i = -\frac{dh}{dx} \tag{13.3}$$

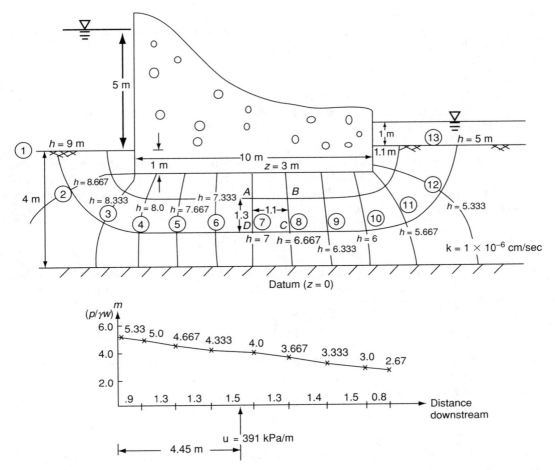

FIGURE 13.1
Illustration of a flow net.

13.2.1 Estimation of the Coefficient of Hydraulic Conductivity

The coefficient of hydraulic conductivity of a soil can be estimated in a number of ways:

1. Using laboratory permeameters (falling-head or constant-head). The readers are referred to Das (2002) for experimental details of these laboratory tests.
2. Using field pumping tests that are discussed in Section 13.5.
3. Using an empirical correlation between k and D_{10} that is listed in Equation (13.38) (Example 13.6).

It can be shown from Equation (13.2a) that the quantity of seepage in the flow domain can also be expressed in terms of the number of equipotential drops (n_e) and the flow conduits (n_f) as

$$q = kH[n_f/n_e] \tag{13.2b}$$

where H is the total head drop.

The following example illustrates the flow net method of seepage analysis and evaluation of uplift pressures.

Example 13.1

Assume that it is necessary to establish the pressure distribution at the bottom and the seepage under the dam as shown in Figure 13.1. Also assume the coefficient of permeability to be 1×10^{-6} cm/sec.

Solution

As the first step in the solution, a flow net has been drawn to scale following the rules given above. Using the bedrock as the datum for the elevation head, total heads have been assigned using Equation (13.1) for all the equipotential lines as shown. It is noted that the head drop between two adjacent equipotential lines is

$$(9m - 5m)/12 = 0.333m$$

Then, by applying Equation (13.1) to the points where the equipotential lines intersect the dam bottom (B_i), the following expression can be obtained for the pressure distribution, which is plotted in Figure 13.1:

$$p = \gamma_w(h - 3.0)$$

The total upthrust can computed from the area of the pressure distribution as 391.34 kPa/m acting at a distance of 4.45 m downstream. Then, by applying Equation (13.3) to the element ABCD, one obtains

$$i = (7.0 - 6.667)/1.1 = 0.302$$

Since $k = 1 \times 10^{-6}$ cm/sec, one can apply Equation (13.2) to obtain the quantity of seepage through ABCD as

$$q_1 = 1 \times (10^{-9})(0.302)(1.3)(1) m^3/\text{sec/m} \quad \text{(since AD} = 1.3 \text{ m)}$$

Since all of the conduits must carry equal flow (rule no. 4 of the flow net construction), the total flow under the dam is given by

$$q = 3 \times (10^{-9})(0.302)(1.3)(1) m^3/\text{sec/m} = 1.18 \times 10^{-9} m^3/\text{sec/m}$$

The following important assumptions made in the above analysis must be noted:

1. The subgrade soil is homogeneous with respect to the coefficient of permeability.
2. Bedrock and concrete dam are free of fault or cracks.
3. There is no free flow under the dam due to piping (or erosion).

Therefore, the design and installation of an adequate pore-pressure monitoring system that can verify the analytical results is an essential part of the design. A piezometer with a geomembrane or sand filter that can be used for monitoring pore pressures is shown in Figure 13.2.

FIGURE 13.2
Piezometer probes. (From Thilakasiri, H.S., 1996, Numerical Simulation of Dynamic Replacement of Florida Organic Soils, Ph.D. Dissertation, University of South Florida. With permission.)

13.3 Numerical Modeling of Groundwater Flow

If it is assumed that the water flow in a saturated soil is laminar, continuous (without any losses or gains in water due to the presence of sinks or sources), and steady with respect to time, the following partial differential equation can be written for continuity of two-dimensional (2D) flow conditions at any given point in the flow domain:

$$\frac{\partial u}{\partial x} + \frac{\partial v}{\partial y} = 0 \tag{13.4}$$

where u and v are the velocities in the X and Y directions.

Using Equation (13.3), the hydraulic gradients in the respective directions can be expressed as

$$i_x = \frac{\partial h}{\partial x} \tag{13.5a}$$

$$i_y = \frac{\partial h}{\partial y} \tag{13.5b}$$

Now if Equation (13.2) is applied per unit flow area, one obtains

$$u = k_x i_x \quad \text{and} \quad v = k_y i_y$$

where k_x and k_y are the hydraulic conductivities in the X and Y directions, respectively, in a generally anisotropic soil.

Then, by substituting in Equation (13.4) along with Equation (13.5), the Laplace equation for 2D flow is obtained as

$$k_x \frac{\partial^2 h}{\partial x^2} + k_y \frac{\partial^2 h}{\partial y^2} = 0 \tag{13.6a}$$

It must be noted that under isotropic conditions Equation (13.6a) reduces to

$$\frac{\partial^2 h}{\partial x^2} + \frac{\partial^2 h}{\partial y^2} = 0 \tag{13.6b}$$

If the second-order differential terms in Equation (13.6a) are replaced by the corresponding forward difference numerical form in Equation (13.7),

$$\frac{\partial^2 h}{\partial x^2} = \frac{h_{x+} + h_{x-} - 2h_0}{l^2} \tag{13.7a}$$

$$\frac{\partial^2 h}{\partial y^2} = \frac{h_{y+} + h_{y-} - 2h_0}{l^2} \tag{13.7b}$$

where the corresponding hydraulic heads hs are defined in Figure 13.3 (where a grid of length interval l is plotted), one obtains the following numerical form of Laplace's equation:

$$k_x \frac{h_{x+} + h_{x-} - 2h_0}{l^2} + k_y \frac{h_{y+} + h_{y-} - 2h_0}{l^2} = 0$$

Then, h_0 can be expressed in Equation (13.8) to obtain the hydraulic head at any point in terms of the hydraulic heads of its neighboring nodes as

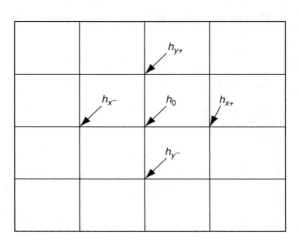

FIGURE 13.3
Finite difference grid for solution of the seepage problem.

$$h_0 = \frac{k_x(h_{x+} + h_{x-}) + k_y(h_{y+} + h_{y-})}{2(k_x + k_y)} \tag{13.8}$$

Knowing the boundary conditions, which are the hydraulic head values of the interfaces of soil with free water (i.e., head water and tail water), and the hydraulic conductivities in each direction, Equation (13.8) can be conveniently coded in the computer to derive the hydraulic heads of the entire flow domain. Then, once the hydraulic heads, h, are available, Equations (13.3) and (13.2) can be used in steps to evaluate the hydraulic gradient at any desired location in the flow domain and the flow within any grid element.

13.4 Analytical Modeling of Groundwater Flow

Groundwater problems can be analytically solved using transformation methods. Generally, the first step in this regard is to mathematically define the family of equipotential lines and flow lines based on the potential function (ϕ) and the stream function (ψ). It is also noted that the potential function (ϕ) is related to the hydraulic head by Equation (13.9)

$$\phi = -kh \tag{13.9}$$

Then, a 2D groundwater regime in the x–y plane can be transformed to the ϕ–ψ domain using an appropriate transformation that accounts for the boundary equipotential lines ($\phi =$ constant) and flow lines ($\psi =$ constant) in the flow domain (x–y plane). This is initiated by defining the above functions in terms of the flow velocities using the Cauchy–Reimann relationships

$$u = \frac{\partial \phi}{\partial x} = \frac{\partial \psi}{\partial y} \tag{13.10a}$$

$$v = \frac{\partial \phi}{\partial y} = -\frac{\partial \psi}{\partial x} \tag{13.10b}$$

It can be shown mathematically that ϕ and ψ are orthogonal to each other.

13.4.1 Conformal Mapping

The actual flow domain in the X–Y plane can be conveniently mapped onto the potential function–flow function domain (ϕ–ψ) using the following complex variables that reduce the amount of mathematical manipulation required:

$$z = x + iy \tag{13.11a}$$

$$\omega = \phi + i\psi \tag{13.11b}$$

Then, one can define appropriate conformal mapping functions in the following formats:

$$\omega = f(z) \tag{13.12}$$

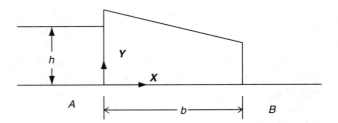

FIGURE 13.4
Unconfined flow under a concrete dam.

Example 13.3
It can be shown that the following transformation is adequate to describe the seepage under a concrete earth dam shown in Figure 13.4

$$z = b \cos \frac{\pi \omega}{kh}$$

since it satisfies the known equipotential surfaces (soil–free water interfaces A and B) and the flow boundary at the dam bottom. k is the hydraulic gradient of the foundation soil. The above expression can be used to plot the flow lines and equipotential lines for the above flow (assume that $b = 5$ m and $h = 6$ m).

Using the following transformation:

$$z = b \cos \frac{\pi \omega}{kh} \tag{13.13}$$

and substituting the complex relations in Equation (13.11),

$$x + iy = b \cos \frac{\pi(\phi + i\psi)}{kh} \tag{13.14}$$

Equation (13.14) can be manipulated using complex algebra to obtain expressions for ϕ and ψ in terms of x and y. Then, by eliminating ϕ from the two equations, the flow lines can be plotted using the following equation (Figure 13.5):

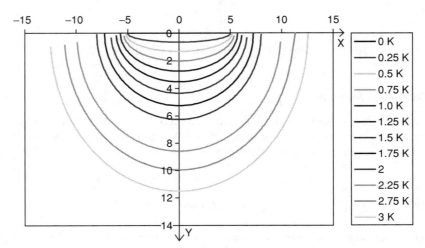

FIGURE 13.5
Flow lines drawn for selected flow quantities indicated in the legend.

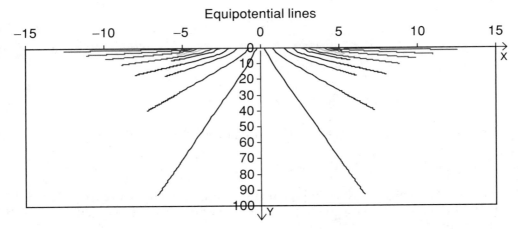

FIGURE 13.6
Equipotential lines drawn for selected hydraulic heads.

$$y = \sqrt{b^2 \sinh^2\left(\frac{\pi\psi}{kh}\right)\left(1 - \frac{x^2}{b^2 \cosh^2(\pi\psi/kh)}\right)} \tag{13.15}$$

Similarly, by eliminating ψ, the equipotential lines can be plotted using the following equation (Figure 13.16):

$$y = \sqrt{b^2 \sin^2(\phi)\left(\frac{x^2}{b^2 \cos^2(\phi)} - 1\right)} \tag{13.16}$$

Also, at the bottom of the dam since $y = 0$, using Equations (13.1) and (13.9), the pressure distribution under the dam can be expressed as (Figure 13.7)

$$\Delta P = \frac{h\gamma_w}{\pi} \cos^{-1}\frac{x}{b} \tag{13.17a}$$

The total uplift force at the bottom of the dam and its moment about the toe of the dam are important parameters for the design of the dam. They can be obtained as follows:

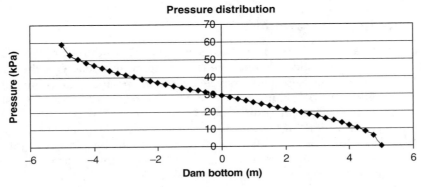

FIGURE 13.7
Pressure distribution under the bottom of the dam.

$$F_U = \int_x^b \frac{h\gamma_w}{\pi} \cos^{-1} \frac{x}{b} \, dx \tag{13.17b}$$

$$M_T = \int_x^b (b-x) \frac{h\gamma_w}{\pi} \cos^{-1} \frac{x}{b} \, dx \tag{13.17c}$$

The above integrals can be evaluated using any numerical integration package.

13.4.2 Complex Flow Velocity

The complex flow velocity ($\bar{V}$) in Equation (13.18) provides another useful relation that can be used to obtain the flow velocity components directly from the derivative of the transformation

$$\bar{V} = \frac{dw}{dz} = u - iv \tag{13.18}$$

Example 13.4

If a certain seepage flow situation can be modeled by the transformation $w = 2z - z^2$, find the resultant velocity at the point (1, 2) represented by

$$z = 1 + 2i$$

Using Equation (13.18),

$$u - iv = \frac{dw}{dz} = 2 - 2z = 2 - 2(1 + 2i) = 0 - 4i$$

where $u = 0$ and $v = 4$. Hence, the magnitude of the flow velocity is 4 units in the $+y$ direction.

When one needs to develop an appropriate transformation for a specific flow situation, one can use the Schwarz–Christoffel transformation technique, which is widely used in such formulations. The reader is referred to Harr (1962) for analytical details of the Schwarz–Christoffel transformation-based groundwater seepage solutions.

13.5 Dewatering of Excavations

Construction in areas of shallow groundwater requires dewatering prior to excavation. Although contractors specialized in such work determine the details of the dewatering program depending on the field performance, a preliminary idea of equipment requirements and feasibility can be obtained by a simplified analysis. Figure 13.8 shows the schematic diagram for such a program.

It also shows the elevations of the depressed water table at various distances from the center of the well. Observation wells (or piezometers) can be placed at intermediate locations, such as those shown at distances of r_1 and r_2, to monitor the water table depression. In analyzing a seepage situation like this, Dupuit (Harr, 1962) assumed that (1) for a small inclination of the line of seepage the flow lines are horizontal and (2) the hydraulic gradient is equal to the slope of the free surface and is invariant with depth.

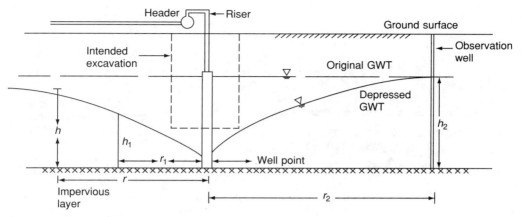

FIGURE 13.8
Dewatering of excavations.

Hence, one can write the expression (13.19) by combining Equations (13.2) and (13.3) for the discharge rate through any general section such as one of the observation well. The underlying assumption is that the entire soil stratum can be considered as homogenous with the average hydraulic conductivity represented by k

$$q = k\left(-\frac{dh}{dr}\right)(2\pi r h)$$ (13.19)

Noting that q is constant throughout the flow regime, Equation (13.19) can be integrated between distances r_1 and r_2 to obtain

$$q = \frac{\pi k \left(h_1^2 - h_2^2\right)}{\ln(r_1/r_2)}$$ (13.20)

One can define the extent of dewatering, using parameters r_1, r_2, h_1, and h_2, and utilize the above expression to determine the capacity requirement of the pump. Alternatively, expression (13.20) can be used to estimate the average permeability coefficient of the soil stratum.

If the site where the dewatering program is executed contains a number of layers with different soil properties, i.e., coefficients of permeability, then Equation (13.20) has to be modified to incorporate properties of all the layers in the area of influence of dewatering. The reader is referred to Cedergreen (1989) for expressions applicable to such complicated site conditions.

13.6 Basic Environmental Geotechnology

Environmental geotechnology is a relatively new civil engineering discipline that is concerned with the design of earthen structures that are utilized in assuring environmental safety. Some examples of such structures include protective clay liners for landfills, soil or soil–fabric filters that control erosion due to groundwater, and earthen barriers against seepage of contaminated groundwater.

The amount of solid waste generated in the United States has exceeded 510 M tons by the year 2000 (Koerner, 1998). Therefore, the immediate need for construction of adequate

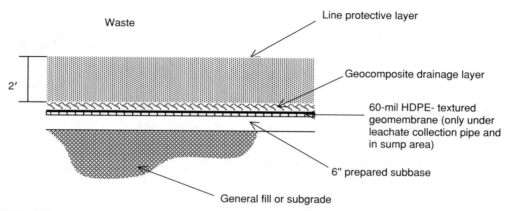

FIGURE 13.9
Typical cross section of a geomembrane-lined landfill.

landfills cannot be overemphasized. Although the construction of landfills involves political and legal issues, properly designed, constructed, and maintained landfills have proven to be secure, especially if they are provided with lined facilities. These are installed at the bottom or sides of a landfill to control groundwater pollution by the liquid mixture (leachate) formed by the interaction of rainwater or snowmelt with waste material.

Types of liners for leachate containment are basically (1) clay liners, (2) geomembranes, and (3) composite liners consisting of geomembranes and clay liners. Of these, until recently, the most frequently used liners were clay liners, which minimized leachate migration by achieving permeability values as low as 5×10^{-8} to 5×10^{-9} cm/sec. However, owing to the large thickness (0.6 to 2 m) requirement and chemical activity in the presence of organic-solvent leachates, geomembranes have been increasingly utilized for landfills.

13.6.1 Design of Landfill Liners

As shown in Figure 13.9 and Figure 13.10, the important components of a solid material containment system are (1) a leachate collection or removal system, (2) a primary leachate barrier, (3) a leachate detection or removal system, (4) a secondary leachate barrier, and (5) a filter above the collection system to prevent clogging. Some of the design criteria (Koerner, 1998) are as follows:

1. The leachate collection system should be capable of maintaining a leachate head of less than 30 cm.

2. Both collection and detection systems should have 3-cm-thick granular drainage layers that are chemically resistant to waste and leachate, and that have

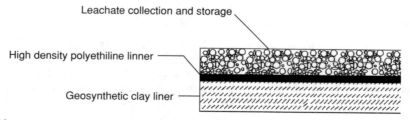

FIGURE 13.10
Typical cross section of a clay/geomembrane-lined composite landfill.

permeability coefficient of not less than 1×10^{-2} cm/sec or an equivalent drainage material.

3. The minimum bottom slope of the facility should be 2%.

13.6.1.1 Design Consideration for Clay Liners

In the case of clay liners, the U.S. Environmental Protection Agency (EPA) requires that the coefficient of permeability be less than 10^{-7} cm/sec. This can be achieved by meeting the following classification criteria:

1. The soil should be at least 20% fine (Section 1.2.1).
2. The plasticity index should have been greater than 10 (Section 1.2.2).
3. The soil should not have more than 10% gravel size (>4.75 mm) particles.
4. The soil should not contain any particles or chunks of rock larger than 50 mm.

It is realized that liner criteria can be satisfied by blending available soil with clay materials like sodium bentonite.

13.6.1.2 Design Considerations for Geomembrane layers

Geomembranes are mainly used in geotechnical engineering to perform the functions of (1) separation, (2) filtration, and (3) stabilization. In this application of geotextiles, the functions of separation and, to a lesser extent, filtration are utilized. Owing to the extreme variation of solid waste leachate composition from landfill to landfill, the candidate liner should be tested for permeability with the actual of synthesized leachate.

In addition to the permeability criterion, other criteria also play a role in geomembrane selection:

1. Resistance to stress-cracking induced by the soil or waste overburden.
2. Different thermal expansion properties in relation to subgrade soil.
3. Coefficient of friction developed with the waste material that governs slope stability criteria.
4. Axisymmetry in tensile elongation when the material is installed in a landfill that is founded on compressible subgrade soils.

In selecting a geomembrane material for a liner, serious consideration should also be given to its durability, which is determined by the possibility of leachate reaction with the geomembrane and premature degradation of the geomembrane. For more details on geomembrane durability and relevant testing, the reader is referenced to Koerner (1998).

According to the U.S. EPA regulations, the minimum required thickness of a geomembrane liner for a hazardous waste pond is 0.75 mm.

13.7 Application of Groundwater Modeling Concepts in Environmental Geotechnology

A major challenge that goetechnical engineers face in the design of earthen structures associated with environmental protection is the evaluation of the effects of pollutant or contaminant migration with groundwater such as the rate of seepage, the extent of the region of contamination, and the seepage forces induced by groundwater flow on such structures.

The analytical techniques introduced in Section 13.4 can certainly provide convenient and effective tools for such evaluations. Hence, the purpose of this section is to illustrate how the above analytical tools can be utilized for the benefit of designing earthen structural elements for geoenvironmental applications.

13.7.1 Analysis of Seepage toward Wells

Analytical consideration of 2-D seepage toward wells is an important starting point for solving problems associated with groundwater contamination.

For pure radial flow in the horizontal plane, the velocity potential ϕ only depends on the radial distance r and not on the transverse position defined by ϕ. Hence, the following mapping function can be conveniently employed to model seepage toward wells:

$$\phi = C_1 \ln(r) + C_2 \tag{13.21a}$$

or

$$\phi = C_1 \ln \sqrt{x^2 + y^2} + C_2 \tag{13.21b}$$

where C_1 and C_2 are constants.

The following check can be performed to verify that the selected velocity potential ϕ in fact satisfies Laplace's equation (Equation 13.6)

$$\phi = C_1 \ln \sqrt{x^2 + y^2} + C_2$$

$$\frac{\partial^2 \phi}{\partial x^2} = \frac{-C_1(x^2 - y^2)}{(x^2 + y^2)^2}$$

$$\frac{\partial^2 \phi}{\partial y^2} = \frac{-C_1(y^2 - x^2)}{(y^2 + x^2)^2}$$

$$\frac{-C_1(x^2 - y^2)}{(x^2 + y^2)^2} + \frac{-C_1(y^2 - x^2)}{(y^2 + x^2)^2} = 0$$

$$\therefore \quad \frac{\partial^2 \phi}{\partial x^2} + \frac{\partial^2 \phi}{\partial y^2} = 0$$

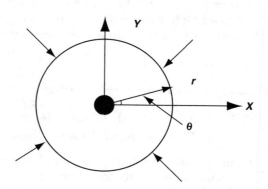

FIGURE 13.11
2-D Seepage toward wells.

Therefore, it is seen that Laplace's equation for isotropic conditions (Equation (13.6b)) is satisfied. Now one can introduce the following boundary conditions to evaluate the constants C_1 and C_2:

1. The hydraulic head and hence the velocity potential (Equation (13.9)) can be assumed as zero at the well where the flow terminates. Then,

   $r = r_w$ when $\phi = 0$

2. The hydraulic head, h, at a known radial distance of R is assumed to initiate the flow toward the well. Then,

   $r = R$ when $\phi = -kh$

By substituting the boundary conditions in Equation (13.21b), one obtains

$$\phi = -\frac{kh}{\ln(R/r_w)} \ln \frac{\sqrt{x^2 + y^2}}{r_w} \tag{13.22}$$

Then, Cauchy–Reimann relationships (Equation (13.10)) can be used to derive the flow velocity components as

$$u = \frac{\partial \phi}{\partial x} = \frac{\partial \psi}{\partial y} \tag{13.10a}$$

$$v = \frac{\partial \phi}{\partial y} = -\frac{\partial \psi}{\partial x} \tag{13.10b}$$

Thus, one obtains

$$u = -\frac{kh}{\ln(R/r_w)} \frac{x}{(x^2 + y^2)} \tag{13.23a}$$

$$v = -\frac{kh}{\ln(R/r_w)} \frac{y}{(x^2 + y^2)} \tag{13.23b}$$

The resultant velocity can be expressed as

$$V = -\frac{kh}{\ln(R/r_w)} \frac{1}{\sqrt{(x^2 + y^2)}} \tag{13.23c}$$

If the quantity of flow is Q, then

$$Q = (2\pi r T)V = (2\pi T)V\sqrt{(x^2 + y^2)}$$

where T is the thickness of the aquifer. It follows that

$$Q = \frac{2\pi T k h}{\ln(R/r_w)} \tag{13.24}$$

Thus, the velocity potential function at any point (x, y) in the flow can be expressed in terms of the intake Q, the aquifer thickness T, and the radius of the well r_w as

$$\phi = -\frac{Q}{2\pi T} \ln \frac{\sqrt{x^2 + y^2}}{r_w} \qquad (13.25)$$

This provides a way to derive the stream function P by using Cauchy–Reimann relationships

$$\psi = -\frac{Q}{2\pi T} \tan^{-1}\left(\frac{y}{x}\right) \qquad (13.26)$$

With the velocity potential and the steam function, the complex potential (Equation (13.11b)) for seepage toward the well can be formulated as

$$\omega = \phi + i\psi = -\frac{Q}{2\pi T}\left[\ln \sqrt{x^2 + y^2} + i\tan^{-1}\left(\frac{y}{x}\right) - \ln r_w\right]$$

$$\omega = -\frac{Q}{2\pi T}\left[\ln \sqrt{x^2 + y^2} + \ln e^{i\tan^{-1}(y/x)} - \ln r_w\right] \qquad (13.27a)$$

$$\omega = -\frac{Q}{2\pi T}\left[\ln\left\{\sqrt{x^2 + y^2}\, e^{i\tan^{-1}(y/x)}\right\} - \ln r_w\right]$$

Using Equation (13.11a) and complex algebraic manipulations, one obtains the following simplified form:

$$\omega = -\frac{Q}{2\pi T}[\ln z - \ln r_w] = -\frac{Q}{2\pi T} \ln \frac{z}{r_w} \qquad (13.27b)$$

A major benefit of possessing Equation (13.27) is that one can conveniently deduce the seepage originating from a source such as a source of contamination by substituting $Q = -Q$.

13.7.2 Uniform Flow in an Aquifer

The regular (uncontaminated) 2-D flow in an aquifer can be described as uniform flow shown in Figure 13.12.

For the above case, the two velocity components can be derived as

$$u = U\cos\theta = U\frac{x}{\sqrt{(x^2 + y^2)}} \qquad (13.28a)$$

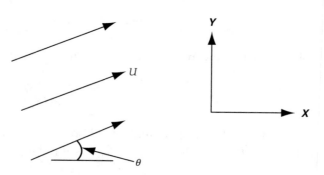

FIGURE 13.12.
Uniform 2-D flow in an aquifer.

$$v = u\sin\theta = U\frac{y}{\sqrt{(x^2 + y^2)}} \tag{13.28b}$$

If the x-axis is chosen along the direction of the flow with the origin of coordinates being at the source of contaminant, then

$$\phi = Ux \tag{13.29a}$$

$$\psi = Uy \tag{13.29b}$$

13.7.3 Transport of Contaminants

The resultant flow caused by the introduction of contaminants into steady flow can be modeled by combining two types of flow: (1) outward flow from a source, which is the opposite scenario of flow into a well (Equation (13.27)) and (2) uniform flow in an aquifer (Equation (13.29)).

Hence, the resultant velocity potential and the stream function can be composed as

$$\phi = \frac{Q}{2\pi T} \ln \frac{\sqrt{x^2 + y^2}}{r_w} + Ux \tag{13.30a}$$

$$\psi = \frac{Q}{2\pi T} \tan^{-1}\left(\frac{y}{x}\right) + Uy \tag{13.30b}$$

(Note that the sign is now $+Q$ to indicate flow out of the source.)

It can be shown that a stagnation point (a point with one component of the flow velocity being zero) is obtained at

$$x_0 = -\frac{Q}{2\pi TU} \tag{13.31}$$

13.7.3.1 *Derivation of the Location of the Stagnation Point*

The resultant x-directional velocity (u) for the contaminated flow can be obtained by superimposing the x-directional velocities for the uniform aquifer flow and the flow emanating from the contaminant source. Then,

$$u = \frac{kh}{\ln (R/r_w)} \frac{x}{(x^2 + y^2)} + U \tag{13.32a}$$

or

$$u = \frac{Q}{2\pi T} \frac{x}{(x^2 + y^2)} + U \tag{13.32b}$$

For $u = 0$ at a stagnation point $(x_0, 0)$

$$0 = \frac{Q}{2\pi T} \frac{x_0}{(x_0^2 + 0^2)} + U$$

$$x_0 = -\frac{Q}{2\pi TU} \tag{13.31}$$

Thus, it is seen that a stagnation point occurs at the above distance *behind* the source of contamination (Figure 13.14).

13.7.3.2 Determination of the Contamination Zone

The streamline passing through the stagnation point must satisfy $(x_0, 0)$.

By substituting $x = x_0$ and $y = 0$ in the stream function (Equation (13.30b)), one obtains

$$\psi = \frac{Q}{2\pi T}\tan^{-1}\left(\frac{0}{x_0}\right) + U(0)$$

However, $\tan^{-1}(0)$ could have multiple values such as;
$-n\pi, 0,$ or $+n\pi$ $(n = 1,2,\ldots)$

If one selects $\tan^{-1}(0)$ to be 0, then $\psi = 0$, which in fact is the first flow line that starts from the contaminant source and extends along the positive x-axis. In order to derive the flow lines of the contaminant boundary that have nonzero ψ, it is obvious that one has to use $-\pi$ or $+\pi$ in the above equation. Then,

$$\psi = -\frac{Q}{2\pi T}(\pm\pi) = \pm\frac{Q}{2T} \tag{13.33}$$

Therefore, the boundary of contamination is given by

$$\pm Q/2T = \frac{Q}{2\pi T}\tan^{-1}\left(\frac{y}{x}\right) + Uy \tag{13.34}$$

Example 13.5

(a) Plot the contaminant flow in a groundwater regime that carries a flow of $10\,\mathrm{m}^3/\mathrm{sec}$ within the flow cross section shown in Figure 13.13.

The values of Q, T, and U are as assigned below:

$$Q = 10\,\mathrm{m}^3/\mathrm{sec}$$

$$T = 15\,\mathrm{m}$$

$$U = 10\mathrm{m}^3/\mathrm{sec}/(15 \times 20) = 0.0333 \ \mathrm{m/s}$$

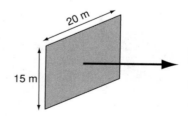

FIGURE 13.13
Illustration for Example 13.5.

By substituting the above values in Equation (13.34), one obtains the following expression for the contamination boundary.

$$\frac{Q}{2T} = \frac{Q}{2\pi T} \tan^{-1}\left(\frac{y}{x}\right) + Uy$$

$$\frac{10}{2 \times 15} = \frac{10}{2\pi \times 15} \tan^{-1}\left(\frac{y}{x}\right) + 0.0333y$$

$$\frac{y}{x} = \tan\left(\frac{0.3333 - 0.0333y}{0.106103}\right)$$

$$x = \frac{y}{\tan(3.142 - 0.3142y)}$$

In order to plot the stream lines for designated flow quantities, vary ψ from $-Q/2T$ to $+Q/2T$ in $0.1Q/T$ increments:

$-Q/2T$:

$$x = \frac{y}{\tan(-3.142 - 0.3142y)}$$

$Q/2T$:

$$x = \frac{y}{\tan(3.142 - 0.3142y)}$$

$-0.4Q/T$:

$$x = \frac{y}{\tan(-2.51328 - 0.3142y)}$$

$0.4Q/T$:

$$x = \frac{y}{\tan(2.51328 - 0.3142y)}$$

$-0.3Q/T$:

$$x = \frac{y}{\tan(-1.885 - 0.3142y)}$$

$0.3Q/T$:

$$x = \frac{y}{\tan(1.885 - 0.3142y)}$$

$0.2Q/T$:

$$x = \frac{y}{\tan(-1.2563 - 0.3142y)}$$

$0.2Q/T$:

$$x = \frac{y}{\tan(1.2563 - 0.3142y)}$$

$-0.1Q/T$:

$$x = \frac{y}{\tan(-0.62769 - 0.3142y)}$$

$0.1Q/T$:

$$x = \frac{y}{\tan(0.62769 - 0.3142y)}$$

These streamlines are graphically illustrated in Figure 13.14.

(b) Determine the limiting streamlines

It can be estimated from the plot that the contaminant flow is bounded by two, horizontal, parallel lines at $y = +10$ and $y = -10$.

The above equations can be derived mathematically by considering the limit of the boundary flow lines as $x \to \infty$

$$x = \frac{y}{\tan(3.142 - 0.3142y)} \quad \text{and} \quad x = \frac{y}{\tan(-3.142 - 0.3142y)}$$

$x \to \infty$ when the denominator is equal to zero, so

$$\tan(3.142 - 0.3142y) = 0 \quad \tan(-3.142 - 0.3142y) = 0$$

$$y = 10 \qquad\qquad\qquad y = -10$$

(c) Locate the stagnation point

Applying Equation (13.31),

$$x_0 = -\frac{10}{2\pi(15)(0.0333)} = 3.18 \text{ m}$$

This is clearly seen in Figure 13.14

(d) Plot the distance traveled by the contaminant versus time on the x axis.

$$u = \frac{dx}{dt}$$

$$t = \int \frac{dx}{u} = \int \frac{dx}{\dfrac{Q}{2\pi T} \cdot \dfrac{x}{x^2 + y^2} + U} = \int \frac{dx}{\dfrac{10}{2\pi \cdot 15} \cdot \dfrac{x}{x^2 + 0} + 0.0333}$$

$$t = -95.68 \ln(x + 3.186) - 0.3138x + C$$

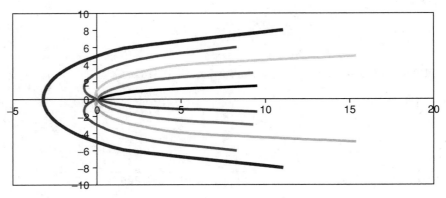

FIGURE 13.14.
Plot of the plume of contamination.

Initial condition

when $t = 0$, $x = 0$

$0 = -95.68 \ln(0 + 3.186) - 0.3138 \cdot 0 + C$

$C = 110.87$

$t = -95.68 \ln(x + 3.186) - 0.3138x + 110.87$

This is plotted in Figure 13.15.

13.8 Design of Filters

Filters are essential for protection of earthen structures from seeping groundwater. A number of empirical criteria for design of filters developed based on experimental studies and past engineering experience are available in the geotechnical literature. In the past, filters were designed primarily using soil layers of different gradation or sizes. However, nowadays geotextile filters are in wide application due to their reduced cost and easy construction.

13.8.1 Design of Soil Filters

The following criteria established by the U.S. Army Corps of Engineers (Huang, 2004) are generally used in the design of soil filters. These criteria are based on particle sizes corresponding to designated percentage of weights of the protected soil and the filter material as reflected in the particle size distribution curve (Chapter 1, Figure 1.3).

Clogging criterion: To ensure that the protected soil does not clog the larger particles of the filter, the following criterion must be satisfied by the relative sizes:

$$\frac{D_{15} \text{ filter}}{D_{85} \text{ soil}} \leq 5 \tag{13.35a}$$

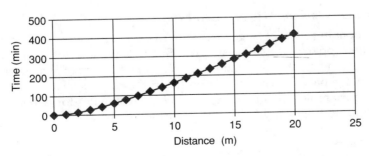

FIGURE 13.15.
Plot of contaminant travel.

Permeability criterion: To ensure that water passes through the filter system without building up excess pressure, the following criterion is recommended:

$$\frac{D_{15} \text{ filter}}{D_{15} \text{ soil}} \geq 5 \tag{13.35b}$$

Additional criterion: U.S. Army Corps of Engineers also recommends the following additional criterion:

$$\frac{D_{50} \text{ filter}}{D_{50} \text{ soil}} \leq 25 \tag{13.35c}$$

Based on Equations (13.35), one can design a satisfactory filter system when the particle size distributions of the relevant soil samples are available.

13.8.2 Design of Geotextile Filters

When filters are designed using geotextiles, a set of unique criteria relevant to water flow through geofabrics and the size interaction between the geofabric and the protected soil have to be considered. These will be briefly introduced in the following sections.

13.8.2.1 *Transmissivity of a Geotextile*

The ultimate transmissivity of a geotextile (θ_{ult}) can be defined in terms of its in-plane coefficient of hydraulic conductivity (k) and the average thickness (t) as (Koerner, 1998)

$$\theta_{\text{ult}} = kt \tag{13.36}$$

13.8.2.2 *Permittivity of a Geotextile*

Permittivity of a geotextile can be defined in terms of its cross-plane coefficient of hydraulic conductivity (k_{n}) and the average thickness (t) as (Koerner, 1998)

$$\psi = \frac{k_{\text{n}}}{t} \tag{13.37}$$

13.8.2.3 *Apparent Opening Size of a Geotextile*

The standard opening size of a geotextile is designated by the U.S. Army Corps of Engineers as the apparent opening size (AOS). AOS (or O_{95}) of a given geotextile is defined as the diameter of a set of uniform size glass beads of which only 5% would pass through that geotextile when it is used to sift the glass beads.

The following examples illustrate the design of a geotextiles for two different applications: (1) underdrain filters and (2) drains in earthen dams

(1) *Underdrain filters*

The following example illustrates the design of a geotextile filter for an underdrain

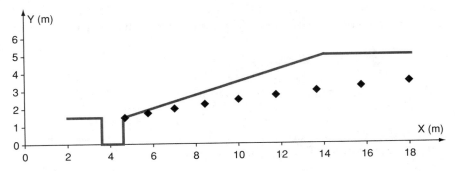

FIGURE 13.16
Illustration for Example 13.6.

Example 13.6

A highway embankment with an open pit edge drain is shown in Figure 13.16. Assuming that the embankment soil adjacent to the drain is well graded with a particle size distribution given in curve A of Figure 13.17, recommend the required properties of a geotextile that can execute the function of a satisfactory drain filter. The seepage surface (phreatic line) under the embankment is shown in a dotted line.

Solution
Step 1. Evaluate the critical nature of the facility and the site conditions
This application has been determined to be of low criticality and low severity.

Step 2. Characterize the soil
Assume the well-graded distribution in curve A of Figure 13.17, from which (curve A) $D_{10} = 0.01\,\text{mm}$; $D_{15} = 0.02\,\text{mm}$, $D_{60} = 2.0\,\text{mm}$; $D_{85} = 19.0\,\text{mm}$

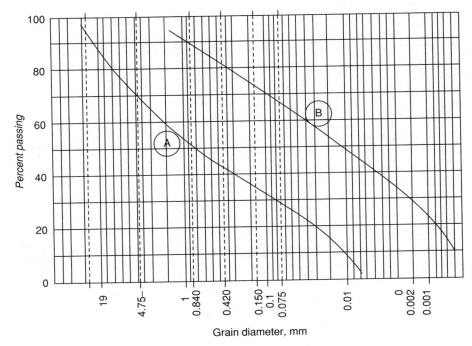

FIGURE 13.17
Soil gradation data for Example 13.6 (same as Figure 1.3).

$C_u = D_{60}/D_{10} = 200$

An approximate value of the coefficient of hydraulic conductivity can be obtained from

$$k = CD_{10}^2 \tag{13.38}$$

when k and D_{10} are expressed in cm/sec and mm, respectively, $C = 1.0$
Hence, $k = 10^{-4}$ cm/sec

Step 3. Estimate the anticipated flow
Assume that the approximate flow can be predicted by

$$q = \frac{\pi k (h_1^2 - h_2^2)}{\ln(r_1/r_2)} \tag{13.20}$$

where h_1 and h_2 are the heights of the phreatic line (free surface) above the impervious base at distances of r_1 and r_2, respectively. Therefore, based on the free surface information in Figure 13.16, the following can be estimated:
$q = 0.414k$ and since $k = 10^{-6}$ m/sec
$q = 4.14 \times 10^{-7}$ m^2/sec
According to Figure 13.16, there is a 1.6 m $\times$ 1.0 m or 1.6 m^2/m area of geotextile to transmit the seeping water.

Step 4. Determine the geotextile requirements based on the retention, permeability, and permittivity criteria in Koerner (1998) as
Retention criterion:
For fine-grained soils with 50% or more passing through No. 200 sieve
$O_{95} < D_{85}$ for woven filters
$O_{95} < 1.8D_{85}$ for nonwoven filters
$O_{95} \geq$ No. 50 sieve
For granular materials with 50% or less passing through No. 200 sieve (Figure 13.17)
$O_{95} < BD_{85}$ where $B = 1$ for $C_u < 2$ or $C_u > 8$
$B = 0.5$ when $2 < C_u < 4$
$B = 8/C_u, 4 < C_u < 8$
For this case, $B = 1$ and $D_{85} = 19.0$ mm; hence, $O_{95} < 19$ mm.
Permeability criterion:
In order to maintain the drainage efficiency after the installation of the geotextile filter,

$$k_{\text{geotextile}} \geq k_{\text{soil}}$$
$$k_{\text{geotextile}} \geq 10^{-6} \text{ m}^2/\text{s}$$

Permittivity criterion:
If the distance and the hydraulic head difference across the geotextile are t and δh, respectively, then by applying Equation (13.2a),

$$q = k_n \frac{\delta h}{t} A$$

By substituting from Equation (13.37)

$$4.41(10)^{-7} = \psi(\delta h)(1.6) \tag{13.39}$$

Assuming that the hydraulic head inside the drain is 0, since the hydraulic head just outside the filter is 1.6 m, then, $\delta h = 1.6$ m.
By substituting in Equation (13.39),
$\psi_{required} = 1.7(10)^{-7} \, \text{sec}^{-1}$
Using an appropriate safety factor, the allowable permittivity can be expressed as

$$F = \frac{\psi_{all}}{\psi_{required}} \tag{13.40}$$

Substituting in Equation (13.40) with $F = 5$,
$\psi_{all} = 8(10)^{-7} \, \text{sec}^{-1}$
However, the permittivity parameter generally specified for geotextiles is the ultimate permittivity value (ψ_{ult}) that can be related to ψ_{all} in the following manner (Koerner, 1998):

$$\psi_{all} = \psi_{ult}\left(\frac{1}{RF_{SCB}RF_{CR}RF_{IN}RF_{CC}RF_{BC}}\right) \tag{13.41}$$

where RF_{SCB}, RF_{CR}, RF_{IN}, RF_{CC}, and RF_{BC} are reduction factors that account for a number of phenomena that reduce the actual flow such as soil clogging and binding, creep reduction of void space, intrusion of adjacent soil into geotextile void space, clogging due to chemical reactions, and clogging due to biological activity, respectively. For underdrain filters, Koerner (1998) recommends the following ranges:

$RF_{SCB} = 5.0 \text{ to } 10.0$
$RF_{CR} = 1.0 \text{ to } 1.5$
$RF_{IN} = 1.0 \text{ to } 1.2$
$RF_{CC} = 1.2 \text{ to } 1.5$
$RF_{BC} = 2.0 \text{ to } 4.0$

By substituting the average values of the above ranges in Equation (13.41), ψ_{ult} is evaluated as

$$\psi_{ult} = 8(10)^{-7}(7.5)(1.25)(1.1)(1.35)(3.0) \, \text{sec}^{-1} = 3.34(10)^{-5} \, \text{sec}^{-1}$$

Step 5. Check for clogging

Clogging criterion:
For soils with $C_u > 3$, $O_{95} > 3(D_{15})$
$O_{95} > 3(0.02) > 0.06$ mm

Geotextile selection criteria:
Select woven geotextile with the following properties

Apparent opening size (O_{95})	Between 0.06 and 19 mm
Permeability ($k_{geotextile}$)	$>4.14 \times 10^{-7} \, m^2/sec$
Permittivity	$>3.34(10)^{-5} \, sec^{-1}$

(2) Design of geotextile drains in earthen dam

Example 13.7

Design a suitable geotextile filter for the "chimney drain" of the earthen dam retaining water at a head difference of 5 m as shown in Figure 13.18. It is given that the filter is inclined at an angle of 60° to the horizontal and the coefficient of hydraulic conductivity of the fine grained soil that is used in constructing the core of the dam (k_{soil}) is 1×10^{-7} m/sec.

(a) Assuming the axes of coordinates to be at the bottom of the filter, the phreatic line can be plotted using the equation, $y^2 = 1.44x$. It must be noted that the mathematical equation of the phreatic line can be derived using a suitable transformation as discussed in Section 13.4.

Then, based on a flow net compute the quantity of seepage using D'Arcy's law (Equation (13.2b)):

$$q = kh\frac{n_f}{n_e} = (1 \times 10^{-7})(5)\frac{3}{9} = 1.67 \times 10^{-7} \, m^2/sec = 1 \times 10^{-5} \, m^2/min$$

(b) Check the transmissivity criterion for the geotextile filter

$$\theta_{ult} = kt$$

where t is the thickness of the geotextile.

Assuming that the thickness and the in-plane permeability of the geotextile are 25 mm and 10^{-3} m/sec,

$$\theta_{ult} = 15 \times 10^{-4} \, m^2/min$$

Calculate the gradient of flow in the geotextile

$$i = \sin(60°) = 0.866$$

Calculate the transmissivity required to handle the given flow.

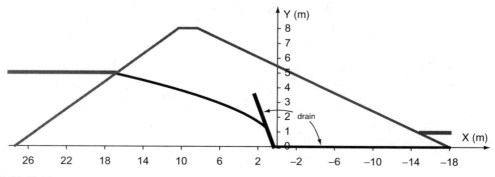

FIGURE 13.18
Illustration for Example 13.7.

This can be done by rearranging D'Arcy's equation (Equation (13.2)) to express the transmissivity (Equation (13.36)) as

$$q = kiA = ki(t \times W) = \theta(i \times W)$$

$$\theta = \frac{q}{i \times W} \Rightarrow \theta_{required} = \frac{1 \times 10^{-5}}{0.866 \times 1.0} = 1.15 \times 10^{-5} \text{ m}^2/\text{min}$$

where W is the width of the geotextile.

Calculate the global factor of safety assuming a reduction factor of 3.0 from θ_{ult} to θ_{all}

$$FS = \frac{\theta_{allow}}{\theta_{required}} = \frac{\theta_{ult}/3.0}{\theta_{required}} = \frac{15 \times 10^{-4}/3.0}{1.15 \times 10^{-5}} = 43.3$$

Since the safety factor is greater than the minimum recommended value of 5, the geotextile can be considered satisfactory.

References

Cedergreen, H.R., 1989, *Seepage, Drainage and Flow Nets*, John Wiley, New York.
Das, B.M., 2002, *Soil Mechanics Laboratory Manual*, Oxford University Press, New York.
Harr, M., 1962, *Groundwater and Seepage*, McGraw-Hill, New York.
Huang, Y.H., 2004, *Pavement Analysis and Design*, Pearson Education Inc., Upper Saddle River, NJ.
Koerner, R., 1998, *Designing with Geosynthetics*, Prentice Hall, Englewood Cliffs, NJ.

Index